等几何有限元与边界元及其耦合方法

董春迎　徐　闯　杨华实　著

科学出版社

北　京

内 容 简 介

本书介绍了等几何分析方法，其内容包括：等几何有限元法的基本理论及其在薄壳裂纹结构、含裂纹和孔洞缺陷的功能梯度薄壁结构和线性黏弹性问题中的应用；瞬态热传导问题的等几何边界元分析；等几何边界元法在含体力的三维黏弹性力学问题和复合材料结构的热弹性-黏弹性力学问题中的应用；三维弹性力学问题等几何有限元-边界元耦合方法中的非相适应界面和对称迭代求解方法，以及与求解问题类型无关的虚拟节点插入技术；混合维度实体-壳结构耦合问题的等几何有限元-边界元耦合方法。

本书可供对等几何分析感兴趣的高等院校教师、研究生和工程技术人员参考，并用其解决实际工程问题。

图书在版编目（CIP）数据

等几何有限元与边界元及其耦合方法 / 董春迎，徐闯，杨华实著. —北京：科学出版社，2024.4

ISBN 978-7-03-078412-4

Ⅰ. ①等… Ⅱ. ①董… ②徐… ③杨… Ⅲ. ①几何–有限元分析 Ⅳ. ①O18

中国国家版本馆 CIP 数据核字（2024）第 080406 号

责任编辑：赵敬伟 田轶静 / 责任校对：杨聪敏
责任印制：张 伟 / 封面设计：无极书装

科学出版社 出版
北京东黄城根北街 16 号
邮政编码：100717
http://www.sciencep.com
北京九州迅驰传媒文化有限公司印刷
科学出版社发行 各地新华书店经销
*
2024 年 4 月第 一 版 开本：720×1000 1/16
2025 年 1 月第二次印刷 印张：22 3/4
字数：459 000

定价：198.00 元

（如有印装质量问题，我社负责调换）

序

计算力学是伴随着电子计算机和新型数值方法的发展而诞生的，是力学学科的重要分支，也是力学与计算机科学和计算数学等的交叉学科。其研究领域和应用范围贯穿于力学学科的其他各个分支和其他相关学科，它们之间相互服务，创新发展并深刻影响。当今时代以计算机为载体的人工智能剧烈冲击着计算力学的发展，催生出以数据驱动为引擎的新的科学研究范式——数据模型和机理模型逐步融合，但其竞争的核心还是要对物理问题进行高效的建模和数值仿真。该书的研究内容——等几何分析方法是数值仿真分析的前沿方向之一，有助于打破计算机辅助设计(CAD)与计算机辅助工程(CAE)之间的壁垒，实现参数化几何建模与数字化仿真的融合。

等几何分析方法在其近二十年的发展历程中，在广大计算力学科技工作者的推动下，在弹性力学、断裂分析、接触力学、声学、电磁学、岩土力学、优化设计等领域得到了广泛的应用探究，这主要得益于两大优势——无几何离散误差的精确几何建模和有利于数值分析的高阶次高连续性样条基函数。“是亦彼也，彼亦是也。彼亦一是非，此亦一是非”，与有限元法等其他传统数值分析方法相比，等几何分析还处于“弱冠”之年，其未来的蓬勃发展仍需研究人员不遗余力地“呵护”。即使在将来随着新兴数值方法的涌现而消逝在计算力学发展的历史长河之中，其“刹那”的辉煌也不应被遗忘。

作为一本学术专著，该书是作者团队近年来在等几何有限元与边界元及其耦合分析领域研究工作的及时梳理和总结，内容涵盖：贯穿裂纹薄壳结构的扩展等几何分析，融合等几何有限胞元法研究含缺陷功能梯度薄壁结构在耦合载荷作用下的自由振动和屈曲问题，线性黏弹性问题的等几何有限元法分析，瞬态热传导问题等几何边界元法分析，热黏弹性问题等几何边界元法分析，非相适应界面等几何有限元-边界元耦合问题，混合维度实体-壳结构耦合问题的等几何有限元-边界元耦合分析等，书籍内容是作者们多年来持之以恒辛勤研究和努力付出的结果，希冀为对等几何分析方法感兴趣的读者提供一些有价值的参考。

该书第一作者从事计算固体力学领域的科学研究工作已达三十多年，学术造诣和治学经验丰厚，所在团队近年来持续关注和致力于等几何数值分析前瞻领域，发表了多篇高水平国际期刊论文，形成了一系列的研究成果。该书是作者团队在继科学出版社出版的《等几何边界元法》中文专著之后，又一部关于等几何分析领域的新作，相信这本新书会继续给从事等几何分析研究的学生、教师和科研人员带来有益的启发和帮助。

姚振汉
清华大学退休教授
Engineering Analysis with Boundary Elements 国际期刊副主编
2023 年 11 月于清华园

前　言

本书总结了作者课题组近年来在等几何有限元法、等几何边界元法及等几何有限元-边界元耦合方法等领域的一些研究成果。等几何分析在结构静动力学、结构优化、弹塑性力学、断裂力学、接触力学、声学、流固耦合等领域得到了广泛的应用，越来越受到研究者的关注。

本书内容共 9 章。第 1 章介绍了等几何分析方法的研究谱系，阐述了与本书内容相关的等几何有限元法、等几何边界元法、等几何有限元-边界元耦合方法的研究进展及发展情况；第 2 章采用基于 PHT(polynomial splines over hierarchical T-meshes)样条的扩展等几何分析方法，研究了含贯穿裂纹的薄壳结构，给出了基于 PHT 样条的扩展等几何分析中位移的富集公式、数值积分的实施方案，以及断裂参数的相关计算公式，通过算例说明了局部细分在分析裂纹问题中的计算优势。第 3 章结合扩展等几何分析和等几何有限胞元法，研究了含裂纹和孔洞缺陷的功能梯度薄壁结构在热、力和热力耦合载荷作用下的自由振动和特征屈曲问题。随后，结合考虑初始几何缺陷的弧长迭代公式，形成了能够有效分析含多孔多向功能梯度薄壁结构后屈曲行为的等几何算法，探讨了各类参数(材料、结构、缺陷几何、边界条件等)对固有频率、临界屈曲载荷和后屈曲路径的影响。第 4 章介绍了线性黏弹性力学问题的等几何有限元法，通过数值算例验证了等几何有限元法在求解线性黏弹性力学问题时的有效性，其中考虑了细化次数、时间步长等因素对计算结果精度的影响。第 5 章给出了二维及三维功能梯度材料瞬态热传导问题的等几何边界元法的解决方案，推导了规则化的边界-域积分方程，讨论了其中的域积分和奇异积分问题，阐述了将精细积分法引入等几何边界元法中求解时域问题的方法，通过算例讨论了网格细化、基函数升阶、内点数等因素对计算结果的影响。第 6 章利用等几何边界元法求解含体力项的三维黏弹性力学问题，推导了黏弹性问题边界积分方程，采用径向积分法将与体力和记忆应力相关的域积分转化为等效的边界积分，通过控制点到配点的变换方法可将刚体位移法用于处理等几何边界元法中的强奇异积分，采用幂级数展开法求解等几何边界元法中的弱奇异积分，采用改进的三维面力恢复法求解模型表面节点的应力和应变，利用正则化的应变和应力积分方程求解内部点的应变和应力。第 7 章利用等几何边界元法分析了多维、多尺度复合结构的热黏弹性力学问题，推导了热黏弹性力学问题的位移边界积分方程和正则化应变边界积分方程，采用径向积分法将与体力、温度和记忆应力有关的域积分转换为等效的边界积分，给出了基于层次四叉树分解算法的自适应积分方法，介绍了热黏弹性力学中的修正面力恢复法，并将其用于求解边界点的应变，利用虚拟结点插入技术构建了非相适应界面网格之间的位移(或面力)耦合约束。第 8 章阐述了三维弹性力学问题等几何有限元-边界元耦合方法中非相适应界面和对称迭代求解方法，给出了一套与求解问题类型无关的虚拟节点插入技术，建立了非相适应(或不完全重合)界面上的位移连续条件，讨论了在选择与等几何边界元子域相关的不同增强矩阵的条件下对称迭代耦合方法的收敛性和准确性。第 9 章介绍了混合维度实体-壳耦合问题的等几何有限元-边界元耦合方法，其实体部分采用等几何边界元模拟，探究了基于壳体变形假设的直接运动耦合方法和基于界面上虚功相等的弱耦合方法的收敛性和稳定性。

作者想对在他之前进入等几何分析领域的所有研究者表示感谢——你们有趣且有创新性的研究工作引导作者进入这个新的研究方向。作者也要感谢他的研究生们在等几何分析领域作出的贡献，特别感谢詹云生硕士研究生对线性黏弹性材料等几何有限元法的贡献。作者还要感谢国家自然科学基金(基金号：12272047, 11972085, 11672038)的资助，使本书得以呈现在读者面前。

等几何分析仍然是一种相对年轻的数值分析方法。由于计算力学领域不断出现新方法，本书的一些内容以后可能会过时。尽管如此，作者还是希望读者在阅读本书时能有所收获。

由于作者水平有限，书中难免有不妥之处，恳请读者批评指正。

董春迎

2023 年 10 月

目　录

第 1 章

绪　论

1.1 引　言

数值分析作为当今公认的四大科学研究范式(理论、试验、计算、数据)之一，起到连接其他三类科学研究范式的桥梁作用，在科学探究之路中持续扮演着重要的角色。等几何分析方法(isogeometric analysis, IGA)作为数值分析领域的前沿方向之一，自 Hughes 及其合作者于 2005 年[1]在计算力学领域顶级期刊 *Computer Methods in Applied Mechanics and Engineering* 上发表以来，因其克服了数值计算中的几何离散误差(其他误差来源主要有模型近似误差、数值计算误差等)，受到众多学者的持续关注和跟进研究，其知识版图和应用领域不断地拓展，近二十年来其发展的研究谱系如图 1.1 所示。

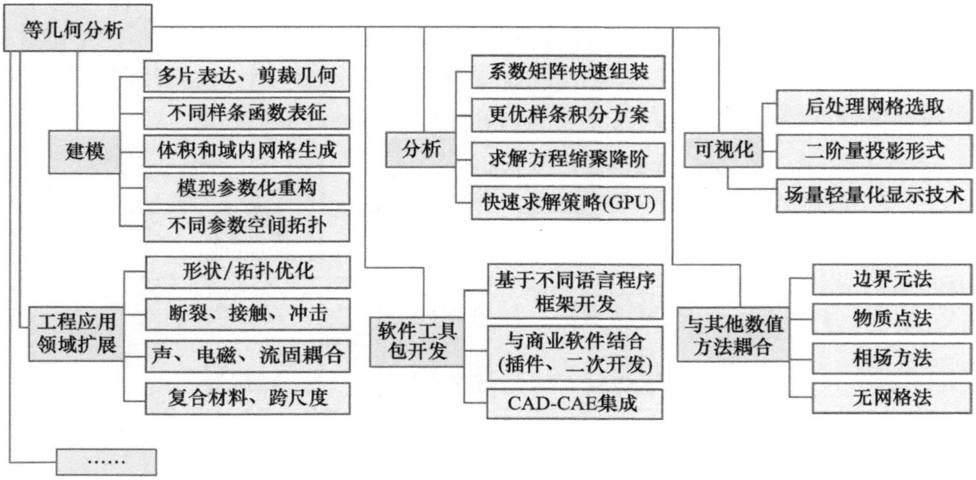

图 1.1　等几何分析方法研究谱系

前期的研究主要集中在将等几何分析方法的思想引入单独一类数值方法中，充分利用等几何模型的精确几何表达、形函数的非负高阶次和高连续性、CAD 与 CAE 之间的无缝连接等优点，对传统数值分析方法进行改进、加强，或者重塑其分析框架，如等几何有限元法(isogeometric analysis finite element method,

IGAFEM)、等几何边界元法(isogeometric analysis boundary element method, IGABEM)、等几何物质点法(isogeometric analysis material point method, IGAMPM)等。然而，正所谓“尺有所短，寸有所长”，每类等几何数值方法都有其发展技术瓶颈和应用局限性。开展各类等几何先进数值方法的耦合算法(如本书中的等几何有限元-边界元耦合方法)研究，可以充分发挥每种方法的优点，达到互补的目的。

本章首先介绍本书中提到的等几何有限元法、等几何边界元法及等几何有限元-边界元耦合方法的相关研究进展和应用，然后介绍本书的章节内容安排。

1.2 等几何有限元法

传统数值方法在分析问题时，首先需要在计算机辅助设计(CAD)软件中建立几何模型，然后将模型导入计算机辅助工程(CAE)软件中进行网格划分。CAD 采用非均匀有理 B 样条(non-uniform rational B-splines, NURBS)函数，CAE 软件采用拉格朗日形函数。CAE 数值计算后不能直接对几何模型进行修改，必须再次在 CAD 软件中修改后重新导入。最终的计算结果可视化展示采用的是双线性插值函数。上述不同几何表达之间频繁的数据传输和交换既麻烦又耗时，极大地增加了分析成本。如图 1.2 所示，在汽车、航空航天和造船工业中，大约 80%的分析时间用于 CAE 网格生成。2005 年，Hughes 等[1]提出了等几何分析(IGA)的概念。IGA 的核心思想是将 CAD 软件几何建模的 NURBS 函数直接用作数值分析的形函数。IGA 完全消除了网格划分的概念，既能精确地描述几何模型，消除几何离散误差和获得更好的数值解，又大大节省了数值计算成本。

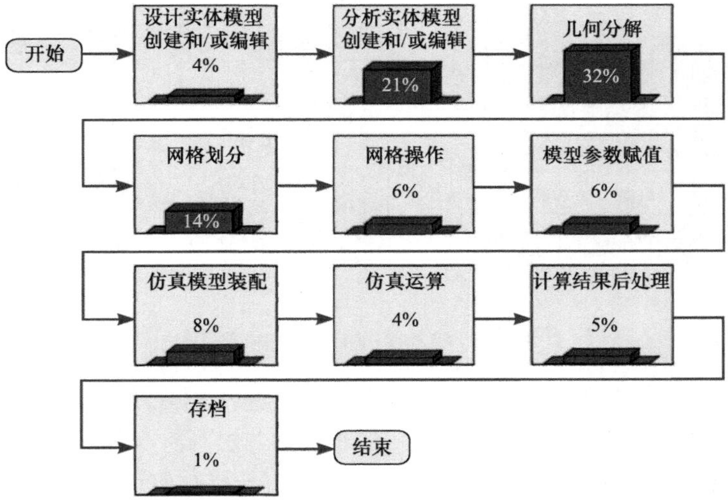

图 1.2 美国桑迪亚(Sandia)国家实验室模型生成和分析过程中每个组成部分的相对时间成本[1]

如图 1.3 所示，IGAFEM 消除了现有数值计算方法中存在的几何离散误差，实现了 CAE 和 CAD 系统的无缝连接和统一，大幅度提高了数值分析和设计的效率[2]。IGA 是将 CAD 中普遍采用的 NURBS 基函数作为形函数用于数值分析，避免了烦琐的网格划分步骤和不同系统间频繁的数据交换，大大缩短了产品的开发周期。等几何概念的提出主要是实现了工程设计与计算模拟之间数学模型表达方式的统一。2017 年，来自挪威科技大学的 Stahl 与其合作者将这一概念拓展到了结果可视化的领域[2]，即相同的数学表达既用于几何描述和数值分析，也用于后处理过程中计算结果的可视化。他们采用 Bézier 分解将结构性样条和非结构性样条单元均转化为 Bézier 单元，进行可视化操作，这样做的好处是利于并行计算和处理局部细分样条下的可视化问题。IGA 采用 CAD 系统中的样条函数当作形函数，同时用于模型几何表示和未知场变量近似，其主要优点如下所述。

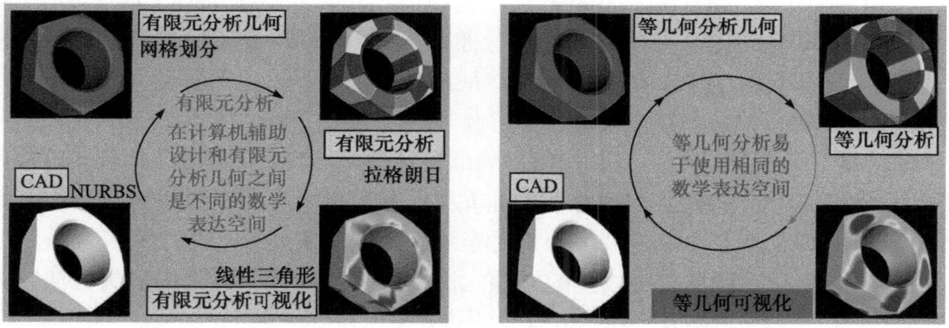

(a)　　　　(b)

图 1.3　传统有限元法和等几何有限元法的区别[2]

(a) 传统有限元法中，设计、分析和可视化被分离，并使用不同的数学表达；(b) 等几何分析中，设计、分析和可视化过程使用相同的数学表达

(1) 消除了传统分析中由拉格朗日(Lagrange)多项式插值引起的几何离散误差，同时也消除了烦琐的网格划分步骤。单元细化过程可以方便地通过插入节点(h 细分)和升阶(p 细分)来实现，而不再需要与 CAD 系统进行额外的数据通信。

(2) 采用很少的单元就能精确描述不同尺度比例结构的几何形状，而且其描述的几何形状不随离散单元的多少而发生改变，并且可以通过其独有的“k 细分”实现单元之间的高阶连续性。

(3) 精确的空间离散使得与曲面几何相关的运动学量(曲率、切线和法向量等)可以在每个计算点处精确确定，而不会产生任何误差。

(4) NURBS 形函数的高阶次和高阶连续性能自然满足各种力学理论中对基本变量 C^1 或更高插值连续性的要求，如薄板壳理论、Cahn-Hilliard 方程[3]等。

(5) 进行 k 细分的高阶 NURBS 基函数在对光滑或者非光滑函数进行插值近

似时不会出现不稳定的振荡现象[4]，如 Runge 现象。高阶 NURBS 形函数由于其非负性和凸包性的特点，在处理大位移和大滑动下的相互作用面问题时具有明显的优势[5]。

(6) 高阶连续性的样条形函数使得应变、应力等二阶张量的计算和可视化不再需要通过在传统有限元分析中广泛应用的节点平均方法来获得光滑的节点变化值[2]。

然而，在应用基于 NURBS 的 IGAFEM 时仍存在一些不足。

(1) NURBS 基函数由于其张量拓扑形式的特点而缺乏有效的局部细分能力，在分析具有局部特征问题时会在非局部区域产生多余的控制点和引入冗余的自由度，因此基于 NURBS 基函数的 IGA 不易于直接分析含缺陷的结构。一些具有固有局部细分特性的样条函数被提出，尽管它们还没有与现有的 CAD 环境完全集成，如 T 样条[6,7]、PHT(polynomial splines over hierarchical T-meshes)样条[8]、LR-B 样条(locally refined-B splines)[9]和细分曲面(subdivision surfaces, SS)[10]等。上述样条不仅克服了 NURBS 在分析局部特征问题时的不足，而且还继承了 NURBS 基函数非负性、单位分解性和局部支撑性等利于数值分析的优点。

(2) 复杂几何体，尤其是那些包含局部特征(如裂纹和孔洞)的几何体，通常需要由多个 NURBS 片组成，这些片之间通常具有不相适应的离散网格。虽然 IGA 可以容易地实现单个片中更高阶的单元连续性，但是实施片与片之间的高阶连续性一直是一个开放且具有挑战性的问题。许多学者提出了一系列的解决方法，如静态凝聚法[11]、罚函数法[12]、Mortar 法[13]和 Nitsche 法[14]等。

(3) IGAFEM 在数值计算时需要将 CAD 系统中几何模型的边界表征向适合于分析的内部区域实体表征进行转换，这种转换的存在使得 IGAFEM 在处理复杂模型时并未能真正实现 CAD 与 CAE 系统间的无缝连接，而且这种转换的精度和有效性目前仍是一个开放性的问题[15-17]。鉴于此，一些学者提出了基于边界样条表示的等几何分析方法(isogeometric B-rep analysis, IBRA)以实现从设计到分析过程的统一[18,19]，允许直接对复杂的 CAD 模型进行数值分析，无须重新建模和进行网格转换。

1.2.1 在壳体分析中的应用

壳体结构分析的两个主要且广泛使用的公式分别来自 Kirchhoff-Love[20]和 Reissner-Mindlin[21]壳体理论。区分薄壳和厚壳的常用标准是壳体曲率半径 R 和厚度 h 的比值。按照这个标准，壳体结构通常分为厚壳($R/h<20$)和薄壳($R/h\geqslant 20$)。Kirchhoff-Love 壳体理论忽略了横向剪切变形，仅需要位移自由度就能描述壳体的变形行为，其仅适用于变形行为主要由弯曲和薄膜应变决定的薄壳结构[22]。该理论的两个关键问题是：①变分公式中涉及形函数的二阶导数，这就要求壳单元

之间至少具有 C^1 连续性，这是传统 FEM 难以实现的[23,24]，这也是仅需要单元间 C^0 连续性的 Reissner-Mindlin 壳单元在传统 FEM 中广泛使用的重要原因；②涉及旋转和力矩的边界条件的施加变得更加复杂，因为它涉及位移自由度的导数，而系统方程中只包含位移自由度[25]。与无旋转自由度的 Kirchhoff-Love 理论相比，Reissner-Mindlin 壳理论考虑了横向剪切应变，可以有效地模拟薄壳和中厚壳。在其壳体的运动学描述中，使用转角作为附加未知量，有助于施加力矩边界条件[26]。此外，平衡方程中只涉及未知场变量的一阶导数。因此，单元间变形满足 C^0 连续性要求就足够了。当然，也有学者利用上述两种壳理论优缺点之间的互补关系提出了混合壳体理论(blended shell theory)，并应用于模拟金属成形过程[27]，即低曲率区域可以使用无旋转的薄壳理论进行有效分析，而高曲率区域可以使用 Reissner-Mindlin 公式进行分析。

近些年来，许多学者结合 IGA 对上述两种壳体理论进行了大量的研究，并从多个方面说明了 IGA 在壳体领域相对于传统方法具有的令人信服的优势，具体可参考等几何 Reissner-Mindlin 壳[21,26,28-30]和等几何 Kirchhoff-Love 壳的文章[20,22,31,32]。一方面，可以在每个积分点处精确计算与壳体基础几何相关的运动学量(曲率、切线和法向量等)。另一方面，NURBS 的高阶光滑性使得壳单元的高阶公式可以直接实现，特别是要求 C^1 连续性的 Kirchhoff-Love 薄壳公式。等几何的出现使得 Kirchhoff-Love 壳体理论经历了一次复苏。尽管如此，与传统方法相比，NURBS 更高的连续性和精确的几何描述仍然为 Reissner-Mindlin 壳单元的实施提供了显著的优势[33]。与传统有限元 Reissner-Mindlin 壳单元相比，等几何 Reissner-Mindlin 壳单元出现剪切自锁问题的可能性要小得多[34]，只需提高 NURBS 基函数的阶次即可简单和有效地减小自锁问题产生的影响。此外，在商业有限元软件 LS-DYNA 中已成功地实现了基于等几何 Reissner-Mindlin 壳单元的数值分析[35]。

1.2.2 在含缺陷问题中的应用

传统 FEM 在处理裂纹问题时主要存在如下几个方面的挑战[36]：①需要相适应的网格拓扑来表示裂纹，裂纹扩展路径必须事先给定且只能沿着单元边界；②基于多项式的插值形函数无法描述裂尖应力奇异场，需要构建特殊类型的单元，如四分之一节点单元[37]，并且在裂纹尖端区域要划分密集的网格，这无疑增加了计算成本；③FEM 网格存在几何离散误差，因此依赖于网格投影的计算误差会不断累积。为了克服传统方法中裂纹求解对网格的严重依赖性和描述裂尖附近应力场的困难，Moës 等[38]利用形函数的单位分解性，在位移插值近似表达式中引入包含裂纹附近解的某些已知特性的富集函数，提出了扩展有限元法(extended finite element method, XFEM)。在该方法中，通过引入阶跃函数和裂尖渐近函数而扩充了原来的形函数，分别对裂纹面的不连续和裂纹尖端的奇异性进行描述，允许离散裂纹的表达独立于底层的求解网格，且富集函数的添加只涉及裂纹附近的局部单元，这大大减轻甚至消除了重新划分网格的负担，同时也提高了求解精度。Dolbow 等[39]首次用 XFEM 模拟了含缺陷的 Reissner-Mindlin 板。随后，Lasry 等[40]用 XFEM 模拟在弯矩载荷作用下含有贯穿裂纹的 Kirchhoff-Love 板，并成功计算了应力强度因子[41]。Bordas 等[42]结合应变光滑技术和 XFEM 求解了 Reissner-Mindlin 板的断裂问题。Areias 和 Belytschko[43]运用 XFEM 研究了含有任意裂纹夹芯壳的非线性变形问题。Bayesteh 和 Mohammadi[44]研究了不同裂纹尖端富集函数对薄壁壳体断裂参数(应力强度因子、J积分)计算精度的影响。Zeng 等[45]基于连续体的(continuum-based, CB)壳单元和 XFEM 模拟了管道中的裂纹扩展问题。

为了消除 XFEM 在求解过程中存在的几何离散误差，进一步提高求解的精度，人们将 XFEM 思想引入 IGA 中形成了新的扩展等几何方法(extended isogeometric analysis, XIGA)[46,47]。高阶的 NURBS 基函数提供了更高的单元间连续性，使得单元的边界处具有连续的应力场，提高了裂尖周围应力场的精度，这对提高裂纹前端应力预测的精度和裂纹扩展方向预判的准确性都尤为重要[48]。另外，NURBS 的高阶连续性也能够有效地解决混合单元带来的误差问题[49]。更详细的 XIGA 在计算断裂力学领域的应用可参考综述文章[50]。XIGA 方法在含裂纹薄壁板壳问题中的研究和应用也得到了广泛关注。Nguyen-Thanh 等[51]运用基于 NURBS 的 XIGA 分析了含贯穿裂纹的 Kirchhoff-Love 薄壳问题。Nguyen-Thanh 等[52]采用基于 NURBS 的 XIGA-无网格耦合方法研究了含裂纹薄壳的静力和自由振动问题。Yang 和 Dong[53]基于局部细分的 PHT 样条和 XIGA 研究了含贯穿裂纹的 Kirchhoff-Love 薄板和薄壳结构。Khatir 等[54]结合粒子群优化算法研究了板结构中裂纹的识别问题，并对 XFEM 和 XIGA 的实施进行了比较。Yu 等[55]基于局部细分的 LR-B 样条和 XIGA 研究了含贯穿裂纹的 Reissner-Mindlin 板结构。

研究含孔洞缺陷功能梯度材料(functionally graded materials, FGM)板的振动和屈曲问题的文献相对较少。Janghorban 和 Rostamsowlat[56]使用 FEM 分析了具有多个圆形和椭圆形孔洞的 FGM 板的自由振动问题。Zhao 等[57]基于一阶剪切变形理论和无网格 Ritz 方法，研究了含有单个圆形和方形孔洞的 FGM 板的热屈曲和力屈曲问题。Saji 等[58]利用 FEM 研究了含单个圆形孔洞的 FGM 板的热屈曲问题，并在分析中考虑了与温度相关的材料特性。Abolghasemi 等[59]利用 FEM 对含有单个椭圆形孔洞的 FGM 板进行了热力耦合屈曲分析。Yang 等[60]使用基于 NURBS 的 XIGA，对含有单个圆形孔洞的 FGM 板进行了自由振动和力屈曲分析。Yu 等[61]利用基于 NURBS 的 XIGA，对含有单个圆形孔洞的 FGM 板进行了热屈曲分析。

上述方法在求解孔洞问题时存在的主要缺陷有：①孔洞边界周围需要密集和

相适应的网格，这增加了网格的划分时间和计算成本；②XFEM/XIGA 需要使用额外的富集函数描述孔洞几何界面对求解的影响，因此增加了富集模式的复杂度，特别是求解同时存在多个孔洞的情况；③高阶剪切变形理论要求单元之间具有 C^1 连续性。传统方法中低阶次的形函数无法自然地满足高阶连续性条件，需要引入额外的方法进行协调处理。在 IGA 中，虽说样条函数可以容易地在单个片中的单元之间实现高阶连续性，但即使只包含一个圆形孔洞的方板，也无法仅由具有 C^1 连续性的单个 NURBS 片所表示。因此，含孔 FGM 板的几何模型需要使用多个 NURBS 片进行建模，并且需要额外的策略来处理片与片之间 C^1 连续性的耦合。

本书将有限胞元法(finite cell method, FCM)[62,63]思想引入等几何分析中，对含有孔洞缺陷的 FGM 板进行动力学分析。FCM 将虚拟域方法与高阶基函数相结合，用于模拟缺陷的几何边界[64,65]。该方法的主要思想是使用非相适应的结构化网格进行数值分析，而复杂缺陷的几何边界则通过在分割单元中的自适应积分来处理，这能够有效地缓解复杂孔洞几何体需要生成边界匹配网格的困难。因此，只需要确定每个积分点是在物理域内还是在物理域外[66]，就能够有效地处理结构中的孔洞缺陷，该方法的精度可由自适应积分中四叉树的层数进行灵活控制。由于其简单、高效的实现方法和显著的性能，FCM 已成功运用到多个领域，如大变形[67]、热弹性[68]、界面问题[69]、剪裁几何[70]、薄壳变形[71]和弹塑性[72]等问题。

1.3 等几何边界元法

上述 IGAFEM 的一些不足的改善和缓解，可通过引入 BEM 来实现。BEM 在数值实施时只需要对实体区域的表面进行离散，从而减少了计算问题的维度。由于在 CAD 系统中，NURBS 和 T 样条等只提供描述物体表面的数据，自然促使了 IGA 和 BEM 的融合。2012 年，Simpson 等将 IGA 引入 BEM 中，形成了 IGABEM[73]，并应用于求解二维弹性力学问题。该方法既继承了 IGA 中几何精确、高阶次和高连续性的优点，又保留了 BEM 中降维及高精度的优点。随后，该方法引起了计算领域学者的持续关注，发展迅速并得到了广泛应用。

与 IGAFEM 相比，IGABEM 提供了许多有吸引力的数值分析优势。

(1) BEM 是半解析形式的数值方法，利用所涉及问题的基本解和格林公式得到对应的边界积分方程，方程的未知量同时包含位移和面力。因此，求解出的应变、应力等二阶量与所求位移量具有相同的精度。

(2) 只需对物体的边界进行离散，从而减少了计算问题的维度，同时也避免了在 IGAFEM 中需要从边界描述到域内体网格转化的复杂过程，更易实现 CAD 与 CAE 系统的无缝集成。

(3) 一般情况下，边界积分方程本质上都满足无限域和半无限域的边界条件[74]。

这使得 BEM 对涉及无限域和半无限域的问题(如声学、电磁学和岩土力学等)具有独特的优势。

然而，IGABEM 仍存在一些局限性。

(1) 由于存在域积分，BEM 在研究非线性材料行为方面缺乏优势。

(2) BEM 的精度主要取决于(拟)奇异积分的精确计算。针对不同阶次的奇异积分需要选择适当的数值积分方法。

(3) BEM 中离散得到的方程组系数矩阵是稠密非对称的满秩矩阵。随着问题自由度的增加，所需的计算存储和计算时间也会急剧增加。这一问题在数值计算时通常表现在两方面。①矩阵存储：如果边界上的总自由度是 N，则系数矩阵的存储需要 $O(N^2)$的量级。②线性方程组的求解：如果使用直接求解技术如高斯消去法，则计算量会达到 $O(N^3)$的量级。

1.3.1 奇异积分及拟奇异积分的计算

由于基本解的特性，在积分单元内部，当源点趋近于场点时，积分会出现不同阶次的奇异性。对于二维问题，当源点处于单元内，且核函数中含有 $\ln(1/r)$ (这里 r 为源点和场点之间的距离)时，积分为弱奇异积分；当核函数中含有 $1/r$ 时，积分则为强奇异积分。对于三维问题，源点在单元内，积分核函数有 $1/r$ 时为弱奇异积分；积分核函数有 $1/r^2$ 时为强奇异积分。当源点在单元外时，积分没有奇异性，可以直接利用高斯积分求解。如何有效地处理奇异积分是保证 BEM 数值结果精确的关键。为了处理不同阶次的奇异积分，Guiggiani 等[75]提出的局部规则化方法已被用于处理各种奇异积分。但这种方法需要将被积函数中的所有量都进行泰勒级数展开，而且只保留至二阶精度，数值实施比较烦琐。高效伟[76]提出的基于径向积分法的高阶奇异曲面积分的直接计算方法，具有被积函数中每个量不需要泰勒级数展开，易于编程实现的优点。

和常规 BEM 相同，IGABEM 在数值分析的过程中也需要处理奇异积分。Simpson 等[73]使用 Telles 变换法[77]处理了弱奇异积分，利用奇异值提取技术(subtraction of singularity technology, SST)[78]处理了强奇异积分。在文献[73]中，Simpson 等详细论述了由于 NURBS 基函数的特殊性质，常规 BEM 中处理强奇异积分最简便的方法(刚体位移法)不能用于 IGABEM。近些年来，学者们采用 SST 或幂级数展开法[79,80]求解强奇异积分。这些方法虽然可以有效地处理强奇异积分，但处理过程相对复杂，不易编程实现。Xu 等[81]通过控制点和配点之间的矩阵变换，保证了剩余奇异积分的个数等于刚体位移的个数，从而使刚体位移法应用于 IGABEM 处理强奇异积分。Gong 等[79]采用幂级数展开法计算了势问题 IGABEM 中的弱奇异积分、强奇异积分和超奇异积分，并与传统 BEM 进行比较，结果表明 IGABEM 具有更高的计算精度。Heltai 和 Arroyo[82]提出了一个三维斯托克斯

(Stokes)流动问题的非奇异等几何边界积分方程，并对其收敛性进行了数值验证。在数值实现中，标准高斯求积法足以对规则化等几何边界积分方程进行积分。Simpson 等[83]针对声学低频问题提出了一个基于规则化的 Burton-Miller 公式，其中所有积分都是弱奇异性的。该公式保证了声学 IGABEM 对所有波数的稳定性。在求解模型表面应力或应变时，如果采用相应的应力或应变积分方程求解，会遇到超强奇异积分。学者此时一般会选用面力恢复法(surface traction recovery method, STRM)[84]来求解模型表面的应力或应变，这样不仅保证了计算精度，还可以有效地回避超强奇异积分。Xu 等[85,86]利用基于 IGABEM 改进的 TRM，求解了黏弹性模型和热黏弹性模型表面的应力和应变。

在分析断裂或薄壁结构问题时，源点与积分单元非常接近，这种积分被称为拟奇异积分[87]。从数学的角度看，拟奇异积分并不具有奇异性[88,89]。然而，从数值积分的角度来看，由于被积函数在积分区间内振荡严重，常规的数值积分法无法准确计算这些积分[90-92]。求解拟奇异积分的常用方法包括全局正则化法[93,94]、单元子分法[95,96]、解析或半解析法[97,98]和坐标转换法[99-101]等。Gong 和 Dong[80]将自适应积分法[79]引入 IGABEM 中求解了三维势问题拟奇异积分。由于自适应积分方法是基于单元细分的思想，随着拟奇异性的增强，通过单元细分得到的子单元数量会急剧增加。因此，对于具有强拟奇异性的积分，该方法的计算效率将大大降低。Xu 等[86]提出了基于四叉树的自适应积分法，其中自适应积分过程由误差公式和四叉树分解方法驱动，在保证计算精度的前提下给出最优高斯点分布，实现了计算精度和效率的平衡。为了进一步提高计算效率，Gong 等[102,103]提出了一种混合积分计算方法，保证了计算精度和效率之间的平衡，并用于求解二维及三维涂层结构。Han 等[104]将 IGABEM 中的拟奇异积分分离为非奇异部分和奇异部分。奇异部分的积分核用泰勒级数多项式表示，其中不同阶导数用 NURBS 样条插值。通过一系列分部积分，导出了具有近似核的奇异部分的解析表达式，而非奇异部分则采用高斯积分求解。该半解析方法能准确计算出更靠近边界的内点位移和应力。

1.3.2 域积分的计算

BEM 的一个初始限制是需要原始偏微分方程的基本解才能得到等效的边界积分方程。但是，对于复杂的问题，很难获得相应的基本解。一般情况下，我们会根据已有的基本解，推导出相应问题的积分方程，其中出现了域积分[105,106]。处理域积分的一般解决方案是将计算域划分为单元。虽然单元积分方案可以得到准确、稳定的结果[107]，但是也使 BEM 失去了只对边界进行离散的优势。近几十年来，学者们研究了很多处理 BEM 域积分的方法。Cruse[108]首次利用散度定理将离心力产生的域积分转换为等效的边界积分。Rizzo 和 Shippy[109]将离心力表示成标量的微分形式，进而提出了一种域积分到边界积分的转换方法。Danson[110]基于伽辽金(Galerkin)向量和高斯-格林(Gauss-Green)公式提出了一种将重力、离心力和热载荷相关的域积分转换为等效边界积分的通用方法，但该方法仅适用于常数或线性的体力项。

尽管上述方法可以在一定程度上解决域积分的问题，但由于适用范围相对较窄，目前还没有一种方法得到广泛推广。1983 年，Nardini 和 Brebbia[111]提出了双重互易边界元法(dual reciprocity method, DRM)。其核心思想是在控制方程的算子上找到域内函数的特解，然后利用格林(Green)公式将域积分转化为等效的边界积分。由于域内函数的特解难以求解，在 DRM 中通常采用径向基函数(radial basis functions, RBF)对域内函数进行插值，从而求得基函数的特解。1992 年，Partridge 等[112]编写的第一本关于 DRM 的著作问世。DRM 在 BEM 分析各类复杂问题中被广泛应用[113-116]，尤其在分析非线性问题上使得 BEM 大放异彩。Yu 等将 DRM 引入 IGABEM 中，用于求解非均质材料稳态热传导[117]、功能梯度材料瞬态热传导[118]、非傅里叶瞬态热传导[119]和热传导反问题[120]等。1989 年，Nowak 和 Brebbia[121]在 DRM 的基础上提出了多重互易方法(multiple reciprocity method, MRM)，将泊松(Poisson)和亥姆霍兹(Helmholtz)方程中的域积分转换为等效的边界积分。在 MRM 中，利用拉普拉斯算子的高阶基本解序列，反复应用互易定理将域积分变换为边界积分。此后，Neves 和 Brebbia[122]利用基于 MRM 的 BEM 求解了纳维(Navier)方程。Ochiai 和 Kobayashi[123]利用基于 MRM 的 BEM 求解了弹塑性力学问题。

DRM 适用的范围比较广泛，但它需要基函数的特解。一些复杂的三维基函数的特解很难求出。此外，即使对于已知的体力问题，该方法仍然需要近似函数。因此，计算结果不如其他方法准确。MRM 是非常强大的，但在有些情况下，它很难计算拉普拉斯算子控制的递归公式中的初始值。此外，该方法还需要在位移基本解中定义一个常数，该常数的不同值可能产生不同的结果。

2002 年，高效伟[124,125]提出了将域积分转化为边界积分的径向积分法(radial integration method, RIM)。RIM 的一个基本特征是它能够将任何复杂的域积分转换为边界积分，而不使用拉普拉斯算子和问题的特解。此外，它还可以降低各种奇异积分[126,127]的奇异性阶数，这自然可以用于求解多边界问题的域积分[128]。对于核函数已知的域积分，利用 RIM 可以精确地将它们转换为等效的边界积分。对于核函数未知的域积分，未知变量需要用基函数逼近，其中可以选择全局近似函数和局部近似函数等多种近似函数[114]。全局近似函数具有精度高、适应性强的特点，但由于计算量大，不适合求解大规模三维问题。局部近似函数计算量小，适用于求解大规模计算问题。一般来说，局部逼近函数的计算精度低于全局逼近函数，但可以通过适当增加节点数量和调整节点分布来保证计算精度。近年来，RIM 在边界元领域引起了广泛的关注，并被用于解决大量的问题，如时域[129,130]、非线性[131,132]和非均质材料[133]等问题。Xu 和 Dong 将 RIM 与 IGABEM 结合而提出了 RI-IGABEM，将其用于分析非均质材料稳态热传导[134]、二维及三维功能梯度材料瞬态热传导[135]、二维及三维非均质材料弹性动力学[81]、三维黏弹性力学[85]，以及多维多层复合材料结构热黏弹性-弹性力学[86]等问题。

1.3.3 时域问题的求解

一般来说，BEM 用于求解瞬态问题的方法分为两类：频域法[136]和时域法[137]。频域法主要是通过傅里叶变换或拉普拉斯变换将时域问题转换到频域上求解，之后再通过逆变换获得时域上的结果。然而，该方法的计算精度取决于变换参数。此外，逆变换不利于大规模计算。时域法是直接在时域上进行求解。对于 BEM，可以通过时域基本解建立积分方程，再通过迭代求解出每一时刻的结果。但是，对于复杂的问题，很难获得时域基本解。还有一种方法是利用已有的稳态问题基本解推导瞬态问题的积分方程，再利用时域方法求解。这种方法应用于求解各类瞬态问题，如瞬态热传导和动力学等问题。瞬态热传导属于抛物型方程初边值问题，弹性动力学属于双曲型方程初边值问题。BEM 在求解不同类型瞬态问题时，选择恰当的时域方法非常重要。下面详细讨论 BEM 在求解瞬态热传导和弹性动力学问题中时域方法的选择。

求解瞬态热传导问题最常用的方法是用差分法代替时间的导数项逐步求解，而欧拉后差分法(backward difference method, BDM)是最常用和最简便的方法，但是该算法对时间步长的选取非常敏感。为了获得更为精确的计算结果，通常需要选取较短的时间步长，这对于复杂的数值算例来说，计算成本较高[138]。此外，从理论上讲，最优的时间步长与模型的材料参数和网格的密度相关，因此很难获得问题的最佳时间步长。1994 年，钟万勰和 Williams[139]提出了精细积分法(precise integration method, PIM)，很好地解决了这个问题。PIM 的核心思想是利用 2^N (N 为给定的正整数)类思想求解指数型矩阵，将每个时间步长划分为 2^N 个精细步长进行求解，因此该算法对时间步长的选取不敏感，同时具有较高的精度[140]。1995 年，钟万勰在 PIM 的基础上又提出了适用于求解大规模问题的子域 PIM[141]。经过近三十年的发展，该方法日趋成熟，并被收录在多本教材[142,143]中，而且已被广泛应用于求解瞬态热传导[144,145]、动力学[146,147]、电磁学[148]等问题。Yu 等[145]将 PIM 与 BEM 结合，求解了变热传导系数的瞬态热传导问题。Xu 等[135]将 PIM 引入 IGABEM 中，求解了功能梯度材料瞬态热传导问题。

在求解弹性动力学问题时，BEM 可以转换为求解一个二阶常微分方程组(ordinary differential equations, ODEs)。一般来说，求解 ODEs 有标准的求解方法，如龙格-库塔(Runge-Kutta)方法。但随着自由度的增大，求解 ODEs 的标准方法变得越来越不经济。因此，在 FEM 和 BEM 分析中，通常采用几种有效的方法，一般分为两类：①振型叠加法；②直接积分法。采用振型叠加法求解方程组时，利用系统的自然模态将方程组转化为 n 个非耦合方程，然后对这些方程进行解析或数值求解，得到各模态的响应。最后，将各模态的响应按一定的方式叠加得到系统响应。直接积分法是在没有任何变换的情况下直接对运动方程进行积分。对于线性问题，振型叠加法和直接积分法是等价的[149]。值得注意的是，由 BEM 或 IGABEM 最终形成的矩阵都是非对称满阵，因此高阶模态振型的计算不够精确。这个问题比 FEM 严重得多。但对于一般的动力学问题，系统的响应主要由低阶模态控制，高阶模态的贡献很小。如果在直接积分中不能有效地过滤掉这些虚假的高阶分量，则会极大地降低计算结果的精度，甚至导致计算失败。FEM 中最常用的求解动力学问题的直接积分法为 Newmark 法[150]，该算法虽然理论上具有二阶精度，但它不能控制数值耗散分布，且在低频区域耗散过大。因此该算法不适用于 BEM 和 IGABEM 求解弹性动力学问题。Gao 等[151]采用了基于 Houbolt 法的 BEM 求解了功能梯度材料热-弹性动力学问题。Xu 等[81]将广义-α 法引入 IGABEM 中求解了非均质材料弹性动力学问题。广义-α 法为 Wilson-θ法[152]和 BA-ρ法[153]的改进形式，当其中参数设置为某个固定值时，该方法可以退化为著名的 HHT-α 法[154]、WBZ-α 法[155]和 Newmark 法[156]。该方法不仅是一种二阶精度的无条件稳定算法，而且能有效地滤除高频的假响应，最小化低频响应的衰减。值得注意的是，这种方法通过调整谱半径来控制高频的数值耗散。

1.4 等几何有限元-边界元耦合方法

IGAFEM 已被广泛地应用于研究结构静动力学、非线性大变形及弹塑性分析等领域，而 IGABEM 在研究应力集中、断裂力学和声场等问题时具有天然的优势，同时又能克服 IGAFEM 在数值计算时需要将 CAD 分析模型的边界表征向适合于分析的实体表征进行转换的困难。因此，研究 IGAFEM 与 IGABEM 的耦合算法可以发挥每种方法各自的优点，达到相辅相成的目的。Zienkiewicz 等[157]首次提出了传统 FEM 与传统 BEM 的耦合方法。这种耦合方法已应用于各种工程问题，如动力分析[158]、弹塑性分析[159]、断裂力学[160]、损伤增长[161]和声-结构相互作用[162,163]等。等几何有限元-边界元(FE-BE)耦合方法的研究则相对有限，主要集中在声-结构或流体-结构耦合领域，包括流体-壳耦合分析[23,164,165]、磁力耦合问题[166]和剪切流中动力分析[167]等。FEM 通过建立结点外力与位移关系的系统方程，得到稀疏对称的正定矩阵。以配点建立的 BEM 则以结点位移和边界上结点面力为基本变量，得到稠密的非对称满秩系数矩阵。因此，耦合分析实施的关键是在耦合界面上建立控制变量的协调方程和面力平衡方程，即建立这两类方程中未知变量之间的关系。

1.4.1　非相适应界面耦合问题

针对求解规模不大且有限元和边界元子域尺度相当的问题，我们可以划分相同的耦合界面网格，轻松实现点对点的强连接。但一般情况下，从 CAD 系统中导出的耦合界面两侧的等几何有限元和等几何边界元网格是不相匹配的(非相适应或非完全重合、界面两侧网格的数量和阶次都不一致)或是基于不同类型的样条函数建立的(如 NURBS 和非结构性 PHT 样条、LR-B 样条等)，甚至耦合界面还可能包含大量的剪裁曲线和剪裁曲面。因此，为了解决这些问题，我们需要研究等几何非相适应界面耦合约束方程建立的稳定算法。

很多学者针对传统有限元与传统边界元的非相适应界面耦合问题，提出了一系列解决方法。Hsiao 等[168,169]利用弱积分意义上的广义相容性条件对不协调的界面网格进行耦合，从而可以放宽对位移场连续性的要求。Schnack 和 Türke[170]通过混合变分公式建立了非相适应界面上的耦合条件。Fischer 和 Gaul[171]提出了一种类似 mortar 方法的耦合算法，用于施加声-结构相互作用问题中的非相适应界面耦合约束。González 等[172,173]构造了子域之间的过渡曲面，并使用局部拉格朗日乘子法来满足界面位移不一致的相容条件。按照同样的思路，Rüberg 和 Schanz[174]将局部拉格朗日乘子法推广应用到动力学领域中的非相适应界面问题。Bazilevs 等[175]提出了增广拉格朗日乘子法对非匹配界面进行耦合，并将其应用到 IGA 流-固耦合问题。上述这些方法都是基于拉格朗日形函数建立的，要么会引入额外的自由度，要么需要推导相关的变分公式。因此，它们不能有效地适用于以样条函数作为形函数的 IGA 中界面耦合的实施。解决等几何 FE-BE 耦合分析中非相适应界面问题的文献很少。Yang 等[176]基于 NURBS 曲面细分的特点和虚功原理，利用虚拟节点插入技术[177]实现了三维弹性力学问题的非相适应界面等几何 FE-BE 强耦合算法，该算法的实施只与界面两侧的 NURBS 控制点和权值有关，易于编程。Zhan 等[178]利用该方法对固体火箭发动机(solid propellant rocket motor, SRT)燃烧室进行了结构完整性分析。Xu 等[86]利用基于非相适应界面的 IGABEM 求解了多层多尺度复合结构的热黏弹性问题。

1.4.2　混合维度实体-壳结构耦合分析

实现混合维度实体-壳耦合分析的主要思想是建立低维单元类型(壳)和高维单元类型(实体)之间耦合界面上控制变量之间的联系，即沿耦合边的位移协调条件、面力和壳体应力合力(法向力矩、扭矩和剪力)的平衡条件。大量的研究主要集中在有限元实体单元与有限元壳单元的耦合分析上。依据壳单元内转角和横向位移表示的边界合反力与三维实体边界上的面力在界面虚位移上所做的虚功相等这一事实，建立了混合维度耦合模型的多点约束方程。McCune 等[179]研究了三维二次 Serendipity 实体单元与 Reissner-Mindlin 板单元的耦合问题。Shim 等[180]建立了弹性问题的实体-壳耦合模型，并考虑了微小偏心率对耦合精度的影响。Cuillière 等[181]采用基于耦合界面网格搭建过渡单元的方法实施了混合维度实体-壳的耦合分析。但致力于使用 FE-BE 耦合方法研究混合维度实体-壳耦合问题的文献较少，Haas 等[182]依据板理论中的直接运动学假设开发了二维有限元板单元与三维对称伽辽金边界元法(symmetric Galerkin boundary element method, SGBEM)的耦合程序。随后，Haas 等[183]又改进了耦合策略，基于不同维度单元的内力在耦合界面上所做功相等的原理，提出了一种弱耦合方法，其中通过压缩耦合界面上未知面力来构建刚度矩阵形式的 BE 子域系数矩阵，然后组装到全局耦合系统方程中。尽管 SGBEM 可以生成对称的半正定系数矩阵，但由于双重面积分和超奇异积分的复杂正则化过程，其计算成本非常高。此外，对等几何实体-壳耦合分析的研究也相对有限。Nguyen 等[184]基于 Nitsche 方法实现了等几何 FE 实体单元和等几何 FE 板单元之间的耦合分析，但是需要根据经验人为设置 Nitsche 方法中的稳定参数。Yang 等[185]采用等几何 FE-BE 耦合方法分析了混合维度实体-壳耦合问题，其中实体区域由 IGABEM 进行模拟，因此整个耦合模型的数值分析只需要提供边界的 CAD 网格信息，在实现耦合分析与 CAD 系统的紧密联系方面具有很大的潜力。

1.5 本书内容安排

本书介绍了作者课题组近年来在等几何有限元、等几何边界元和等几何有限元-边界元耦合方法等领域的一些研究成果，旨在为对等几何分析方法感兴趣的读者提供一些参考。本书的内容安排如下所述。

第 1 章，首先介绍等几何分析方法作为数值分析领域前沿方向的研究谱系，接着分析与本书研究内容相关的等几何有限元法、等几何边界元法、等几何有限元-边界元耦合方法的发展情况，最后给出本书章节内容安排。

第 2 章，利用基于 PHT 样条的扩展等几何分析方法研究含贯穿裂纹的薄壳结构，给出基于 PHT 样条的扩展等几何分析中位移的富集公式、数值积分的实施方案和断裂参数的相关计算公式，通过数值算例展示局部细分在分析含裂纹问题时的计算优势。

第 3 章，结合扩展等几何分析和等几何有限胞元法，研究同时含裂纹和孔洞缺陷的功能梯度薄壁结构在热、力和热力耦合载荷作用下的自由振动和特征屈曲问题。之后，结合考虑初始几何缺陷的弧长迭代公式，形成能够有效分析含多孔多向功能梯度薄壁结构后屈曲行为的等几何算法，并探究各类参数(材料、结构、缺陷几何、边界条件等)对固有频率、临界屈曲载荷和后屈曲路径的影响。

第 4 章，介绍线性黏弹性力学问题的等几何有限元法，通过数值算例验证了等几何有限元法在求解线性黏弹性力学问题时的有效性，其中考虑了细化次数、时间步长等因素对计算结果精度的影响。

第 5 章，针对二维及三维功能梯度材料瞬态热传导问题，给出了等几何边界元法的解决方案。为了避免积分方程中温度梯度的存在，推导了规则化的边界-域积分方程，详细讨论了其中遇到的域积分问题和奇异积分问题。将精细积分法引入等几何边界元法中求解时域问题，降低了时间步长对计算结果的影响。通过数值算例讨论网格细化、基函数升阶、内点数等因素对计算结果的影响。

第 6 章，利用等几何边界元法求解含体力项的三维黏弹性力学问题。详细推导黏弹性问题边界积分方程。为了降低计算成本，将随时间变化的剪切模量用普罗尼(Prony)级数展开，同时利用遗传积分计算记忆应力。由于采用 Kelvin 基本解推导出边界积分方程，所以积分方程中存在域积分。为了保证等几何边界元法仅对边界离散化的优点，采用径向积分法将与体力和记忆应力相关的域积分转化为等效的边界积分。通过控制点到配点的变换方法，可以用刚体位移法处理等几何边界元法中的强奇异积分，并采用幂级数展开法求解等几何边界元法中的弱奇异积分。利用改进的三维面力恢复法求解模型表面节点的应力和应变。利用正则化的应变和应力积分方程来求解内部点的应变和应力。

第 7 章，利用等几何边界元法分析多维、多尺度复合结构的热弹性-黏弹性力学问题。首先我们详细推导了热黏弹性力学问题的位移边界积分方程和正则化应变边界积分方程。其次，将随时间变化的剪切模量用 Prony 级数表示，采用遗传积分计算记忆应力，降低了计算成本；采用径向积分法将与体力、温度和记忆应力有关的域积分转换为等效的边界积分，不仅使等几何边界元法保留了只在边界上离散的优势，又降低了积分的奇异性。提出了基于层次四叉树分解算法的自适应积分方法，在计算成本最优的情况下更加灵活有效地处理拟奇异积分。提出了热黏弹性力学中的修正面力恢复法，并将其用于求解边界点的应变，从而避免了求解超奇异积分。在此基础上，利用虚拟结点插入技术，构建了非相适应界面网格之间的位移(或面力)耦合约束。该方法的主要优点是操作简便和鲁棒性强，因为它与研究的问题无关，只依赖于界面两侧的 NURBS 网格。

第 8 章，研究了三维弹性力学问题等几何有限元-边界元耦合方法中非相适应界面和对称迭代求解问题。针对非相适应耦合界面，提出了一套与求解问题类型无关的虚拟节点插入技术，有效地建立了非相适应(或不完全重合)界面上的位移连续条件，其实施仅依赖于界面两侧的 NURBS 信息。其次，探究了在选择与等几何边界元子域相关的不同增强矩阵的条件下，对称迭代耦合方法的收敛性和准确性。

第 9 章，建立了混合维度实体-壳结构耦合问题的等几何有限元-边界元耦合方法，其实体部分采用等几何边界元模拟，能避免实体内部体网格的划分，壳体区域采用等几何壳单元进行分析。因此，耦合分析仅需提供整个模型边界的 CAD 网格信息，易于实现等几何耦合方法与 CAD 系统的紧密融合；提出和探究了基于壳体变形假设的直接运动耦合方法和基于界面上虚功相等的弱耦合方法的收敛性和稳定性。

参 考 文 献

[1] Hughes T J R, Cottrell J A, Bazilevs Y. Isogeometric analysis: CAD, finite elements, NURBS, exact geometry and mesh refinement[J]. Computer Methods in Applied Mechanics and Engineering, 2005, 194(39): 4135-4195.

[2] Stahl A, Kvamsdal T, Schellewald C. Post-processing and visualization techniques for isogeometric analysis results[J]. Computer Methods in Applied Mechanics and Engineering, 2017, 316: 880-943.

[3] Gómez H, Calo V M, Bazilevs Y, et al. Isogeometric analysis of the Cahn-Hilliard phase-field model[J]. Computer Methods in Applied Mechanics and Engineering, 2008, 197(49): 4333-4352.

[4] Beer G, Marussig B, Duenser C. The Isogeometric Boundary Element Method[M]. Switzerland: Springer Nature Switzerland AG, 2020.

[5] De Lorenzis L, Wriggers P, Hughes T J R. Isogeometric contact: a review[J]. GAMM-Mitteilungen, 2014, 37(1): 85-123.

[6] Bazilevs Y, Calo V M, Cottrell J A, et al. Isogeometric analysis using T-splines[J]. Computer Methods in Applied Mechanics and Engineering, 2010, 199(5): 229-263.

[7] Dörfel M R, Jüttler B, Simeon B. Adaptive isogeometric analysis by local h-refinement with T-splines[J]. Computer Methods in Applied Mechanics and Engineering, 2010, 199(5): 264-275.

[8] Deng J, Chen F, Li X, et al. Polynomial splines over hierarchical T-meshes[J]. Graphical Models, 2008, 70(4): 76-86.

[9] Johannessen K A, Kvamsdal T, Dokken T. Isogeometric analysis using LR B-splines[J]. Computer Methods in Applied Mechanics and Engineering, 2014, 269: 471-514.

[10] Liu Z, Majeed M, Cirak F, et al. Isogeometric FEM-BEM coupled structural-acoustic analysis of shells using subdivision surfaces[J]. International Journal for Numerical Methods in Engineering, 2018, 113(9): 1507-1530.

[11] Lei Z, Gillot F, Jezequel L. A C0/G1 multiple patches connection method in isogeometric analysis[J]. Applied Mathematical Modelling, 2015, 39(15): 4405-4420.

[12] Herrema A J, Johnson E L, Proserpio D, et al. Penalty coupling of non-matching isogeometric Kirchhoff-Love shell patches with application to composite wind turbine blades[J]. Computer Methods in Applied Mechanics and Engineering, 2019, 346: 810-840.

[13] Dornisch W, Vitucci G, Klinkel S. The weak substitution method—an application of the mortar method for patch coupling in NURBS-based isogeometric analysis[J]. International Journal for Numerical Methods in Engineering, 2015, 103(3): 205-234.

[14] Nguyen V P, Kerfriden P, Brino M, et al. Nitsche's method for two and three dimensional NURBS patch coupling[J]. Computational Mechanics, 2014, 53(6): 1163-1182.

[15] Xu G, Mourrain B, Duvigneau R, et al. Constructing analysis-suitable parameterization of computational domain from CAD boundary by variational harmonic method[J]. Journal of Computational Physics, 2013, 252: 275-289.

[16] Xu G, Mourrain B, Galligo A, et al. High-quality construction of analysis-suitable trivariate NURBS solids by reparameterization methods[J]. Computational Mechanics, 2014, 54(5): 1303-1313.

[17] Al Akhras H, Elguedj T, Gravouil A, et al. Isogeometric analysis-suitable trivariate NURBS models from standard B-Rep models[J]. Computer Methods in Applied Mechanics and Engineering, 2016, 307: 256-274.

[18] Breitenberger M, Apostolatos A, Philipp B, et al. Analysis in computer aided design: nonlinear isogeometric B-Rep analysis of shell structures[J]. Computer Methods in Applied Mechanics and Engineering, 2015, 284: 401-457.

[19] Leidinger L F, Breitenberger M, Bauer A M, et al. Explicit dynamic isogeometric B-Rep analysis of penalty-coupled trimmed NURBS shells[J]. Computer Methods in Applied Mechanics and Engineering, 2019, 351: 891-927.

[20] Kiendl J, Bletzinger K U, Linhard J, et al. Isogeometric shell analysis with Kirchhoff-Love elements[J]. Computer Methods in Applied Mechanics and Engineering, 2009, 198(49-52): 3902-3914.

[21] Benson D J, Bazilevs Y, Hsu M C, et al. Isogeometric shell analysis: the Reissner-Mindlin shell[J]. Computer Methods in Applied Mechanics and Engineering, 2010, 199(5): 276-289.

[22] Kiendl J, Hsu M C, Wu M C H, et al. Isogeometric Kirchhoff-Love shell formulations for general hyperelastic materials[J]. Computer Methods in Applied Mechanics and Engineering, 2015, 291: 280-303.

[23] Kiendl J, Bazilevs Y, Hsu M C, et al. The bending strip method for isogeometric analysis of Kirchhoff-Love shell structures comprised of multiple patches[J]. Computer Methods in Applied Mechanics and Engineering, 2010, 199(37-40): 2403-2416.

[24] Guo Y, Ruess M. Nitsche's method for a coupling of isogeometric thin shells and blended shell structures[J]. Computer Methods in Applied Mechanics and Engineering, 2015, 284: 881-905.

[25] Ivannikov V, Tiago C, Pimenta P M. On the boundary conditions of the geometrically nonlinear Kirchhoff-Love shell theory[J]. International Journal of Solids and Structures, 2014, 51(18): 3101-3112.

[26] Dornisch W, Müller R, Klinkel S. An efficient and robust rotational formulation for isogeometric Reissner-Mindlin shell elements[J]. Computer Methods in Applied Mechanics and Engineering, 2016, 303: 1-34.

[27] Benson D J, Hartmann S, Bazilevs Y, et al. Blended isogeometric shells[J]. Computer Methods in Applied Mechanics and Engineering, 2013, 255: 133-146.

[28] Dornisch W, Klinkel S, Simeon B. Isogeometric Reissner-Mindlin shell analysis with exactly calculated director vectors[J]. Computer Methods in Applied Mechanics and Engineering, 2013, 253: 491-504.

[29] Sobota P M, Dornisch W, Müller R, et al. Implicit dynamic analysis using an isogeometric

Reissner-Mindlin shell formulation[J]. International Journal for Numerical Methods in Engineering, 2017, 110(9): 803-825.

[30] Kikis G, Ambati M, De Lorenzis L, et al. Phase-field model of brittle fracture in Reissner-Mindlin plates and shells[J]. Computer Methods in Applied Mechanics and Engineering, 2021, 373: 113490.

[31] Kiendl J, Ambati M, De Lorenzis L, et al. Phase-field description of brittle fracture in plates and shells[J]. Computer Methods in Applied Mechanics and Engineering, 2016, 312: 374-394.

[32] Zareh M, Qian X. Kirchhoff-Love shell formulation based on triangular isogeometric analysis[J]. Computer Methods in Applied Mechanics and Engineering, 2019, 347: 853-873.

[33] Kiendl J, Marino E, De lorenzis L. Isogeometric collocation for the Reissner-Mindlin shell problem[J]. Computer Methods in Applied Mechanics and Engineering, 2017, 325: 645-665.

[34] Adam C, Bouabdallah S, Zarroug M, et al. Improved numerical integration for locking treatment in isogeometric structural elements. Part II: Plates and shells[J]. Computer Methods in Applied Mechanics and Engineering, 2015, 284: 106-137.

[35] Benson D J, Bazilevs Y, De Luycker E, et al. A generalized finite element formulation for arbitrary basis functions: from isogeometric analysis to XFEM[J]. International Journal for Numerical Methods in Engineering, 2010, 83(6): 765-785.

[36] Egger A, Pillai U, Agathos K, et al. Discrete and phase field methods for linear elastic fracture mechanics: a comparative study and state-of-the-art review[J]. Applied Sciences, 2019, 9(12): 2436-2500.

[37] Barsoum R S. On the use of isoparametric finite elements in linear fracture mechanics[J]. International Journal for Numerical Methods in Engineering, 1976, 10(1): 25-37.

[38] Moës N, Dolbow J, Belytschko T. A finite element method for crack growth without remeshing[J]. International Journal for Numerical Methods in Engineering, 1999, 46(1): 131-150.

[39] Dolbow J, Moës N, Belytschko T. Modeling fracture in Mindlin-Reissner plates with the extended finite element method[J]. International Journal of Solids and Structures, 2000, 37(48): 7161-7183.

[40] Lasry J, Pommier J, Renard Y, et al. Extended finite element methods for thin cracked plates with Kirchhoff-Love theory[J]. International Journal for Numerical Methods in Engineering, 2010, 84(9): 1115-1138.

[41] Lasry J, Renard Y, Salaün M. Stress intensity factors computation for bending plates with extended finite element method[J]. International Journal for Numerical Methods in Engineering, 2012, 91(9): 909-928.

[42] Bordas S P A, Rabczuk T, Hung N X, et al. Strain smoothing in FEM and XFEM[J]. Computers & Structures, 2010, 88(23): 1419-1443.

[43] Areias P M A, Belytschko T. Non-linear analysis of shells with arbitrary evolving cracks using XFEM[J]. International Journal for Numerical Methods in Engineering, 2005, 62(3): 384-415.

[44] Bayesteh H, Mohammadi S. XFEM fracture analysis of shells: the effect of crack tip enrichments[J]. Computational Materials Science, 2011, 50(10): 2793-2813.

[45] Zeng Q, Liu Z, Xu D, et al. Modeling arbitrary crack propagation in coupled shell/solid structures with X-FEM[J]. International Journal for Numerical Methods in Engineering, 2016, 106(12):

1018-1040.
[46] De Luycker E, Benson D J, Belytschko T, et al. X-FEM in isogeometric analysis for linear fracture mechanics[J]. International Journal for Numerical Methods in Engineering, 2011, 87(6): 541-565.
[47] Ghorashi S S, Valizadeh N, Mohammadi S. Extended isogeometric analysis for simulation of stationary and propagating cracks[J]. International Journal for Numerical Methods in Engineering, 2012, 89(9): 1069-1101.
[48] Fathi F, Chen L, de Borst R. Extended isogeometric analysis for cohesive fracture[J]. International Journal for Numerical Methods in Engineering, 2020, 121(20): 4584-4613.
[49] Nguyen-Thanh N, Valizadeh N, Nguyen M N, et al. An extended isogeometric thin shell analysis based on Kirchhoff-Love theory[J]. Computer Methods in Applied Mechanics and Engineering, 2015, 284: 265-291.
[50] Yadav A, Godara R K, Bhardwaj G. A review on XIGA method for computational fracture mechanics applications[J]. Engineering Fracture Mechanics, 2020, 230: 107001.
[51] Nguyen-Thanh N, Valizadeh N, Nguyen M N, et al. An extended isogeometric thin shell analysis based on Kirchhoff-Love theory[J]. Computer Methods in Applied Mechanics and Engineering, 2015, 284: 265-291.
[52] Nguyen-Thanh N, Li W, Zhou K. Static and free-vibration analyses of cracks in thin-shell structures based on an isogeometric-meshfree coupling approach[J]. Computational Mechanics, 2018, 62(6): 1287-1309.
[53] Yang H S, Dong C Y. Adaptive extended isogeometric analysis based on PHT-splines for thin cracked plates and shells with Kirchhoff-Love theory[J]. Applied Mathematical Modelling, 2019, 76: 759-799.
[54] Khatir S, Abdel Wahab M. A computational approach for crack identification in plate structures using XFEM, XIGA, PSO and Jaya algorithm[J]. Theoretical and Applied Fracture Mechanics, 2019, 103: 102240.
[55] Yu T, Yuan H, Gu J, et al. Error-controlled adaptive LR B-splines XIGA for assessment of fracture parameters in through-cracked Mindlin-Reissner plates[J]. Engineering Fracture Mechanics, 2020, 229: 106964.
[56] Janghorban M, Rostamsowlat I. Free vibration analysis of functionally graded plates with multiple circular and non-circular cutouts[J]. Chinese Journal of Mechanical Engineering, 2012, 25(2): 277-284.
[57] Zhao X, Lee Y Y, Liew K M. Mechanical and thermal buckling analysis of functionally graded plates[J]. Composite Structures, 2009, 90(2): 161-171.
[58] Saji D, Varughese B, Pradhan S C. Finite element analysis for thermal buckling behavior on functionally graded plates with cut-outs[C]//Proceedings of the International Conference on Aerospace Science and Technology, Bangalore, India, 26-28 June, 2008.
[59] Abolghasemi S, Shaterzadeh A R, Rezaei R. Thermo-mechanical buckling analysis of functionally graded plates with an elliptic cutout[J]. Aerospace Science and Technology, 2014, 39: 250-259.
[60] Yang H S, Dong C Y, Qin X C, et al. Vibration and buckling analyses of FGM plates with multiple internal defects using XIGA-PHT and FCM under thermal and mechanical loads[J]. Applied

Mathematical Modelling, 2020, 78: 433-481.

[61] Yu T, Bui T Q, Yin S, et al. On the thermal buckling analysis of functionally graded plates with internal defects using extended isogeometric analysis[J]. Composite Structures, 2016, 136: 684-695.

[62] Parvizian J, Düster A, Rank E. Finite cell method[J]. Computational Mechanics, 2007, 41(1): 121-133.

[63] Düster A, Parvizian J, Yang Z, et al. The finite cell method for three-dimensional problems of solid mechanics[J]. Computer Methods in Applied Mechanics and Engineering, 2008, 197(45): 3768-3782.

[64] Schillinger D, Ruess M, Zander N, et al. Small and large deformation analysis with the p- and B-spline versions of the finite cell method[J]. Computational Mechanics, 2012, 50(4): 445-478.

[65] Schillinger D, Ruess M. The finite cell method: a review in the context of higher-order structural analysis of CAD and image-based geometric models[J]. Archives of Computational Methods in Engineering, 2015, 22(3): 391-455.

[66] Schillinger D, Dedè L, Scott M A, et al. An isogeometric design-through-analysis methodology based on adaptive hierarchical refinement of NURBS, immersed boundary methods, and T-spline CAD surfaces[J]. Computer Methods in Applied Mechanics and Engineering, 2012, 249-252: 116-150.

[67] Schillinger D, Ruess M, Zander N, et al. Small and large deformation analysis with the p- and B-spline versions of the finite cell method[J]. Computational Mechanics, 2012, 50(4): 445-478.

[68] Zander N, Kollmannsberger S, Ruess M, et al. The finite cell method for linear thermoelasticity[J]. Computers & Mathematics with Applications, 2012, 64(11): 3527-3541.

[69] Schillinger D, Rank E. An unfitted hp-adaptive finite element method based on hierarchical B-splines for interface problems of complex geometry[J]. Computer Methods in Applied Mechanics and Engineering, 2011, 200(47): 3358-3380.

[70] Ruess M, Schillinger D, Özcan A I, et al. Weak coupling for isogeometric analysis of non-matching and trimmed multi-patch geometries[J]. Computer Methods in Applied Mechanics and Engineering, 2014, 269: 46-71.

[71] Rank E, Kollmannsberger S, Sorger C, et al. Shell finite cell method: a high order fictitious domain approach for thin-walled structures[J]. Computer Methods in Applied Mechanics and Engineering, 2011, 200(45): 3200-3209.

[72] Taghipour A, Parvizian J, Heinze S, et al. The finite cell method for nearly incompressible finite strain plasticity problems with complex geometries[J]. Computers & Mathematics with Applications, 2018, 75(9): 3298-3316.

[73] Simpson R N, Bordas S P A, Trevelyan J, et al. A two-dimensional isogeometric boundary element method for elastostatic analysis[J]. Computer Methods in Applied Mechanics and Engineering, 2012, 209-212: 87-100.

[74] Beer G, Smith I, Duenser C. The Boundary Element Method with Programming: For Engineers and Scientists[M]. Germany, Mörlenbach: Springer-Verlag/Wien, 2008.

[75] Guiggiani M, Krishnasamy G, Rudolphi T J, et al. A general algorithm for the numerical solution

of hypersingular boundary integral equations[J]. Journal of Applied Mechanics-Transactions of the Asme, 1992, 59: 604-614.

[76] Gao X W. An effective method for numerical evaluation of general 2D and 3D high order singular boundary integrals[J]. Computer Methods in Applied Mechanics and Engineering, 2010, 199: 2856-2864.

[77] Telles J C F. A self-adaptive co-ordinate transformation for efficient numerical evaluation of general boundary integrals[J]. International Journal for Numerical Methods in Engineering, 1987, 24: 959-973.

[78] Guiggiani M, Casalini P. Direct computation of Cauchy principal value integrals in advanced boundary elements[J]. International Journal for Numerical Methods in Engineering, 1987, 24 : 1711-1720.

[79] Gong Y P, Dong C Y, Qin X C, An isogeometric boundary element method for three dimensional potential problems[J]. Journal of Computational and Applied Mathematics, 2017, 313: 454-468.

[80] Gong Y P, Dong C Y. An isogeometric boundary element method using adaptive integral method for 3D potential problems[J]. Journal of Computational and Applied Mathematics, 2017, 319: 141-158.

[81] Xu C, Dai R, Dong C Y, et al. RI-IGABEM based on generalized-α method in 2D and 3D elastodynamic problems[J]. Computer Methods in Applied Mechanics and Engineering, 2021, 383 : 113890.

[82] Heltai L, Arroyo M A. DeSimone M. Nonsingular isogeometric boundary element method for stokes flows in 3D[J]. Computer Methods in Applied Mechanics and Engineering, 2014, 268 : 514-539.

[83] Simpson R N, Scott M A, Taus M, et al. Acoustic isogeometric boundary element analysis[J]. Computer Methods in Applied Mechanics and Engineering, 2014, 269: 265-290.

[84] Banerjee P K, Raveendra S T. Advanced boundary element analysis of two- and three-dimensional problems of elasto-plasticity[J]. International Journal for Numerical Methods in Engineering, 1986, 23: 985-1002.

[85] Xu C, Zhan Y S, Dai R, et al. RI-IGABEM for 3D viscoelastic problems with body force[J]. Computer Methods in Applied Mechanics and Engineering, 2022, 394: 114911.

[86] Xu C, Yang H, Zhan Y, et al. Non-conforming coupling RI-IGABEM for solving multidimensional and multiscale thermoelastic-viscoelastic problems[J]. Computer Methods in Applied Mechanics and Engineering, 2023, 403: 115725.

[87] 郑保敬. 功能梯度材料动态热力耦合分析的径向积分边界元法研究及其应用[D]. 大连: 大连理工大学, 2015.

[88] Cruse T A. Boundary Element Analysis in Computational Fracture Mechanics[M]. Boston: Kluwer Academic Publishers, 1988.

[89] Cheng A H D, Cheng D T. Heritage and early history of the boundary element method[J]. Engineering Analysis with Boundary Elements, 2005, 29: 268-302.

[90] Cruse T A, Aithal R. Non-singular boundary integral equation implementation[J]. International Journal for Numerical Methods in Engineering, 1993, 36: 237-254.

[91] Sladek V, Sladek J. Singular integrals and boundary elements[J]. Computer Methods in Applied Mechanics and Engineering, 1998, 157 : 251-266.

[92] Zhang Y M, Gu Y. An effective method in BEM for potential problems of thin bodies[J]. Journal of Materials Science & Technology, 2010, 18:137-144.

[93] Sladek V, Sladek J, Tanaka M. Regularization of hypersingular and nearly singular integrals in the potential theory and elasticity[J]. International Journal for Numerical Methods in Engineering, 1993, 36: 1609-1628.

[94] Schulz H, Schwab C, Wendl W L. The computation of potentials near and on the boundary by an extraction technique for boundary element methods[J]. Computer Methods in Applied Mechanics and Engineering, 1998, 157: 225-238.

[95] Telles J C F. A self-adaptive coordinate transformation for efficient numerical evaluation of general boundary element integral[J]. International Journal for Numerical Methods in Engineering, 1987, 24 : 959-973.

[96] Gao X W, Yang K, Wang J. An adaptive element subdivision technique for evaluation of various 2D singular boundary integrals[J]. Engineering Analysis with Boundary Elements, 2008, 32: 692-696.

[97] Niu Z R, Wendland W L, Wang X X, et al. A semi-analytical algorithm for the evaluation of the nearly singular integrals in three dimensional boundary element methods[J]. Computer Methods in Applied Mechanics and Engineering, 2005, 194: 1057-1074.

[98] Zhang Y M, Gu Y, Chen J T. Stress analysis for multilayered coating systems using semi-analytical BEM with geometric non-linearities[J]. Computational Mechanics, 2011, 47: 493-504.

[99] Sladek V, Sladek J, Tanaka M. Optimal transformations of the integration variables in computation of singular integrals in BEM[J].International Journal for Numerical Methods in Engineering, 2000, 47: 1263-1283.

[100] Ye W J. A new transformation technique for evaluating nearly singular integrals[J]. Computational Mechanics, 2008, 42: 457-466.

[101] Zhang Y M, Qu W Z, Chen J T. BEM analysis of thin structures for thermoelastic problems [J]. Engineering Analysis with Boundary Elements, 2013, 37: 441-452.

[102] Gong Y P, Trevelyan J, Hattori G, et al. Hybrid nearly singular integration for isogeometric boundary element analysis of coatings and other thin 2D structures[J]. Computer Methods in Applied Mechanics and Engineering, 2019, 346: 642-673.

[103] Gong Y P, Dong C Y, Qin F, et al. Hybrid nearly singular integration for three-dimensional isogeometric boundary element analysis of coatings and other thin structures[J]. Computer Methods in Applied Mechanics and Engineering, 2020, 367: 113099.

[104] Han Z L, Cheng C Z, Hu Z, et al. The semi-analytical evaluation for nearly singular integrals in isogeometric elasticity boundary element method[J]. Engineering Analysis with Boundary Elements, 2018, 95: 286-296.

[105] Aliabadi M H, Martin D. Boundary element hyper-singular formulation for elastoplastic contact problems[J]. International Journal for Numerical Methods in Engineering, 2000, 48: 995-1014.

[106] Mukherjee S. Boundary Element Methods in Creep and Fracture[M]. London: Applied Science

Publishers, 1982.
[107] Gao X W, Davies T G. Boundary Element Programming in Mechanics[M]. Cambridge: Cambridge University Press, 2002.
[108] Cruse T A. Boundary Integral Equation Method for Three-Dimensional Elastic Fracture Mechanics[M]. AFOSR-TR-75-0813, ADA 011660, Pratt and Whitney Aircraft, Connecticut, 1975.
[109] Rizzo F J, Shippy D J. An advanced boundary integral equation method for three-dimensional thermoelasticity[J]. International Journal for Numerical Methods in Engineering, 1977, 11: 1753-1768.
[110] Danson D J. A boundary element formulation for problems in linear isotropic elasticity with body forces[C]//Brebbia C A. Boundary Element Methods, 1981: 105-122.
[111] Nardini D, Brebbia C A. A new approach for free vibration analysis using boundary elements [J]. Applied Mathematical Modelling, 1983, 7: 157-162.
[112] Partridge P W, Brebbia C A, Wrobel L C. The Dual Reciprocity Boundary Element Method[M]. Southampton: Computational Mechanics Publications, 1992.
[113] Cheng A H D, Young D L, Tsai C C. Solution of Poisson's equation by iterative DRBEM using compactly supported, positive definite radial basis function[J]. Engineering Analysis with Boundary Elements, 2000, 24: 549-557.
[114] Golberg M A, Chen C S, Bowman H. Some recent results and proposals for the use of radial basis functions in the BEM[J]. Engineering Analysis with Boundary Elements, 1999, 23: 285-296.
[115] Power H, Mingo R. The DRM subdomain decomposition approach to solve the two-dimensional Navier-Stokes system of equations[J]. Engineering Analysis with Boundary Elements, 2000, 24: 107-119.
[116] Wen P H, Aliabadi M H, Rooke D P. A new method for transformation of domain integrals to boundary integrals in boundary element method[J]. International Journal for Numerical Methods in Engineering, 1998, 14: 1055-1065.
[117] Yu B, Cao G Y, Huo W D, et al. Isogeometric dual reciprocity boundary element method for solving transient heat conduction problems with heat sources[J]. Journal of Computational and Applied Mathematics, 2020, 385 : 113197.
[118] Yu B, Cao G Y, Gong Y P, et al. IG-DRBEM of three-dimensional transient heat conduction problems[J]. Engineering Analysis with Boundary Elements, 2021, 128: 298-309.
[119] Cao G Y, Yu B, Chen L L, et al. Isogeometric dual reciprocity BEM for solving non-Fourier transient heat transfer problems in FGMs with uncertainty analysis[J]. International Journal of Heat And Mass Transfer, 2023, 203: 123783.
[120] Yu B, Cao G Y, Ren S H, et al. An isogeometric boundary element method for transient heat transfer problems in inhomogeneous materials and the non-iterative inversion of loads[J]. Applied Thermal Engineering, 2022, 212: 118600.
[121] Nowak A J, Brebbia C A. The multiple-reciprocity method. A new approach for transforming B.E.M. domain integrals to the boundary[J]. Engineering Analysis with Boundary Elements,

1989, 6: 164-186.

[122] Neves A C, Brebbia C A. The multiple reciprocity boundary element method in elasticity: a new approach for transforming domain integral to the boundary[J]. International Journal for Numerical Methods in Engineering, 1991, 31 : 709-727.

[123] Ochiai Y, Kobayashi T. Initial stress formulation for elastoplastic analysis by improved multiple-reciprocity boundary element method[J]. Engineering Analysis with Boundary Elements, 1999, 23: 167-173.

[124] Gao X W. A boundary element method without internal cells for two dimensional and three-dimensional elastoplastic problems[J]. Journal of Applied Mechanics-Transactions of the ASME, 2002, 69: 154-160.

[125] Gao X W. The radial integration method for evaluation of domain integrals with boundary-only discretization[J]. Engineering Analysis with Boundary Elements, 2002, 26: 905-916.

[126] Feng W Z, Gao L F, Du J M, et al. A meshless interface integral BEM for solving heat conduction in multi-non-homogeneous media with multiple heat sources[J]. International Communications in Heat and Mass Transfer, 2019, 104 : 70-82.

[127] Feng W Z, Li H Y, Gao L F, et al. Hypersingular flux interface integral equation for multi-medium heat transfer analysis[J]. International Journal of Heat And Mass Transfer, 2019, 138: 852-865.

[128] Dong C Y, Lo S H, Cheung Y K. Numerical solution for elastic inclusion problems by domain integral equation with integration by means of radial basis functions[J]. Engineering Analysis with Boundary Elements, 2004, 28 : 623-632.

[129] Yang K, Gao X W. Radial integration BEM for transient heat conduction problems[J]. Engineering Analysis with Boundary Elements, 2010, 34 : 557-563.

[130] Yao W A, Yu B, Gao X W, et al. A precise integration boundary element method for solving transient heat conduction problems[J]. International Journal of Heat And Mass Transfer, 2014, 78 : 883-891.

[131] Gao X W, Zhang C, Guo L. Boundary-only element solutions of 2D and 3D nonlinear and nonhomogeneous elastic problems[J]. Engineering Analysis with Boundary Elements, 2007, 31 : 974-982.

[132] Yang K, Peng H F, Wang J, et al. Radial integration BEM for solving transient nonlinear heat conduction with temperature-dependent conductivity[J]. International Journal of Heat And Mass Transfer, 2017, 108 : 1551-1559.

[133] 高效伟, 周魏巍, 吕军, 等. 基于多边形网格的变系数问题边界单元法[J]. 应用力学学报, 2016, 33: 188-194.

[134] Xu C, Dong C Y. RI-IGABEM in inhomogeneous heat conduction problems [J]. Engineering Analysis with Boundary Elements, 2021, 124: 21-236.

[135] Xu C, Dong C Y, Dai R. RI-IGABEM based on PIM in transient heat conduction problems of FGMs [J]. Computer Methods in Applied Mechanics and Engineering, 2020, 374: 113601.

[136] Guo S P, Zhang J M, Li G Y, et al. Three dimensional transient heat conduction analysis by Laplace transformation and multiple reciprocity boundary face method[J]. Engineering Analysis with Boundary Elements, 2013, 37 : 15-22.

[137] Wang C H, Grigoriev M M, Dargush G F. A fast multi-level convolution boundary element method for transient diffusion problems[J]. International Journal for Numerical Methods in Engineering, 2005, 62 : 1895-1926.
[138] 徐闯. 功能梯度材料瞬态热传导问题边界条件和几何形状的直接反演方法研究[D]. 合肥: 合肥工业大学, 2019.
[139] Zhong W X, Williams F W. A precise time step integration method[J]. Proceedings of the Institution of Mechanical Engineers Part C-Journal of Mechanical Engineering SCIE, 1994, 208: 427-430.
[140] 余波. 非稳态热传导问题分析的时域径向积分边界元法[D]. 大连: 大连理工大学, 2014.
[141] 钟万勰. 子域精细积分及偏微分方程数值解[J]. 计算结构力学及其应用, 1995, 3: 253-260.
[142] 施吉林, 张宏伟, 金光日. 计算机科学计算[M]. 北京: 高等教育出版社, 2005.
[143] 林家浩,张亚辉. 结构动力学基础[M]. 大连: 大连理工大学出版社, 2007.
[144] Chen B S, Tong L Y, Gu Y X, et al. transient heat transfer analysis of functionally graded materials using adaptive precise time integration and graded finite elements[J]. Numerical Heat Transfer Part B-Fundamentals, 2004, 45: 181-200.
[145] Yu B, Yao W A, Gao Q. A precise integration boundary element method for solving transient heat conduction problems with variable thermal conductivity[J]. Numerical Heat Transfer Part B-Fundamentals, 2014, 65: 472-493.
[146] 李红云,沈为平. 结构动力响应的精细时程积分并行算法[J]. 上海交通大学学报, 1998, 8: 83-87.
[147] Tang B. Combined dynamic stiffness matrix and precise time integration method for transient forced vibration response analysis of beams[J]. Journal of Sound and Vibration, 2008, 309 : 868-876.
[148] 方宏远. 基于辛算法的层状结构探地雷达检测正反演研究[D]. 大连: 大连理工大学, 2012.
[149] 张雄,王天舒, 刘岩. 计算动力学[M]. 北京: 清华大学出版社, 2015.
[150] Newmark N M. A method of computation for structural dynamics[J]. Journal of Engineering Mechanics, 1959, 85 : 67-94.
[151] Gao X W, Zheng B J, Yang K, et al. Radial integration BEM for dynamic coupled thermoelastic analysis under thermal shock loading[J]. Computers & Structures, 2015, 158 : 140-147.
[152] Wilson E L. Division of Structural Engineering and Structural Mechanics[M]. Berkeley, CA: University of California, 1968.
[153] Bazzi G, Anderheggen E. The ρ-family of algorithms for time-step integration with improved numerical dissipation[J]. Earthquake Engineering & Structural Dynamics, 1982, 10: 537-550.
[154] Hilber H M, Hughes T J, Taylor R L. Improved numerical dissipation for time integration algorithms in structural dynamics[J]. Earthquake Engineering & Structural Dynamics, 1977, 5 : 283-292.
[155] Wood W, Bossak M, Zienkiewicz O. An alpha modification of newmark's method[J]. International Journal for Numerical Methods in Engineering, 1980, 15: 1562-1566.
[156] 吴月. 动载荷识别的非迭代法研究[D]. 合肥: 合肥工业大学, 2020.
[157] Zienkiewicz O C, Kelly D W, Bettess P. The coupling of the finite element method and boundary

solution procedures[J]. International Journal for Numerical Methods in Engineering, 1977, 11(2): 355-375.
[158] Coda H B, Venturini W S, Aliabadi M H. A general 3D BEM/FEM coupling applied to elastodynamic continua/frame structures interaction analysis[J]. International Journal for Numerical Methods in Engineering, 1999, 46(5): 695-712.
[159] Elleithy W, Grzhibovskis R. An adaptive domain decomposition coupled finite element-boundary element method for solving problems in elasto-plasticity[J]. International Journal for Numerical Methods in Engineering, 2009, 79(8): 1019-1040.
[160] Aour B, Rahmani O, Nait-Abdelaziz M. A coupled FEM/BEM approach and its accuracy for solving crack problems in fracture mechanics[J]. International Journal of Solids and Structures, 2007, 44(7): 2523-2539.
[161] Mobasher M E, Waisman H. Adaptive modeling of damage growth using a coupled FEM/BEM approach[J]. International Journal for Numerical Methods in Engineering, 2016, 105(8): 599-619.
[162] Soares D, Jr. Acoustic modelling by BEM-FEM coupling procedures taking into account explicit and implicit multi-domain decomposition techniques[J]. International Journal for Numerical Methods in Engineering, 2009, 78(9): 1076-1093.
[163] Zhao W, Chen L, Chen H, et al. Topology optimization of exterior acoustic-structure interaction systems using the coupled FEM-BEM method[J]. International Journal for Numerical Methods in Engineering, 2019, 119(5): 404-431.
[164] Heltai L, Kiendl J, Desimone A, et al. A natural framework for isogeometric fluid-structure interaction based on BEM-shell coupling[J]. Computer Methods in Applied Mechanics and Engineering, 2017, 316: 522-546.
[165] Yildizdag M E, Ardic I T, Kefal A, et al. An isogeometric FE-BE method and experimental investigation for the hydroelastic analysis of a horizontal circular cylindrical shell partially filled with fluid[J]. Thin-Walled Structures, 2020, 151: 106755.
[166] May S, Kästner M, Müller S, et al. A hybrid IGAFEM/IGABEM formulation for two-dimensional stationary magnetic and magneto-mechanical field problems[J]. Computer Methods in Applied Mechanics and Engineering, 2014, 273: 161-180.
[167] Maestre J, Pallares J, Cuesta I, et al. A 3D isogeometric BE-FE analysis with dynamic remeshing for the simulation of a deformable particle in shear flows[J]. Computer Methods in Applied Mechanics and Engineering, 2017, 326: 70-101.
[168] Hsiao G C, Schnack E, Wendland W L. A hybrid coupled finite-boundary element method in elasticity[J]. Computer Methods in Applied Mechanics and Engineering, 1999, 173(3): 287-316.
[169] Hsiao G C, Schnack E, Wendland W L. Hybrid coupled finite-boundary element methods for elliptic systems of second order[J]. Computer Methods in Applied Mechanics and Engineering, 2000, 190(5): 431-485.
[170] Schnack E, Türke K. Domain decomposition with BEM and FEM[J]. International Journal for Numerical Methods in Engineering, 1997, 40(14): 2593-2610.
[171] Fischer M, Gaul L. Fast BEM-FEM mortar coupling for acoustic-structure interaction[J].

International Journal for Numerical Methods in Engineering, 2005, 62(12): 1677-1690.

[172] González J A, Park K C, Felippa C A. FEM and BEM coupling in elastostatics using localized Lagrange multipliers[J]. International Journal for Numerical Methods in Engineering, 2007, 69(10): 2058-2074.

[173] González J A, Rodríguez-Tembleque L, Park K C, et al. The nsBETI method: an extension of the FETI method to non-symmetrical BEM-FEM coupled problems[J]. International Journal for Numerical Methods in Engineering, 2013, 93(10): 1015-1039.

[174] Rüberg T, Schanz M. Coupling finite and boundary element methods for static and dynamic elastic problems with non-conforming interfaces[J]. Computer Methods in Applied Mechanics and Engineering, 2008, 198(3): 449-458.

[175] Bazilevs Y, Hsu M C, Scott M A. Isogeometric fluid-structure interaction analysis with emphasis on non-matching discretizations, and with application to wind turbines[J]. Computer Methods in Applied Mechanics and Engineering, 2012, 249-252: 28-41.

[176] Yang H S, Dong C Y, Wu Y H. Non-conforming interface coupling and symmetric iterative solution in isogeometric FE-BE analysis[J]. Computer Methods in Applied Mechanics and Engineering, 2021, 373: 113561.

[177] Coox L, Greco F, Atak O, et al. A robust patch coupling method for NURBS-based isogeometric analysis of non-conforming multipatch surfaces[J]. Computer Methods in Applied Mechanics and Engineering, 2017, 316: 235-260.

[178] Zhan Y, Xu C, Yang H, et al. Isogeometric FE-BE method with non-conforming coupling interface for solving elasto-thermoviscoelastic problems[J]. Engineering Analysis with Boundary Elements, 2022, 141 :199-221.

[179] McCune R W, Armstrong C G, Robinson D J. Mixed-dimensional coupling in finite element models[J]. International Journal for Numerical Methods in Engineering, 2000, 49(6): 725-750.

[180] Shim K W, Monaghan D J, Armstrong C G. Mixed dimensional coupling in finite element stress analysis[J]. Engineering with Computers, 2002, 18(3): 241-252.

[181] Cuillière J C, Bournival S, François V. A mesh-geometry-based solution to mixed-dimensional coupling[J]. Computer-Aided Design, 2010, 42(6): 509-522.

[182] Haas M, Kuhn G. Mixed-dimensional, symmetric coupling of FEM and BEM[J]. Engineering Analysis with Boundary Elements, 2003, 27(6): 575-582.

[183] Haas M, Helldörfer B, Kuhn G. Improved coupling of finite shell elements and 3D boundary elements[J]. Archive of Applied Mechanics, 2006, 75(10): 649.

[184] Nguyen V P, Kerfriden P, Claus S, et al. Nitsche's method for mixed dimensional analysis: conforming and non-conforming continuum-beam and continuum-plate coupling[J]. arXiv, http://arxiv.org/abs/1308.2910, 2013.

[185] Yang H S, Dong C Y, Wu Y H, et al. Mixed dimensional isogeometric FE-BE coupling analysis for solid-shell structures[J]. Computer Methods in Applied Mechanics and Engineering, 2021, 382: 113841.

第 2 章

含裂纹薄壳的等几何有限元分析

2.1 引　　言

飞机机身、压力容器和管道等大量工程结构都可以用薄壳结构进行模拟，这些结构在循环载荷作用下往往会产生贯穿厚度方向的裂纹。由于几何、壳体运动学、材料本构关系和断裂现象描述之间的复杂相互作用，所以准确计算相应的断裂参数对含裂纹壳体的建模和数值分析至关重要。传统有限元法在处理裂纹问题时存在的挑战性问题[1,2]和相应的处理方法[3-10]，以及 XIGA[11-18]在断裂力学问题中的应用已在 1.2.2 节中进行了描述。

基于 NURBS 基函数的等几何分析不容易直接处理含缺陷的结构，因为 NURBS 基函数的张量拓扑形式导致产生多余的控制点和引入冗余的自由度。为此，一些具有固有局部细分特性的样条函数[19-23]被提出。本章采用基于 PHT 样条[21]的 XIGA 来研究含贯穿裂纹的薄壳结构。

2.1.1 分层网格

T 网格是参数空间中包含 T 结点的非张量积拓扑形式的矩形网格。分层 T 网格开始于张量积拓扑形式的 NURBS 或 B 样条参数网格，从 k 层到 k+1 层，分层 T 网格中的一部分参数单元通过连接其相对边中点的方式被划分为四个子单元。图 2.1 所示为分层 T 网格逐层细分的示意图。

PHT 样条基函数是在分层 T 网格上逐层构造的。分层 T 网格上的网格点根据位置不同可以分为边界点、十字交叉点和 T 结点。在图 2.1 中，蓝色方点、红色圆点和绿色三角形点分别表示边界点、十字交叉点和 T 结点。根据文献[1]，PHT 样条空间的维度仅与十字交叉点和边界点的数量有关。因此，十字交叉点和边界点也称作基点。在本章中，仅涉及样条空间为 $S(3,3,1,1,\mathscr{T})$ 的情形，其空间维度表达式为[21]

$$\dim S(3,3,1,1,\mathscr{T}) = 4\left(V^{\mathrm{b}} + V^{+}\right) \tag{2.1}$$

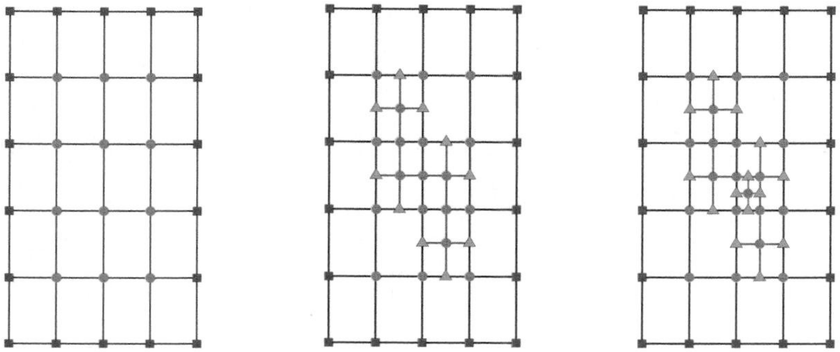

(a)　(b)　(c)

图 2.1　分层 T 网格示意图

蓝色方点、红色圆点和绿色三角形点分别表示边界基点、十字交叉基点和 T 结点。(a) 初始网格；(b) 第一次细分；(c) 第二次细分

其中，$\mathcal{T}$ 代表样条函数空间，括号内数字 3 和 1 表示样条为 3 阶次和 C^1 连续性；V^{b} 和 V^+ 分别代表分层 T 网格上边界点和十字交叉点的数量。通过维度公式可知，每个基点只与 4 个基函数相关联。

2.1.2　样条函数构造

PHT 样条基函数的构造按照在参数空间中逐层的方式进行，从 k 层到 k+1 层，基于分层策略其过程被分为两个步骤：第一步是在 k 层上修改原始的基函数；第二步是生成与 k+1 层上的新基点相关的新基函数。

1. k 层上原始基函数的修改

首先，与待细分单元相关的基函数需要在该单元参数空间中用 16 个 Bézier 坐标(图 2.2(a))表示

$$R(\xi,\eta)=\sum_{j=1}^{4}\sum_{i=1}^{4}c_{ij}B_{i,3}(\xi)B_{j,3}(\eta) \tag{2.2}$$

其中，c_{ij} 为对应的 Bézier 坐标[2]。当 PHT 样条空间的初始层为 NURBS 基函数网格时(控制点的权值不全为 1)，可将 NURBS 样条转化为齐次坐标形式的 B 样条，仍按照上述方法获取 Bézier 坐标，待细分完成后再转换回去。$B_{i,3}$ 和 $B_{j,3}$ 为定义在 $[-1,1]$ 上的三次 Bernstein 基函数，其表达式为[22]

$$B_{m,3}(\xi)=\frac{1}{2^3}\frac{3!}{(m-1)!(4-m)!}(1-\xi)^{4-m}(1+\xi)^{m-1}\quad(1\leqslant m\leqslant 4,\ m=i,j) \tag{2.3}$$

然后利用 de Casteljau 算法[3]，得到该基函数在细分后的四个子单元内的 Bézier 坐标，如图 2.2(b)所示。值得注意的是，此时的原始基函数还未发生改变，只是被定义在 k+1 层的四个子单元内。单元细分后会有新的基点生成，通过将与新基点关联位置处的 Bézier 坐标设置为零来完成对原始基函数的修改，如图 2.2(c)所示。

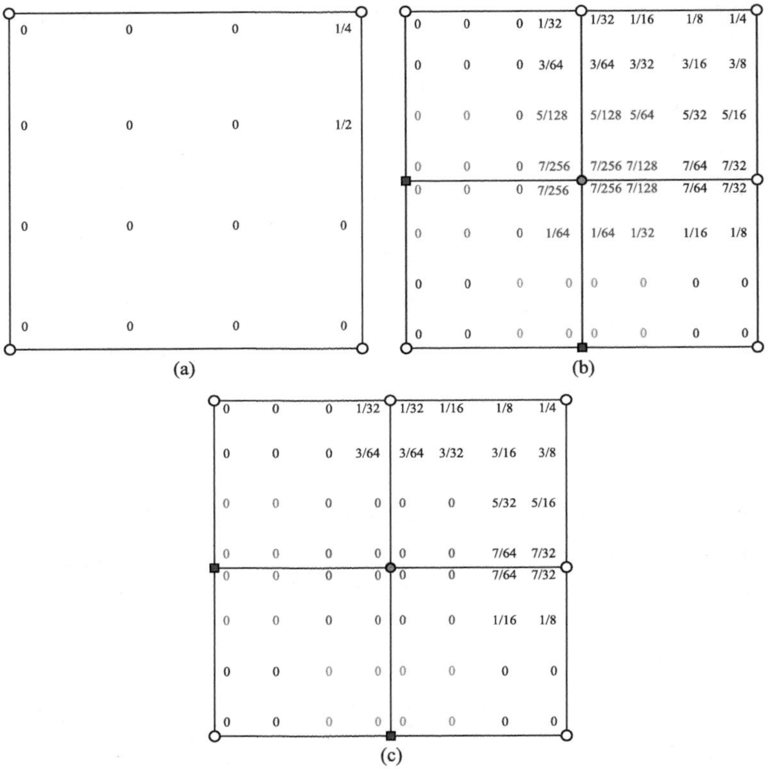

图 2.2 基函数修改过程

(a) 通过在单元中指定 16 个 Bézier 坐标表示基函数；(b) 单元被细分为四个子单元，蓝色方点和红色圆点分别表示新的边界基点和十字交叉基点，然后利用 de Casteljau 算法计算在四个子单元中的 Bézier 坐标；(c) 将新基点周围的 Bézier 坐标设置为零

2. k+1 层上新基函数的生成

接下来是生成与新基点相关的新基函数。与新基点 $\left(\xi_i, \eta_j\right)$ 相关的 4 个新基函数有着共同的支持域 $(\xi_{i-1}, \xi_{i+1}) \times (\eta_{j-1}, \eta_{j+1})$，如图 2.3 所示。与 4 个新的 PHT 样条基函数相关的局部节点矢量为

$$\begin{aligned}&\{\xi_{i-1},\xi_{i-1},\xi_i,\xi_i,\xi_{i+1}\}\times\{\eta_{j-1},\eta_{j-1},\eta_j,\eta_j,\eta_{j+1}\}\\&\{\xi_{i-1},\xi_i,\xi_i,\xi_{i+1},\xi_{i+1}\}\times\{\eta_{j-1},\eta_{j-1},\eta_j,\eta_j,\eta_{j+1}\}\\&\{\xi_{i-1},\xi_{i-1},\xi_i,\xi_i,\xi_{i+1}\}\times\{\eta_{j-1},\eta_j,\eta_j,\eta_{j+1},\eta_{j+1}\}\\&\{\xi_{i-1},\xi_{i-1},\xi_i,\xi_i,\xi_{i+1}\}\times\{\eta_{j-1},\eta_j,\eta_j,\eta_{j+1},\eta_{j+1}\}\end{aligned}\tag{2.4}$$

已知局部节点矢量后，新的基函数就可通过 Cox-de Boor 递推公式[4]或 Bézier 提取算子进行计算。当该细分单元为边界单元时，需将局部节点矢量中对应的节点值设为零，即 $\xi_{i-1}=0$ (左边界)、$\xi_{i+1}=0$ (右边界)、$\eta_{j-1}=0$ (下边界)和 $\eta_{j+1}=0$ (上边界)。

记 k 层上修改的和未修改的基函数分别为 $\overline{R}_j^k$ 和 R_j^k，以及 k+1 层上新增加的基函数为 $\overline{R}_j^{k+1}$，我们可以得到细分后样条空间内 PHT 样条基函数为

$$R^{k+1}(\xi,\eta)=\sum_{j=1}^{N_{k+1}}\overline{R}_j^{k+1}(\xi,\eta)+\sum_{j=1}^{N_k}\overline{R}_j^k(\xi,\eta)+\sum_{j=1}^{N_k^*}R_j^k(\xi,\eta)\tag{2.5}$$

其中，N_k^*、N_k 和 N_{k+1} 分别代表 k 层上未修改的基函数、k 层上需要修改的基函数和 k+1 层上新增加的基函数的数量。

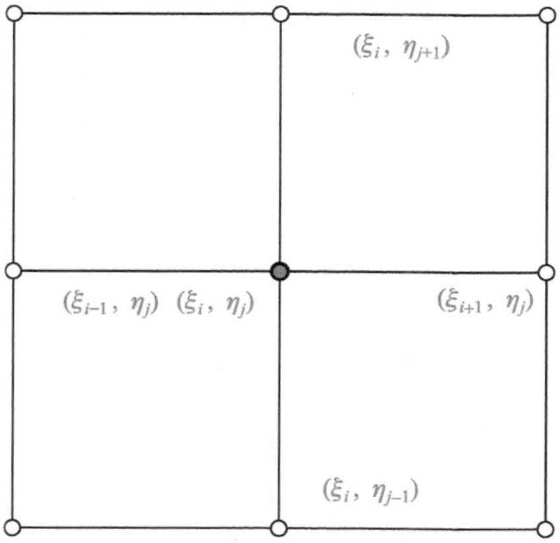

图 2.3　新十字交叉基点与相邻单元

上述局部细分的过程依然能够保证基函数单位分解性、线性无关性和局部支撑性的优点[4]。PHT 样条网格的粗化过程相对复杂且存在一定误差[21]，因此在本章中规定相邻单元的层数之差不能大于 1。为了更好地展示 PHT 样条细分过程中基函数的变化，此处给出一个示例进行说明。

考虑如图 2.4 所示的定义在节点矢量 $\Xi=\left\{0,0,0,0,\frac{1}{3},\frac{1}{3},\frac{2}{3},\frac{2}{3},1,1,1,1\right\}$ 和

$\boldsymbol{\Psi}=\left\{0,0,0,0,\frac{1}{3},\frac{1}{3},\frac{2}{3},\frac{2}{3},1,1,1,1\right\}$ 上第 0 层(k 层)的 NURBS 参数网格，该层上包含 9 个三次单元、64 个基函数，并且单元间保持 C^1 连续性。

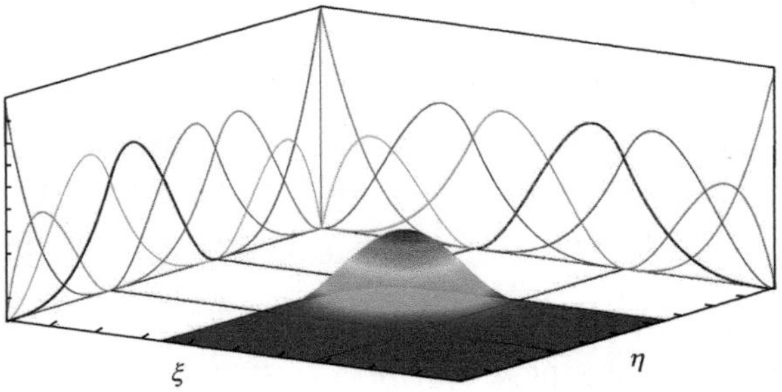

图 2.4　第 0 层上的 NURBS 参数网格及两个参数方向上的基函数曲线

初始层上单元的编号如图 2.5(a)所示。本例中将对初始层上的第 1、第 2 和第 5 个单元进行局部细分。局部细分后第 1 层(k+1 层)上子单元的编号如图 2.5(b)所示。考虑第 29 个基函数(由 ξ 方向上第 5 个基函数与 η 方向上第 4 个基函数的张量积形成，如图 2.4 所示)在初始层第 5 个单元被细分后的修改过程。

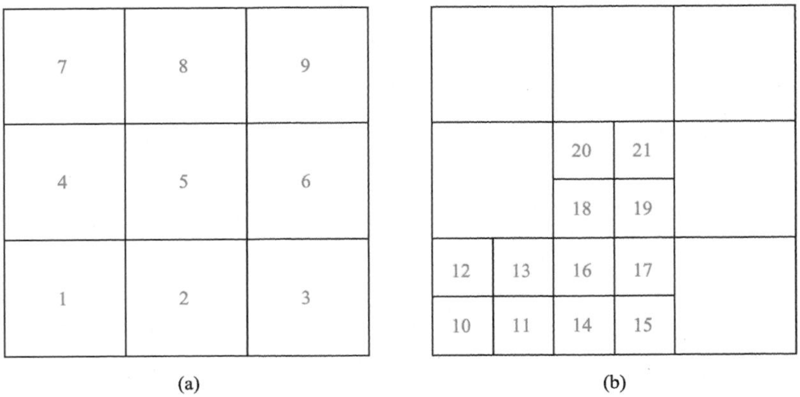

图 2.5　初始层单元和第 1 层上子单元的编号

(a) 初始层；(b) 第 1 层上子单元

首先将该基函数（其在初始层中第 5 个单元内的图像如图 2.6(a)所示）通过式(2.2)表示为第 5 个单元内的 Bézier 坐标(图 2.6(b))，当在该单元内插入十字交叉节点后(crossing insertion)，该基函数在 4 个子单元内的图像如图 2.7(a)所示，通过 de Casteljau 算法得到 4 个子单元内的 Bézier 坐标(图 2.7(b))。

此例中单元 5 局部细分后多增加了两个十字交叉基点(如图 2.8(b)中的红色圆点所示)。然后将与新基点相关联的 Bézier 坐标设置为零，如图 2.8(b)所示，可以

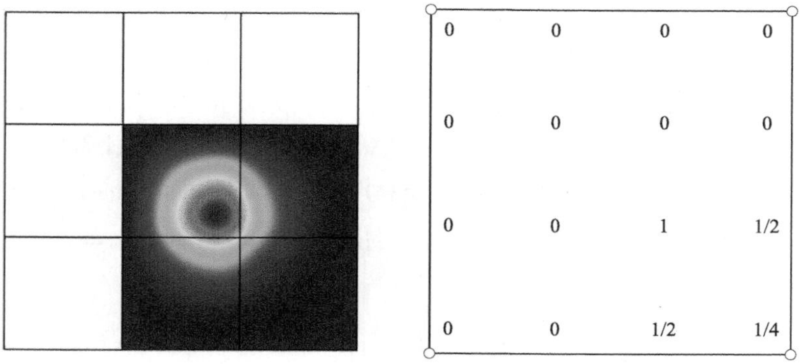

(a) (b)

图 2.6 选定的基函数在初始层中第 5 个单元内的图像和 Bézier 坐标

(a) 基函数图像；(b) 单元内的 Bézier 坐标

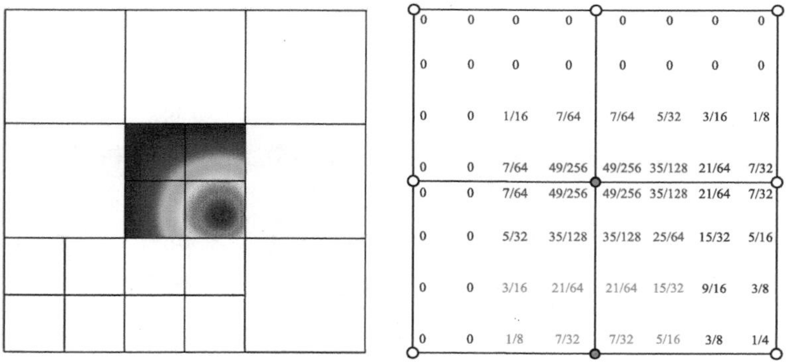

(a) (b)

图 2.7 利用 de Casteljau 算法，选定的基函数在 4 个子单元内的图像和 Bézier 坐标

(a) 基函数图像；(b) 子单元内的 Bézier 坐标

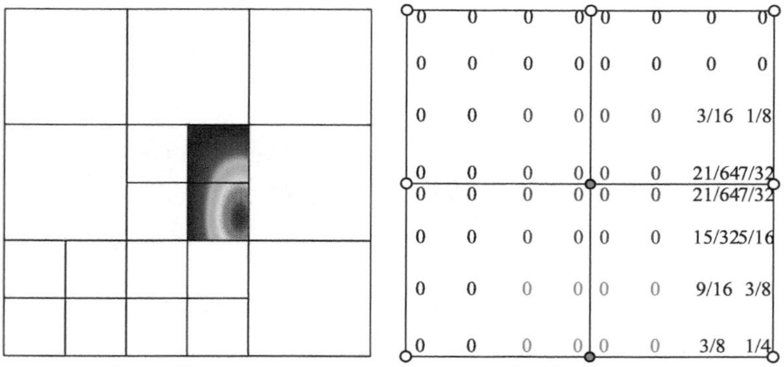

(a) (b)

图 2.8 将与新基点相关的 Bézier 坐标设置为零后，基函数在 4 个子单元内的图像和 Bézier 坐标

(a) 基函数图像；(b) 单元内的 Bézier 坐标

看到修改后的基函数支持域(图 2.8(a))不再包含子单元 18 和子单元 20。上述修改过程需遍历完初始层中支持单元 5 的所有基函数。

接下来是第 1 层上新基函数的生成。如图 2.9 所示，给出了与单元 5 中新增加的十字交叉基点在局部节点矢量 $\{1/3,1/2,1/2,2/3,2/3\}\times\{1/3,1/3,1/2,1/2,2/3\}$ 下新基函数在子单元 19 中的示意图(图 2.9(a))及相关的 Bézier 坐标(图 2.9(b))。

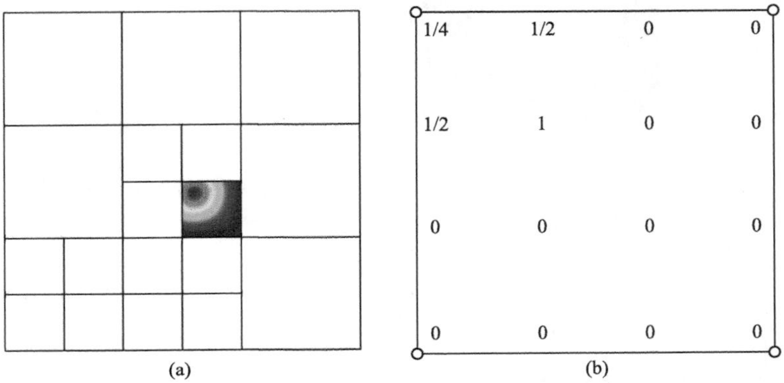

图 2.9 与单元 5 中新增加的十字交叉基点相关的新基函数在子单元 19 中的图像和 Bézier 坐标
(a) 基函数图像；(b) 单元内的 Bézier 坐标

求解出所有的与新基函数相关的 Bézier 坐标后，与初始层中修改后基函数的 Bézier 坐标合并，即可得到子单元 19 的 Bézier 提取算子矩阵。

2.1.3 样条曲面

PHT 曲面是 PHT 样条基函数的线性组合，组合系数为控制点，其表示为

$$\boldsymbol{S}(\xi,\eta)=\sum_{i}R_i^{k+1}(\xi,\eta)\boldsymbol{P}_i \tag{2.6}$$

式中，$R_i^{k+1}(\xi,\eta)$是 k+1 层上的 PHT 样条基函数；$\boldsymbol{P}_i$ 是相关的控制点。

局部细化过程中，曲面的几何和参数特性保持不变。上一层中与旧基点相关的控制点保持不变，当前层中与新基点相关的控制点通过以下方式确定[21]。

定义与 PHT 样条基函数相关的几何信息算子

$$\mathcal{L}R(\xi,\eta)=\left\{R(\xi,\eta),R_{,\xi}(\xi,\eta),R_{,\eta}(\xi,\eta),R_{,\xi\eta}(\xi,\eta)\right\} \tag{2.7}$$

式中，算子 $\mathcal{L}$ 包含 PHT 样条基函数 $R(\xi,\eta)$、对两个参数坐标方向上的偏导数 $R_{,\xi}(\xi,\eta)$ 和 $R_{,\eta}(\xi,\eta)$ 以及混合二阶偏导数 $R_{,\xi\eta}(\xi,\eta)$。

记与新基点$\left(\xi_{i^*},\eta_{j^*}\right)$相关的新基函数的编号为 j_1、j_2、j_3 和 j_4，则 PHT 样条

曲面 $\boldsymbol{S}$ 在新基点处的几何信息表示为

$$\begin{aligned}\mathcal{L}\boldsymbol{S}\left(\xi_{i^*},\eta_{j^*}\right)&=\left(\boldsymbol{S}\left(\xi_{i^*},\eta_{j^*}\right),\boldsymbol{S}_{,\xi}\left(\xi_{i^*},\eta_{j^*}\right),\boldsymbol{S}_{,\eta}\left(\xi_{i^*},\eta_{j^*}\right),\boldsymbol{S}_{,\xi\eta}\left(\xi_{i^*},\eta_{j^*}\right)\right)\\&=\sum_i \mathcal{L}R_i\left(\xi_{i^*},\eta_{j^*}\right)\boldsymbol{P}=\sum_{k=1}^{4}\mathcal{L}R_{j_k}\left(\xi_{i^*},\eta_{j^*}\right)\boldsymbol{P}_{j_k}=\boldsymbol{P}\cdot\boldsymbol{R}\end{aligned} \tag{2.8}$$

式中，$\boldsymbol{R}=\left[\mathcal{L}R_{j_1}\left(\xi_{i^*},\eta_{j^*}\right)\ \mathcal{L}R_{j_2}\left(\xi_{i^*},\eta_{j^*}\right)\ \mathcal{L}R_{j_3}\left(\xi_{i^*},\eta_{j^*}\right)\ \mathcal{L}R_{j_4}\left(\xi_{i^*},\eta_{j^*}\right)\right]$ 为 4×4 矩阵，其具体表达式为[21]

$$\boldsymbol{R}=\begin{bmatrix}(1-\lambda)(1-\mu) & \lambda(1-\mu) & \lambda\mu & \mu(1-\lambda)\\ -\alpha(1-\mu) & \alpha(1-\mu) & \alpha\mu & -\alpha\mu\\ -\beta(1-\lambda) & -\beta\lambda & \beta\lambda & \beta(1-\lambda)\\ \alpha\beta & -\alpha\beta & \alpha\beta & -\alpha\beta\end{bmatrix} \tag{2.9}$$

式中，

$$\alpha=\frac{1}{\Delta\xi_1+\Delta\xi_2},\quad \beta=\frac{1}{\Delta\eta_1+\Delta\eta_2},\quad \lambda=\alpha\Delta\xi_1,\quad \mu=\beta\Delta\eta_1 \tag{2.10}$$

其中，由图 2.3 可得

$$\Delta\xi_1=\xi_i-\xi_{i-1},\quad \Delta\xi_2=\xi_{i+1}-\xi_i,\quad \Delta\eta_1=\eta_j-\eta_{j-1},\quad \Delta\eta_2=\eta_{j+1}-\eta_j \tag{2.11}$$

式(2.8)中的 $\boldsymbol{P}=\left[\boldsymbol{P}_{j_1}\ \ \boldsymbol{P}_{j_2}\ \ \boldsymbol{P}_{j_3}\ \ \boldsymbol{P}_{j_4}\right]$ 为 3×4 矩阵。

为了进行比较，图 2.10(a)表示通过 k 细分策略后具有 C^1 连续性的三次 NURBS 等几何单元的控制点分布，而图 2.10(b)表示通过 PHT 样条局部细分后等几何单元的控制点分布。可以明显地看出，在 PHT 样条局部细化情况下，单元的控制点向新基点周围聚集。

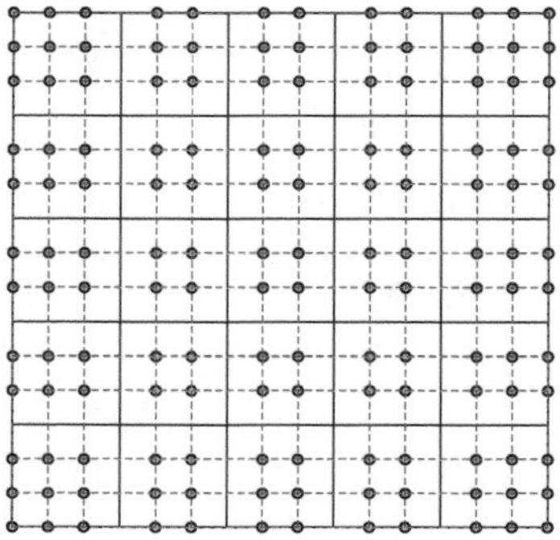

(a)

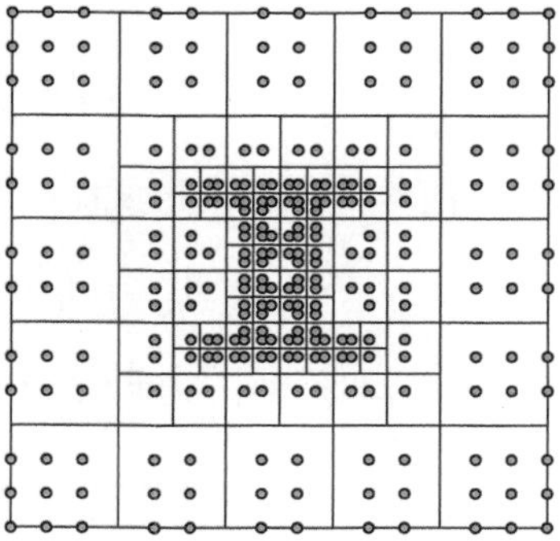

(b)

图 2.10 不同细分方式下，控制点分布的差异

(a) NURBS 细分；(b) PHT 局部细分

2.2 等几何薄壳公式

在 Kirchhoff-Love 壳理论中，横向剪切变形可以忽略不计，变形后的横截面依然垂直于壳体的中性面。因此，三维连续壳体可以用其中性面及其法向量来描述。壳体中性面的映射由其两个曲线坐标表示。本章使用如下的符号约定，即大写字母代表变形前的变量，小写字母代表变形后的变量。拉丁字母的取值范围为{1,2,3}，而希腊字母的取值范围为{1,2}。

记壳体两个参数方向的曲线坐标为$\theta^1=\xi$ 、$\theta^2=\eta$，以及厚度方向上的参数坐标为$\theta^3=\zeta$，则壳体变形前后任意一点处的位置矢量$\boldsymbol{X}(\xi,\eta,\zeta)$和$\boldsymbol{x}(\xi,\eta,\zeta)$表示为

$$\begin{cases}\boldsymbol{X}(\xi,\eta,\zeta)=\boldsymbol{R}(\xi,\eta)+\zeta\boldsymbol{N}(\xi,\eta)\\ \boldsymbol{x}(\xi,\eta,\zeta)=\boldsymbol{r}(\xi,\eta)+\zeta\boldsymbol{n}(\xi,\eta)\end{cases}\tag{2.12}$$

其中，$\boldsymbol{R}(\xi,\eta)$和$\boldsymbol{r}(\xi,\eta)$为该点在壳中性面上投影点处的位置矢量；厚度坐标ζ的取值范围为$(-h/2,h/2)$，这里h为壳体的厚度；$\boldsymbol{N}(\xi,\eta)$和$\boldsymbol{n}(\xi,\eta)$为中性面上投影点处的单位法向量。

壳体在该点变形前后的协变基矢量分别为

$$\begin{cases}\boldsymbol{G}_1=\boldsymbol{X}_{,1}=\boldsymbol{R}_{,1}+\zeta\boldsymbol{N}_{,1}=\boldsymbol{A}_1+\zeta\boldsymbol{N}_{,1}\\ \boldsymbol{G}_2=\boldsymbol{X}_{,2}=\boldsymbol{R}_{,2}+\zeta\boldsymbol{N}_{,2}=\boldsymbol{A}_2+\zeta\boldsymbol{N}_{,2}\\ \boldsymbol{G}_3=\boldsymbol{X}_{,3}=\boldsymbol{N}\end{cases},\quad \begin{cases}\boldsymbol{g}_1=\boldsymbol{x}_{,1}=\boldsymbol{r}_{,1}+\zeta\boldsymbol{n}_{,1}=\boldsymbol{a}_1+\zeta\boldsymbol{n}_{,1}\\ \boldsymbol{g}_2=\boldsymbol{x}_{,2}=\boldsymbol{r}_{,2}+\zeta\boldsymbol{n}_{,2}=\boldsymbol{a}_2+\zeta\boldsymbol{n}_{,2}\\ \boldsymbol{g}_3=\boldsymbol{x}_{,3}=\boldsymbol{n}\end{cases}\tag{2.13}$$

因此，法向量的计算式为

$$\boldsymbol{N}=\frac{\boldsymbol{A}_1\times\boldsymbol{A}_2}{\left\|\boldsymbol{A}_1\times\boldsymbol{A}_2\right\|},\quad \boldsymbol{n}=\frac{\boldsymbol{a}_1\times\boldsymbol{a}_2}{\left\|\boldsymbol{a}_1\times\boldsymbol{a}_2\right\|}\tag{2.14}$$

壳体中性面的一阶基本形式(fundamental form)为[5]

$$A_{\alpha\beta}=\boldsymbol{A}_\alpha\cdot\boldsymbol{A}_\beta,\quad a_{\alpha\beta}=\boldsymbol{a}_\alpha\cdot\boldsymbol{a}_\beta\tag{2.15}$$

则可得到壳体中性面在该点处的逆变基矢量为

$$\boldsymbol{A}^\alpha=A^{\alpha\beta}\cdot\boldsymbol{A}_\beta,\quad \boldsymbol{a}^\alpha=a^{\alpha\beta}\cdot\boldsymbol{a}_\beta\tag{2.16}$$

式中，$A^{\alpha\beta}=\left(A_{\alpha\beta}\right)^{-1}$，$a^{\alpha\beta}=\left(a_{\alpha\beta}\right)^{-1}$。

根据壳体的基本假设，薄壳的变形完全可以用其中性面的变形来描述。中性面上任意一点处的位移矢量表示为

$$\boldsymbol{u}(\xi,\eta)=\boldsymbol{r}(\xi,\eta)-\boldsymbol{R}(\xi,\eta) \tag{2.17}$$

壳体变形分析中的格林–拉格朗日(Green-Lagrange)应变张量 $\boldsymbol{E}$ 定义为

$$\boldsymbol{E}=\frac{1}{2}\left(\boldsymbol{F}^{\mathrm{T}}\boldsymbol{F}-\boldsymbol{I}\right) \tag{2.18}$$

其中，$\boldsymbol{F}=\partial\boldsymbol{x}/\partial\boldsymbol{X}$ 为材料的变形梯度；$\boldsymbol{I}$ 为单位张量。由于 Kirchhoff-Love 薄壳假设，可以得到 $\boldsymbol{N}\cdot\boldsymbol{A}_\alpha=0$、$\boldsymbol{n}\cdot\boldsymbol{a}_\alpha=0$、$\boldsymbol{N}\cdot\boldsymbol{N}_{,\alpha}=0$ 和 $\boldsymbol{n}\cdot\boldsymbol{n}_{,\alpha}=0$。因此，只有中性面内的应变张量分量不为零。忽略高阶小量后，应变张量 $\boldsymbol{E}$ 的面内分量表示为[6]

$$E_{\alpha\beta}=\frac{1}{2}\left[\boldsymbol{a}_\alpha\cdot\boldsymbol{a}_\beta-\boldsymbol{A}_\alpha\cdot\boldsymbol{A}_\beta+2\zeta\left(\boldsymbol{A}_{\alpha,\beta}\cdot\boldsymbol{N}-\boldsymbol{a}_{\alpha,\beta}\cdot\boldsymbol{n}\right)\right] \tag{2.19}$$

上式推导过程中利用到如下的关系：

$$\begin{aligned}\boldsymbol{A}_\alpha\cdot\boldsymbol{N}_{,\beta}=\boldsymbol{A}_\beta\cdot\boldsymbol{N}_{,\alpha}=-\boldsymbol{N}\cdot\boldsymbol{A}_{\alpha,\beta}\\ \boldsymbol{a}_\alpha\cdot\boldsymbol{n}_{,\beta}=\boldsymbol{a}_\beta\cdot\boldsymbol{n}_{,\alpha}=-\boldsymbol{n}\cdot\boldsymbol{a}_{\alpha,\beta}\end{aligned} \tag{2.20}$$

式(2.19)可以简记为

$$E_{\alpha\beta}=\varepsilon_{\alpha\beta}+\zeta\kappa_{\alpha\beta} \tag{2.21}$$

式中，$\varepsilon_{\alpha\beta}$ 为薄膜应变分量，其表达式为

$$\varepsilon_{\alpha\beta}=\frac{1}{2}\left(\boldsymbol{a}_\alpha\cdot\boldsymbol{a}_\beta-\boldsymbol{A}_\alpha\cdot\boldsymbol{A}_\beta\right) \tag{2.22}$$

$\kappa_{\alpha\beta}$ 为弯曲应变分量，其表达式为

$$\kappa_{\alpha\beta}=\boldsymbol{A}_{\alpha,\beta}\boldsymbol{N}-\boldsymbol{a}_{\alpha,\beta}\boldsymbol{n} \tag{2.23}$$

将式(2.13)、式(2.17)和式(2.19)代入式(2.22)和式(2.23)中，并忽略非线性项，可得到

$$\varepsilon_{\alpha\beta}=\frac{1}{2}\left(\boldsymbol{A}_\alpha\boldsymbol{u}_{,\beta}+\boldsymbol{A}_\beta\boldsymbol{u}_{,\alpha}\right) \tag{2.24}$$

$$\kappa_{\alpha\beta}=\boldsymbol{A}_{\alpha,\beta}\boldsymbol{N}-\left(\boldsymbol{A}_{\alpha,\beta}+\boldsymbol{u}_{,\alpha\beta}\right)\boldsymbol{n} \tag{2.25}$$

由于式(2.25)中包含位移的二阶导数，所以形函数至少需要 C^1 连续性。这对于基于拉格朗日多项式函数的传统 FEM 来说实施比较困难，而 IGA 则可以轻易地构造高阶次和高连续性的形函数。

二阶 Piola-Kirchhoff 应力张量 $\boldsymbol{S}$ 表示为

$$\boldsymbol{S}=\boldsymbol{C}:\boldsymbol{E} \tag{2.26}$$

其中，$\boldsymbol{C}$ 为四阶弹性张量。考虑薄壳厚度方向上的平面应力假设[7]，可得其分量为

$$C^{\alpha\beta\gamma\delta}=\frac{E}{1-\nu^2}\left[\nu A^{\alpha\beta}A^{\gamma\delta}+\frac{1}{2}(1-\nu)\left(A^{\alpha\gamma}A^{\delta\beta}+A^{\alpha\delta}A^{\gamma\beta}\right)\right] \tag{2.27}$$

式中，E 和 ν 分别为材料的弹性模量和泊松比。

因此，应力张量 $\boldsymbol{S}$ 的分量可以表示为

$$S^{\alpha\beta}=C^{\alpha\beta\gamma\delta}E_{\gamma\delta} \tag{2.28}$$

应用虚功原理，我们可以得到

$$\delta\mathcal{W}=\delta\mathcal{W}_{\text{int}}+\delta\mathcal{W}_{\text{ext}}=0 \tag{2.29}$$

其中，$\delta\mathcal{W}$ 为总势能的变分；$\delta\mathcal{W}_{\text{int}}$ 和 $\delta\mathcal{W}_{\text{ext}}$ 分别表示内力虚功和外力虚功。

内力虚功的变分 $\delta\mathcal{W}_{\text{int}}$ 表示为

$$\begin{aligned}\delta\mathcal{W}_{\text{int}}&=-\int_{\mathcal{V}}\boldsymbol{S}:\delta\boldsymbol{E}\,\mathrm{d}\mathcal{V}=-\int_{\mathcal{V}}\delta E_{\alpha\beta}C^{\alpha\beta\gamma\delta}E_{\gamma\delta}\,\mathrm{d}\mathcal{V}\\&=-\int_{\mathcal{V}}\left(\delta\varepsilon_{\alpha\beta}+\zeta\delta\kappa_{\alpha\beta}\right)C^{\alpha\beta\gamma\delta}\left(\varepsilon_{\gamma\delta}+\zeta\kappa_{\gamma\delta}\right)\mathrm{d}\mathcal{V}\\&=-\int_{\mathcal{A}}(\delta\varepsilon_{\alpha\beta}\underbrace{hC^{\alpha\beta\gamma\delta}\varepsilon_{\gamma\delta}}_{n^{\alpha\beta}}+\delta\kappa_{\alpha\beta}\underbrace{\frac{h^3}{12}C^{\alpha\beta\gamma\delta}\kappa_{\gamma\delta}}_{m^{\alpha\beta}})\mathrm{d}\mathcal{A}\end{aligned} \tag{2.30}$$

式中，$\delta\varepsilon_{\alpha\beta}$ 为薄膜应变的变分；$\delta\kappa_{\alpha\beta}$ 为弯曲应变的变分；$\mathcal{V}$ 代表整个壳体区域；$\mathcal{A}$ 代表壳体中性面的区域。应力可以分为两类：一是对应于薄膜应变的薄膜应力 $n^{\alpha\beta}$，二是对应于弯曲应变的弯曲应力 $m^{\alpha\beta}$，它们的表达式为

$$n^{\alpha\beta}=hC^{\alpha\beta\gamma\delta}\varepsilon_{\gamma\delta},\quad m^{\alpha\beta}=\frac{h^3}{12}C^{\alpha\beta\gamma\delta}\kappa_{\gamma\delta} \tag{2.31}$$

为了便于计算机编程，我们使用 Voigt 符号来表示上述变量，即将应力与应变的关系记为

$$\boldsymbol{n}=\frac{Eh}{1-\nu^2}\bar{\boldsymbol{D}}\boldsymbol{\varepsilon},\quad \boldsymbol{m}=\frac{Eh^3}{12\left(1-\nu^2\right)}\bar{\boldsymbol{D}}\boldsymbol{\kappa} \tag{2.32}$$

其中，

$$
\begin{aligned}
\boldsymbol{n}&=\left\{n_{11}, n_{22}, n_{12}\right\}^{\mathrm{T}}, \quad \boldsymbol{\varepsilon}=\left\{\varepsilon_{11}, \varepsilon_{22}, 2\varepsilon_{12}\right\}^{\mathrm{T}} \\
\boldsymbol{m}&=\left\{m_{11}, m_{22}, m_{12}\right\}^{\mathrm{T}}, \quad \boldsymbol{\kappa}=\left\{\kappa_{11}, \kappa_{22}, 2\kappa_{12}\right\}^{\mathrm{T}}
\end{aligned} \tag{2.33}
$$

和

$$
\bar{\boldsymbol{D}}=\begin{bmatrix}
\left(A^{11}\right)^2 & \nu A^{11}A^{22}+(1-\nu)\left(A^{12}\right)^2 & A^{11}A^{12} \\
\vdots & \left(A^{22}\right)^2 & A^{22}A^{12} \\
\text{对称} & \cdots & \dfrac{1}{2}\left[(1-\nu)A^{11}A^{22}-(1-\nu)\left(A^{12}\right)^2\right]
\end{bmatrix} \tag{2.34}
$$

因此，内力虚功的变分可以更简单地记作

$$
\delta\mathcal{W}_{\text{int}}=-\int_{\mathcal{A}}\left(\delta\boldsymbol{\varepsilon}^{\mathrm{T}}\cdot\boldsymbol{n}+\delta\boldsymbol{\kappa}^{\mathrm{T}}\cdot\boldsymbol{m}\right)\mathrm{d}\mathcal{A}=-\frac{E}{1-\nu^2}\int_{\mathcal{A}}\left(h\delta\boldsymbol{\varepsilon}^{\mathrm{T}}\bar{\boldsymbol{D}}\boldsymbol{\varepsilon}+\frac{h^3}{12}\delta\boldsymbol{\kappa}^{\mathrm{T}}\bar{\boldsymbol{D}}\boldsymbol{\kappa}\right)\mathrm{d}\mathcal{A} \tag{2.35}
$$

外力虚功的变分 $\delta\mathcal{W}_{\text{ext}}$ 表示为

$$
\begin{aligned}
\delta\mathcal{W}_{\text{ext}}&=\int_{-h/2}^{h/2}\left(\int_{\mathcal{A}}\delta\boldsymbol{u}^{\mathrm{T}}\cdot\rho\boldsymbol{F}_{\mathrm{b}}\mathrm{d}\mathcal{A}\right)\mathrm{d}\zeta+\int_{-h/2}^{h/2}\left(\int_{\Gamma_{\mathrm{t}}}\delta\boldsymbol{u}^{\mathrm{T}}\cdot\boldsymbol{F}_{\mathrm{t}}\mathrm{d}\Gamma_{\mathrm{t}}\right)\mathrm{d}\zeta \\
&=\int_{\mathcal{A}}\delta\boldsymbol{u}^{\mathrm{T}}\cdot\rho\left(h\boldsymbol{F}_{\mathrm{b}}\right)\mathrm{d}\mathcal{A}+\int_{\Gamma_{\mathrm{t}}}\delta\boldsymbol{u}^{\mathrm{T}}\cdot\left(h\boldsymbol{F}_{\mathrm{t}}\right)\mathrm{d}\Gamma_{\mathrm{t}}
\end{aligned} \tag{2.36}
$$

式中，ρ 为密度；$\delta\boldsymbol{u}$ 为虚位移矢量；Γ_{t} 是面力作用的边界；$\boldsymbol{F}_{\mathrm{b}}$ 和 $\boldsymbol{F}_{\mathrm{t}}$ 分别表示单位面积上的体力矢量和单位长度上的面力矢量。

PHT 样条基函数的 C^1 连续性能够直接满足上述变分公式中对形函数高阶连续性的要求，而不需要额外的操作。在 IGA 中，PHT 样条不仅用于几何精确描述，同时也作为数值分析中的形函数进行未知场变量的插值近似。

壳体中性面几何在单元 e 内通过 PHT 样条基函数插值表示为

$$
\boldsymbol{X}^e(\xi,\eta)=\sum_{i=1}^{n_e}R_i(\xi,\eta)\boldsymbol{P}_i \tag{2.37}
$$

其中，R_i 为 PHT 样条基函数；n_e 为支持单元 e 的控制点个数；$\boldsymbol{P}_i$ 为控制点。基于等参思想，单元 e 的位移场通过 PHT 样条基函数插值近似为

$$
\boldsymbol{u}^e(\xi,\eta)=\sum_{i=1}^{n_e}R_i(\xi,\eta)\boldsymbol{u}_i \tag{2.38}
$$

将式(2.37)和式(2.38)代入式(2.24)中可得到单元的薄膜应变向量为

$$\boldsymbol{\varepsilon}^e = \left\{\varepsilon_{11}, \varepsilon_{22}, 2\varepsilon_{12}\right\}_e^{\mathrm{T}} = \sum_{i=1}^{n_e} \boldsymbol{B}_{i\mathrm{m}}^e \boldsymbol{u}_i \tag{2.39}$$

式中，$\boldsymbol{B}_{i\mathrm{m}}^e$ 为薄膜应变–位移矩阵，其表达式为

$$\boldsymbol{B}_{i\mathrm{m}}^e = \begin{bmatrix} R_{i,1}\boldsymbol{A}_1 \cdot \boldsymbol{e}_1 & R_{i,1}\boldsymbol{A}_1 \cdot \boldsymbol{e}_2 & R_{i,1}\boldsymbol{A}_1 \cdot \boldsymbol{e}_3 \\ R_{i,2}\boldsymbol{A}_2 \cdot \boldsymbol{e}_1 & R_{i,2}\boldsymbol{A}_2 \cdot \boldsymbol{e}_2 & R_{i,2}\boldsymbol{A}_2 \cdot \boldsymbol{e}_3 \\ \left(R_{i,2}\boldsymbol{A}_1 + R_{i,1}\boldsymbol{A}_2\right)\cdot \boldsymbol{e}_1 & \left(R_{i,2}\boldsymbol{A}_1 + R_{i,1}\boldsymbol{A}_2\right)\cdot \boldsymbol{e}_2 & \left(R_{i,2}\boldsymbol{A}_1 + R_{i,1}\boldsymbol{A}_2\right)\cdot \boldsymbol{e}_3 \end{bmatrix} \tag{2.40}$$

其中，$\boldsymbol{e}_1$、$\boldsymbol{e}_2$ 和 $\boldsymbol{e}_3$ 是整体坐标系下的基矢量。

同理，将式(2.37)和式(2.38)代入式(2.25)中可得到单元的弯曲应变矢量为

$$\boldsymbol{\kappa}^e = \left\{\kappa_{11}, \kappa_{22}, 2\kappa_{12}\right\}_e^{\mathrm{T}} = \sum_{i=1}^{n_e} \boldsymbol{B}_{i\mathrm{b}}^e \boldsymbol{u}_i \tag{2.41}$$

式中，$\boldsymbol{B}_{i\mathrm{b}}^e$ 为弯曲应变–位移矩阵，其表达式为

$$\boldsymbol{B}_{i\mathrm{b}}^e = \begin{bmatrix} \boldsymbol{B}_{i\mathrm{b}}^{11} \cdot \boldsymbol{e}_1 & \boldsymbol{B}_{i\mathrm{b}}^{11} \cdot \boldsymbol{e}_2 & \boldsymbol{B}_{i\mathrm{b}}^{11} \cdot \boldsymbol{e}_3 \\ \boldsymbol{B}_{i\mathrm{b}}^{22} \cdot \boldsymbol{e}_1 & \boldsymbol{B}_{i\mathrm{b}}^{22} \cdot \boldsymbol{e}_2 & \boldsymbol{B}_{i\mathrm{b}}^{22} \cdot \boldsymbol{e}_3 \\ 2\boldsymbol{B}_{i\mathrm{b}}^{12} \cdot \boldsymbol{e}_1 & 2\boldsymbol{B}_{i\mathrm{b}}^{12} \cdot \boldsymbol{e}_2 & 2\boldsymbol{B}_{i\mathrm{b}}^{12} \cdot \boldsymbol{e}_3 \end{bmatrix} \tag{2.42}$$

其中，

$$\begin{aligned} \boldsymbol{B}_{i\mathrm{b}}^{11} &= -R_{i,11} + \frac{1}{\sqrt{g}}\left[R_{i,1}\cdot\left(\boldsymbol{A}_{1,1}\times\boldsymbol{A}_2\right) + R_{i,2}\cdot\left(\boldsymbol{A}_1\times\boldsymbol{A}_{1,1}\right)\right] + \frac{\boldsymbol{A}_3\times\boldsymbol{A}_{1,1}}{\sqrt{g}}\left[R_{i,1}\cdot\left(\boldsymbol{A}_2\times\boldsymbol{A}_3\right) + R_{i,2}\cdot\left(\boldsymbol{A}_3\times\boldsymbol{A}_1\right)\right] \\ \boldsymbol{B}_{i\mathrm{b}}^{22} &= -R_{i,22} + \frac{1}{\sqrt{g}}\left[R_{i,1}\cdot\left(\boldsymbol{A}_{2,2}\times\boldsymbol{A}_2\right) + R_{i,2}\cdot\left(\boldsymbol{A}_1\times\boldsymbol{A}_{2,2}\right)\right] + \frac{\boldsymbol{A}_3\times\boldsymbol{A}_{2,2}}{\sqrt{g}}\left[R_{i,1}\cdot\left(\boldsymbol{A}_2\times\boldsymbol{A}_3\right) + R_{i,2}\cdot\left(\boldsymbol{A}_3\times\boldsymbol{A}_1\right)\right] \\ \boldsymbol{B}_{i\mathrm{b}}^{12} &= -R_{i,12} + \frac{1}{\sqrt{g}}\left[R_{i,1}\cdot\left(\boldsymbol{A}_{1,2}\times\boldsymbol{A}_2\right) + R_{i,2}\cdot\left(\boldsymbol{A}_1\times\boldsymbol{A}_{1,2}\right)\right] + \frac{\boldsymbol{A}_3\times\boldsymbol{A}_{1,2}}{\sqrt{g}}\left[R_{i,1}\cdot\left(\boldsymbol{A}_2\times\boldsymbol{A}_3\right) + R_{i,2}\cdot\left(\boldsymbol{A}_3\times\boldsymbol{A}_1\right)\right] \end{aligned} \tag{2.43}$$

式中，$g = \left\|\boldsymbol{A}_1 \times \boldsymbol{A}_2\right\|$。

因此，应变矢量的变分为

$$\delta\boldsymbol{\varepsilon}^e = \sum_{i=1}^{n_e} \boldsymbol{B}_{i\mathrm{m}}^e \delta\boldsymbol{u}_i,\quad \delta\boldsymbol{\kappa}^e = \sum_{i=1}^{n_e} \boldsymbol{B}_{i\mathrm{b}}^e \delta\boldsymbol{u}_i \tag{2.44}$$

由于变分变量的任意性，可以得到线性方程组

$$\boldsymbol{K}\boldsymbol{u} = \boldsymbol{f} \tag{2.45}$$

其中，

$$K_{ij}=\frac{E}{1-\nu^2}\int_{\mathcal{A}}\left[h\left(\boldsymbol{B}_i^{\mathrm{m}}\right)^{\mathrm{T}}\bar{\boldsymbol{D}}\boldsymbol{B}_j^{\mathrm{m}}+\frac{h^3}{12}\left(\boldsymbol{B}_j^{\mathrm{b}}\right)^{\mathrm{T}}\bar{\boldsymbol{D}}\boldsymbol{B}_j^{\mathrm{b}}\right]\mathrm{d}\mathcal{A} \tag{2.46}$$

式中，$\boldsymbol{B}_i^{\mathrm{m}}=\sum_e \boldsymbol{B}_{i\mathrm{m}}^e$；$\boldsymbol{B}_i^{\mathrm{b}}=\sum_e \boldsymbol{B}_{i\mathrm{b}}^{\mathrm{e}}$；$\mathrm{d}\mathcal{A}=\sqrt{g}\mathrm{d}\xi\mathrm{d}\eta$ 为中性面微元的面积。

外力矢量表示为

$$\boldsymbol{f}_i=\int_{\mathcal{A}}R_i\cdot\rho h\boldsymbol{F}_{\mathrm{b}}\mathrm{d}\mathcal{A}+\int_{\Gamma_{\mathrm{t}}}R_i\cdot h\boldsymbol{F}_{\mathrm{t}}\,\mathrm{d}\Gamma_{\mathrm{t}} \tag{2.47}$$

2.3 基于 PHT 样条的扩展等几何分析

2.3.1 富集模式

通过在位移模式中添加富集函数的形式来模拟由裂纹引起的位移不连续性和裂尖应力奇异性，且对扩展等几何分析(XIGA)中的位移近似只需进行局部富集，即根据裂纹的位置，只有受到裂纹影响区域的控制点才被富集。在裂纹面和裂纹尖端的影响下，控制点分别由 Heaviside 函数和裂尖渐近函数进行富集。含裂纹 Kirchhoff-Love 薄壳的 XIGA 位移插值近似表达式为[8]

$$\begin{gathered}\begin{Bmatrix}u_1\\u_2\end{Bmatrix}=\sum_{i=1}^{NI}R_i\begin{Bmatrix}u_{i1}\\u_{i2}\end{Bmatrix}+\sum_{j=1}^{NJ}R_j\left[H(\xi)-H\left(\xi_j\right)\right]\begin{Bmatrix}a_{j1}\\a_{j2}\end{Bmatrix}+\sum_{k=1}^{NK}R_k\sum_{L=1}^{4}\left[\psi_L(\xi)-\psi_L(\xi_k)\right]\begin{Bmatrix}b_{k1}^L\\b_{k2}^L\end{Bmatrix}\\u_3=\sum_{i=1}^{NI}R_iu_{i3}+\sum_{j=1}^{NJ}R_j\left[H(\xi)-H\left(\xi_j\right)\right]a_{j3}+\sum_{k=1}^{NK}R_k\sum_{L=1}^{4}\left[\phi_L(\xi)-\phi_L(\xi_k)\right]b_{k3}^L\end{gathered} \tag{2.48}$$

其中，R_i、R_j 和 R_k 为 PHT 样条基函数；u_{i1}、u_{i2} 和 u_{i3} 为标准的位移自由度；a_{j1}、a_{j2}、a_{j3}、b_{k1}^L、b_{k2}^L 和 b_{k3}^L 为相应的富集自由度；ξ_j 和 ξ_k 为与控制点相关的参数坐标；NI、NJ 和 NK 分别为 PHT 曲面控制点的个数、受 Heaviside 函数富集控制点的个数和受裂尖渐近函数富集控制点的个数；$H(\xi)$ 为 Heaviside 函数，其值在裂纹处发生阶跃，一般情况下在裂纹两侧取值为 ±1，其分布如图 2.11(a)所示。

ψ 和 ϕ 分别是面内和面外裂尖渐近函数，其表达式为[15]

$$\begin{gathered}\psi(r,\theta)=\left\{\sqrt{r}\sin\left(\frac{\theta}{2}\right),\sqrt{r}\cos\left(\frac{\theta}{2}\right),\sqrt{r}\sin\left(\frac{\theta}{2}\right)\sin(\theta),\sqrt{r}\cos\left(\frac{\theta}{2}\right)\sin(\theta)\right\}\\\phi(r,\theta)=\left\{r^{\frac{3}{2}}\sin\left(\frac{\theta}{2}\right),r^{\frac{3}{2}}\cos\left(\frac{\theta}{2}\right),r^{\frac{3}{2}}\sin\left(\frac{3\theta}{2}\right),r^{\frac{3}{2}}\cos\left(\frac{3\theta}{2}\right)\right\}\end{gathered} \tag{2.49}$$

式中，(r,θ) 为坐标原点位于裂尖的局部圆柱坐标系下的空间分量(图 2.11(b))。裂尖渐近函数对物理空间坐标的一阶及二阶导数表达式见文献[14]，[15]。

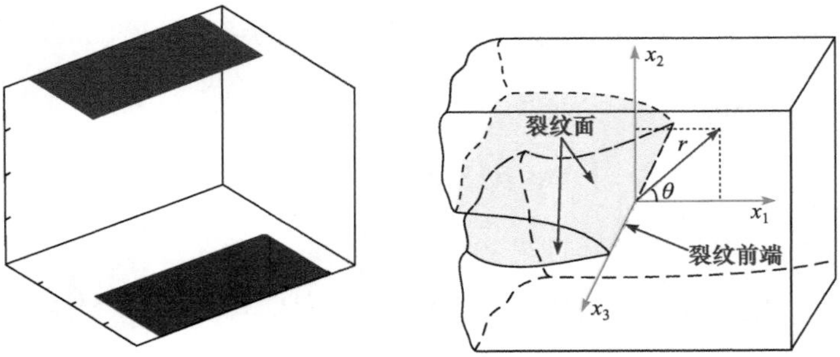

(a) (b)

图 2.11 Heaviside 富集函数分布和裂尖局部坐标系

(a) Heaviside 富集函数；(b) 裂尖局部坐标系

裂尖渐近函数的分布如图 2.12 所示。可以清楚地看到，只有第一个面内渐近函数，以及第一个和第三个面外渐近函数表征出了裂纹面的不连续特性，而其他函数则可以用来提高富集近似的精度。

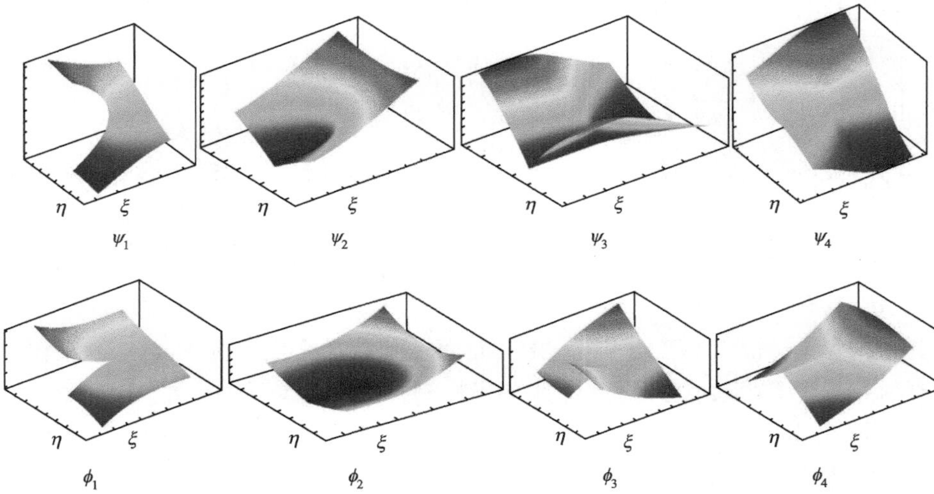

图 2.12 裂尖渐近函数的分布

利用代表位移不连续性的 Heaviside 函数和相关的裂尖渐近函数，可以计算沿裂纹面的张开位移$[\boldsymbol{u}](\boldsymbol{x})$，其表达式为

$$[\boldsymbol{u}](\boldsymbol{x})=\begin{Bmatrix}\boldsymbol{u}_1\\ \boldsymbol{u}_2\\ \boldsymbol{u}_3\end{Bmatrix}=\begin{Bmatrix}2\sum_{j=1}^{NJ}R_j a_{j1}+2\sqrt{r}\sum_{k}^{NK}R_k b_{k1}^1\\ 2\sum_{j=1}^{NJ}R_j a_{j2}+2\sqrt{r}\sum_{k=1}^{NK}R_k b_{k2}^1\\ 2\sum_{j=1}^{NJ}R_j a_{j3}+2r^{3/2}\sum_{k=1}^{NK}R_k b_{k3}^1-2r^{3/2}\sum_{k=1}^{NK}R_k b_{k3}^3\end{Bmatrix} \tag{2.50}$$

为了更好地展示基函数被相关富集函数富集前后的变化，此处我们给出一个例子进行说明。考虑如图 2.13 所示的包含一个边裂纹的 5×5 三次 C^1 连续性的 PHT 样条初始网格，与每个单元相关的基函数的个数为 16。裂纹分割单元(整体编号为 11)中局部编号为 6 的基函数被 Heaviside 函数富集前后的变化如图 2.14 所示。裂尖单元(整体编号为 13)中局部编号为 6 的基函数被裂尖渐近函数富集前后的变化如图 2.15 所示。

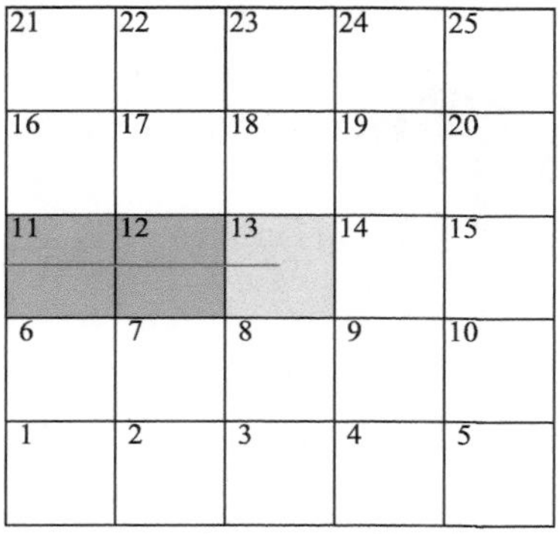

图 2.13　含边裂纹的 PHT 样条初始网格

红线代表裂纹，深蓝和浅蓝单元分别为裂纹分割单元和裂尖单元，图中数字为单元的整体编号

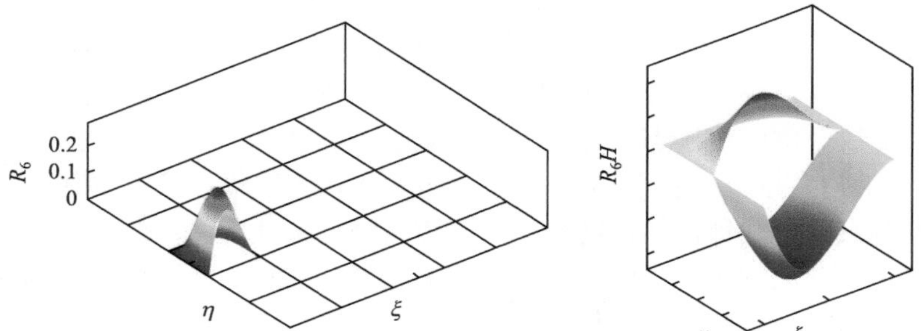

(a)　　(b)

图 2.14　基函数被 Heaviside 函数富集前后的变化

(a) 富集前；(b) 富集后

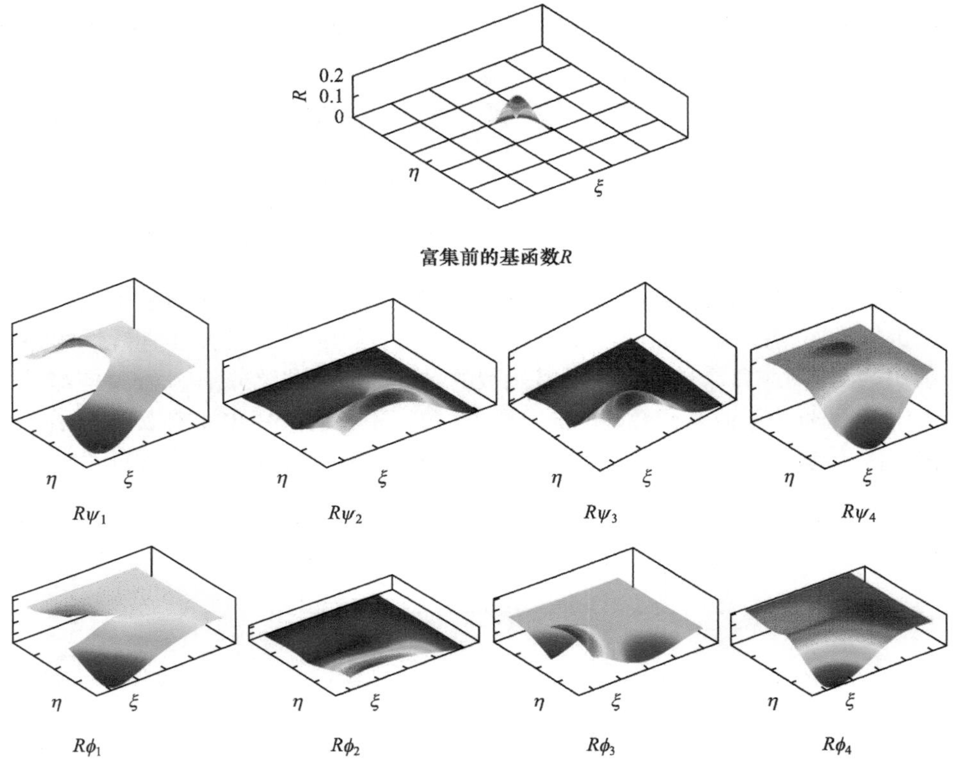

图 2.15　基函数被裂尖渐近函数富集前后的变化

2.3.2　富集控制点及单元选择

由于裂纹的影响，求解单元被分为四种类型：标准单元(与单元相关的所有基函数均未被富集)、裂纹分割单元(与单元相关的所有基函数均被 Heaviside 函数富集)、裂尖单元(与单元相关的所有基函数均被裂尖渐近函数富集)和混合单元(与单元相关的基函数部分被富集)。在等几何分析(IGA)中，每个基函数都与唯一的控制点相对应，且 PHT 样条函数的局部支持性也有利于富集控制点的选择。其支持域完全被裂纹分割的控制点用 Heaviside 函数进行富集。采用裂尖渐近函数富集控制点的选择存在两种主要方式[9]：拓扑富集和几何富集。仅富集其支持域包含裂尖的控制点的方式称作拓扑富集，其富集区域的尺寸随着网格局部细分的过程而逐渐变小。几何富集方式中富集范围则为离裂尖的距离在预设值内的整个周围区域，其中提前预设的距离称作富集半径。几何富集方式虽说能获得更优的收敛率，但其会导致系数矩阵条件数差的问题[10]，这是因为远离裂尖的富集单元内富集后的形函数与未被富集的形函数之间差异性逐渐减少，线性相关性逐渐增强。因此，本章采用拓扑富集的方式。图 2.16 给出了单元类型分布和富集控制点选择

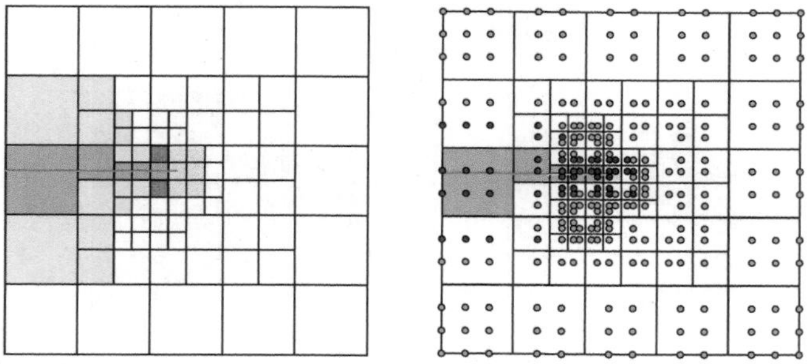
(a)　(b)

图 2.16　单元类型和富集控制点选择

(a) 单元类型分布，深蓝和浅蓝单元分别为裂纹分割单元和裂尖单元，绿色单元为仅部分控制点被 Heaviside 函数富集的混合单元，灰色单元为仅部分控制点被裂尖渐近函数富集的混合单元，红色单元为同时有部分控制点被 Heaviside 函数或裂尖渐近函数富集的混合单元；(b) 富集控制点分布，绿点表示非富集控制点，洋红点表示 Heaviside 富集控制点，蓝点表示裂尖富集控制点

的一个示例。值得注意的是，同时被选择为 Heaviside 函数和裂尖渐近函数富集的控制点仅使用裂尖渐近函数富集。

2.3.3　数值积分方案

不连续的和奇异的富集函数的引入使得系统方程中的积分包含非光滑项，因此标准高斯求积公式不再适用。在裂纹分割单元中，本章采用单元划分的方式进行数值积分，即根据裂纹的几何信息将单元划分为子三角形区域，然后在每个子单元内利用三角形积分点进行计算，而且每个子三角形内使用 7 个积分点。在裂尖单元中，需要结合单元划分和积分空间转换方法来同时消除积分中的不连续性和奇异性。首先将裂尖单元划分为 6 个子三角形单元，并使裂尖位于每个子三角形单元的顶点处，这样做可以使更多的积分点聚集在裂尖附近从而提高求解精度，然后利用几乎极性积分(almost polar integration)[11]进行积分空间转换来消除 $r^{-1/2}$ 的奇异性。此外，包含裂尖渐近函数富集的混合单元则采用 6×6 高斯积分点进行计算。图 2.17 给出了在不同类型单元中积分点分布的示意图。整个 XIGA 中涉及的积分空间转换总结于图 2.18 中。

图 2.18 中的 $\boldsymbol{J}_1$ 和 $\boldsymbol{J}_2$ 分别为物理空间到参数空间，以及参数空间到积分空间的转换矩阵，它们的表达式和含义与 IGA 中的保持一致。$\boldsymbol{J}_3$ 和 $\boldsymbol{J}_4$ 为非连续单元内数值积分所引入的新转换矩阵，其转换关系和表达式为

$$\begin{cases}\xi=\xi_1\left(1-\tilde{\xi}-\tilde{\eta}\right)+\xi_2\tilde{\xi}+\xi_3\tilde{\eta}\\ \eta=\eta_1\left(1-\tilde{\xi}-\tilde{\eta}\right)+\eta_2\tilde{\xi}+\eta_3\tilde{\eta}\end{cases},\quad \boldsymbol{J}_3=\begin{bmatrix}-\xi_1+\xi_2 & -\xi_1+\xi_3\\ -\eta_1+\eta_2 & -\eta_1+\eta_3\end{bmatrix} \tag{2.51}$$

和

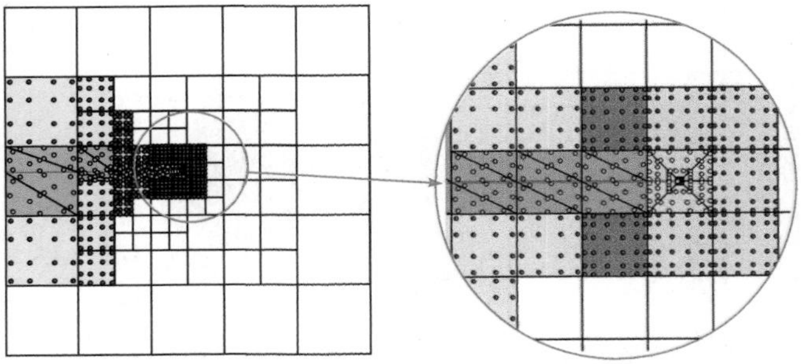

图 2.17　积分点在不同类型单元中的分布及局部放大图

$$\begin{cases}\tilde{\xi}=1/4\cdot\left(1+\overline{\xi}-\overline{\eta}-\overline{\xi}\overline{\eta}\right)\\ \tilde{\eta}=1/2\cdot\left(1+\overline{\eta}\right)\end{cases},\quad \boldsymbol{J}_4=\begin{bmatrix}1/4\cdot\left(1-\overline{\eta}\right) & 1/4\cdot\left(-1-\overline{\xi}\right)\\ 0 & 1/2\end{bmatrix} \tag{2.52}$$

式中，ξ_1、ξ_2、ξ_3、η_1、η_2、η_3、$\tilde{\xi}$、$\tilde{\eta}$、$\overline{\xi}$、$\overline{\eta}$ 的含义见图 2.18；$\boldsymbol{J}_3$、$\boldsymbol{J}_4$ 表示不同空间坐标转换的雅可比矩阵。

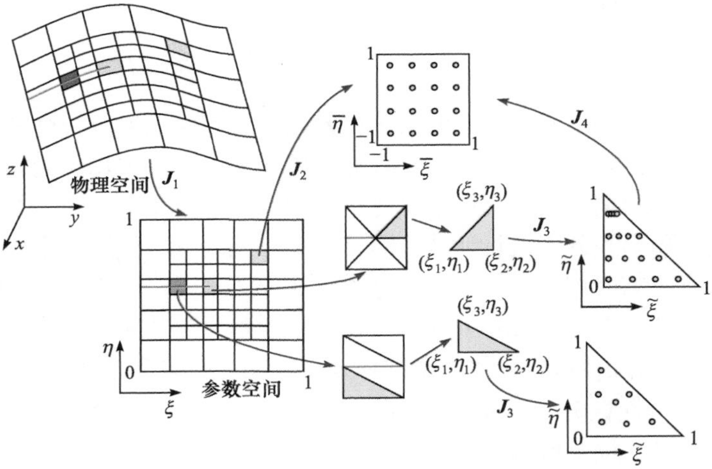

图 2.18　整个 XIGA 中的积分空间转换

2.3.4　求解方程

XIGA 中富集后的离散化系统方程可以化为如下代数方程组的形式：

$$\boldsymbol{K}^{\mathrm{enr}}\boldsymbol{U}^{\mathrm{enr}}=\boldsymbol{F}^{\mathrm{enr}} \tag{2.53}$$

其中，$\boldsymbol{U}^{\mathrm{enr}}$ 为包含标准位移自由度和富集自由度的全局广义位移矢量，其扩展形式为

$$\boldsymbol{U}^{\text{enr}}=\left[\boldsymbol{u}^{\mathrm{T}}\quad \boldsymbol{a}^{\mathrm{T}}\quad \boldsymbol{b}^{\mathrm{T}}\right]^{\mathrm{T}} \tag{2.54}$$

式中，$\boldsymbol{u}$ 是位移矢量；$\boldsymbol{a}$ 是 Heaviside 富集自由度矢量；$\boldsymbol{b}$ 是与裂尖渐近函数相关的富集自由度矢量；$\boldsymbol{K}^{\text{enr}}$ 是富集后的全局刚度矩阵，表示为

$$\boldsymbol{K}_{ij}^{\text{enr}}=\frac{E}{1-\nu^2}\int_{\mathcal{A}}\left[h\left(\boldsymbol{B}_{i\mathrm{m}}^{\text{enr}}\right)^{\mathrm{T}}\bar{\boldsymbol{D}}\boldsymbol{B}_{j\mathrm{m}}^{\text{enr}}+\frac{h^3}{12}\left(\boldsymbol{B}_{i\mathrm{b}}^{\text{enr}}\right)^{\mathrm{T}}\bar{\boldsymbol{D}}\boldsymbol{B}_{j\mathrm{b}}^{\text{enr}}\right]\mathrm{d}\mathcal{A} \tag{2.55}$$

其中，$\boldsymbol{B}_{i\mathrm{m}}^{\text{enr}}=\sum\limits_{e}\boldsymbol{B}_{i\mathrm{m}}^{\text{enr}(e)}$，$\boldsymbol{B}_{i\mathrm{b}}^{\text{enr}}=\sum\limits_{e}\boldsymbol{B}_{i\mathrm{b}}^{\text{enr}(e)}$。

根据控制点是否被富集，或富集形式的不同，$\boldsymbol{B}_{i\mathrm{m}}^{\text{enr}(e)}$ 和 $\boldsymbol{B}_{i\mathrm{b}}^{\text{enr}(e)}$ 的表示形式如下所述。

如果控制点 $\boldsymbol{P}_i$ 没有被富集，则可得

$$\boldsymbol{B}_{i\mathrm{m}}^{\text{enr}(e)}=\boldsymbol{B}_{i\mathrm{m}}^{e},\quad \boldsymbol{B}_{i\mathrm{b}}^{\text{enr}(e)}=\boldsymbol{B}_{i\mathrm{b}}^{e} \tag{2.56}$$

如果控制点 $\boldsymbol{P}_i$ 被 Heaviside 函数富集，则

$$\boldsymbol{B}_{i\mathrm{m}}^{\text{enr}(e)}=\left[\boldsymbol{B}_{i\mathrm{m}}^{e}\quad \boldsymbol{B}_{i\mathrm{m}}^{\text{Henr}(e)}\right],\quad \boldsymbol{B}_{i\mathrm{b}}^{\text{enr}(e)}=\left[\boldsymbol{B}_{i\mathrm{b}}^{e}\quad \boldsymbol{B}_{i\mathrm{b}}^{\text{Henr}(e)}\right] \tag{2.57}$$

式中，$\boldsymbol{B}_{i\mathrm{m}}^{\text{Henr}(e)}$ 和 $\boldsymbol{B}_{i\mathrm{b}}^{\text{Henr}(e)}$ 分别称作 Heaviside 富集薄膜应变-位移矩阵和弯曲应变-位移矩阵，其表达式为

$$\boldsymbol{B}_{i\mathrm{m}}^{\text{Henr}(e)}=\left[H(\boldsymbol{x})-H(\boldsymbol{x}_i)\right]\cdot\begin{bmatrix} R_{i,1}\boldsymbol{A}_1\cdot\boldsymbol{e}_1 & R_{i,1}\boldsymbol{A}_1\cdot\boldsymbol{e}_2 & R_{i,1}\boldsymbol{A}_1\cdot\boldsymbol{e}_3 \\ R_{i,2}\boldsymbol{A}_2\cdot\boldsymbol{e}_1 & R_{i,2}\boldsymbol{A}_2\cdot\boldsymbol{e}_1 & R_{i,2}\boldsymbol{A}_2\cdot\boldsymbol{e}_3 \\ \left(R_{i,2}\boldsymbol{A}_1+R_{i,1}\boldsymbol{A}_2\right)\cdot\boldsymbol{e}_1 & \left(R_{i,2}\boldsymbol{A}_1+R_{i,1}\boldsymbol{A}_2\right)\cdot\boldsymbol{e}_2 & \left(R_{i,2}\boldsymbol{A}_1+R_{i,1}\boldsymbol{A}_2\right)\cdot\boldsymbol{e}_3 \end{bmatrix} \tag{2.58}$$

和

$$\boldsymbol{B}_{i\mathrm{b}}^{\text{Henr}(e)}=\left[H(\boldsymbol{x})-H(\boldsymbol{x}_i)\right]\cdot\begin{bmatrix} \boldsymbol{B}_{i\mathrm{b}}^{11}\cdot\boldsymbol{e}_1 & \boldsymbol{B}_{i\mathrm{b}}^{11}\cdot\boldsymbol{e}_2 & \boldsymbol{B}_{i\mathrm{b}}^{11}\cdot\boldsymbol{e}_3 \\ \boldsymbol{B}_{i\mathrm{b}}^{22}\cdot\boldsymbol{e}_1 & \boldsymbol{B}_{i\mathrm{b}}^{22}\cdot\boldsymbol{e}_2 & \boldsymbol{B}_{i\mathrm{b}}^{22}\cdot\boldsymbol{e}_3 \\ 2\boldsymbol{B}_{i\mathrm{b}}^{12}\cdot\boldsymbol{e}_1 & 2\boldsymbol{B}_{i\mathrm{b}}^{12}\cdot\boldsymbol{e}_2 & 2\boldsymbol{B}_{i\mathrm{b}}^{12}\cdot\boldsymbol{e}_3 \end{bmatrix} \tag{2.59}$$

其中，$\boldsymbol{B}_{i\mathrm{b}}^{11}$、$\boldsymbol{B}_{i\mathrm{b}}^{22}$ 和 $\boldsymbol{B}_{i\mathrm{b}}^{12}$ 的定义见式(2.43)。

采用如下的记号：

$$
\begin{aligned}
&R_{i,1}^{\mathrm{TL}}=R_{i,1}\left[\mathscr{R}_L(\boldsymbol{x})-\mathscr{R}_L(\boldsymbol{x}_i)\right]+R_i\mathscr{R}_{L,1}(\boldsymbol{x})\quad(L=1,2,3,4)\\
&R_{i,2}^{\mathrm{TL}}=R_{i,2}\left[\mathscr{R}_L(\boldsymbol{x})-\mathscr{R}_L(\boldsymbol{x}_i)\right]+R_i\mathscr{R}_{L,2}(\boldsymbol{x})\\
&R_{i,11}^{\mathrm{TL}}=R_{i,11}\left[\mathscr{R}_L(\boldsymbol{x})-\mathscr{R}_L(\boldsymbol{x}_i)\right]+2R_{i,1}\mathscr{R}_{L,1}(\boldsymbol{x})+R_i\mathscr{R}_{L,11}(\boldsymbol{x})\\
&R_{i,22}^{\mathrm{TL}}=R_{i,22}\mathscr{R}\left[\phi_L(\boldsymbol{x})-\mathscr{R}_L(\boldsymbol{x}_i)\right]+2R_{i,2}\mathscr{R}_{L,2}(\boldsymbol{x})+R_i\mathscr{R}_{L,22}(\boldsymbol{x})\\
&R_{i,12}^{\mathrm{TL}}=R_{i,12}\mathscr{R}\left[\phi_L(\boldsymbol{x})-\mathscr{R}_L(\boldsymbol{x}_i)\right]+R_{i,1}\mathscr{R}_{L,2}(\boldsymbol{x})+R_{i,2}\mathscr{R}_{L,1}(\boldsymbol{x})+R_i\mathscr{R}_{L,12}(\boldsymbol{x})
\end{aligned}
\tag{2.60}
$$

式中，$\mathscr{R}$代表式(2.49)中的裂尖渐近函数，即ψ和ϕ，它们对参数坐标的导数可参见文献[13],[15]。

如果控制点$\boldsymbol{P}_i$被裂尖渐近函数富集，则

$$
\begin{aligned}
&\boldsymbol{B}_{i\mathrm{m}}^{\mathrm{enr}(e)}=\begin{bmatrix}\boldsymbol{B}_{i\mathrm{m}}^{e} & \boldsymbol{B}_{i\mathrm{m}1}^{\mathrm{Tenr}(e)} & \boldsymbol{B}_{i\mathrm{m}2}^{\mathrm{Tenr}(e)} & \boldsymbol{B}_{i\mathrm{m}3}^{\mathrm{Tenr}(e)} & \boldsymbol{B}_{i\mathrm{m}4}^{\mathrm{Tenr}(e)}\end{bmatrix}\\
&\boldsymbol{B}_{i\mathrm{b}}^{\mathrm{enr}(e)}=\begin{bmatrix}\boldsymbol{B}_{i\mathrm{b}}^{e} & \boldsymbol{B}_{i\mathrm{b}1}^{\mathrm{Tenr}(e)} & \boldsymbol{B}_{i\mathrm{b}2}^{\mathrm{Tenr}(e)} & \boldsymbol{B}_{i\mathrm{b}3}^{\mathrm{Tenr}(e)} & \boldsymbol{B}_{i\mathrm{b}4}^{\mathrm{Tenr}(e)}\end{bmatrix}
\end{aligned}
\tag{2.61}
$$

其中，$\boldsymbol{B}_{i\mathrm{m}L}^{\mathrm{Tenr}(e)}$和$\boldsymbol{B}_{i\mathrm{b}L}^{\mathrm{Tenr}(e)}$ $(L=1,2,3,4)$分别称作裂尖富集薄膜应变–位移矩阵和弯曲应变–位移矩阵，其表达式为

$$
\boldsymbol{B}_{i\mathrm{m}L}^{\mathrm{Tenr}(e)}=\begin{bmatrix}
R_{i,1}^{\mathrm{TL}}\boldsymbol{A}_1\cdot\boldsymbol{e}_1 & R_{i,1}^{\mathrm{TL}}\boldsymbol{A}_1\cdot\boldsymbol{e}_2 & R_{i,1}^{\mathrm{TL}}\boldsymbol{A}_1\cdot\boldsymbol{e}_3\\
R_{i,2}^{\mathrm{TL}}\boldsymbol{A}_2\cdot\boldsymbol{e}_1 & R_{i,2}^{\mathrm{TL}}\boldsymbol{A}_2\cdot\boldsymbol{e}_1 & R_{i,2}^{\mathrm{TL}}\boldsymbol{A}_2\cdot\boldsymbol{e}_3\\
\left(R_{i,2}^{\mathrm{TL}}\boldsymbol{A}_1+R_{i,1}^{\mathrm{TL}}\boldsymbol{A}_2\right)\cdot\boldsymbol{e}_1 & \left(R_{i,2}^{\mathrm{TL}}\boldsymbol{A}_1+R_{i,1}^{\mathrm{TL}}\boldsymbol{A}_2\right)\cdot\boldsymbol{e}_2 & \left(R_{i,2}^{\mathrm{TL}}\boldsymbol{A}_1+R_{i,1}^{\mathrm{TL}}\boldsymbol{A}_2\right)\cdot\boldsymbol{e}_3
\end{bmatrix}
\tag{2.62}
$$

和

$$
\boldsymbol{B}_{i\mathrm{b}L}^{\mathrm{Tenr}(e)}=\begin{bmatrix}
\boldsymbol{B}_{i\mathrm{b}}^{11*}\cdot\boldsymbol{e}_1 & \boldsymbol{B}_{i\mathrm{b}}^{11*}\cdot\boldsymbol{e}_2 & \boldsymbol{B}_{i\mathrm{b}}^{11*}\cdot\boldsymbol{e}_3\\
\boldsymbol{B}_{i\mathrm{b}}^{22*}\cdot\boldsymbol{e}_1 & \boldsymbol{B}_{i\mathrm{b}}^{22*}\cdot\boldsymbol{e}_2 & \boldsymbol{B}_{i\mathrm{b}}^{22*}\cdot\boldsymbol{e}_3\\
2\boldsymbol{B}_{i\mathrm{b}}^{12*}\cdot\boldsymbol{e}_1 & 2\boldsymbol{B}_{i\mathrm{b}}^{12*}\cdot\boldsymbol{e}_2 & 2\boldsymbol{B}_{i\mathrm{b}}^{12*}\cdot\boldsymbol{e}_3
\end{bmatrix}
\tag{2.63}
$$

将矩阵$\boldsymbol{B}_{i\mathrm{b}}^{11}$、$\boldsymbol{B}_{i\mathrm{b}}^{22}$和$\boldsymbol{B}_{i\mathrm{b}}^{12}$中的形函数、形函数对参数坐标的一阶和二阶导数用式(2.60)中的表达式进行替换，即可得到富集矩阵$\boldsymbol{B}_{i\mathrm{b}}^{11*}$、$\boldsymbol{B}_{i\mathrm{b}}^{22*}$和$\boldsymbol{B}_{i\mathrm{b}}^{12*}$。

式(2.53)中的$\boldsymbol{F}^{\mathrm{enr}}$是全局外力矢量，其分量扩展形式为

$$
\boldsymbol{F}_i^{\mathrm{enr}}=\left[(\boldsymbol{f}_i^{\boldsymbol{u}})^{\mathrm{T}}\quad(\boldsymbol{f}_i^{\boldsymbol{a}})^{\mathrm{T}}\quad(\boldsymbol{f}_i^{\boldsymbol{b}})^{\mathrm{T}}\right]^{\mathrm{T}}
\tag{2.64}
$$

结合式(2.48)、式(2.47)和式(2.60)可得

$$
\begin{aligned}
\boldsymbol{f}_i^{\boldsymbol{u}} &= \int_{\mathcal{A}} R_i \cdot \rho h \boldsymbol{F}_{\mathrm{b}} \mathrm{d}\mathcal{A} + \int_{\Gamma_{\mathrm{t}}} R_i \cdot h \boldsymbol{F}_{\mathrm{t}} \mathrm{d}\Gamma_{\mathrm{t}} \\
\boldsymbol{f}_i^{\boldsymbol{a}} &= \int_{\mathcal{A}} R_i \left[H(\boldsymbol{x}) - H(\boldsymbol{x}_i) \right] \cdot \rho h \boldsymbol{F}_{\mathrm{b}} \mathrm{d}\mathcal{A} + \int_{\Gamma_{\mathrm{t}}} R_i \left[H(\boldsymbol{x}) - H(\boldsymbol{x}_i) \right] \cdot h \boldsymbol{F}_{\mathrm{t}} \mathrm{d}\Gamma_{\mathrm{t}} \\
\boldsymbol{f}_i^{\boldsymbol{b}} &= \int_{\mathcal{A}} R_i \left[\mathcal{R}_L(\boldsymbol{x}) - \mathcal{R}_L(\boldsymbol{x}_i) \right] \cdot \rho h \boldsymbol{F}_{\mathrm{b}} \mathrm{d}\mathcal{A} + \int_{\Gamma_{\mathrm{t}}} R_i \left[\mathcal{R}_L(\boldsymbol{x}) - \mathcal{R}_L(\boldsymbol{x}_i) \right] \cdot h \boldsymbol{F}_{\mathrm{t}} \mathrm{d}\Gamma_{\mathrm{t}}
\end{aligned}
\tag{2.65}
$$

2.3.5 断裂参数计算

本章中采用等效域积分形式的交互作用积分来获取含贯穿裂纹薄壳结构的应力强度因子。考虑如图 2.19 所示的裂纹前端垂直于壳体中性面的裂纹模型，我们在裂纹前端附近截取一小段圆柱区域，V 和 V_ε 分别为由圆柱面 $\mathcal{A}$ 和 $\mathcal{A}_\varepsilon$ 所围成的体积域，$\mathcal{A}_1$ 和 $\mathcal{A}_2$ 分别为圆柱体前后两端的表面。裂尖局部坐标系 (x_1, x_2, x_3) 定义为：x_1 位于裂纹面内且与裂尖垂直，x_3 位于裂纹面内且与裂尖相切，x_2 与 x_1 和 x_3 相互垂直。

裂纹前端的 J 积分定义为[12]

$$
J \cdot f = -\int_{\mathcal{A}+\mathcal{A}_\varepsilon} \left(\mathcal{W} n_1 - \sigma_{ij} \frac{\partial u_i}{\partial x_1} n_j \right) q \mathrm{d}\mathcal{A} \tag{2.66}
$$

式中，$\mathcal{W} = \frac{1}{2}\sigma_{ij}\varepsilon_{ij}$ 为应变能密度，其中 σ_{ij} 和 ε_{ij} 分别为应力和应变张量的分量；u_i 为位移矢量的分量；n_j 为表面 $\mathcal{A}$ 和 $\mathcal{A}_\varepsilon$ 的单位法向量的分量；q 为选取的连续函数，其值在表面 $\mathcal{A}_\varepsilon$ 处取 1，在表面 $\mathcal{A}$ 处取 0，在数值实施中积分点处的 q 值可由单元控制点处的值通过基函数插值而得；f 为沿裂纹前端连续函数 q 所围成区

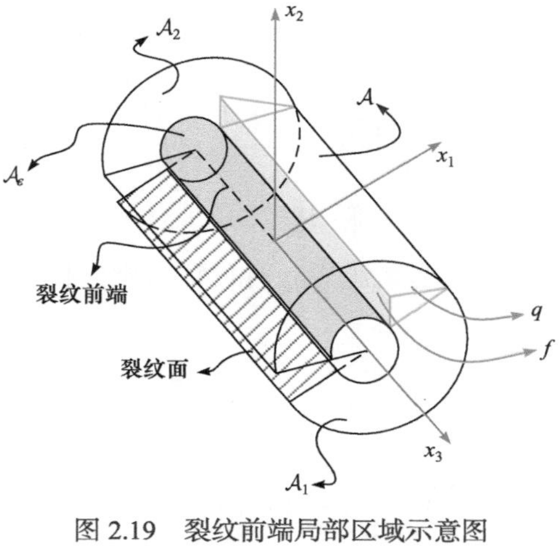

图 2.19 裂纹前端局部区域示意图

域的面积。当圆柱内表面趋近于裂纹前端时，$\mathcal{A}_\varepsilon \to 0$。利用散度定理，可以得到等效域积分形式的$J$积分为

$$
\begin{aligned}
J \cdot f &= -\int_V \frac{\partial\left[\left(\mathcal{W}\delta_{1j} - \sigma_{ij}\dfrac{\partial u_i}{\partial x_1}\right)q\right]}{\partial x_j}\mathrm{d}V + \int_{\mathcal{A}_1+\mathcal{A}_2}\left(\mathcal{W}n_1 - \sigma_{ij}\frac{\partial u_i}{\partial x_1}n_j\right)q\mathrm{d}\mathcal{A} \\
&= -\int_V \left(\mathcal{W}\delta_{1j} - \sigma_{ij}\frac{\partial u_i}{\partial x_1}\right)\frac{\partial q}{\partial x_j}\mathrm{d}V - \int_V\left[\frac{\partial \mathcal{W}}{\partial x_1} - \frac{\partial}{\partial x_j}\left(\sigma_{ij}\frac{\partial u_i}{\partial x_1}\right)\right]q\mathrm{d}V \\
&\quad + \int_{\mathcal{A}_1+\mathcal{A}_2}\left(\mathcal{W}n_1 - \sigma_{ij}\frac{\partial u_i}{\partial x_1}n_j\right)q\mathrm{d}\mathcal{A}
\end{aligned}
\tag{2.67}
$$

其中，δ为克罗内克(Kronecker)符号。线弹性材料的应力、应变和位移分量均能满足几何方程和平衡方程，则式(2.67)中的第二项为零。等厚度的薄壳满足$f = h$。因此，式(2.67)中的J积分可转化为

$$
J = -\frac{1}{h}\int_V \left(\mathcal{W}\delta_{1j} - \sigma_{ij}\frac{\partial u_i}{\partial x_1}\right)\frac{\partial q}{\partial x_j}\mathrm{d}V + \frac{1}{h}\int_{\mathcal{A}_1+\mathcal{A}_2}\left(\mathcal{W}n_1 - \sigma_{ij}\frac{\partial u_i}{\partial x_1}n_j\right)q\mathrm{d}\mathcal{A} \tag{2.68}
$$

$\mathcal{A}_1$和$\mathcal{A}_2$面积相等且与裂纹前端相互垂直，故可以得到：在表面$\mathcal{A}_1$上，$n_1 = n_2 = 0$和$n_3 = 1$；在表面$\mathcal{A}_2$上，$n_1 = n_2 = 0$和$n_3 = -1$。则式(2.68)中的第二项可以化为

$$
\frac{1}{h}\int_{\mathcal{A}_1+\mathcal{A}_2}\left(\mathcal{W}n_1 - \sigma_{ij}\frac{\partial u_i}{\partial x_1}n_j\right)q\mathrm{d}\mathcal{A} = -\frac{1}{h}\int_{\mathcal{A}_1+\mathcal{A}_2}\sigma_{i3}\frac{\partial u_i}{\partial x_1}n_3 q\mathrm{d}\mathcal{A} \tag{2.69}
$$

将式(2.69)代入式(2.68)，得到J积分的最终表达式为

$$
J = -\frac{1}{h}\int_V \left(\mathcal{W}\frac{\partial q}{\partial x_1} - \sigma_{ij}\frac{\partial u_i}{\partial x_1}\frac{\partial q}{\partial x_j}\right)\mathrm{d}V - \frac{1}{h}\int_{\mathcal{A}_1+\mathcal{A}_2}\sigma_{i3}\frac{\partial u_i}{\partial x_1}n_3 q\mathrm{d}\mathcal{A} \tag{2.70}
$$

给定两种状态：当前求解状态变量$\left(\sigma_{ij}, \varepsilon_{ij}, u_i\right)$和虚拟辅助状态$\left(\sigma_{ij}^{\mathrm{aux}}, \varepsilon_{ij}^{\mathrm{aux}}, u_i^{\mathrm{aux}}\right)$(具体表达式见文献[9],[13],[15])，则两种状态下的混合J积分为

$$
\begin{aligned}
J^{\sup} &= -\frac{1}{h}\int_V\left[\frac{1}{2}\left(\sigma_{ij} + \sigma_{ij}^{\mathrm{aux}}\right)\left(\varepsilon_{ij} + \varepsilon_{ij}^{\mathrm{aux}}\right)\frac{\partial q}{\partial x_1} - \left(\sigma_{ij} + \sigma_{ij}^{\mathrm{aux}}\right)\frac{\partial\left(u_i + u_i^{\mathrm{aux}}\right)}{\partial x_1}\frac{\partial q}{\partial x_j}\right]\mathrm{d}V \\
&\quad - \frac{1}{h}\int_{\mathcal{A}_1+\mathcal{A}_2}\left(\sigma_{i3} + \sigma_{i3}^{\mathrm{aux}}\right)\frac{\partial\left(u_i + u_i^{\mathrm{aux}}\right)}{\partial x_1}n_3 q\mathrm{d}\mathcal{A} = J + J^{\mathrm{aux}} + I
\end{aligned}
\tag{2.71}
$$

式中，J 和 J^{aux} 分别是当前状态和辅助状态下的 J 积分；I 为交互作用积分，表示为

$$I = \frac{1}{h}\int_V \left(\sigma_{ij} \frac{\partial u_i^{\text{aux}}}{\partial x_1} + \sigma_{ij}^{\text{aux}} \frac{\partial u_i}{\partial x_1} - \sigma_{ij}\varepsilon_{ij}^{\text{aux}}\delta_{1j} \right) \frac{\partial q}{\partial x_j} \mathrm{d}V - \frac{1}{h}\int_{\mathcal{A}_1+\mathcal{A}_2} \left(\sigma_{i3} \frac{\partial u_i^{\text{aux}}}{\partial x_1} + \sigma_{i3}^{\text{aux}} \frac{\partial u_i}{\partial x_1} \right) n_3 q \mathrm{d}\mathcal{A} \tag{2.72}$$

交互作用积分与应力强度因子之间的关系为[13]

$$I = \frac{2}{E}\left(K_{\text{I}}K_{\text{I}}^{\text{aux}} + K_{\text{II}}K_{\text{II}}^{\text{aux}} \right) + \frac{2\pi}{3E}\left(\frac{1+\nu}{3+\nu} \right)\left(k_1 k_1^{\text{aux}} + k_2 k_2^{\text{aux}} \right) \tag{2.73}$$

其中，E 和 ν 分别为弹性模量和泊松比；K_{I} 和 K_{II} 分别为面内薄膜载荷作用下的拉伸和剪切应力强度因子；k_1 和 k_2 分别为面外载荷作用下的弯曲和扭转应力强度因子。这些应力强度因子在裂尖局部极坐标系下的定义为[13]

$$\begin{aligned} &K_{\text{I}} = \lim_{r\to 0}\sqrt{2\pi r}\,\sigma_{\theta\theta}(r,0), \quad K_{\text{II}} = \lim_{r\to 0}\sqrt{2\pi r}\,\sigma_{r\theta}(r,0) \\ &k_1 = \lim_{r\to 0}\sqrt{2r}\,\sigma_{\theta\theta}\left(r,0,\frac{h}{2}\right), \quad k_2 = \lim_{r\to 0}\frac{3+\nu}{1+\nu}\sqrt{2r}\,\sigma_{r\theta}\left(r,0,\frac{h}{2}\right) \end{aligned} \tag{2.74}$$

如选取弯曲应力强度因子下的辅助状态，即 $k_1^{\text{aux}}=1$，则其他类型的应力强度因子全为零。然后，通过式(2.73)可提取出当前状态下的弯曲应力强度因子为

$$k_1 = \frac{3E}{2\pi}\left(\frac{3+\nu}{1+\nu} \right) I \tag{2.75}$$

采取同样的方式即可得到其他应力强度因子。应注意的是，交互作用积分的计算涉及位移、应变和应力的坐标变换，这是因为当前状态下的量是在整体坐标系中获得的，而辅助状态下的量是在裂尖局部坐标系中定义的。

2.4 数值算例

2.4.1 含边裂纹的薄板

含边裂纹的一端固支薄板受到在相反方向上的两个点载荷作用，如图 2.20 所示。载荷的大小为 $P=1\text{N}$。材料参数为：弹性模量 $E=1000\,\text{Pa}$ 和泊松比 ν=0.3。板的尺寸为 $b\times 2b(b=1\text{m})$，其厚度为 h=0.1m。图 2.21 给出了在裂纹长度 a=1m 情况下 PHT 样条网格局部细分的过程，其网格在裂尖附近逐渐加密。

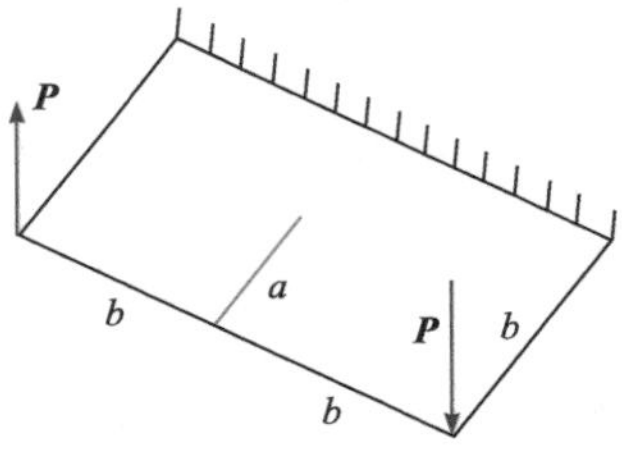

图 2.20　含边裂纹薄板的几何与边界信息

最后一次局部细分状态下的控制点分布如图 2.22 所示。该网格下控制点总数为 3412，单元数量为 874，富集之后总自由度数为 11004，富集自由度占总自由度的比例仅为 6.98%。针对不同单元类型选择不同的积分方案，其积分点的分布如图 2.23 所示。不同裂纹长度下的应力强度因子 K_{II} 与参考解[14]的比较如图 2.24 所示。结果表明，本方法的数值解与参考结果吻合得很好。图 2.25 和图 2.26 展示了在裂纹长度 $a=1\,\text{m}$ 时板沿作用力方向的位移分布和弯曲应力的分布。

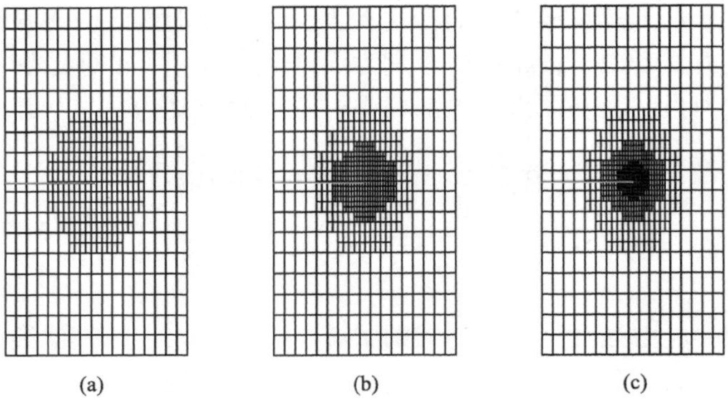

图 2.21　PHT 样条网格的局部细分过程

(a) 第一次细分；(b) 第二次细分；(c) 第三次细分

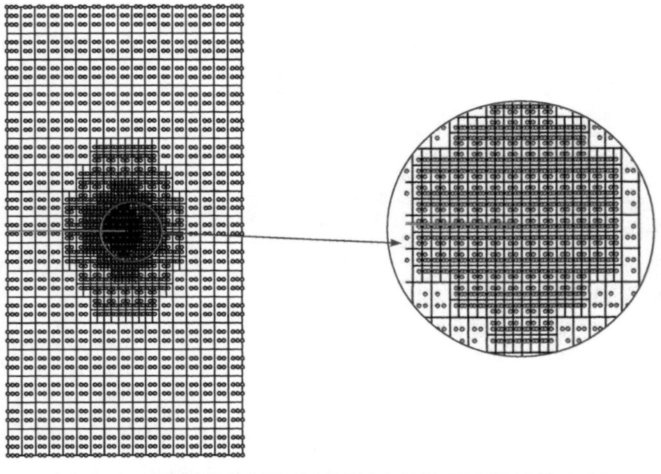

图 2.22　局部细分网格下的控制点分布及局部放大图

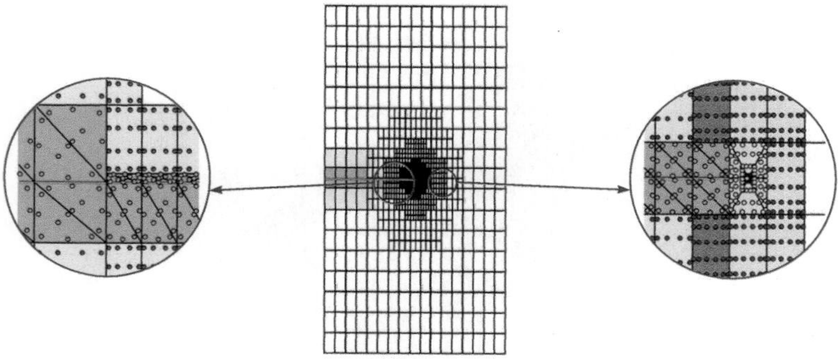

图 2.23　积分点在不同类型单元中的分布及局部放大图

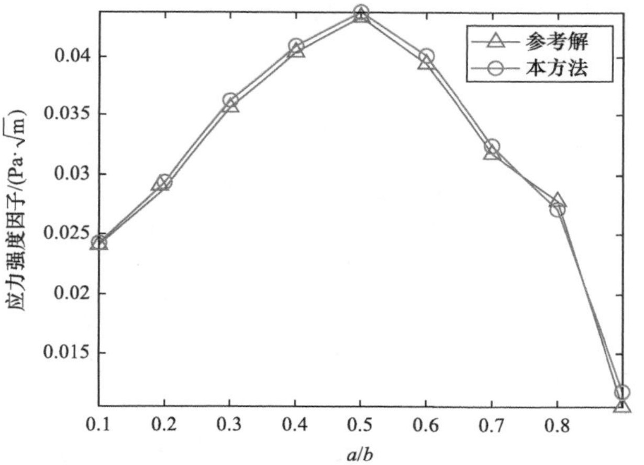

图 2.24　不同裂纹长度下，应力强度因子与参考解[14]的比较

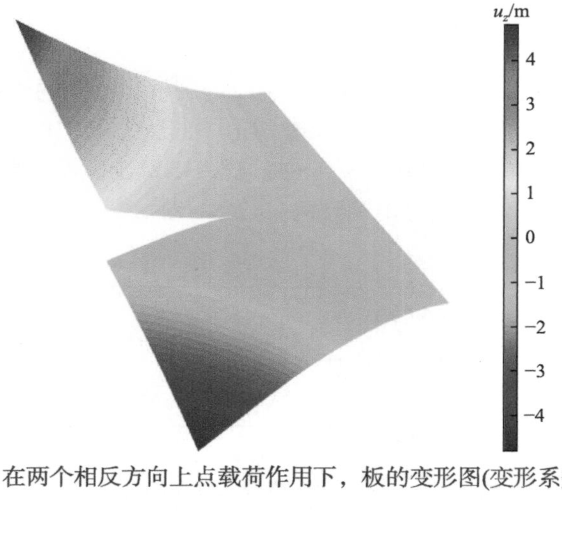

图 2.25　在两个相反方向上点载荷作用下，板的变形图(变形系数为 0.1)

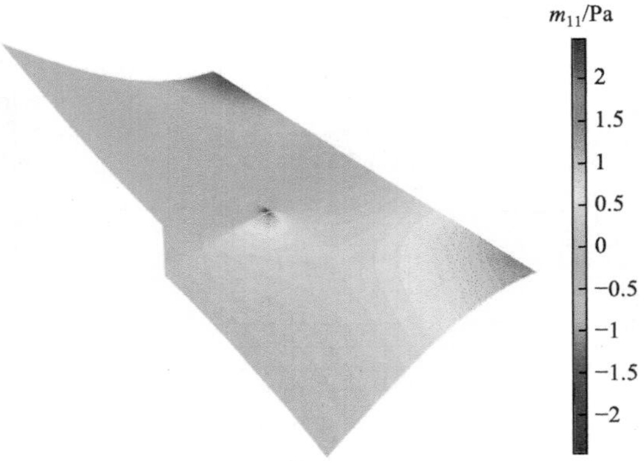

(a)

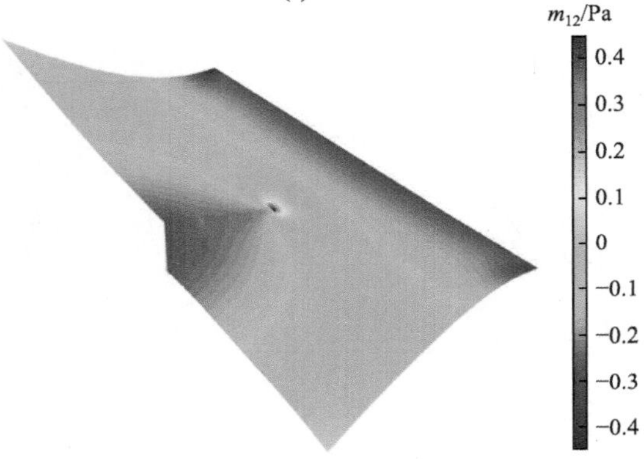

(b)

图 2.26 在两个相反方向上点载荷作用下，板弯曲应力的分布

(a) 应力 m_{11}；(b) 应力 m_{12}

2.4.2 含轴向裂纹的圆柱壳

如图 2.27 所示，考虑含有轴向贯穿裂纹的圆柱壳，壳的半径为 $R=20\text{m}$，厚度为 $h=0.25\text{m}$，长度为 $L=100\text{m}$，承受均匀内部压力 $p=10\text{Pa}$。材料参数为：弹性模量 $E=1000\,\text{Pa}$ 和泊松比 $\nu=0.3$。PHT 样条的基函数至少要求为三阶次和 C^1 连续性的样条函数，但直接对半圆弧的二阶 NURBS 模型进行升阶，并不能得到

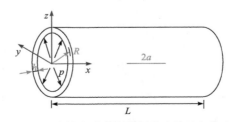

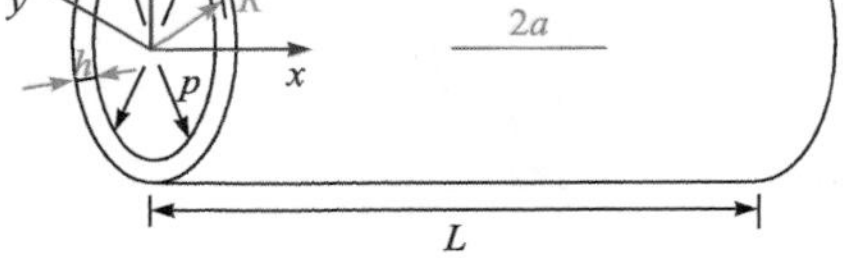

图 2.27 含轴向贯穿裂纹圆柱壳的几何信息

整体均为 C^1 连续性的几何。本算例中圆弧参数方向上选取的初始节点矢量即为三阶的，如表 2.1 所示，因此在后续局部细分过程中能一直保持 C^1 连续性。圆柱壳几何的初始控制点及相关权值如表 2.2 所示。

表 2.1　圆柱壳的节点矢量

参数方向	阶次	节点矢量
ξ	3	$\Xi=\{0,0,0,0,1,1,1,1\}$
η	1	$\Psi=\{0,0,1,1\}$

表 2.2　初始控制点坐标及权值

编号	坐标	权值	编号	坐标	权值
1	$(0,-R,0)$	1	5	$(L,-R,0)$	1
2	$(0,-R,2R)$	1/3	6	$(L,-R,2R)$	1/3
3	$(0,R,2R)$	1/3	7	$(L,R,2R)$	1/3
4	$(0,R,0)$	1	8	$(L,R,0)$	1

裂纹长度 $2a=14.76\,\text{m}$ 时的 PHT 网格局部细分过程如图 2.28 所示，可以看出，模型的 PHT 样条网格在裂纹尖端附近逐步加密。最后一次局部细分状态下的控制点分布如图 2.29 所示，该网格下控制点总数为 3444，单元数量为 888，富集之后总自由度数为 11124，富集自由度占总自由度的比例为 7.12%。积分点在不同类型单元中的分布如图 2.30 所示。不同裂纹长度下裂尖的参数坐标值如表 2.3 所示。

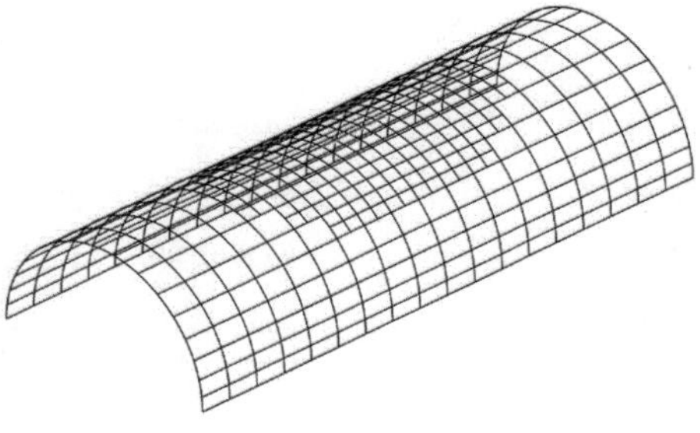
(a)

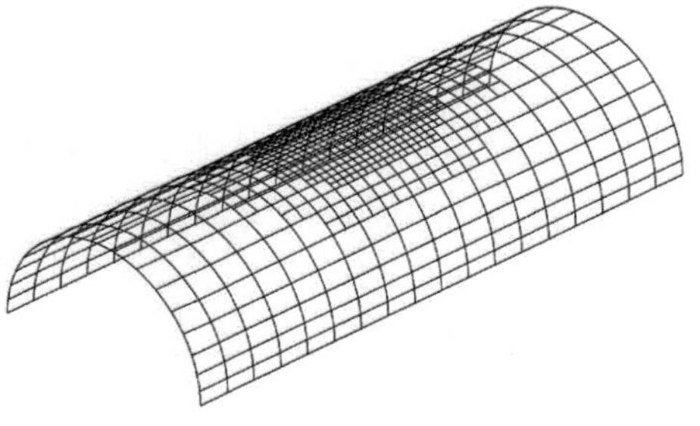
(b)

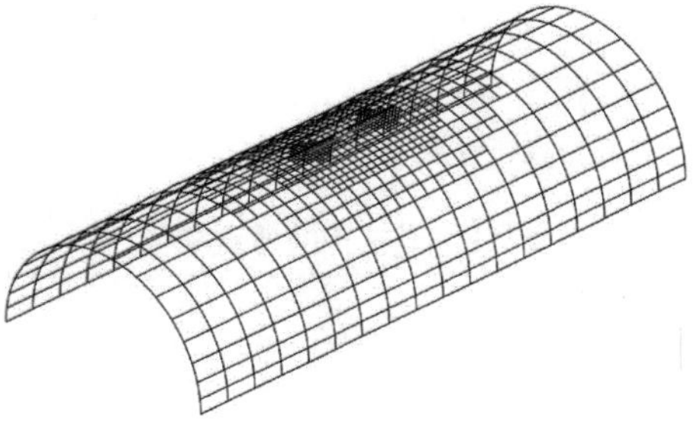

(c)

图 2.28 PHT 样条网格的局部细分过程

(a) 第一次细分；(b) 第二次细分；(c) 第三次细分

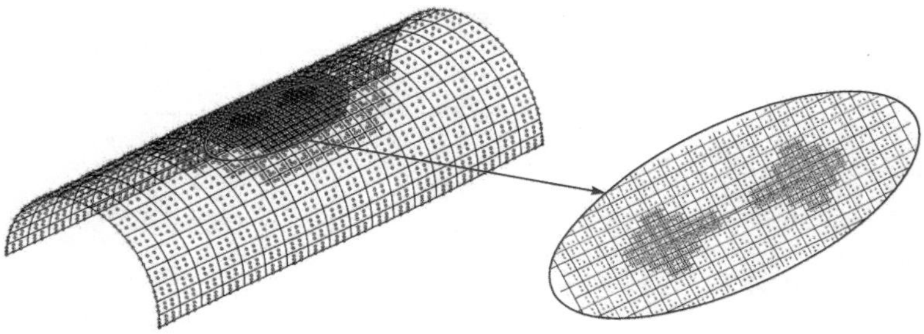

图 2.29 局部细分网格下的控制点分布及局部放大图

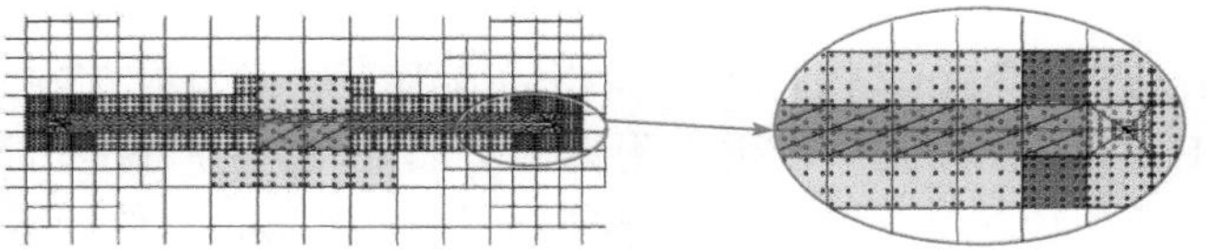

图 2.30 积分点在不同类型单元中的分布及局部放大图

表 2.3 不同裂纹长度下，裂尖的参数坐标

半裂纹长度/m	裂尖 1	裂尖 2
1.23	(0.5, 0.4877)	(0.5, 0.5123)
2.46	(0.5, 0.4754)	(0.5, 0.5246)
4.92	(0.5, 0.4508)	(0.5, 0.5492)
7.38	(0.5, 0.4262)	(0.5, 0.5738)
9.841	(0.5, 0.4016)	(0.5, 0.5984)

接下来，我们讨论富集之后刚度矩阵的稀疏特性。在本章算例中，富集自由度的编号是在标准自由度编号完成之后才开始的，因此在富集单元内，存在小编

号值的基函数与大编号值的基函数之间互相影响并被组装到单元刚度矩阵中。从图 2.31 中可以看出，这会导致在刚度矩阵的非对角线部分区域产生密集的行或列，虽然其破坏了刚度矩阵带状结构的特点，但是仍然保留了刚度矩阵的稀疏特性。

不同轴向裂纹长度下圆柱壳的位移场分布如图 2.32 所示。裂纹长度 $2a$=9.84m 时的薄膜应力和弯曲应力的分布如图 2.33 所示。利用交互作用积分计算了面外弯曲应力强度因子 K_{I}。表 2.4 给出了不同裂纹长度下，本章所得的数值结果与基于 NURBS 的 XIGA 计算的参考解[8]、虚拟节点法(phantom-node method)所得的结果[15]和 Folias 近似解[16]的比较，可以看出，本章所得的数值解与参考解吻合得很好。

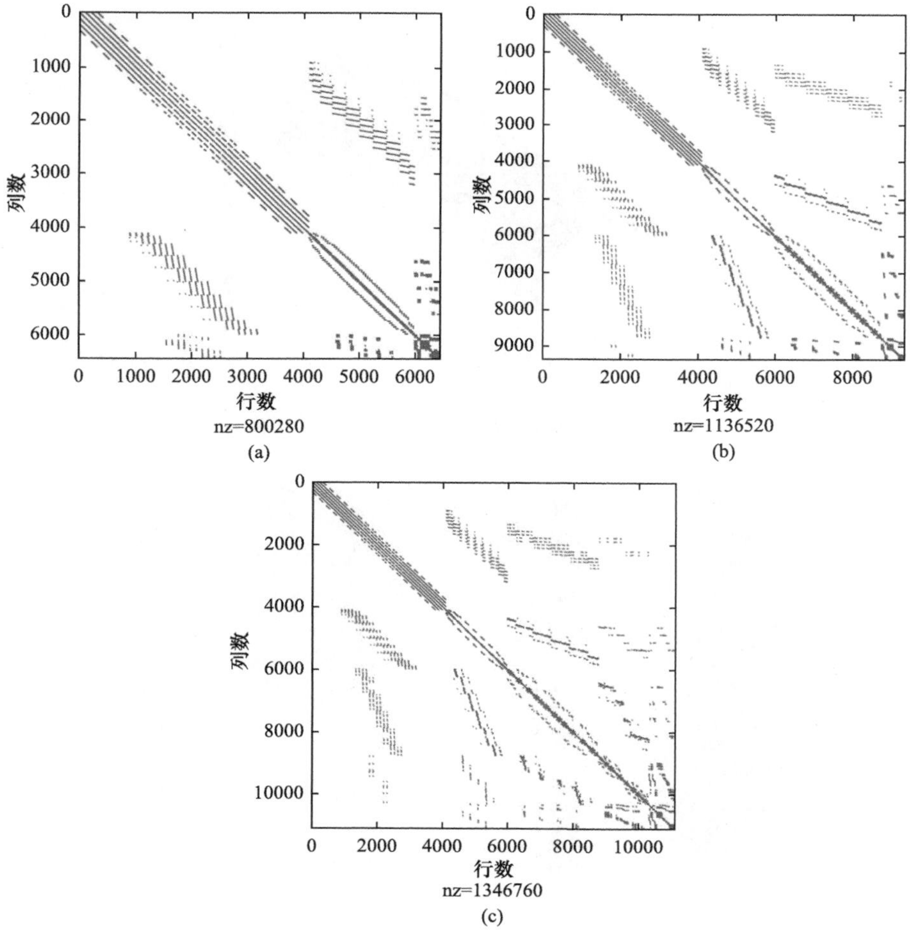

图 2.31　刚度矩阵结构(nz 是非零元素的个数)
(a) 第一次细分；(b) 第二次细分；(c) 第三次细分

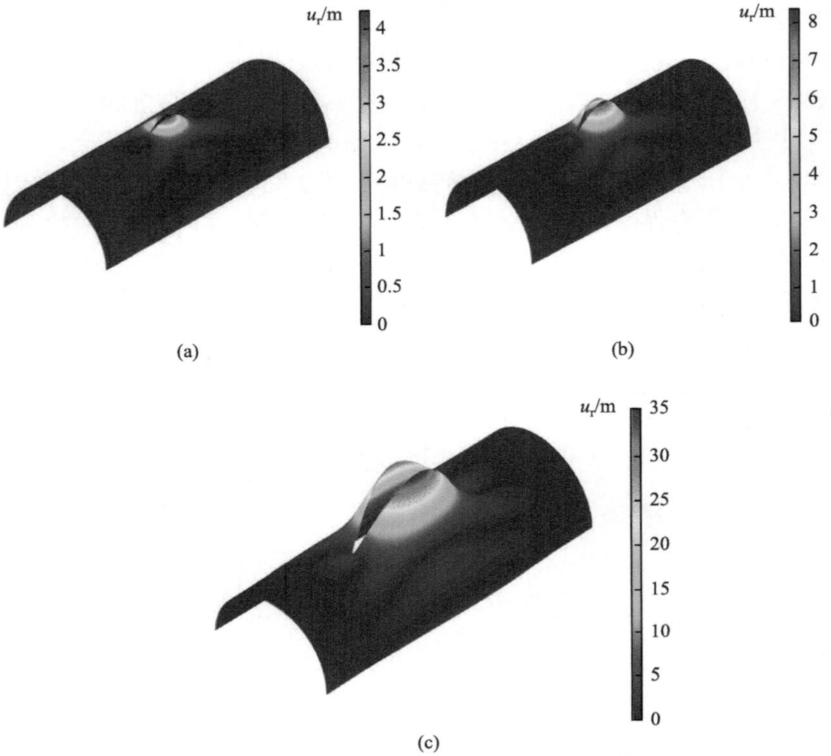

图 2.32　不同轴向裂纹长度下，圆柱壳的位移场(变形因子为 0.05)

(a) 裂纹长度 4.92m；(b) 裂纹长度 9.84m；(c) 裂纹长度 19.682m

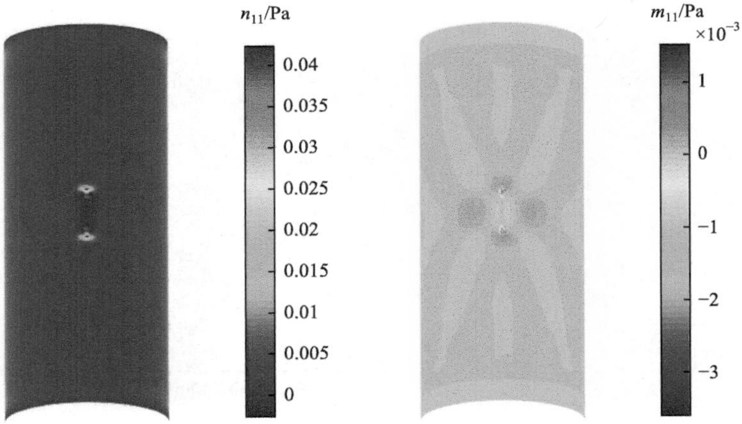

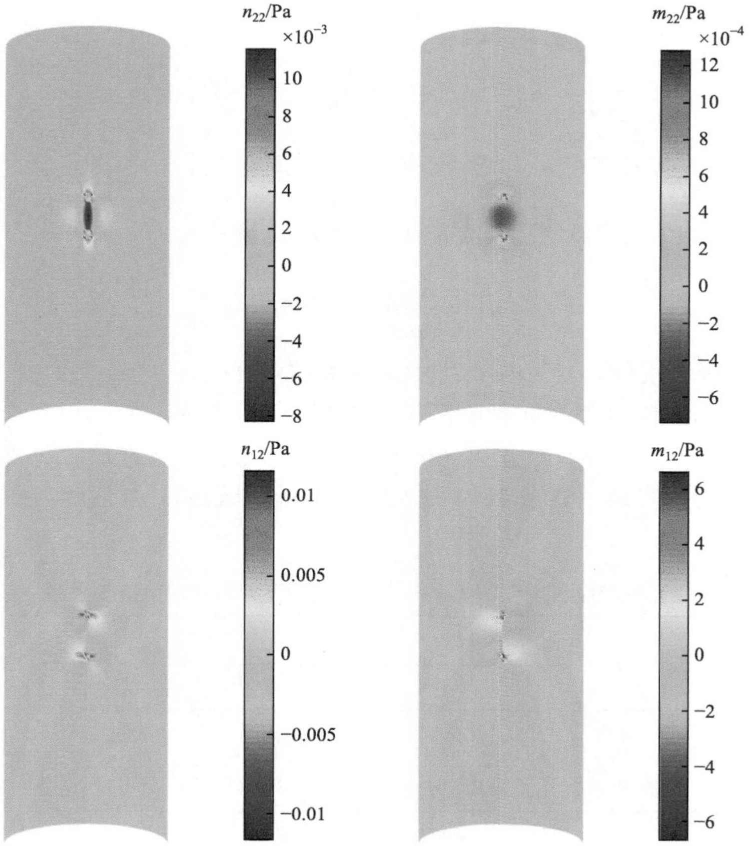

图 2.33 含轴向裂纹圆柱壳的薄膜应力和弯曲应力分布

表 2.4 不同裂纹长度下，弯曲应力强度因子与参考解的比较

裂纹长度/m	弯曲应力强度因子 $K_I/\left(\text{Pa}\cdot\sqrt{\text{m}}\right)$			
	本章	虚拟节点法[15]	基于 NURBS 的 XIGA[8]	Folias 近似解[16]
2.46	188.76	186.85	189.73	188.60
4.92	361.83	360.81	363.92	359.77
9.84	815.68	808.71	811.82	817.72
14.76	1365.62	1323.03	1343.71	1374.45
19.682	1989.24	1934.68	1947.12	1997.02

为了验证本方法的准确性，计算了裂纹中点处的张开位移。首先定义无量纲

的裂纹长度指标 ρ 为[17]

$$\rho = \frac{a}{\sqrt{Rh}} \tag{2.76}$$

式中，a 为半裂纹长度；R 为壳的半径；h 为壳体厚度。

根据文献[17]，无量纲的裂纹张开位移 $[\boldsymbol{u}]^*$ 计算式为

$$[\boldsymbol{u}]^* = \frac{E \cdot [\boldsymbol{u}]}{4a\sigma^{\infty}}, \quad \sigma^{\infty} = \frac{pR}{2h} \tag{2.77}$$

式中，E 为弹性模量；p 为内压值；$[\boldsymbol{u}]$ 为裂纹张开位移。

含轴向裂纹圆柱壳在四种不同无量纲裂纹长度指标 $\rho = 0.5, 1, 2, 3$，以及不同半径与厚度比值 $R/h = 5, 10, 20, 30$ 情况下，裂纹中点处的无量纲裂纹张开位移数值结果与参考解[17]的比较如表 2.5 所示。结果表明，数值解与参考结果吻合良好。

表 2.5　不同裂纹长度和不同半径与厚度比值下，裂纹中点处的张开位移

R/h	方法	ρ			
		0.5	1	2	3
5	本章	2.6477	3.9091	8.5795	15.3977
	参考解[17]	2.6304	3.9222	8.5814	15.4170
10	本章	2.4873	3.7056	8.2437	15.0964
	参考解[17]	2.5074	3.7219	8.2543	15.0810
20	本章	2.4498	3.6192	8.1169	14.9235
	参考解[17]	2.4459	3.6249	8.0929	14.9187
30	本章	2.4255	3.6239	8.0536	14.8342
	参考解[17]	2.4253	3.5939	8.0384	14.8597

2.4.3　含环向裂纹的圆柱壳

考虑如图 2.34 所示的受拉伸载荷作用的含环向裂纹圆柱壳的算例。圆柱半径为 R=0.0529 m。环向裂纹的长度为 $2a = 2\theta R$，其中 θ 为半裂纹跨越的角度。拉伸载荷的大小为 $P = 90\,\text{Pa}$。材料参数设置为：弹性模量 $E = 2.07 \times 10^{11}\,\text{Pa}$，泊松比 $\nu = 0.3$。圆柱壳初始节点矢量和控制点及相关权值与上例中选取的一样。裂纹跨

越角度为 $2\theta=45°$ 时的 PHT 样条网格局部细分过程如图 2.35 所示。最后一次局部细分状态下的控制点分布如图 2.36 所示，可以看出，控制点在裂纹尖端附近逐

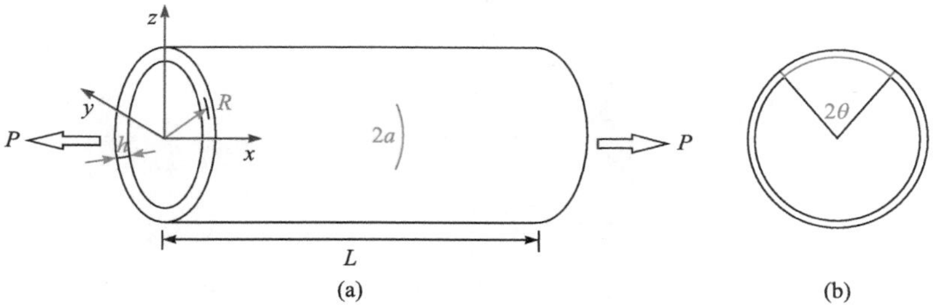

图 2.34　含环向贯穿裂纹圆柱壳

(a) 几何信息；(b) 横截面示意图

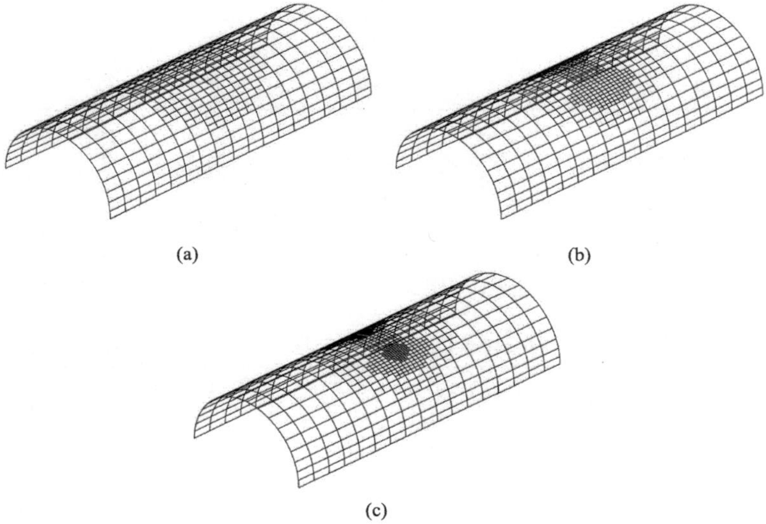

图 2.35　PHT 样条网格的局部细分过程

(a) 第一次细分；(b) 第二次细分；(c) 第三次细分

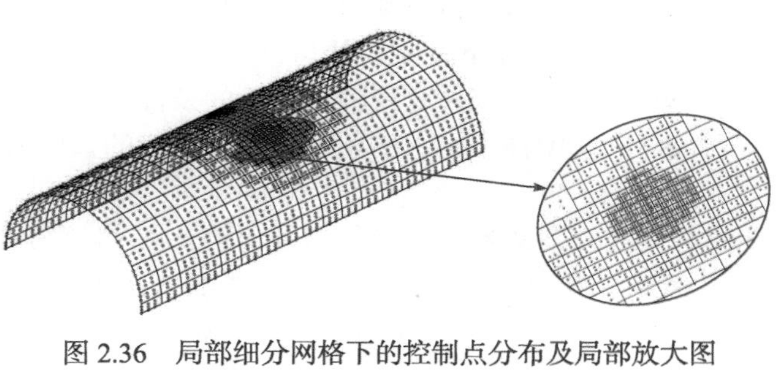

图 2.36　局部细分网格下的控制点分布及局部放大图

步聚集。该网格下控制点总数为 3468，单元数量为 900，富集之后总自由度数为 11292，富集自由度占总自由度的比例为 7.86%。积分点在不同类型单元中的分布如图 2.37 所示。不同裂纹长度下裂尖的参数坐标值如表 2.6 所示。

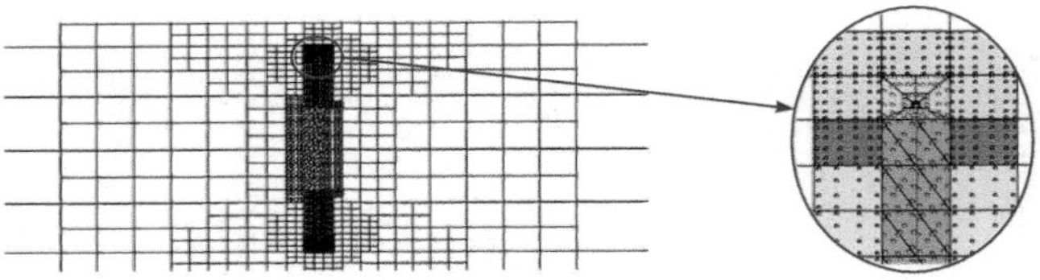

图 2.37 积分点在不同类型单元中的分布及局部放大图

表 2.6 不同裂纹长度下，裂尖的参数坐标

2θ	裂尖 1	裂尖 2
45°	(0.4005, 0.5)	(0.5995, 0.5)
90°	(0.2929, 0.5)	(0.7071, 0.5)

利用等效域积分形式的公式计算了断裂参数 J 积分的值。表 2.7 给出了在不同裂纹长度和不同壳体厚度条件下，本章所得的数值结果与虚拟节点法所得的结果[15]、虚拟裂纹扩展方法所得的结果[18]、基于 NURBS 的 XIGA 计算的参考解[8]和 Folias 近似解[16]的比较。可以看出，本章的数值解与参考解相一致。

表 2.7 不同裂纹长度和不同壳体厚度下，J 积分的值

2θ	R/h	本章	基于 NURBS 的 XIGA[8]	虚拟节点法[15]	虚拟裂纹扩展方法[18]	Folias 近似解[16]
45°	40	3.072×10^{-2}	3.056×10^{-2}	3.03×10^{-2}	3.24×10^{-2}	3.09×10^{-2}
90°	20	2.465×10^{-2}	2.423×10^{-2}	2.40×10^{-2}	2.57×10^{-2}	2.48×10^{-2}

此外，网格局部细分对刚度矩阵的条件数也会产生很大的影响。在 NURBS 全局细分和 PHT 样条局部细分下，刚度矩阵条件数随自由度数的变化趋势如图 2.38 所示。基于 NURBS 的 XIGA 所得刚度矩阵的条件数随自由度数增加呈现线性增长的趋势，而基于 PHT 样条的 XIGA 所得刚度矩阵的条件数则增长缓慢且增长率小。

在半裂纹跨越角度 $\theta=22.5°$ 和壳体半径厚度比值 $R/h=40$ 情况下，圆柱壳的位移分布如图 2.39 所示，其薄膜应力和弯曲应力的分布如图 2.40 所示。

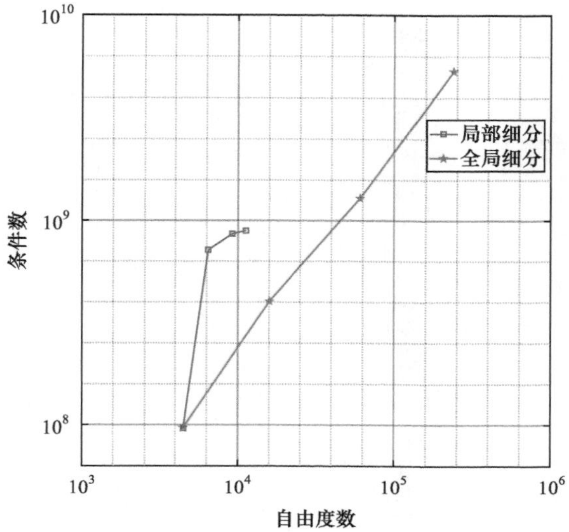

图 2.38　刚度矩阵的条件数随自由度数的变化趋势

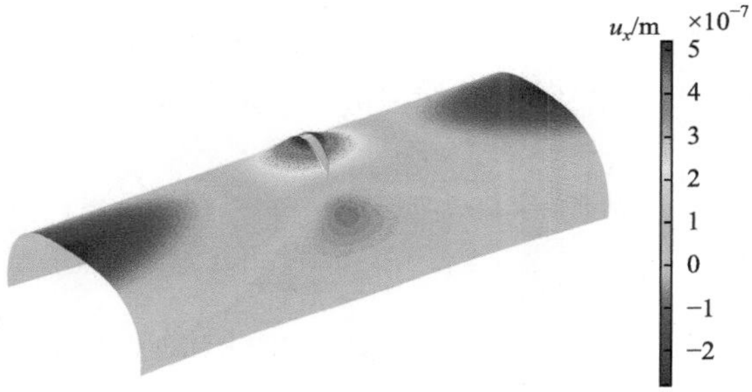

图 2.39　含环向裂纹圆柱壳的位移分布(变形因子为 0.01)

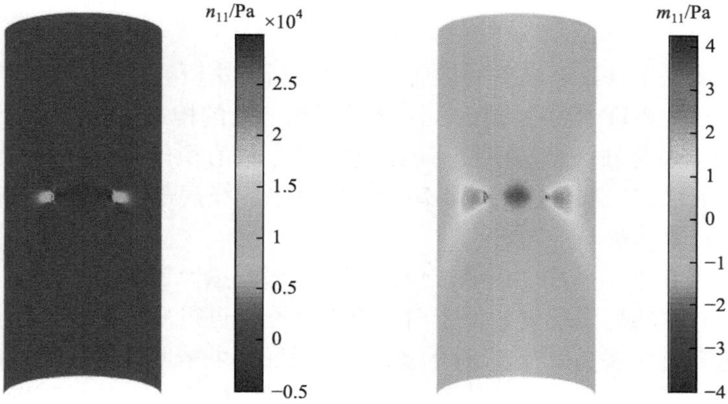

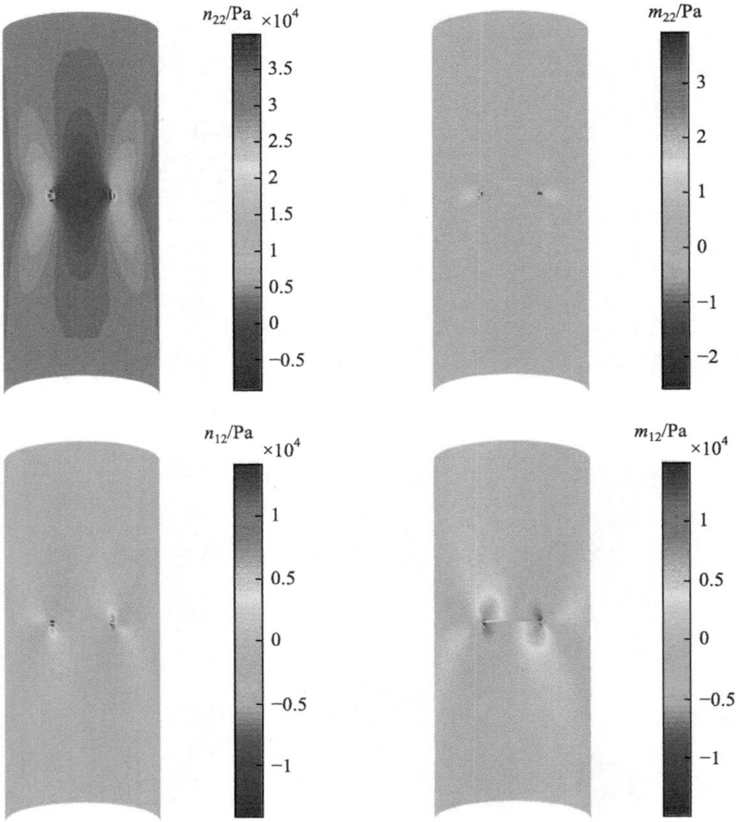

图 2.40 含环向裂纹圆柱壳的薄膜应力和弯曲应力分布

2.5 小 结

本章利用基于 PHT 样条的扩展等几何分析方法研究了含贯穿裂纹的薄壳结构：首先介绍了 PHT 样条基函数在网格局部细分时的构造过程；然后给出了基于 PHT 样条的扩展等几何分析中位移的富集公式、数值积分的实施方案和断裂参数的相关计算公式；接着通过数值算例展示了 PHT 样条局部细分的优点在分析含裂纹问题时的计算优势。

PHT 样条在继承 NURBS 单位分解性和局部支撑性等优点的基础上，还能够进行方便和有效的局部细分，便于对裂纹附近网格进行局部加密，而且能自动满足薄壳理论中对位移场 C^1 连续性的要求，在保证结果精度的同时能提高求解效率和减少求解规模。裂纹引起的位移不连续性和裂尖应力的奇异性可以通过阶跃函数和裂尖渐近函数对样条基函数进行局部富集的方式准确模拟，同时利用分割单元中的子单元划分技术和裂尖单元中的几乎极性积分可有效地消除数值积分中的非光滑项和 $r^{-1/2}$ 奇异性。

参 考 文 献

[1] Deng J, Chen F, Li X, et al. Polynomial splines over hierarchical T-meshes[J]. Graphical Models, 2008, 70(4): 76-86.

[2] Borden M J, Scott M A, Evans J A, et al. Isogeometric finite element data structures based on Bézier extraction of NURBS[J]. International Journal for Numerical Methods in Engineering, 2011, 87(1-5): 15-47.

[3] Farin G. Curves and Surfaces for CAGD A Practical Guide[M]. 5th ed. New York: Academic Press, 2002.

[4] Cottrell J A, Hughes T J R, Bazilevs Y. Isogeometric Analysis: Toward Integration of CAD and FEA[M]. New York: John Wiley & Sons, Inc., 2009.

[5] Kiendl J, Hsu M C, Wu M C H, et al. Isogeometric Kirchhoff-Love shell formulations for general hyperelastic materials[J]. Computer Methods in Applied Mechanics and Engineering, 2015, 291: 280-303.

[6] Kiendl J, Bletzinger K U, Linhard J, et al. Isogeometric shell analysis with Kirchhoff-Love elements[J]. Computer Methods in Applied Mechanics and Engineering, 2009, 198(49-52): 3902-3914.

[7] Bischoff M, Ramm E, Irslinger J. Models and finite elements for thin-walled structures//Stein E, Borst R, Hughes T J R. Encyclopedia of Computational Mechanics[M]. 2nd ed. New York: John Wiley & Sons, Inc., 2017: 1-86.

[8] Nguyen-Thanh N, Valizadeh N, Nguyen M N, et al. An extended isogeometric thin shell analysis based on Kirchhoff-Love theory[J]. Computer Methods in Applied Mechanics and Engineering, 2015, 284: 265-291.

[9] Laborde P, Pommier J, Renard Y, et al. High-order extended finite element method for cracked domains[J]. International Journal for Numerical Methods in Engineering, 2005, 64(3): 354-381.

[10] Agathos K, Chatzi E, Bordas S P A. A unified enrichment approach addressing blending and conditioning issues in enriched finite elements[J]. Computer Methods in Applied Mechanics and Engineering, 2019, 349: 673-700.

[11] Ghorashi S S, Valizadeh N, Mohammadi S. Extended isogeometric analysis for simulation of stationary and propagating cracks[J]. International Journal for Numerical Methods in Engineering, 2012, 89(9): 1069-1101.

[12] Nikishkov G P, Atluri S N. Calculation of fracture mechanics parameters for an arbitrary three-dimensional crack, by the 'equivalent domain integral' method[J]. International Journal for Numerical Methods in Engineering, 1987, 24(9): 1801-1821.

[13] Zehnder A T, Viz M J. Fracture mechanics of thin plates and shells under combined membrane, bending, and twisting loads[J]. Applied Mechanics Reviews, 2005, 58(1): 37-48.

[14] Nguyen-Thanh N, Valizadeh N, Nguyen M N, et al. An extended isogeometric thin shell analysis

based on Kirchhoff-Love theory[J]. Computer Methods in Applied Mechanics and Engineering, 2015, 284: 265-291.

[15] Chau-Dinh T, Zi G, Lee P S, et al. Phantom-node method for shell models with arbitrary cracks[J]. Computers & Structures, 2012, 92-93: 242-256.

[16] Folias E S. On the effect of initial curvature on cracked flat sheets[J]. International Journal of Fracture Mechanics, 1969, 5(4): 327-346.

[17] Huh N S, Shim D J, Choi S, et al. Stress intensity factors and crack opening displacements for slanted axial through-wall cracks in pressurized pipes[J]. Fatigue & Fracture of Engineering Materials & Structures, 2008, 31(6): 428-440.

[18] Lefort P, Delorenzi H G, Kumar V, et al. Virtual crack extension method for energy release rate calculations in flawed thin shell structures[J]. Journal of Pressure Vessel Technology, 1987, 109(1): 101-107.

[19] Bazilevs Y, Calo V M, Cottrell J A, et al. Isogeometric analysis using T-splines[J]. Computer Methods in Applied Mechanics and Engineering, 2010, 199(5): 229-263.

[20] Dörfel M R, Jüttler B, Simeon B. Adaptive isogeometric analysis by local h-refinement with T-splines[J]. Computer Methods in Applied Mechanics and Engineering, 2010, 199(5): 264-275.

[21] Deng J, Chen F, Li X, et al. Polynomial splines over hierarchical T-meshes[J]. Graphical Models, 2008, 70(4): 76-86.

[22] Johannessen K A, Kvamsdal T, Dokken T. Isogeometric analysis using LR B-splines[J]. Computer Methods in Applied Mechanics and Engineering, 2014, 269: 471-514.

[23] Liu Z, Majeed M, Cirak F, et al. Isogeometric FEM-BEM coupled structural-acoustic analysis of shells using subdivision surfaces[J]. International Journal for Numerical Methods in Engineering, 2018, 113(9): 1507-1530.

第 3 章

含缺陷功能梯度材料的等几何有限元分析

3.1 引　言

功能梯度材料(FGM)是一类新型的先进复合材料，其材料组分按照规定的梯度分布规律在空间方向上发生连续和光滑的变化，由于继承了各组分材料的优良性能，其材料性能优于传统的层状复合材料，因此被广泛应用于飞机、航空航天和核电站等工程领域，但其在制造或服役过程中，不可避免地会出现裂纹或孔洞等缺陷。这些缺陷的存在改变了材料性能在结构中的分布，对其在热力载荷作用下的动态响应产生了重要的影响。针对含裂纹或含孔洞缺陷 FGM 薄壁结构在热、力和热力耦合载荷作用下的动态问题（振动和屈曲），许多学者已基于不同的板壳理论和 XIGA/XFEM 进行了一些研究。

有限胞元法(FCM)[1,2]旨在将虚拟域方法与高阶基函数相结合用于模拟缺陷的几何边界[3,4]。该方法的主要思想是使用非相适应结构化网格进行数值分析，而通过分割单元中的自适应积分来处理复杂缺陷的几何边界，可以有效地缓解复杂孔洞几何形状生成边界匹配网格的困难。因此，只需要确定每个积分点是在物理域内还是在物理域外[5,6]，就可以有效地处理结构中的孔洞缺陷。该方法的精度可以通过自适应积分中四叉树的层数灵活控制。FCM 由于其简单高效的实现方法和显著的性能，已成功应用于多个领域[7-12]。

由于工程领域对先进材料优良性能的诉求越来越高，有必要对材料性能在多个方向(同时沿厚度方向和面内方向)发生变化的 FGM 进行含孔洞缺陷失效分析的研究。非线性屈曲分析是开孔 FGM 板的重要设计准则之一。当压力载荷的微小变化导致变形显著增加时，FGM 板表现出屈曲，这是由压力载荷所引起的薄膜应变能转化为弯曲应变能导致的，且在分岔点后还能继续承受相当大的附加压力载荷。而且在某些情况下，这些极限载荷是临界载荷的数倍。因此，线性屈曲分析可能无法准确地评估 FGM 板的极限承载能力，有必要研究多孔多向 FGM 在压缩载荷作用下的后屈曲响应。

一些学者采用不同的方法对不含缺陷的面内双向 FGM 板的静态弯曲、自由振动和特征屈曲行为进行了研究，包括 FEM[1]、径向基配置法[2,3]、Ritz 方法[4]、无网格局部 Petrov-Galerkin 方法[5]、比例边界有限元法(scaled boundary finite element method, SBFEM)[6]以及微分求积法[7]等。近年来，利用 CAD 中样条函数几何精确描述、高阶次和高连续性的优点，IGA 也被逐渐应用到面内非均匀 FGM 板的自由振动和特征屈曲领域[8-13]。利用 IGA 预测 FGM 板壳结构后屈曲行为的文献则相对较少。Thai 等[14]基于高阶剪切变形理论(higher-order shear deformation theory, HSDT)和修正的应变梯度理论，利用 IGA 对含孔隙 FGM 微板在热载荷和力载荷作用下的后屈曲行为进行了研究。Nguyen 等[15,16]利用基于 NURBS 的一阶剪切变形双曲壳理论，研究了功能梯度碳纳米管增强复合圆柱壳的静态、自由振动和后屈曲问题。因此，本章的其中一项工作是结合 FCM 和 IGA 研究含多孔多向 FGM 板在面内边缘压力载荷作用下的后屈曲行为，并探究初始几何缺陷、孔洞几何形状和材料分布等因素对后屈曲路径的影响。

3.2 功能梯度材料的分布及等效参数

FGM 是由两种不同材料混合而成的复合材料。假设顶面和底面分别为纯陶瓷和纯金属，板中性面位于 xOy 平面内。梯度特性沿厚度方向变化，如图 3.1 所示。

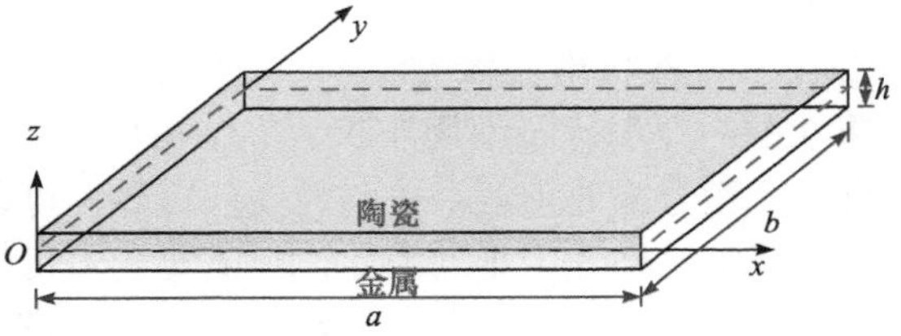

图 3.1 FGM 板

陶瓷和金属相的体积分数分别用 V_c 和 V_m 表示，其表达式为

$$V_c(z)=\left(\frac{1}{2}+\frac{z}{h}\right)^n,\quad V_m(z)=1-V_c(z)\quad \left(-\frac{h}{2}\leqslant z\leqslant\frac{h}{2}\right) \tag{3.1}$$

其中，n 为材料梯度指数；$n=0$ 和 $n=\infty$ 分别代表纯陶瓷和纯金属材质。

根据混合法则[17]，任意一点处的等效材料特性(P_e)计算如下：

$$P_e = P_c V_c(z) + P_m V_m(z) \tag{3.2}$$

式中，P_c 和 P_m 分别代表陶瓷成分和金属成分的材料特性；$P_s(s=e,c,m)$ 可以代表材料弹性模量 E、泊松比 ν、密度 ρ、热传导系数 κ 和热膨胀系数 α 等。

混合规则没有考虑两种成分之间的相互影响。基于 Mori-Tanaka 均匀化技术[17]，等效材料特性是通过其等效体积模量和等效剪切模量获取的。任意一点处的材料

9 体积模量 $K(z)$ 和剪切模量 $\mu(z)$ 的计算表达式为

$$K(z)=\frac{E(z)}{3\left[1-2\nu(z)\right]},\quad \mu(z)=\frac{E(z)}{2\left[1+\nu(z)\right]} \tag{3.3}$$

通过以下公式计算等效体积模量 $K_{\rm e}$ 和等效剪切模量 $\mu_{\rm e}$：

$$\frac{K_{\rm e}-K_{\rm m}}{K_{\rm c}-K_{\rm m}}=\frac{V_{\rm c}}{1+V_{\rm m}\dfrac{3(K_{\rm c}-K_{\rm m})}{3K_{\rm m}+4\mu_{\rm m}}},\qquad \frac{\mu_{\rm e}-\mu_{\rm m}}{\mu_{\rm c}-\mu_{\rm m}}=\frac{V_{\rm c}}{1+V_{\rm m}\dfrac{\mu_{\rm c}-\mu_{\rm m}}{\mu_{\rm m}+f_1}} \tag{3.4}$$

其中，$f_1=\dfrac{\mu_{\rm m}(9K_{\rm m}+8\mu_{\rm m})}{6(K_{\rm m}+2\mu_{\rm m})}$。

因此，在此情况下等效弹性模量和等效泊松比的计算表达式为

$$E_{\rm e}=\frac{9K_{\rm e}\mu_{\rm e}}{3K_{\rm e}+\mu_{\rm e}},\quad \nu_{\rm e}=\frac{3K_{\rm e}-2\mu_{\rm e}}{2(3K_{\rm e}+\mu_{\rm e})} \tag{3.5}$$

等效热传导系数和等效热膨胀系数的计算表达式为

$$\frac{\kappa_{\rm e}-\kappa_{\rm m}}{\kappa_{\rm c}-\kappa_{\rm m}}=\frac{V_{\rm c}}{1+V_{\rm m}\dfrac{\kappa_{\rm c}-\kappa_{\rm m}}{3\kappa_{\rm m}}},\qquad \frac{\alpha_{\rm e}-\alpha_{\rm m}}{\alpha_{\rm c}-\alpha_{\rm m}}=\frac{1/\kappa_{\rm e}-1/\kappa_{\rm m}}{1/\kappa_{\rm c}-1/\kappa_{\rm m}} \tag{3.6}$$

假定 FGM 板的材料性能在两个或多个方向上连续变化，则称为多向 FGM 板[18]。除非另有说明，只考虑以下两种情况。

考虑如图 3.2 所示的具有长度 a、宽度 b 和厚度 h 的多向 FGM 矩形板，且板的中性面位于 xOy 平面内。

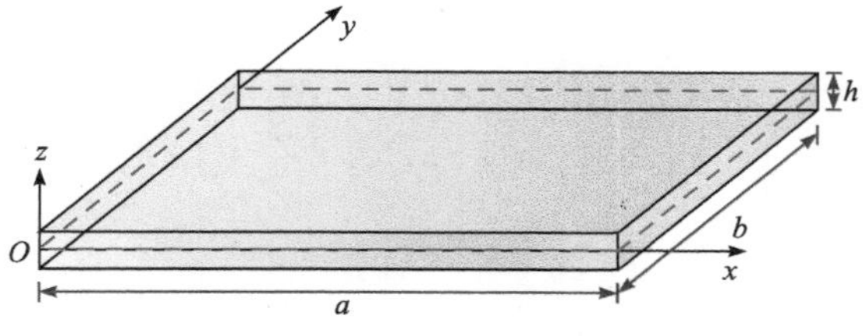

图 3.2　多向 FGM 矩形板

假设该板由三种不同的材料复合组成，且每种材料成分沿厚度和面内方向均发生变化，它们的体积分数用 V_1、V_2 和 V_3 来表示。首先定义三类面内分布函数[10]为

$$\phi_1=\left(\frac{x}{a}\right)^{k_1}\left(\frac{y}{b}\right)^{k_2},\quad 0\leqslant x\leqslant a,\ 0\leqslant y\leqslant b \tag{3.7}$$

$$\phi_2=\begin{cases}\left(\dfrac{2x}{a}\right)^{k_1}\left(\dfrac{2y}{b}\right)^{k_2}, & 0\leqslant x\leqslant\dfrac{a}{2},\ 0\leqslant y\leqslant\dfrac{b}{2}\\ \left(2-\dfrac{2x}{a}\right)^{k_1}\left(\dfrac{2y}{b}\right)^{k_2}, & \dfrac{a}{2}\leqslant x\leqslant a,\ 0\leqslant y\leqslant\dfrac{b}{2}\\ \left(\dfrac{2x}{a}\right)^{k_1}\left(2-\dfrac{2y}{b}\right)^{k_2}, & 0\leqslant x\leqslant\dfrac{a}{2},\ \dfrac{b}{2}\leqslant y\leqslant b\\ \left(2-\dfrac{2x}{a}\right)^{k_1}\left(2-\dfrac{2y}{b}\right)^{k_2}, & \dfrac{a}{2}\leqslant x\leqslant a,\ \dfrac{b}{2}\leqslant y\leqslant b\end{cases}\tag{3.8}$$

$$\phi_3=\begin{cases}\left(1-\dfrac{2x}{a}\right)^{k_1}\left(1-\dfrac{2y}{b}\right)^{k_2}, & 0\leqslant x\leqslant\dfrac{a}{2},\ 0\leqslant y\leqslant\dfrac{b}{2}\\ \left(\dfrac{2x}{a}-1\right)^{k_1}\left(1-\dfrac{2y}{b}\right)^{k_2}, & \dfrac{a}{2}\leqslant x\leqslant a,\ 0\leqslant y\leqslant\dfrac{b}{2}\\ \left(1-\dfrac{2x}{a}\right)^{k_1}\left(\dfrac{2y}{b}-1\right)^{k_2}, & 0\leqslant x\leqslant\dfrac{a}{2},\ \dfrac{b}{2}\leqslant y\leqslant b\\ \left(\dfrac{2x}{a}-1\right)^{k_1}\left(\dfrac{2y}{b}-1\right)^{k_2}, & \dfrac{a}{2}\leqslant x\leqslant a,\ \dfrac{b}{2}\leqslant y\leqslant b\end{cases}\tag{3.9}$$

根据文献[1],[10]，在三类面内分布函数下，不同类型材料的体积分数计算如下所述。

类型 A：

$$V_3=\left(\frac{z}{h}+\frac{1}{2}\right)^{k_3},\quad V_2=(1-V_3)\phi_1,\quad V_1=(1-V_3)(1-\phi_1)\tag{3.10}$$

类型 B：

$$V_3=\left(\frac{z}{h}+\frac{1}{2}\right)^{k_3},\quad V_2=(1-V_3)\phi_2,\quad V_1=(1-V_3)(1-\phi_2)\tag{3.11}$$

类型 C：

$$V_3=\left(\frac{z}{h}+\frac{1}{2}\right)^{k_3},\quad V_2=(1-V_3)\phi_3,\quad V_1=(1-V_3)(1-\phi_3)\tag{3.12}$$

式(3.7)～(3.12)中，k_1、k_2和k_3分别是控制各个组分沿坐标轴方向变化的材料梯度指数。可以发现，当$k_1=k_2=0$时，多向 FGM 板退化为由材料 3 和材料 2 组合而成的单方向 FGM 板。

材料在任意一点处的等效材料特性P_e根据混合法则计算如下：

$$P_e(x,y,z)=P_1V_1(x,y,z)+P_2V_2(x,y,z)+P_3V_3(x,y,z)\tag{3.13}$$

其中，P_1、P_2和P_3分别代表三种成分的材料特性；P可以代表材料的弹性模量E、

泊松比 ν 和密度 ρ 等。

材料 3 在不同 k_3 值下沿厚度方向的分布曲线如图 3.3 所示。另外，图 3.4 给出了在 $k_1 = 2$ 和 $k_2 = 0.2$ 情况下的三种面内分布函数的云图。

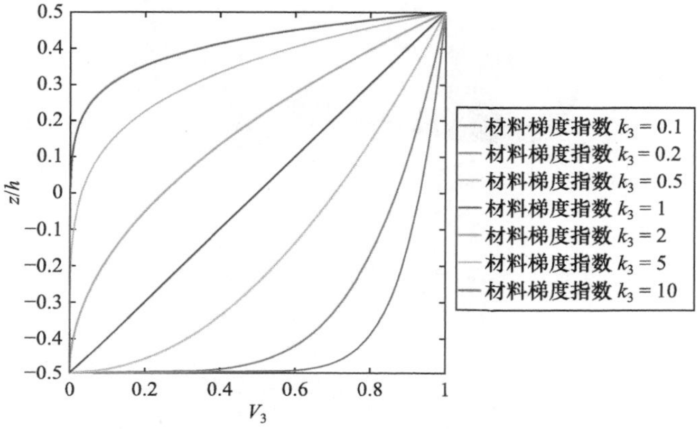

图 3.3　材料 3 在不同 k_3 值下沿厚度方向的分布曲线

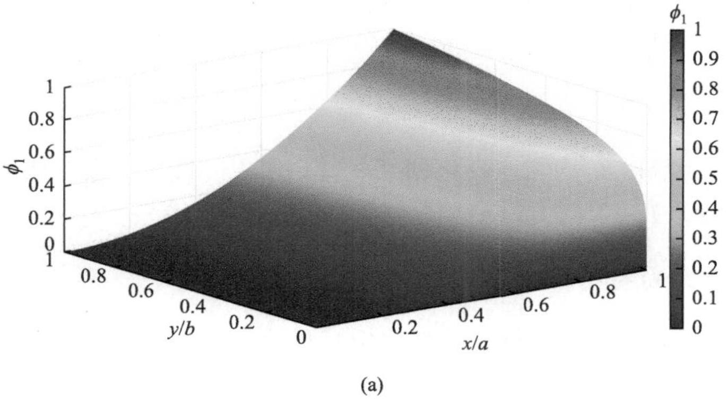

(a)

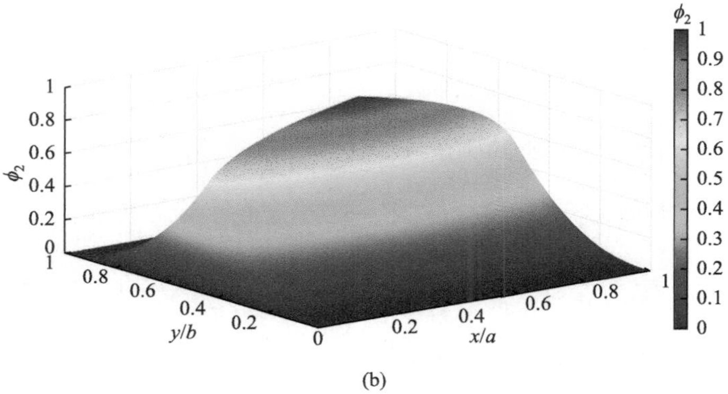

(b)

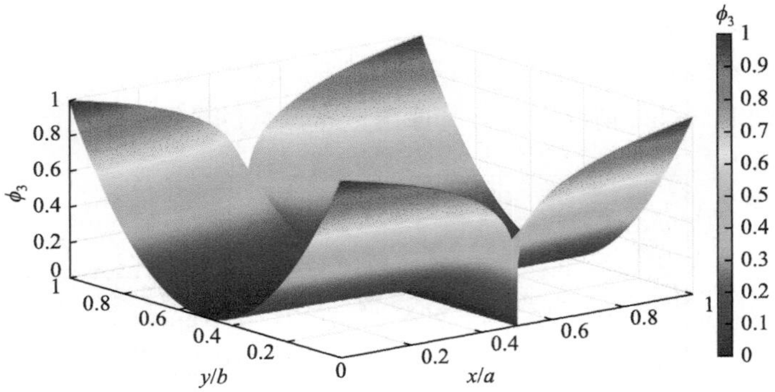

(c)

图 3.4 在 $k_1 = 2$ 和 $k_2 = 0.2$ 情况下，三种面内分布函数的云图

(a) ϕ_1 的分布；(b) ϕ_2 的分布；(c) ϕ_3 的分布

3.3 材料厚度方向上的温度分布

假设温度场在 xOy 平面内为常值，仅在厚度方向上发生变化。本章考虑三种情况下的温度分布。

1. 均匀温度分布

温度从参考温度 T_0 均匀地变化到某一特定值 T_f。因此，板上任意一点处的温度变化都是 $\Delta T = T_f - T_0$。

2. 沿厚度方向的线性温度分布

温度沿厚度方向线性变化，其变化量表示为

$$T(z) = T_m + (T_c - T_m)\left(\frac{z}{h} + \frac{1}{2}\right) \tag{3.14}$$

其中，T_c 和 T_m 分别是板顶面和底面上的温度。

3. 沿厚度方向的非线性温度分布

温度沿厚度方向的分布可通过求解一维稳态热传导方程获得。

$$-\frac{d}{dz}\left[\kappa(z)\frac{dT}{dz}\right] = 0 \qquad \left(T = T_c 在\ z = \frac{h}{2},\quad T = T_m 在\ z = -\frac{h}{2}\right) \tag{3.15}$$

式(3.15)的解通过泰勒级数表示为[19]

$$T(z) = T_m + (T_c - T_m)\eta(z,h) \tag{3.16}$$

式中，

$$\eta(z,h)=\frac{1}{C}\left[\left(\frac{z}{h}+\frac{1}{2}\right)-\frac{\kappa_{\mathrm{cm}}}{(n+1)\kappa_{\mathrm{m}}}\left(\frac{z}{h}+\frac{1}{2}\right)^{n+1}+\frac{\kappa_{\mathrm{cm}}^2}{(2n+1)\kappa_{\mathrm{m}}^2}\left(\frac{z}{h}+\frac{1}{2}\right)^{2n+1}\right.$$
$$\left.-\frac{\kappa_{\mathrm{cm}}^3}{(3n+1)\kappa_{\mathrm{m}}^3}\left(\frac{z}{h}+\frac{1}{2}\right)^{3n+1}+\frac{\kappa_{\mathrm{cm}}^4}{(4n+1)\kappa_{\mathrm{m}}^4}\left(\frac{z}{h}+\frac{1}{2}\right)^{4n+1}-\frac{\kappa_{\mathrm{cm}}^5}{(5n+1)\kappa_{\mathrm{m}}^5}\left(\frac{z}{h}+\frac{1}{2}\right)^{5n+1}\right] \tag{3.17}$$

其中，

$$C=1-\frac{\kappa_{\mathrm{cm}}}{(n+1)\kappa_{\mathrm{m}}}+\frac{\kappa_{\mathrm{cm}}^2}{(2n+1)\kappa_{\mathrm{m}}^2}-\frac{\kappa_{\mathrm{cm}}^3}{(3n+1)\kappa_{\mathrm{m}}^3}+\frac{\kappa_{\mathrm{cm}}^4}{(4n+1)\kappa_{\mathrm{m}}^4}-\frac{\kappa_{\mathrm{cm}}^5}{(5n+1)\kappa_{\mathrm{m}}^5} \tag{3.18}$$

$$\kappa_{\mathrm{cm}}=\kappa_{\mathrm{c}}-\kappa_{\mathrm{m}}$$

3.4 基于样条函数的高阶功能梯度结构公式

根据 Reddy 高阶剪切板理论[20]，板上任意一点处的位移模式为

$$\begin{aligned}
u(x,y,z)&=u_0+z\beta_x-\frac{4z^3}{3h^2}(\beta_x+w_{,x})\\
v(x,y,z)&=v_0+z\beta_y-\frac{4z^3}{3h^2}(\beta_y+w_{,y}) \quad \left(-\frac{h}{2}\leqslant z\leqslant\frac{h}{2}\right)\\
w(x,y,z)&=w_0
\end{aligned} \tag{3.19}$$

其中，(u_0,v_0,w_0)代表中性面的位移；β_x 和 β_y 分别代表绕 yOz 平面和 xOz 平面的转角。

由于高阶剪切变形量的引入，该理论能够自动满足剪应变沿厚度方向的分布，而不再需要剪切修正因子。小应变和中等旋转假设下的应变与位移关系为

$$\begin{Bmatrix}\varepsilon_{\mathrm{x}}\\ \varepsilon_y\\ \gamma_{xy}\\ \gamma_{xz}\\ \gamma_{yz}\end{Bmatrix}=\begin{Bmatrix}u_{,x}\\ v_{,y}\\ u_{,y}+v_{,x}\\ u_{,z}+w_{,x}\\ v_{,z}+w_{,y}\end{Bmatrix} \tag{3.20}$$

将式(3.19)代入式(3.20)可得应变为

$$\begin{aligned}
\boldsymbol{\varepsilon}&=\left[\varepsilon_{\mathrm{xx}},\varepsilon_{yy},\gamma_{xy}\right]^{\mathrm{T}}=\boldsymbol{\varepsilon}_{\mathrm{m}}+z\boldsymbol{\kappa}_{\mathrm{b1}}+z^3\boldsymbol{\kappa}_{\mathrm{b2}}\\
\gamma&=\left[\gamma_{xz},\gamma_{yz}\right]^{\mathrm{T}}=\boldsymbol{\varepsilon}_{\mathrm{s}}+z^2\boldsymbol{\kappa}_{\mathrm{bs}}
\end{aligned} \tag{3.21}$$

式中，

$$\boldsymbol{\varepsilon}_{\mathrm{m}}=\begin{Bmatrix} u_{0,x} \\ v_{0,y} \\ u_{0,y}+v_{0,x} \end{Bmatrix},\quad \boldsymbol{\kappa}_{\mathrm{b1}}=\begin{Bmatrix} \beta_{x,x} \\ \beta_{y,y} \\ \beta_{x,y}+\beta_{y,x} \end{Bmatrix},\quad \boldsymbol{\kappa}_{\mathrm{b2}}=-\frac{4}{3h^2}\begin{Bmatrix} \beta_{x,x}+w_{0,xx} \\ \beta_{y,y}+w_{0,yy} \\ \beta_{x,y}+\beta_{y,x}+2w_{0,xy} \end{Bmatrix}$$

$$\boldsymbol{\varepsilon}_{\mathrm{s}}=\begin{Bmatrix} \beta_x+w_{0,x} \\ \beta_y+w_{0,y} \end{Bmatrix},\quad \boldsymbol{\kappa}_{\mathrm{bs}}=-\frac{4}{h^2}\begin{Bmatrix} \beta_x+w_{0,x} \\ \beta_y+w_{0,y} \end{Bmatrix} \tag{3.22}$$

考虑热效应，其热应变矢量定义为

$$\boldsymbol{\varepsilon}^{\mathrm{th}}=\alpha_{\mathrm{e}}(z)\Delta T(z)\{1,1,0\}^{\mathrm{T}} \tag{3.23}$$

其中，$\alpha_{\mathrm{e}}(z)$ 是等效热膨胀系数；$\Delta T(z)$ 表示温度的改变，其计算式为 $\Delta T(z)=T(z)-T_0$。此处的 $T(z)$ 和 T_0 分别代表当前温度和无热应变的参考温度。

根据广义胡克定律，假设法向应力 σ_z 为零，板的应力表示为

$$\begin{Bmatrix} \boldsymbol{\sigma} \\ \boldsymbol{\tau} \end{Bmatrix}=\begin{bmatrix} \boldsymbol{C} & 0 \\ 0 & \boldsymbol{G} \end{bmatrix}\begin{Bmatrix} \boldsymbol{\varepsilon}-\boldsymbol{\varepsilon}^{\mathrm{th}} \\ \boldsymbol{\gamma} \end{Bmatrix} \tag{3.24}$$

式中，材料矩阵 $\boldsymbol{C}$ 和 $\boldsymbol{G}$ 表示为

$$\boldsymbol{C}=\frac{E_{\mathrm{e}}(z)}{1-\nu_{\mathrm{e}}^2(z)}\begin{bmatrix} 1 & \nu_{\mathrm{e}}(z) & 0 \\ \nu_{\mathrm{e}}(z) & 1 & 0 \\ 0 & 0 & [1-\nu_{\mathrm{e}}(z)]/2 \end{bmatrix},\ \boldsymbol{G}=\frac{E_{\mathrm{e}}(z)}{2[1+\nu_{\mathrm{e}}(z)]}\begin{bmatrix} 1 & 0 \\ 0 & 1 \end{bmatrix} \tag{3.25}$$

其中，$E_{\mathrm{e}}(z)$ 和 $\nu_{\mathrm{e}}(z)$ 分别为沿板厚度方向的等效弹性模量和等效泊松比。

通过厚度方向积分，面内拉力、弯矩和剪力合力分别表示为

$$\begin{Bmatrix} \boldsymbol{N} \\ \boldsymbol{M} \\ \boldsymbol{P} \end{Bmatrix}=\int_{-\frac{h}{2}}^{\frac{h}{2}}\boldsymbol{\sigma}\begin{Bmatrix} 1 \\ z \\ z^3 \end{Bmatrix}\mathrm{d}z,\quad \begin{Bmatrix} \boldsymbol{Q} \\ \boldsymbol{R} \end{Bmatrix}=\int_{-\frac{h}{2}}^{\frac{h}{2}}\boldsymbol{\tau}\begin{Bmatrix} 1 \\ z^2 \end{Bmatrix}\mathrm{d}z \tag{3.26}$$

将式(3.24)代入式(3.26)，应力合力和应变之间的关系以矩阵形式表示为

$$\begin{Bmatrix} \boldsymbol{N} \\ \boldsymbol{M} \\ \boldsymbol{P} \\ \boldsymbol{Q} \\ \boldsymbol{R} \end{Bmatrix}=\begin{bmatrix} \boldsymbol{A} & \boldsymbol{B} & \boldsymbol{E} & 0 & 0 \\ \boldsymbol{B} & \boldsymbol{D} & \boldsymbol{F} & 0 & 0 \\ \boldsymbol{E} & \boldsymbol{F} & \boldsymbol{H} & 0 & 0 \\ 0 & 0 & 0 & \boldsymbol{A}_{\mathrm{s}} & \boldsymbol{B}_{\mathrm{s}} \\ 0 & 0 & 0 & \boldsymbol{B}_{\mathrm{s}} & \boldsymbol{D}_{\mathrm{s}} \end{bmatrix}\begin{Bmatrix} \boldsymbol{\varepsilon}_{\mathrm{m}} \\ \boldsymbol{\kappa}_{\mathrm{b1}} \\ \boldsymbol{\kappa}_{\mathrm{b2}} \\ \boldsymbol{\varepsilon}_{\mathrm{s}} \\ \boldsymbol{\kappa}_{\mathrm{bs}} \end{Bmatrix}-\begin{Bmatrix} \boldsymbol{N}^{\mathrm{th}} \\ \boldsymbol{M}^{\mathrm{th}} \\ \boldsymbol{P}^{\mathrm{th}} \\ 0 \\ 0 \end{Bmatrix} \tag{3.27}$$

式中，

$$
\begin{aligned}
&[\boldsymbol{A}\ \ \boldsymbol{B}\ \ \boldsymbol{D}\ \ \boldsymbol{E}\ \ \boldsymbol{F}\ \ \boldsymbol{H}]=\int_{-\frac{h}{2}}^{\frac{h}{2}}\boldsymbol{C}\left\{1,z,z^2,z^3,z^4,z^6\right\}\mathrm{d}z\\
&[\boldsymbol{A}_\mathrm{s}\ \ \boldsymbol{B}_\mathrm{s}\ \ \boldsymbol{D}_\mathrm{s}]=\int_{-\frac{h}{2}}^{\frac{h}{2}}\boldsymbol{G}\left\{1,z^2,z^4\right\}\mathrm{d}z
\end{aligned}
\tag{3.28}
$$

和热应变产生的热应力合力表示为

$$
\left[\boldsymbol{N}^\mathrm{th}\ \ \boldsymbol{M}^\mathrm{th}\ \ \boldsymbol{P}^\mathrm{th}\right]=\int_{-\frac{h}{2}}^{\frac{h}{2}}\boldsymbol{C}\begin{Bmatrix}\alpha_\mathrm{e}(z)\\ \alpha_\mathrm{e}(z)\\ 0\end{Bmatrix}\left\{1,z,z^3\right\}\mathrm{d}z \tag{3.29}
$$

FGM 板静态分析的弱形式表示为[8]

$$
\int_\Omega \delta\boldsymbol{\varepsilon}^\mathrm{T}\boldsymbol{D}\boldsymbol{\varepsilon}\,\mathrm{d}\Omega+\int_\Omega \delta\boldsymbol{\gamma}^\mathrm{T}\boldsymbol{D}^\mathrm{s}\boldsymbol{\gamma}\,\mathrm{d}\Omega=\int_\Omega \delta wp\,\mathrm{d}\Omega+\int_\Omega \delta\boldsymbol{\varepsilon}^\mathrm{T}\boldsymbol{\sigma}^\mathrm{th}\,\mathrm{d}\Omega \tag{3.30}
$$

其中，p 为单位面积上的横向载荷；Ω 为板的中性面；$\boldsymbol{D}$ 、$\boldsymbol{D}^\mathrm{s}$ 和 $\boldsymbol{\sigma}^\mathrm{th}$ 的形式如下：

$$
\boldsymbol{D}=\begin{bmatrix}\boldsymbol{A}&\boldsymbol{B}&\boldsymbol{E}\\ \boldsymbol{B}&\boldsymbol{D}&\boldsymbol{F}\\ \boldsymbol{E}&\boldsymbol{F}&\boldsymbol{H}\end{bmatrix},\quad \boldsymbol{D}^\mathrm{s}=\begin{bmatrix}\boldsymbol{A}_\mathrm{s}&\boldsymbol{B}_\mathrm{s}\\ \boldsymbol{B}_\mathrm{s}&\boldsymbol{D}_\mathrm{s}\end{bmatrix},\quad \boldsymbol{\sigma}^\mathrm{th}=\begin{Bmatrix}\boldsymbol{N}^\mathrm{th}\\ \boldsymbol{M}^\mathrm{th}\\ \boldsymbol{P}^\mathrm{th}\end{Bmatrix} \tag{3.31}
$$

FGM 板自由振动的弱形式表示为[8]

$$
\int_\Omega \delta\boldsymbol{\varepsilon}^\mathrm{T}\boldsymbol{D}\boldsymbol{\varepsilon}\,\mathrm{d}\Omega+\int_\Omega \delta\boldsymbol{\gamma}^\mathrm{T}\boldsymbol{D}^\mathrm{s}\boldsymbol{\gamma}\,\mathrm{d}\Omega=\int_\Omega \delta\boldsymbol{u}^\mathrm{T}\boldsymbol{m}\ddot{\boldsymbol{u}}\,\mathrm{d}\Omega \tag{3.32}
$$

式中，

$$
\boldsymbol{u}=\begin{Bmatrix}\boldsymbol{u}_1\\ \boldsymbol{u}_2\\ \boldsymbol{u}_3\end{Bmatrix},\quad \boldsymbol{u}_1=\begin{Bmatrix}u_0\\ \beta_x\\ \beta_x+w_{0,x}\end{Bmatrix},\quad \boldsymbol{u}_2=\begin{Bmatrix}v_0\\ \beta_y\\ \beta_y+w_{0,y}\end{Bmatrix},\quad \boldsymbol{u}_3=\begin{Bmatrix}w_0\\ 0\\ 0\end{Bmatrix} \tag{3.33}
$$

其中，$\boldsymbol{m}$ 为一致形式的质量矩阵，其表达式为

$$
\boldsymbol{m}=\begin{bmatrix}\boldsymbol{I}_0&0&0\\ 0&\boldsymbol{I}_0&0\\ 0&0&\boldsymbol{I}_0\end{bmatrix} \tag{3.34}
$$

式中，

$$\boldsymbol{I}_0 = \begin{bmatrix} I_1 & I_2 & -\dfrac{4}{3h^2}I_4 \\ I_2 & I_3 & -\dfrac{4}{3h^2}I_5 \\ -\dfrac{4}{3h^2}I_4 & -\dfrac{4}{3h^2}I_5 & \dfrac{16}{9h^4}I_6 \end{bmatrix} \tag{3.35}$$

$$\{I_1,I_2,I_3,I_4,I_5,I_6\} = \int_{-\frac{h}{2}}^{\frac{h}{2}} \rho_e(z)\left\{1,z,z^2,z^3,z^4,z^6\right\} \mathrm{d}z$$

其中，$\rho_e(z)$为沿板厚度方向的等效密度。

FGM 板特征屈曲的弱形式表示为[21]

$$\begin{aligned}&\int_{\Omega}\delta\boldsymbol{\varepsilon}^{\mathrm{T}}\boldsymbol{D}\boldsymbol{\varepsilon}\mathrm{d}\Omega+\int_{\Omega}\delta\boldsymbol{\gamma}^{\mathrm{T}}\boldsymbol{D}^{\mathrm{s}}\boldsymbol{\gamma}\mathrm{d}\Omega+h\int_{\Omega}\nabla^{\mathrm{T}}\delta w\boldsymbol{\sigma}_0\nabla w\mathrm{d}\Omega\\&+h\int_{\Omega}\begin{bmatrix}\nabla^{\mathrm{T}}\delta u_0 & \nabla^{\mathrm{T}}\delta v_0\end{bmatrix}\begin{bmatrix}\boldsymbol{\sigma}_0 & 0\\0 & \boldsymbol{\sigma}_0\end{bmatrix}\begin{bmatrix}\nabla u_0\\\nabla v_0\end{bmatrix}\mathrm{d}\Omega\\&+\frac{h^3}{12}\int_{\Omega}\begin{bmatrix}\nabla^{\mathrm{T}}\delta\beta_x & \nabla^{\mathrm{T}}\delta\beta_y\end{bmatrix}\begin{bmatrix}\boldsymbol{\sigma}_0 & 0\\0 & \boldsymbol{\sigma}_0\end{bmatrix}\begin{bmatrix}\nabla\beta_x\\\nabla\beta_y\end{bmatrix}\mathrm{d}\Omega=0\end{aligned} \tag{3.36}$$

式中，$\nabla=\begin{bmatrix}\dfrac{\partial}{\partial x} & \dfrac{\partial}{\partial y}\end{bmatrix}^{\mathrm{T}}$为梯度算子；$\boldsymbol{\sigma}_0=\begin{bmatrix}\sigma_x^0 & \tau_{xy}^0\\\tau_{xy}^0 & \sigma_y^0\end{bmatrix}$为预屈曲载荷。

在面内力载荷屈曲状态下，沿板边的屈曲应力表示为

$$\sigma_x^0=\frac{N_x^0}{h},\quad \sigma_y^0=\frac{N_y^0}{h},\quad \tau_{xy}^0=\frac{N_{xy}^0}{h} \tag{3.37}$$

其中，N_x^0、N_y^0和N_{xy}^0为预屈曲合力。

在热屈曲下，预屈曲合力为

$$N_x^0=N_y^0=\int_{-\frac{h}{2}}^{\frac{h}{2}}\frac{E_e(z)}{1-\nu_e(z)}\alpha_e(z)\Delta T(z)\mathrm{d}z,\quad N_{xy}^0=0 \tag{3.38}$$

值得注意的是，热屈曲公式中计算的临界载荷为$N_{\mathrm{cr}}^{\mathrm{th}}$，需要进行转化才能得到临界温度改变量$\Delta T_{\mathrm{cr}}$，由不同温度分布情况可得如下 3 种情况。

1. 均匀温度分布

由式(3.29)和均匀温度分布下温度变化$\Delta T(z)=T_{\mathrm{f}}-T_0$可得

$$N_{\mathrm{cr}}^{\mathrm{th}}=\Delta T_{\mathrm{cr}}\int_{-\frac{h}{2}}^{\frac{h}{2}}\frac{E_e(z)}{1-\nu_e(z)}\alpha_e(z)\mathrm{d}z \tag{3.39}$$

令 $T_{\mathrm{p}}^{1}=\int_{-\frac{h}{2}}^{\frac{h}{2}}\frac{E_{\mathrm{e}}(z)}{1-\nu_{\mathrm{e}}(z)}\alpha_{\mathrm{e}}(z)\mathrm{d}z$，可得 $\Delta T_{\mathrm{cr}}=\frac{N_{\mathrm{cr}}^{\mathrm{th}}}{T_{\mathrm{p}}^{1}}$。

2. 沿厚度方向的线性温度分布

由式(3.29)和式(3.14)可得

$$\Delta T(z)=T(z)-T_0$$

$$N_{\mathrm{cr}}^{\mathrm{th}}=\Delta T_{\mathrm{cr}}\int_{-\frac{h}{2}}^{\frac{h}{2}}\frac{E_{\mathrm{e}}(z)}{1-\nu_{\mathrm{e}}(z)}\alpha_{\mathrm{e}}(z)\left(\frac{z}{h}+\frac{1}{2}\right)\mathrm{d}z+(T_{\mathrm{m}}-T_0)\int_{-\frac{h}{2}}^{\frac{h}{2}}\frac{E_{\mathrm{e}}(z)}{1-\nu_{\mathrm{e}}(z)}\alpha_{\mathrm{e}}(z)\,\mathrm{d}z \tag{3.40}$$

令 $T_{\mathrm{p}}^{2}=\int_{-\frac{h}{2}}^{\frac{h}{2}}\frac{E_{\mathrm{e}}(z)}{1-\nu_{\mathrm{e}}(z)}\alpha_{\mathrm{e}}(z)\left(\frac{z}{h}+\frac{1}{2}\right)\mathrm{d}z$，可得 $\Delta T_{\mathrm{cr}}=T_{\mathrm{c}}-T_{\mathrm{m}}=\frac{N_{\mathrm{cr}}^{\mathrm{th}}-T_{\mathrm{p}}^{1}(T_{\mathrm{m}}-T_0)}{T_{\mathrm{p}}^{2}}$。

3. 沿厚度方向的非线性温度分布

由式(3.29)和式(3.16)可得

$$\Delta T(z)=T_{\mathrm{f}}-T_0$$

$$N_{\mathrm{cr}}^{\mathrm{th}}=\Delta T_{\mathrm{cr}}\int_{-\frac{h}{2}}^{\frac{h}{2}}\frac{E_{\mathrm{e}}(z)}{1-\nu_{\mathrm{e}}(z)}\alpha_{\mathrm{e}}(z)\eta(z)\mathrm{d}z+(T_{\mathrm{m}}-T_0)\int_{-\frac{h}{2}}^{\frac{h}{2}}\frac{E_{\mathrm{e}}(z)}{1-\nu_{\mathrm{e}}(z)}\alpha_{\mathrm{e}}(z)\,\mathrm{d}z \tag{3.41}$$

令 $T_{\mathrm{p}}^{3}=\int_{-\frac{h}{2}}^{\frac{h}{2}}\frac{E_{\mathrm{e}}(z)}{1-\nu_{\mathrm{e}}(z)}\alpha_{\mathrm{e}}(z)\eta(z)\mathrm{d}z$，可得 $\Delta T_{\mathrm{cr}}=T_{\mathrm{c}}-T_{\mathrm{m}}=\frac{N_{\mathrm{cr}}^{\mathrm{th}}-T_{\mathrm{p}}^{1}(T_{\mathrm{m}}-T_0)}{T_{\mathrm{p}}^{3}}$。

3.5 缺陷的等几何求解

3.5.1 裂纹分析方法

含裂纹高阶 FGM 板的 XIGA 位移插值近似表达式为

$$\begin{Bmatrix}u_0\\v_0\\w_0\end{Bmatrix}=\sum_{i=1}^{NI}R_i\begin{Bmatrix}u_i\\v_i\\w_i\end{Bmatrix}+\sum_{j=1}^{NJ}R_j\left[H(\boldsymbol{\xi})-H(\boldsymbol{\xi}_j)\right]\begin{Bmatrix}a_{uj}\\a_{vj}\\a_{wj}\end{Bmatrix}+\sum_{k=1}^{NK}R_k\sum_{L=1}^{4}\left[\phi_L(\boldsymbol{\xi})-\phi_L(\boldsymbol{\xi}_k)\right]\begin{Bmatrix}b_{uk}^L\\b_{vk}^L\\b_{wk}^L\end{Bmatrix}$$

$$\begin{Bmatrix}\beta_x\\\beta_y\end{Bmatrix}=\sum_{i=1}^{NI}R_i\begin{Bmatrix}\beta_{xi}\\\beta_{yi}\end{Bmatrix}+\sum_{j=1}^{NJ}R_j\left[H(\boldsymbol{\xi})-H(\boldsymbol{\xi}_j)\right]\begin{Bmatrix}a_{\beta_x j}\\a_{\beta_y j}\end{Bmatrix}+\sum_{k=1}^{NK}R_k\sum_{L=1}^{4}\left[\psi_L(\boldsymbol{\xi})-\psi_L(\boldsymbol{\xi}_k)\right]\begin{Bmatrix}b_{\beta_x k}^L\\b_{\beta_y k}^L\end{Bmatrix} \tag{3.42}$$

式中，R_i、R_j 和 R_k 为 PHT 样条基函数；u_i、v_i、w_i、β_{xi} 和 β_{yi} 为标准的位移和

旋转自由度；a_{uj}、a_{vj}、a_{wj}、$a_{\beta_x j}$、$a_{\beta_y j}$、b_{uk}^L、b_{vk}^L、b_{wk}^L、$b_{\beta_x k}^L$和$b_{\beta_y k}^L$为相应的富集自由度；$\boldsymbol{\xi}$为计算点处的参数坐标；$\boldsymbol{\xi}_j$和$\boldsymbol{\xi}_k$为与控制点相关的参数坐标；NI、NJ和NK分别为PHT曲面控制点的个数、受Heaviside函数富集控制点的个数和受裂尖渐近函数富集控制点的个数；$H(\boldsymbol{\xi})$为Heaviside阶跃函数，其值根据计算点位于裂纹面下方或上方设置为−1或+1。

ϕ和ψ分别为平移和旋转裂尖渐近函数，其表达式为[17]

$$
\begin{aligned}
\phi(r,\theta) &= \left\{ r^{\frac{3}{2}}\sin\left(\frac{\theta}{2}\right), r^{\frac{3}{2}}\cos\left(\frac{\theta}{2}\right), r^{\frac{3}{2}}\sin\left(\frac{3\theta}{2}\right), r^{\frac{3}{2}}\cos\left(\frac{3\theta}{2}\right) \right\} \\
\psi(r,\theta) &= \left\{ \sqrt{r}\sin\left(\frac{\theta}{2}\right), \sqrt{r}\cos\left(\frac{\theta}{2}\right), \sqrt{r}\sin\left(\frac{\theta}{2}\right)\sin(\theta), \sqrt{r}\cos\left(\frac{\theta}{2}\right)\sin(\theta) \right\}
\end{aligned}
\tag{3.43}
$$

其中，(r,θ)为坐标原点位于裂尖的局部极坐标系下的空间分量，其对物理空间坐标的一阶和二阶导数见文献[22]。另外，富集单元及富集控制点的选择，以及非连续单元内积分规则的选取均与第2章介绍的内容保持一致，此处不再赘述。

3.5.2 孔洞分析方法

FCM将非匹配结构网格下的虚拟域方法与高阶基函数相结合，用于模拟孔洞缺陷对求解位移场的影响。该方法的主要思想是使用简单的边界非相适应的结构化网格进行数值分析，而复杂缺陷的几何边界则通过在分割单元中的自适应积分进行处理。FCM将结构化的网格区域Ω分为两个部分：实际的物理求解区域Ω_{phy}和孔洞缺陷占据的虚拟域Ω_{fict}，如图3.5所示。

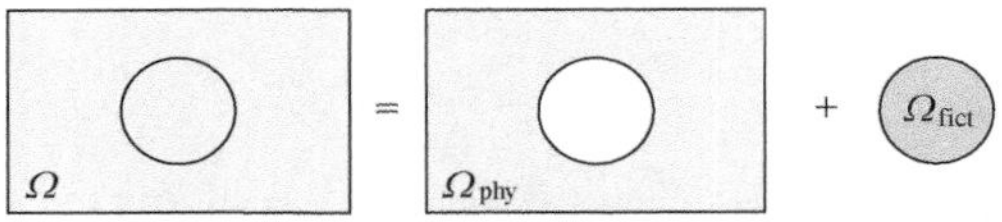

图3.5 虚拟域方法示意图

该方法通过对本构方程中的材料张量乘以相应的罚因子来表征孔洞带来的影响，因此虚拟域中的材料本构关系变为

$$\boldsymbol{\sigma}=\chi\boldsymbol{C}:\boldsymbol{\varepsilon} \tag{3.44}$$

式(3.44)中，$\boldsymbol{\sigma}$和$\boldsymbol{\varepsilon}$分别为应力和应变张量；χ是罚因子，其取值为

$$\chi=\begin{cases}1, & \forall \boldsymbol{x}\in\Omega_{\text{phy}} \\ \gamma, & \forall \boldsymbol{x}\in\Omega_{\text{fict}}\end{cases} \tag{3.45}$$

由文献[23]可知，对于三次样条函数，γ 的值可以直接设置为零。

根据孔洞与结构化背景网格的位置关系，区域 Ω 内的单元可以分为如图 3.6 所示三种类型：标准单元(单元区域完全位于 Ω_{phy} 内)、空单元(单元区域完全位于 Ω_{fict} 内，图中浅蓝色填充区域)和孔洞分割单元(单元区域被孔洞边界分为两部分，图中深灰色填充区域)。标准单元内的积分无需额外操作，而空单元内的数值积分则直接舍弃。

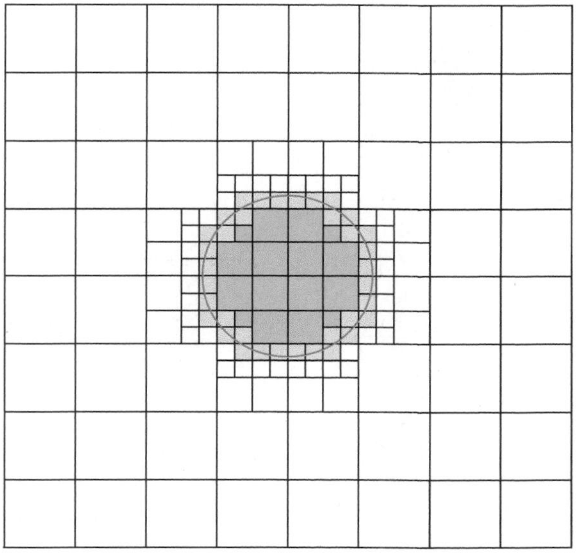

图 3.6　孔洞影响下的单元类型划分

分割单元内的数值计算则结合分层四叉树分解算法和自适应积分策略进行，即将分割单元内的积分胞元逐层分解为 4 个积分子胞元，每个子胞元内使用 $(p+1)\times(q+1)$ 个积分点，其中 p 和 q 为两个参数方向上样条基函数的阶次。数值实施时每个子胞元中落在 Ω_{fict} 内的积分点的计算直接舍弃。在 IGA 中，孔洞边界的几何是通过参数表达的形式精确描述的，积分点对应的整体坐标可直接用于判断其位置是落在求解域内还是域外，该过程可将积分点对应的参数坐标通过样条函数精确的几何离散映射到物理空间中来实现。分割单元内必须保证有足够多的积分点参与计算才能确保求解的精度，该目标可通过设置自适应积分中四叉树的最大层数来灵活控制。

此处通过一个简单的例子来说明自适应积分的实施过程。考虑如图 3.7 所示的仅包含一个圆形孔洞的矩形区域，在初始层中预先通过 PHT 样条对孔洞边界附近的单元进行局部细分，在该例中单元被局部细分了两次。针对在 k 层上的每个子胞元，首先检查其与孔洞边界的位置关系，若属于标准胞元则使用

$(p+1)\times(q+1)$ 个积分点计算，若属于空胞元则直接舍弃，若属于分割胞元则将其分解为 k+1 层的四个子胞元。然后重复此过程，直到所有分割单元的自适应积分层数都达到预设的最大值。另外，单个分割单元内的四叉树分层结构也展示在图 3.8 中。

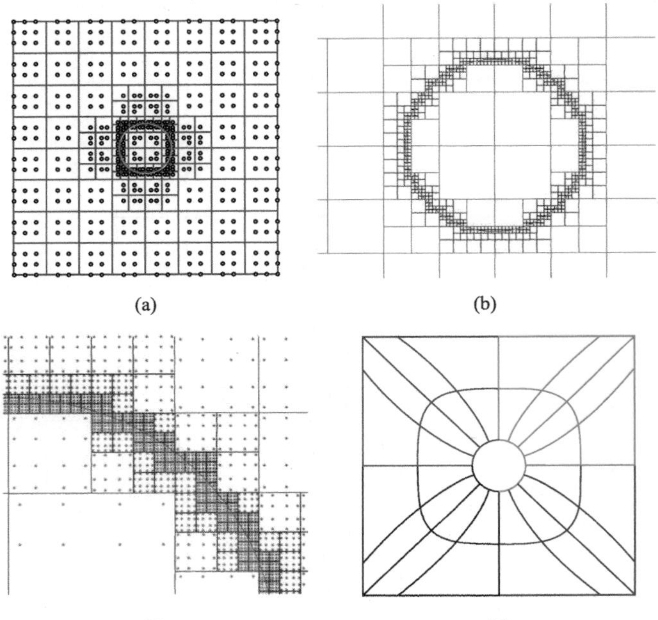

(a) (b) (c) (d)

图 3.7 自适应积分法的实施过程

(a) 初始层 PHT 样条网格和孔洞边界；(b) 自适应积分子胞元的分布，黑线代表积分子胞元的边界；(c) 积分点的分布，洋红色点位于求解域内，红色点位于虚拟域内；(d) 含孔洞几何的多片表达

积分子胞元的划分是独立于底层结构化样条网格的，其功能仅用于数值积分。样条函数仍然是在初始层网格上定义的，因此不会引入额外的自由度。其支持域完全落在 Ω_{fict} 内的基函数对应控制点处的自由度将直接从系统方程中去除，如图 3.9 所示的白色填充控制点。FCM 能够有效地缓解在描述复杂孔洞几何时需要生成边界匹配网格的困难，同时也能避免需要使用多个样条片(patch)对复杂孔洞结构的等几何建模，以及实施片与片之间 C^1 连续性的困难。如图 3.7(d)所示，仅包含一个圆形孔洞的几何描述就至少需要四个具有 C^1 连续性的样条片。

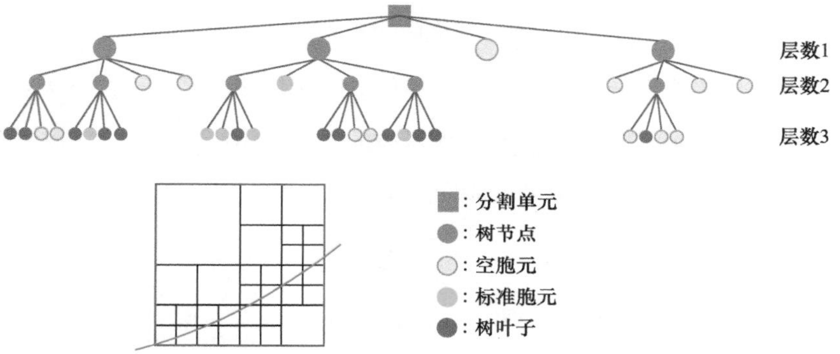

图 3.8　单个分割单元内的四叉树分层结构

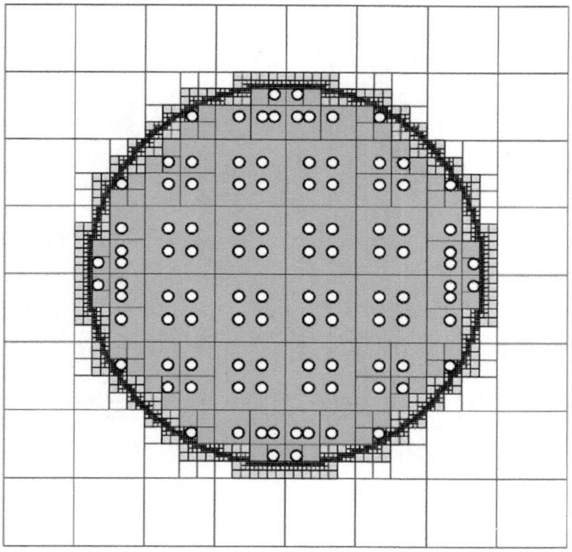

图 3.9　基函数支持域完全落在虚拟域内的控制点

3.5.3　振动和屈曲求解公式

基于 IGA 的概念，FGM 板的位移场用 PHT 样条函数近似表示为

$$\boldsymbol{u}=\begin{Bmatrix}u_0\\v_0\\w_0\\\beta_x\\\beta_y\end{Bmatrix}=\sum_{i=1}^{n\times m}\begin{bmatrix}\tilde{R}_i&0&0&0&0\\0&\tilde{R}_i&0&0&0\\0&0&\tilde{R}_i&0&0\\0&0&0&\tilde{R}_i&0\\0&0&0&0&\tilde{R}_i\end{bmatrix}\begin{Bmatrix}u_{0i}\\v_{0i}\\w_{0i}\\\beta_{xi}\\\beta_{yi}\end{Bmatrix}=\sum_{i=1}^{n\times m}\tilde{\boldsymbol{R}}_i\boldsymbol{u}_i \tag{3.46}$$

其中，n、m 为两个参数方向上基函数的个数；根据控制点 $\boldsymbol{P}_i$ 的富集情况，$\tilde{R}_i$ 可以为

PHT 样条基函数 R_i、Heaviside 函数富集后的形函数 $R_j\left[H(\xi)-H(\xi_j)\right]$、平移或旋转裂尖渐近函数富集后的形函数 $R_k\left[\phi_L(\xi)-\phi_L(\xi_k)\right]$ 或 $R_k\left[\psi_L(\xi)-\psi_L(\xi_k)\right]$；$\boldsymbol{u}_i=\left\{u_{0i},v_{0i},w_{0i},\beta_{xi},\beta_{yi}\right\}^{\mathrm{T}}$ 是控制点 $\boldsymbol{P}_i$ 处对应的广义自由度。

将式(3.46)代入式(3.22)中可得应变矢量为

$$\tilde{\boldsymbol{\varepsilon}}=\begin{bmatrix}\boldsymbol{\varepsilon}_{\mathrm{m}}^{\mathrm{T}} & \boldsymbol{\kappa}_{\mathrm{b1}}^{\mathrm{T}} & \boldsymbol{\kappa}_{\mathrm{b2}}^{\mathrm{T}} & \boldsymbol{\varepsilon}_{\mathrm{s}}^{\mathrm{T}} & \boldsymbol{\kappa}_{\mathrm{bs}}^{\mathrm{T}}\end{bmatrix}^{\mathrm{T}}=\boldsymbol{B}^{\mathrm{L}}\boldsymbol{u}_i \tag{3.47}$$

式中，$\boldsymbol{B}^{\mathrm{L}}=\left[\left(\boldsymbol{B}_i^{\mathrm{m}}\right)^{\mathrm{T}}\left(\boldsymbol{B}_i^{\mathrm{b1}}\right)^{\mathrm{T}}\left(\boldsymbol{B}_i^{\mathrm{b2}}\right)^{\mathrm{T}}\left(\boldsymbol{B}_i^{\mathrm{s}}\right)^{\mathrm{T}}\left(\boldsymbol{B}_i^{\mathrm{bs}}\right)^{\mathrm{T}}\right]^{\mathrm{T}}$ 为线性应变矩阵，其各个分量表示为

$$\begin{gathered}\boldsymbol{B}_i^{\mathrm{m}}=\begin{bmatrix}\tilde{R}_{i,x} & 0 & 0 & 0 & 0\\ 0 & \tilde{R}_{i,y} & 0 & 0 & 0\\ \tilde{R}_{i,y} & \tilde{R}_{i,x} & 0 & 0 & 0\end{bmatrix},\quad \boldsymbol{B}_i^{\mathrm{b1}}=\begin{bmatrix}0 & 0 & 0 & \tilde{R}_{i,x} & 0\\ 0 & 0 & 0 & 0 & \tilde{R}_{i,y}\\ 0 & 0 & 0 & \tilde{R}_{i,y} & \tilde{R}_{i,x}\end{bmatrix}\\ \boldsymbol{B}_i^{\mathrm{b2}}=-\frac{4}{3h^2}\begin{bmatrix}0 & 0 & \tilde{R}_{i,xx} & \tilde{R}_{i,x} & 0\\ 0 & 0 & \tilde{R}_{i,yy} & 0 & \tilde{R}_{i,y}\\ 0 & 0 & 2\tilde{R}_{i,xy} & \tilde{R}_{i,y} & \tilde{R}_{i,x}\end{bmatrix}\\ \boldsymbol{B}_i^{\mathrm{s}}=\begin{bmatrix}0 & 0 & \tilde{R}_{i,x} & \tilde{R}_i & 0\\ 0 & 0 & \tilde{R}_{i,y} & 0 & \tilde{R}_i\end{bmatrix},\quad \boldsymbol{B}_i^{\mathrm{bs}}=-\frac{4}{h^2}\begin{bmatrix}0 & 0 & \tilde{R}_{i,x} & \tilde{R}_i & 0\\ 0 & 0 & \tilde{R}_{i,y} & 0 & \tilde{R}_i\end{bmatrix}\end{gathered} \tag{3.48}$$

应变矩阵分量 $\boldsymbol{B}_i^{\mathrm{b2}}$ 中包含形函数的二阶导数，这就要求有限元近似求解中单元之间至少保持 C^1 连续性。这是传统 FEM 难以实现的。PHT 样条基函数的高阶连续性的特点可以自然地满足 C^1 连续性条件。应变的变分定义为

$$\delta\boldsymbol{\varepsilon}=\left(\boldsymbol{B}^{\mathrm{L}}\right)\delta\boldsymbol{u} \tag{3.49}$$

将式(3.48)和式(3.49)代入式(3.32)和式(3.36)，可得 FGM 板在热、力和热力耦合载荷作用下的自由振动和特征屈曲问题的离散公式(以矩阵形式表示)，如表 3.1 所示。表中各个矩阵的含义解释如下所述。

表 3.1 FGM 板的振动和屈曲求解公式

问题类型	求解公式
自由振动	$\left(\boldsymbol{K}-\omega^2\boldsymbol{M}\right)\boldsymbol{u}=0$
热振动	$\left(\boldsymbol{K}+\boldsymbol{K}_{\mathrm{g}}^{\mathrm{t}}-\omega^2\boldsymbol{M}\right)u=0$
力屈曲	$\left(\boldsymbol{K}-\lambda_{\mathrm{cr}}\boldsymbol{K}_{\mathrm{g}}^{\mathrm{m}}\right)\boldsymbol{u}=0$

续表

问题类型	求解公式
热屈曲	$\left(\boldsymbol{K}-\Delta T_{\mathrm{cr}}\boldsymbol{K}_{\mathrm{g}}^{\mathrm{t}}\right)\boldsymbol{u}=0$
热力耦合屈曲	$\begin{cases}\left(\boldsymbol{K}-\boldsymbol{K}_{\mathrm{g}}^{\mathrm{m}}-\Delta T_{\mathrm{cr}}\boldsymbol{K}_{\mathrm{g}}^{\mathrm{t}}\right)\boldsymbol{u}=0\\ \left(\boldsymbol{K}-\boldsymbol{K}_{\mathrm{g}}^{\mathrm{t}}-\lambda_{\mathrm{cr}}\boldsymbol{K}_{\mathrm{g}}^{\mathrm{m}}\right)\boldsymbol{u}=0\end{cases}$

$\boldsymbol{K}$ 是线性刚度矩阵，其表达式为

$$\boldsymbol{K}=\int_{\Omega}\begin{Bmatrix}\boldsymbol{B}_i^{\mathrm{m}}\\ \boldsymbol{B}_i^{\mathrm{b1}}\\ \boldsymbol{B}_i^{\mathrm{b2}}\\ \boldsymbol{B}_i^{\mathrm{s}}\\ \boldsymbol{B}_i^{\mathrm{bs}}\end{Bmatrix}^{\mathrm{T}}\begin{bmatrix}\boldsymbol{A}&\boldsymbol{B}&\boldsymbol{E}&0&0\\ \boldsymbol{B}&\boldsymbol{D}&\boldsymbol{F}&0&0\\ \boldsymbol{E}&\boldsymbol{F}&\boldsymbol{H}&0&0\\ 0&0&0&\boldsymbol{A}_{\mathrm{s}}&\boldsymbol{B}_{\mathrm{s}}\\ 0&0&0&\boldsymbol{B}_{\mathrm{s}}&\boldsymbol{D}_{\mathrm{s}}\end{bmatrix}\begin{Bmatrix}\boldsymbol{B}_i^{\mathrm{m}}\\ \boldsymbol{B}_i^{\mathrm{b1}}\\ \boldsymbol{B}_i^{\mathrm{b2}}\\ \boldsymbol{B}_i^{\mathrm{s}}\\ \boldsymbol{B}_i^{\mathrm{bs}}\end{Bmatrix}\mathrm{d}\Omega=\int_{\Omega}\left(\boldsymbol{B}^{\mathrm{L}}\right)^{\mathrm{T}}\tilde{\boldsymbol{D}}\boldsymbol{B}^{\mathrm{L}}\mathrm{d}\Omega \tag{3.50}$$

$\boldsymbol{M}$ 是一致质量矩阵，其表达式为

$$\boldsymbol{M}=\int_{\Omega}\tilde{\boldsymbol{N}}^{\mathrm{T}}\boldsymbol{m}\tilde{\boldsymbol{N}}\mathrm{d}\Omega \tag{3.51}$$

其中，

$$\tilde{\boldsymbol{N}}=\begin{bmatrix}\tilde{R}_i&0&0&0&0&0&0&0&0\\ 0&0&0&\tilde{R}_i&0&0&0&0&0\\ 0&0&\tilde{R}_{i,x}&0&0&\tilde{R}_{i,y}&\tilde{R}_i&0&0\\ 0&\tilde{R}_i&\tilde{R}_i&0&0&0&0&0&0\\ 0&0&0&0&\tilde{R}_i&\tilde{R}_i&0&0&0\end{bmatrix}^{\mathrm{T}} \tag{3.52}$$

$\boldsymbol{K}_{\mathrm{g}}^{\mathrm{t}}$ 和 $\boldsymbol{K}_{\mathrm{g}}^{\mathrm{m}}$ 分别是在初始力屈曲载荷和初始温度屈曲载荷作用下的几何刚度矩阵，它们的表达式为

$$\begin{aligned}\boldsymbol{K}_{\mathrm{g}}^{\mathrm{t}}&=\int_{\Omega}\left(\boldsymbol{B}^{\mathrm{g}}\right)^{\mathrm{T}}\boldsymbol{\Theta}^{\mathrm{t}}\boldsymbol{B}^{\mathrm{g}}\mathrm{d}\Omega\\ \boldsymbol{K}_{\mathrm{g}}^{\mathrm{m}}&=\int_{\Omega}\left(\boldsymbol{B}^{\mathrm{g}}\right)^{\mathrm{T}}\boldsymbol{\Theta}^{\mathrm{m}}\boldsymbol{B}^{\mathrm{g}}\mathrm{d}\Omega\end{aligned} \tag{3.53}$$

其中，

$$\boldsymbol{B}^{\mathrm{g}}=\begin{bmatrix}\tilde{R}_{i,x}&\tilde{R}_{i,y}&0&0&0&0&0&0&0&0\\ 0&0&\tilde{R}_{i,x}&\tilde{R}_{i,y}&0&0&0&0&0&0\\ 0&0&0&0&\tilde{R}_{i,x}&\tilde{R}_{i,y}&0&0&0&0\\ 0&0&0&0&0&0&\tilde{R}_{i,x}&\tilde{R}_{i,y}&0&0\\ 0&0&0&0&0&0&0&0&\tilde{R}_{i,x}&\tilde{R}_{i,y}\end{bmatrix}^{\mathrm{T}} \tag{3.54}$$

$$\boldsymbol{\Theta}^{\mathrm{m}}=\begin{bmatrix} h\boldsymbol{\sigma}_0^{\mathrm{m}} & 0 & 0 & 0 & 0 \\ 0 & h\boldsymbol{\sigma}_0^{\mathrm{m}} & 0 & 0 & 0 \\ 0 & 0 & h\boldsymbol{\sigma}_0^{\mathrm{m}} & 0 & 0 \\ 0 & 0 & 0 & \dfrac{h^3}{12}\boldsymbol{\sigma}_0^{\mathrm{m}} & 0 \\ 0 & 0 & 0 & 0 & \dfrac{h^3}{12}\boldsymbol{\sigma}_0^{\mathrm{m}} \end{bmatrix},\quad \boldsymbol{\Theta}^{\mathrm{t}}=\begin{bmatrix} h\boldsymbol{\sigma}_0^{\mathrm{t}} & 0 & 0 & 0 & 0 \\ 0 & h\boldsymbol{\sigma}_0^{\mathrm{t}} & 0 & 0 & 0 \\ 0 & 0 & h\boldsymbol{\sigma}_0^{\mathrm{t}} & 0 & 0 \\ 0 & 0 & 0 & \dfrac{h^3}{12}\boldsymbol{\sigma}_0^{\mathrm{t}} & 0 \\ 0 & 0 & 0 & 0 & \dfrac{h^3}{12}\boldsymbol{\sigma}_0^{\mathrm{t}} \end{bmatrix} \tag{3.55}$$

式(3.55)中，预应力 $\boldsymbol{\sigma}_0^{\mathrm{m}}$ 和 $\boldsymbol{\sigma}_0^{\mathrm{t}}$ 通过式(3.37)和式(3.38)求得。

3.5.4 后屈曲求解公式

在 3.5.3 节高阶板公式的基础上，引入 von Kármán 非线性理论[24]，应变和位移的非线性关系表示如下：

$$\begin{Bmatrix} \varepsilon_x \\ \varepsilon_y \\ \gamma_{xy} \\ \gamma_{xz} \\ \gamma_{yz} \end{Bmatrix}=\begin{Bmatrix} u_{,x} \\ v_{,y} \\ u_{,y}+v_{,x} \\ u_{,z}+w_{,x} \\ v_{,z}+w_{,y} \end{Bmatrix}+\frac{1}{2}\begin{Bmatrix} w_{,x}^2 \\ w_{,y}^2 \\ 2w_{,x}^2w_{,y}^2 \\ 0 \\ 0 \end{Bmatrix} \tag{3.56}$$

将式(3.19)代入式(3.56)可得非线性关系下的应变分量，其区别为面内薄膜应变发生了改变，其余分量仍保持一致。非线性面内薄膜应变表示为

$$\boldsymbol{\varepsilon}_{\mathrm{m}}=\begin{Bmatrix} u_{0,x} \\ v_{0,y} \\ u_{0,y}+v_{0,x} \end{Bmatrix}+\frac{1}{2}\begin{Bmatrix} w_{0,x}^2 \\ w_{0,y}^2 \\ w_{0,x}^2w_{0,y}^2 \end{Bmatrix}=\boldsymbol{\varepsilon}_{\mathrm{l}}+\boldsymbol{\varepsilon}_{\mathrm{nl}} \tag{3.57}$$

式中，$\boldsymbol{\varepsilon}_{\mathrm{l}}$ 为线性应变部分，其形式与式(3.22)一样，非线性面内薄膜应变 $\boldsymbol{\varepsilon}_{\mathrm{nl}}$ 可表示为

$$\boldsymbol{\varepsilon}_{\mathrm{nl}}=\frac{1}{2}\begin{Bmatrix} w_{0,x}^2 \\ w_{0,y}^2 \\ w_{0,x}^2w_{0,y}^2 \end{Bmatrix}=\frac{1}{2}\boldsymbol{H\Theta} \tag{3.58}$$

其中，

$$\boldsymbol{H}=\begin{bmatrix} w_{0,x} & 0 \\ 0 & w_{0,y} \\ w_{0,y} & w_{0,x} \end{bmatrix},\quad \boldsymbol{\Theta}=\begin{Bmatrix} w_{0,x} \\ w_{0,y} \end{Bmatrix} \tag{3.59}$$

对于多向 FGM 板，材料矩阵 $\boldsymbol{C}$ 和 $\boldsymbol{G}$ 更新如下：

$$\begin{aligned} &\boldsymbol{C}=\frac{E_{\mathrm{e}}(x,y,z)}{1-\nu_{\mathrm{e}}^{2}(x,y,z)}\begin{bmatrix} 1 & \nu_{\mathrm{e}}(x,y,z) & 0 \\ \nu_{\mathrm{e}}(x,y,z) & 1 & 0 \\ 0 & 0 & \left[1-\nu_{\mathrm{e}}(x,y,z)\right]/2 \end{bmatrix} \\ &\boldsymbol{G}=\frac{E_{\mathrm{e}}(x,y,z)}{2\left[1+\nu_{\mathrm{e}}(x,y,z)\right]}\begin{bmatrix} 1 & 0 \\ 0 & 1 \end{bmatrix} \end{aligned} \tag{3.60}$$

式中，$E_{\mathrm{e}}(x,y,z)$ 和 $\nu_{\mathrm{e}}(x,y,z)$ 分别为等效弹性模量和泊松比，其值是每个积分点处的函数，从而体现出材料性能的连续变化。在程序实施中，首先根据每个积分点在物理空间中的 x 和 y 坐标计算出面内分布函数，其次根据积分点的 z 坐标确定材料的体积分数，然后根据式(3.13)获得该积分点处的等效弹性模量和泊松比，最后依据式(3.28)和式(3.31)计算出材料矩阵 $\boldsymbol{D}$ 和 $\boldsymbol{D}^{\mathrm{s}}$ 。

基于虚功原理，非线性稳定性问题的平衡方程表示为[24]

$$\delta\mathcal{W}_{\mathrm{int}}(\boldsymbol{u},\delta\boldsymbol{u})-\delta\mathcal{W}_{\mathrm{ext}}(\boldsymbol{u},\delta\boldsymbol{u},\lambda)=0 \tag{3.61}$$

其中，$\delta\mathcal{W}_{\mathrm{int}}$ 和 $\delta\mathcal{W}_{\mathrm{ext}}$ 分别为内力虚功和外力虚功，它们的表达式为

$$\begin{aligned} &\delta\mathcal{W}_{\mathrm{int}}=\int_{\Omega}\delta\boldsymbol{\varepsilon}^{\mathrm{T}}\boldsymbol{D}\boldsymbol{\varepsilon}\,\mathrm{d}\Omega+\int_{\Omega}\delta\boldsymbol{\gamma}^{\mathrm{T}}\boldsymbol{D}^{\mathrm{s}}\boldsymbol{\gamma}\,\mathrm{d}\Omega \\ &\delta\mathcal{W}_{\mathrm{ext}}=\lambda\left(\int_{\Omega}\boldsymbol{f}\cdot\delta\boldsymbol{u}\,\mathrm{d}\Omega+\int_{\Gamma_{\mathrm{t}}}\overline{\boldsymbol{t}}\cdot\delta\boldsymbol{u}\,\mathrm{d}\Gamma_{\mathrm{t}}\right) \end{aligned} \tag{3.62}$$

式中，$\boldsymbol{f}$ 是单位体力；$\overline{\boldsymbol{t}}$ 是在自然边界上作用的单位面力；λ 是载荷因子；$\boldsymbol{D}$ 和 $\boldsymbol{D}^{\mathrm{s}}$ 的定义见式(3.31)。从板的特征屈曲振型中提取初始几何缺陷，并用于后续的稳定性分析。

引入等几何样条函数离散公式，非线性分析下的应变矢量为

$$\tilde{\boldsymbol{\varepsilon}}=\begin{Bmatrix} \boldsymbol{\varepsilon} \\ \boldsymbol{\gamma} \end{Bmatrix}=\begin{Bmatrix} \boldsymbol{\varepsilon}_{\mathrm{m}} \\ \boldsymbol{\kappa}_{\mathrm{b1}} \\ \boldsymbol{\kappa}_{\mathrm{b2}} \\ \boldsymbol{\varepsilon}_{\mathrm{s}} \\ \boldsymbol{\kappa}_{\mathrm{bs}} \end{Bmatrix}=\sum_{i=1}^{n\times m}\left(\boldsymbol{B}^{\mathrm{L}}+\frac{1}{2}\boldsymbol{B}^{\mathrm{NL}}\right)\boldsymbol{u}_{i} \tag{3.63}$$

其中，n 、m 为两个参数方向上基函数的个数；线性应变矩阵 $\boldsymbol{B}^{\mathrm{L}}$ 各个分量的表达

式见式(3.48)；非线性应变矩阵 $\boldsymbol{B}^{\mathrm{NL}}$ 表示为

$$\boldsymbol{B}^{\mathrm{NL}} = \boldsymbol{H}\boldsymbol{B}_i^{\mathrm{g}} \tag{3.64}$$

式中，

$$\boldsymbol{B}_i^{\mathrm{g}} = \begin{bmatrix} 0 & 0 & R_{i,x} & 0 & 0 \\ 0 & 0 & R_{i,y} & 0 & 0 \end{bmatrix} \tag{3.65}$$

应变的变分为

$$\delta\boldsymbol{\varepsilon} = \sum_{i=1}^{n\times m}\left(\boldsymbol{B}^{\mathrm{L}} + \boldsymbol{B}^{\mathrm{NL}}\right)\delta\boldsymbol{u}_i \tag{3.66}$$

将式(3.66)代入式(3.61)和式(3.62)中，可得非线性稳定性问题的代数方程为

$$\left[\boldsymbol{K}^{\mathrm{L}} + \boldsymbol{K}^{\mathrm{NL}}(\boldsymbol{u})\right]\boldsymbol{u} = \boldsymbol{F} \tag{3.67}$$

其中，

$$\begin{aligned}
&\boldsymbol{K}^{\mathrm{L}} = \int_{\Omega}\left(\boldsymbol{B}^{\mathrm{L}}\right)^{\mathrm{T}}\tilde{\boldsymbol{D}}\boldsymbol{B}^{\mathrm{L}}\mathrm{d}\Omega \\
&\boldsymbol{K}^{\mathrm{NL}} = \frac{1}{2}\int_{\Omega}\left(\boldsymbol{B}^{\mathrm{L}}\right)^{\mathrm{T}}\tilde{\boldsymbol{D}}\boldsymbol{B}^{\mathrm{NL}}\mathrm{d}\Omega + \int_{\Omega}\left(\boldsymbol{B}^{\mathrm{NL}}\right)^{\mathrm{T}}\tilde{\boldsymbol{D}}\boldsymbol{B}^{\mathrm{L}}\mathrm{d}\Omega + \frac{1}{2}\int_{\Omega}\left(\boldsymbol{B}^{\mathrm{NL}}\right)^{\mathrm{T}}\tilde{\boldsymbol{D}}\boldsymbol{B}^{\mathrm{NL}}\mathrm{d}\Omega \\
&\boldsymbol{F} = \lambda\left(\int_{\Omega}\boldsymbol{R}^{\mathrm{T}}\boldsymbol{f}\,\mathrm{d}\Omega + \int_{\Gamma_{\mathrm{t}}}\boldsymbol{R}^{\mathrm{T}}\bar{\boldsymbol{t}}\,\mathrm{d}\Gamma_{\mathrm{t}}\right)
\end{aligned} \tag{3.68}$$

由于 $\boldsymbol{K}^{\mathrm{NL}}$ 与位移的非线性关系，式(3.67)无法直接求解，因此需要采用增量迭代法进行求解。上述非线性平衡方程可以重新写作

$$\mathfrak{R}(\boldsymbol{u},\lambda) = \left[\boldsymbol{K}^{\mathrm{L}} + \boldsymbol{K}^{\mathrm{NL}}(\boldsymbol{u})\right]\boldsymbol{u} - \lambda\boldsymbol{F}_0 = 0 \tag{3.69}$$

式中，$\boldsymbol{F}_0 = \int_{\Omega}\boldsymbol{R}^{\mathrm{T}}\boldsymbol{f}\,\mathrm{d}\Omega + \int_{\Gamma_{\mathrm{t}}}\boldsymbol{R}^{\mathrm{T}}\bar{\boldsymbol{t}}\,\mathrm{d}\Gamma_{\mathrm{t}}$。

随着外力载荷从 $\lambda\boldsymbol{F}_0$ 增加到 $(\lambda + \Delta\lambda)\boldsymbol{F}_0$，物体达到新的平衡状态，表示为

$$\mathfrak{R}(\boldsymbol{u} + \Delta\boldsymbol{u}, \lambda + \Delta\lambda) = 0 \tag{3.70}$$

采用泰勒级数对其进行展开，并忽略高阶小量，式(3.70)可表示为

$$\mathfrak{R}(\boldsymbol{u} + \Delta\boldsymbol{u}, \lambda + \Delta\lambda) = \mathfrak{R}(\boldsymbol{u},\lambda) + \boldsymbol{K}_{\mathrm{T}}(\boldsymbol{u})\Delta\boldsymbol{u} - \Delta\lambda\boldsymbol{F}_0 = 0 \tag{3.71}$$

其中，$\Delta\boldsymbol{u}$ 为位移增量；$\boldsymbol{K}_{\mathrm{T}}(\boldsymbol{u})$ 为切线刚度矩阵，其表达式为

$$\boldsymbol{K}_{\mathrm{T}}(\boldsymbol{u})=\frac{\partial \boldsymbol{\mathfrak{R}}}{\partial \boldsymbol{u}}=\int_{\Omega}\left(\boldsymbol{B}^{\mathrm{L}}+\boldsymbol{B}^{\mathrm{NL}}\right)^{\mathrm{T}} \tilde{\boldsymbol{D}}\left(\boldsymbol{B}^{\mathrm{L}}+\boldsymbol{B}^{\mathrm{NL}}\right) \mathrm{d} \Omega + \int_{\Omega}\left(\boldsymbol{B}^{\mathrm{g}}\right)^{\mathrm{T}}\begin{bmatrix} N_{x} & N_{xy} \\ N_{xy} & N_{y} \end{bmatrix} \boldsymbol{B}^{\mathrm{g}} \mathrm{d} \Omega \tag{3.72}$$

式中，$\left\{N_x, N_y, N_{xy}\right\}^{\mathrm{T}}=[\boldsymbol{A} \quad \boldsymbol{B} \quad \boldsymbol{E}]\boldsymbol{\varepsilon}_{\mathrm{m}}$。

路径跟踪法的核心思想是[25]，将总载荷步分解为许多小的增量步，在每个增量步中通过迭代找到平衡解，向非线性平衡方程添加额外约束，即将载荷增量作为附加变量。圆柱形弧长迭代过程的示意图如图 3.10 所示。

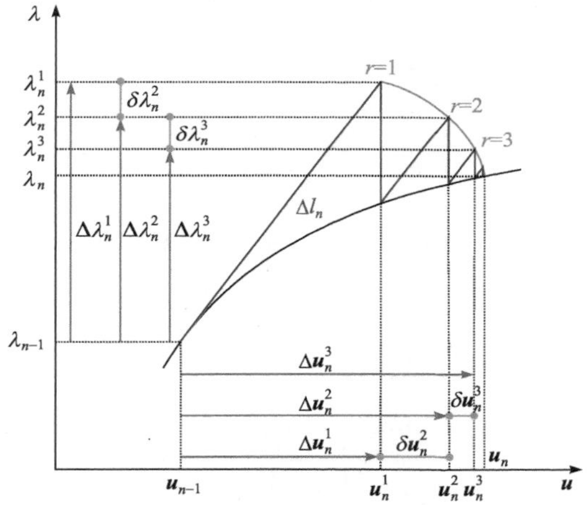

图 3.10　路径跟踪法的迭代过程

假设在第 n 个载荷步的第 $r-1$ 次迭代的解 $\left(\boldsymbol{u}_n^{r-1}, \lambda_n^{r-1}\right)$ 已获得，接下来要获取下一迭代步的解 $\left(\boldsymbol{u}_n^{r}, \lambda_n^{r}\right)$，根据图 3.10，我们可以得到

$$\begin{aligned}\boldsymbol{\mathfrak{R}}\left(\boldsymbol{u}_n^r, \lambda_n^r\right) &= \boldsymbol{\mathfrak{R}}\left(\boldsymbol{u}_n^{r-1}, \lambda_n^{r-1}\right)+\frac{\partial \boldsymbol{\mathfrak{R}}}{\partial \lambda} \delta \lambda_n^r+\frac{\partial \boldsymbol{\mathfrak{R}}}{\partial \boldsymbol{u}} \delta \boldsymbol{u}_n^r=0 \\ &= \boldsymbol{\mathfrak{R}}\left(\boldsymbol{u}_n^{r-1}, \lambda_n^{r-1}\right)-\delta \lambda_n^r \boldsymbol{F}_0+\boldsymbol{K}_{\mathrm{T}}\left(\boldsymbol{u}_n^{r-1}\right) \delta \boldsymbol{u}_n^r \\ &= \boldsymbol{\mathfrak{R}}\left(\boldsymbol{u}_n^{r-1}, \lambda_n^{r-1}\right)-\delta \lambda_n^r \boldsymbol{F}_0+\boldsymbol{K}_{\mathrm{T}}\left(\boldsymbol{u}_{n-1}\right) \delta \boldsymbol{u}_n^r\end{aligned} \tag{3.73}$$

所对应的位移校正量为

$$\begin{aligned}\delta \boldsymbol{u}_n^r &= \delta \lambda_n^r\left[\boldsymbol{K}_{\mathrm{T}}\left(\boldsymbol{u}_{n-1}\right)\right]^{-1} \boldsymbol{F}_0-\left[\boldsymbol{K}_{\mathrm{T}}\left(\boldsymbol{u}_{n-1}\right)\right]^{-1} \boldsymbol{\mathfrak{R}}\left(\boldsymbol{u}_n^{r-1}, \lambda_n^{r-1}\right) \\ &= \delta \lambda_n^r \delta \hat{\boldsymbol{u}}_n+\delta \overline{\boldsymbol{u}}_n^r\end{aligned} \tag{3.74}$$

其中，

$$\begin{aligned}\delta\hat{\boldsymbol{u}}_n &= \left[\boldsymbol{K}_{\mathrm{T}}\left(\boldsymbol{u}_{n-1}\right)\right]^{-1}\boldsymbol{F}_0\\ \delta\overline{\boldsymbol{u}}_n^r &= -\left[\boldsymbol{K}_{\mathrm{T}}\left(\boldsymbol{u}_{n-1}\right)\right]^{-1}\boldsymbol{\mathfrak{R}}\left(\boldsymbol{u}_n^{r-1},\lambda_n^{r-1}\right)\\ &= -\left[\boldsymbol{K}_{\mathrm{T}}\left(\boldsymbol{u}_{n-1}\right)\right]^{-1}\left\{\left[\boldsymbol{K}^{\mathrm{L}}+\boldsymbol{K}^{\mathrm{NL}}\left(\boldsymbol{u}_n^{r-1}\right)\right]\boldsymbol{u}_n^{r-1}-\lambda_n^{r-1}\boldsymbol{F}_0\right\}\end{aligned} \tag{3.75}$$

增加的弧长公式为[25]

$$\left(\Delta\boldsymbol{u}_n^r\right)^{\mathrm{T}}\Delta\boldsymbol{u}_n^r+\left(\Delta\lambda_n^r\right)^2\left(\boldsymbol{F}_0\right)^{\mathrm{T}}\boldsymbol{F}_0=\left(\Delta l_n\right)^2 \tag{3.76}$$

式中，Δl_n 是第 n 个载荷步的弧长。

将式(3.74)代入式(3.76)可得求解载荷因子校正量的方程为

$$\left(\Delta\boldsymbol{u}_n^{r-1}+\delta\overline{\boldsymbol{u}}_n^r+\delta\lambda_n^r\delta\hat{\boldsymbol{u}}_n\right)^{\mathrm{T}}\left(\Delta\boldsymbol{u}_n^{r-1}+\delta\overline{\boldsymbol{u}}_n^r+\delta\lambda_n^r\delta\hat{\boldsymbol{u}}_n\right)+\left(\Delta\lambda_n^{r-1}+\delta\lambda_n^r\right)^2\left(\boldsymbol{F}_0\right)^{\mathrm{T}}\boldsymbol{F}_0=\left(\Delta l_n\right)^2 \tag{3.77}$$

式(3.77)简化为

$$a\left(\delta\lambda_n^r\right)^2+b\delta\lambda_n^r+c=0 \tag{3.78}$$

其中，

$$\begin{aligned}a&=\left(\delta\hat{\boldsymbol{u}}_n\right)^{\mathrm{T}}\delta\hat{\boldsymbol{u}}_n+\left(\boldsymbol{F}_0\right)^{\mathrm{T}}\boldsymbol{F}_0\\ b&=2\left(\delta\hat{\boldsymbol{u}}_n\right)^{\mathrm{T}}\left(\Delta\boldsymbol{u}_n^{r-1}+\delta\overline{\boldsymbol{u}}_n^r\right)+2\Delta\lambda_n^{r-1}\left(\boldsymbol{F}_0\right)^{\mathrm{T}}\boldsymbol{F}_0\\ c&=\left(\Delta\boldsymbol{u}_n^{r-1}+\delta\overline{\boldsymbol{u}}_n^r\right)^{\mathrm{T}}\left(\Delta\boldsymbol{u}_n^{r-1}+\delta\overline{\boldsymbol{u}}_n^r\right)-\left(\Delta l_n\right)^2+\left(\Delta\lambda_n^{r-1}\right)^2\left(\boldsymbol{F}_0\right)^{\mathrm{T}}\boldsymbol{F}_0\end{aligned} \tag{3.79}$$

式(3.78)是一个一元二次方程。若方程根为复数，则将当前载荷步的弧长缩减为原始值的一半；若两根均为正，则取 $\delta\lambda_n^r=-\dfrac{c}{b}$；若根为一正一负，则取能保证 $\left(\Delta\boldsymbol{u}_n^{r-1}\right)^{\mathrm{T}}\Delta\boldsymbol{u}_n^r$ 为正的根为载荷因子的校正量。

增量和校正量之间的关系为

$$\begin{aligned}&\Delta\boldsymbol{u}_n^r=\sum_{i=1}^{r}\delta\boldsymbol{u}_n^r,\quad \Delta\lambda_n^r=\sum_{i=1}^{r}\delta\lambda_n^r\\ &\boldsymbol{u}_n^r=\boldsymbol{u}_{n-1}+\Delta\boldsymbol{u}_n^r,\quad \lambda_n^r=\lambda_{n-1}+\Delta\lambda_n^r\\ &\boldsymbol{u}_n^r=\boldsymbol{u}_n^{r-1}+\delta\boldsymbol{u}_n^r,\quad \lambda_n^r=\lambda_{n-1}+\Delta\lambda_n^r\end{aligned} \tag{3.80}$$

然后，此载荷步的最终位移和载荷因子更新为 $\boldsymbol{u}_n=\boldsymbol{u}_n^r$ 和 $\lambda_n=\lambda_n^r$。另外，上述迭代过程中需要注意的几个问题如下所述。

1. 当前载荷步的弧长确定

$$\frac{\Delta l_n}{\Delta l_{n-1}}=\left(\frac{N_d}{N_{n-1}}\right)^{\zeta} \tag{3.81}$$

式中，N_{n-1} 为上一载荷步的迭代次数；N_d 为本载荷步期望的迭代次数，一般取值为 3～5；ζ 为阻尼或放大指数，一般取值 0.5[26]。

2. 当前载荷步的初始载荷因子增量(也为校正量)的确定(符号和大小)

初始步也满足弧长约束

$$\begin{gathered}\left(\Delta\boldsymbol{u}_n^1\right)^{\mathrm{T}}\Delta\boldsymbol{u}_n^1+\left(\Delta\lambda_n^1\right)^2\left(\boldsymbol{F}_0\right)^{\mathrm{T}}\boldsymbol{F}_0=\left(\Delta l_n\right)^2\\ \Delta\boldsymbol{u}_n^1\approx\Delta\lambda_n^1\left[\boldsymbol{K}_{\mathrm{T}}\left(\boldsymbol{u}_n^{r-1}\right)\right]^{-1}\boldsymbol{F}_0=\Delta\lambda_n^1\delta\hat{\boldsymbol{u}}_n\end{gathered} \tag{3.82}$$

由式(3.82)可得 $\Delta\lambda_n^1$ 为

$$\Delta\lambda_n^1=\frac{\Delta l_n}{\sqrt{\left(\delta\hat{\boldsymbol{u}}_n\right)^{\mathrm{T}}\delta\hat{\boldsymbol{u}}_n+\left(\boldsymbol{F}_0\right)^{\mathrm{T}}\boldsymbol{F}_0}} \tag{3.83}$$

为了使迭代过程收敛，应给定载荷步的最小值 $\Delta\lambda_{\min}$ 和最大值 $\Delta\lambda_{\max}$，同时也应给定每个载荷步的最大迭代次数(一般取 12～20)。

在每个载荷步的初始迭代步，满足

$$\Delta\boldsymbol{u}_n^1=\delta\boldsymbol{u}_n^1,\quad \Delta\lambda_n^1=\delta\lambda_n^1 \tag{3.84}$$

其次，确定载荷因子的符号

$$\Delta\lambda_n^1=\begin{cases}+\left|\Delta\lambda_n^1\right|, & \text{如果}\left(\Delta\boldsymbol{u}_{n-1}^r\right)^{\mathrm{T}}\delta\hat{\boldsymbol{u}}_n>0\\ -\left|\Delta\lambda_n^1\right|, & \text{如果}\left(\Delta\boldsymbol{u}_{n-1}^r\right)^{\mathrm{T}}\delta\hat{\boldsymbol{u}}_n<0\end{cases} \tag{3.85}$$

3. 收敛标准的确定

连续迭代步的位移相对差值的 L_2 范数小于预先给定的允许值 η (prescribed convergence tolerance)

$$\sqrt{\frac{\left(\boldsymbol{u}_n^r-\boldsymbol{u}_n^{r-1}\right)^{\mathrm{T}}\left(\boldsymbol{u}_n^r-\boldsymbol{u}_n^{r-1}\right)}{\left(\boldsymbol{u}_n^r\right)^{\mathrm{T}}\left(\boldsymbol{u}_n^r\right)}}\leqslant\eta \tag{3.86}$$

4. 初始载荷步的弧长及初始位移增量的确定

初始载荷因子 $\Delta\lambda_1^1$ 可以自行设定，比如设置为总载荷因子的 0.1 或 0.001 等，一般选择期望最大载荷的 10%～30%。则初始位移增量为

$$\Delta\boldsymbol{u}_1^1=\Delta\lambda_1^1\left[\boldsymbol{K}_{\mathrm{T}}\left(\boldsymbol{u}_0\right)\right]^{-1}\boldsymbol{F}_0=\Delta\lambda_1^1\delta\hat{\boldsymbol{u}}_1 \tag{3.87}$$

因此，初始弧长求得为

$$\begin{aligned}\Delta l_1&=\sqrt{\left(\Delta\boldsymbol{u}_1^1\right)^{\mathrm{T}}\Delta\boldsymbol{u}_1^1+\left(\Delta\lambda_1^1\right)^2\left(\boldsymbol{F}_0\right)^{\mathrm{T}}\boldsymbol{F}_0}\\&=\left|\Delta\lambda_1^1\right|\sqrt{\left(\delta\hat{\boldsymbol{u}}_1\right)^{\mathrm{T}}\delta\hat{\boldsymbol{u}}_1+\left(\boldsymbol{F}_0\right)^{\mathrm{T}}\boldsymbol{F}_0}\end{aligned} \tag{3.88}$$

3.6 算法验证及计算比较

为了验证本章方法的可靠性和准确性，考虑了含多个缺陷的均质和非均质 FGM 板在热载荷和力载荷作用下的自由振动和特征屈曲问题，并将所提方法得到的数值结果与文献中的参考解进行了比较。本章中使用的 FGM 各组分的性能参数见表 3.2。不同类型边界条件下需要约束的自由度见表 3.3，其中我们约定字母 S、C 和 F 分别代表简支、固支和自由边界类型。在固支边界条件下，通过将定义板边缘切面的两行控制点处的横向位移设置为零，来实现对 $w_{0,x}$ 和 $w_{0,y}$ 约束条件的施加。

表 3.2　FGM 各组分的性能参数

材料	弹性模量 E/GPa	泊松比 ν	热传导系数 κ/(W/(m·K))	热膨胀系数 α/(×10^{-6}/℃)	密度 ρ/(kg/m^3)
Al	70	0.3	204	23	2707
ZrO_2-1	200	0.3	2.09	10	5700
ZrO_2-2	151	0.3	2.09	10	3000
Al_2O_3	380	0.3	10.4	7.4	3800

表 3.3　边界条件的约束类型

类型		约束的自由度
简支	SSSS	$\begin{cases}v_0=w_0=\beta_y=0, & 在\ x=0,a\\u_0=w_0=\beta_x=0, & 在\ y=0,b\end{cases}$

续表

类型		约束的自由度
其他类型	CCSS	$\begin{cases} u_0 = v_0 = w_0 = \beta_x = \beta_y = w_{0,x} = w_{0,y} = 0, & 在\ x = 0, a \\ u_0 = w_0 = \beta_x = 0, & 在\ y = 0, b \end{cases}$
	CSCS	$\begin{cases} v_0 = w_0 = \beta_y = 0, & 在\ x = 0 \\ u_0 = w_0 = \beta_x = 0, & 在\ y = b \\ u_0 = v_0 = w_0 = \beta_x = \beta_y = w_{0,x} = w_{0,y} = 0, & 在\ x = a,\ y = 0 \end{cases}$
	CCFF	$u_0 = v_0 = w_0 = \beta_x = \beta_y = w_{0,x} = w_{0,y} = 0$，在 $x = a,\ y = 0$
	CFCF	$u_0 = v_0 = w_0 = \beta_x = \beta_y = w_{0,x} = w_{0,y} = 0$，在 $y = 0, b$
固支	CCCC	$u_0 = v_0 = w_0 = \beta_x = \beta_y = w_{0,x} = w_{0,y} = 0$，在所有边

在 XIGA 中，求解域由裂纹分割单元、裂尖单元、混合单元和标准(非富集)单元组成。在对这些富集单元进行数值积分时，由于富集函数中包含非连续项和奇异项，使用标准高斯求积会产生较大的误差。本章仍采用子三角形划分技术和几乎极性积分转换方法计算裂纹分割单元和裂尖单元内的数值积分。表 3.4 列出了不同类型单元中的求积规则。与温度相关的材料特性计算式为

$$P = P_0\left(P_{-1}T^{-1} + 1 + P_1T + P_2T^2 + P_3T^3\right) \tag{3.89}$$

其中，P_{-1}、P_0、P_1、P_2 和 P_3 是与温度相关的系数，其值见表 3.5。

表 3.4　不同类型单元中的积分规则

规则	单元类型	积分点数目
高斯求积	标准单元	4×4
	自适应积分子胞元	4×4
	混合单元	6×6
子三角形划分	裂纹分割单元	每个子三角形内 7 个
几乎极性积分	裂尖单元	每个子三角形内 4×4

表 3.5　温度相关的材料系数

材料	参数	P_0	P_{-1}	P_1	P_2	P_3
Si_3N_4	E/Pa	3.4843×10^{11}	0	3.07×10^{-4}	2.16×10^{-7}	-8.946×10^{-11}
	α/(1/K)	5.8723×10^{-6}	0	9.095×10^{-4}	0	0
SUS304	E/Pa	2.0104×10^{11}	0	3.079×10^{-4}	-6.534×10^{-7}	0
	α/(1/K)	1.233×10^{-5}	0	8.086×10^{-4}	0	0

3.6.1 收敛性分析与开销对比

考虑如图 3.11 所示含边裂纹的四边简支 Al/Al_2O_3 板，其边长和厚度比值为 $a/h=10$，裂纹与边长的比值为 $c/a=0.5$，等效材料特性通过混合法则计算。无量纲固有频率定义为 $\bar{\omega}=\omega\dfrac{a^2}{h}\sqrt{\dfrac{\rho_c}{E_c}}$。PHT 样条网格局部细化过程如图 3.11 所示。不同材料梯度指数下，板一阶无量纲固有频率随网格变化的情况如表 3.6 所示。数据表明，随着网格的逐渐加密，与参考解相比，数值结果具有良好的收敛性。

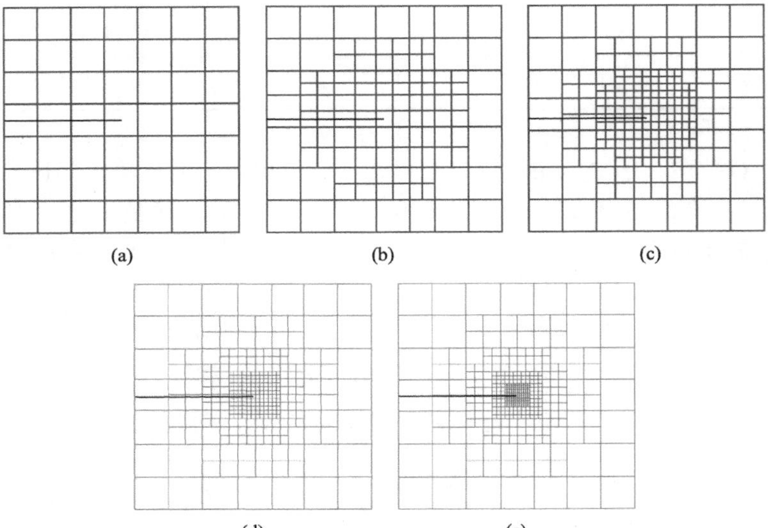

图 3.11　含边裂纹板的 PHT 样条网格局部细化过程

黑线代表裂纹，不同网格的控制点数量为：(a) 256; (b)468; (c) 804; (d) 1188; (e) 1572

表 3.6　不同材料梯度指数和网格下，FGM 板的一阶无量纲固有频率

n	控制点数量(本章)					文献[27]	文献[28]	文献[29]
	256	468	804	1188	1572			
0	5.3029	5.3414	5.3495	5.3551	5.3573	5.379	5.387	5.373
1	4.0654	4.0956	4.1001	4.1037	4.1051	4.122	4.122	4.117
5	3.4608	3.4876	3.4910	3.4940	3.4951	3.511	3.626	3.526
10	3.3373	3.3630	3.3671	3.3703	3.3715	3.388	3.409	3.403

无量纲固有频率的相对误差定义为 $\left|(\bar{\omega}-\bar{\omega}_{ref})/\bar{\omega}_{ref}\right|$，其中 $\bar{\omega}_{ref}$ 为参考值，取自三个参考文献中结果的平均值。在材料梯度指数为 5 的情况下，前三阶无量纲固

有频率的相对误差随控制点数的变化趋势如图 3.12 所示。相对误差的变化趋势与参考解相比表现出了很好的收敛性，且控制点数为 1188 时的数值结果已达到精度要求，表明 PHT 样条局部细化的优势。

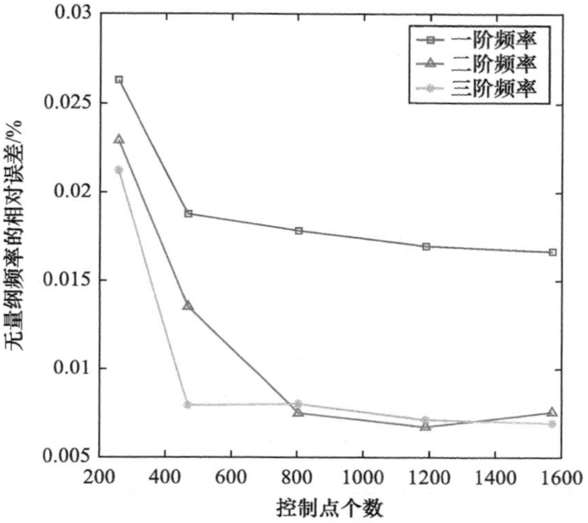

图 3.12　材料梯度指数为 5 时，前三阶无量纲固有频率的相对误差随控制点数的变化

接下来讨论利用 FCM 处理孔洞问题的收敛性。考虑如图 3.13 所示含中心圆孔的四边简支 Al/ZrO_2-2 板的单轴特征屈曲问题，其边长和厚度比值为 $a/h=100$，圆孔与板边长的比值为 $r/a=0.1$，等效材料特性通过混合法则计算。图 3.13 展示了不同自适应积分层数下 PHT 样条网格和自适应积分子胞元。如图 3.13(a)所示，初始层中圆孔周围的网格利用 PHT 样条进行了两次局部细分。临界屈曲载荷参数定义为 $\bar{N}_{cr}=\dfrac{N_{cr}a^2}{\pi^2 D_c}$，其中 $D_c=\dfrac{E_c h^3}{12\left(1-\nu_c^2\right)}$。临界屈曲载荷参数与参考解的比较见表 3.7，数值解与参考解吻合得很好，且自适应积分层数为 3 时就可以获得很好的收敛结果。因此，本章的后续分析中自适应积分的层数均设置为 3。

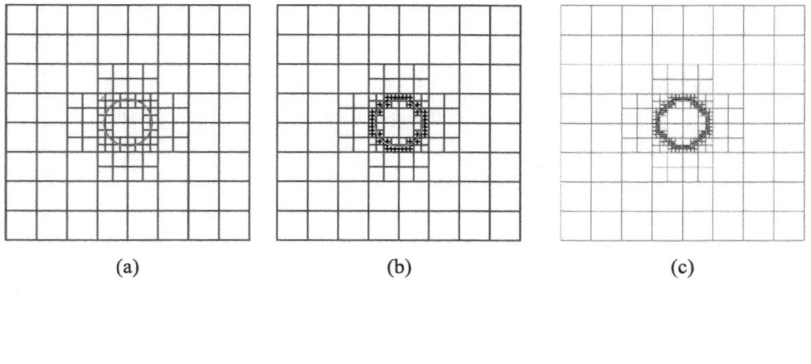

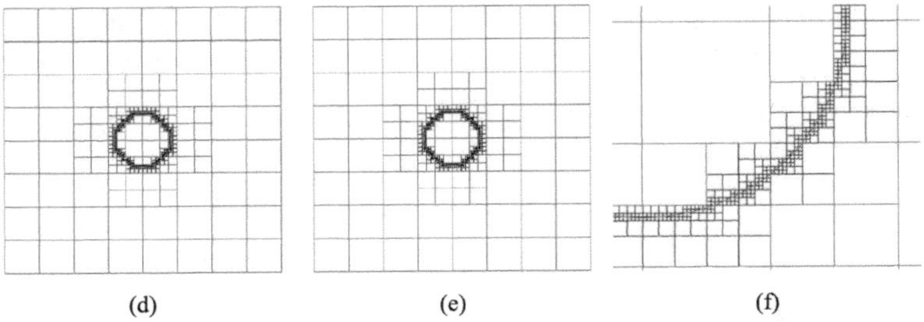

图 3.13　在不同自适应积分层数下，含中心圆孔的四边简支方板的 PHT 样条网格和积分子胞元

(a) 层数 0；(b) 层数 1；(c) 层数 2；(d) 层数 3；(e) 层数 4；(f) 图(e)的局部放大图

表 3.7　临界屈曲载荷参数与参考解的比较

n	自适应积分层数					文献[30]
	0	1	2	3	4	
0	6.9833	6.9932	7.0030	7.0048	7.0046	6.9999
1	4.8817	4.8885	4.8953	4.8966	4.8964	4.8933
5	4.1525	4.1586	4.1644	4.1656	4.1654	4.1625

然后，对 PHT 样条局部细分网格和 NURBS 全局细分网格的计算开销进行了比较。板的几何参数和材料特性的设置与上述算例相同。如图 3.13(d)所示，具有 532 个控制点且自适应层数为 3 时的 PHT 样条网格计算的结果用作参考。图 3.14 显示了不同网格尺寸下的 NURBS 离散网格和自适应积分子胞元。临界屈曲载荷参数的相对误差定义为 $\left|\dfrac{\bar{N}_{\mathrm{cr}}-\bar{N}_{\mathrm{cr}}^{\mathrm{ref}}}{\bar{N}_{\mathrm{cr}}^{\mathrm{ref}}}\right|$，其中 $\bar{N}_{\mathrm{cr}}^{\mathrm{ref}}$ 为文献[30]中的参考值。图 3.15 显示了不同 NURBS 网格尺寸下，临界屈曲参数的相对误差随控制点数量和每边单元数量的变化趋势，其中红色虚线表示 PHT 样条网格下的相对误差。如图 3.15 所示，与仅包含 532 个控制点的局部细化 PHT 样条网格相比，NURBS 模型每边至少需要 17 个单元或 1296 个控制点才能达到相同的精度。

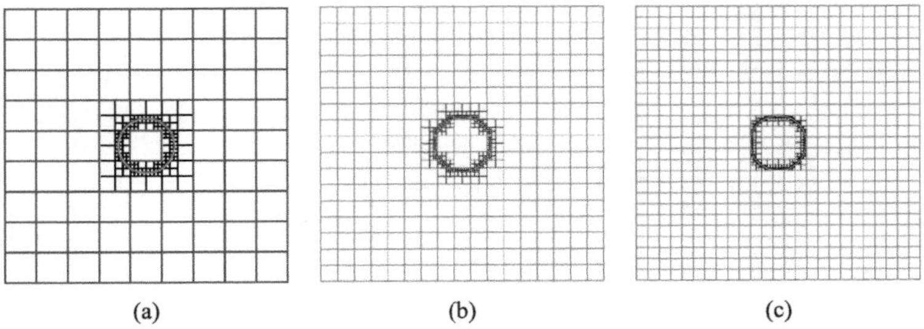

图 3.14　含中心圆孔的简支方板在不同网格尺寸下的 NURBS 单元和自适应积分子胞元

网格规模为：(a) 9×9；(b) 17×17；(c) 25×25

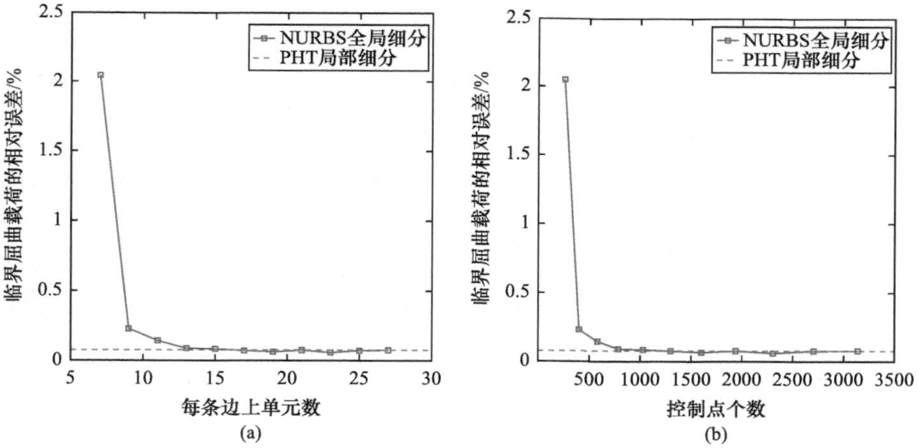

图 3.15 不同 NURBS 网格下临界屈曲参数的相对误差

(a) 随每条边单元数的变化(PHT 局部细分每条边单元数为 8)；(b) 随控制点数量的变化(PHT 局部细分控制点数为 532)

此外，还对计算时间进行了比较。不同 NURBS 网格和控制点数量下的无量纲计算费用T/T_{ref}如图 3.16 所示，其中T_{ref}代表 PHT 样条网格所需的时间且在图中用红色虚线表示。与 PHT 样条网格的计算时间相比，NURBS 模型的无量纲计算费用至少等于 1.8657。综上所述，本章所采用的方法需要更少的自由度和更短的计算时间就能达到所需的精度，并且还可以灵活地控制缺陷周围网格的分布。

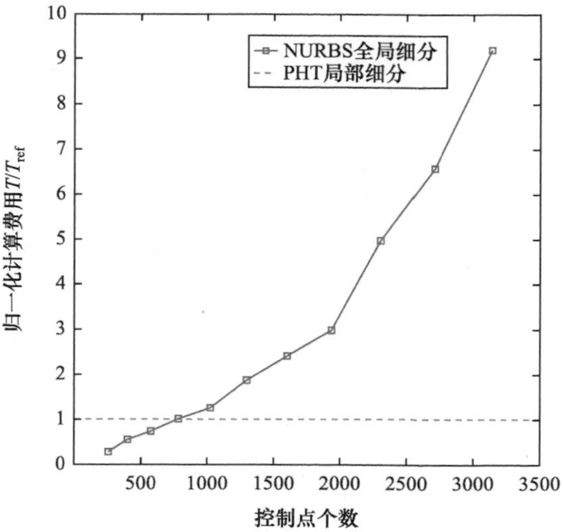

图 3.16 不同 NURBS 网格和控制点数量下的无量纲计算开销(PHT 局部细分控制点数为 532)

3.6.2 含缺陷振动分析验证

1. 含中心裂纹的方板的自由振动

考虑含中心裂纹的四边简支 Al/Al_2O_3 板，其边长和厚度比值表示为 a/h，裂纹长度与板边长的比值为 $c/a = 0.5$，等效材料特性通过混合法则计算。PHT 样条离散网格如图 3.17 所示。无量纲固有频率定义为 $\bar{\omega} = \omega \dfrac{a^2}{h}\sqrt{\dfrac{\rho_c}{E_c}}$。表 3.8 给出了不同材料梯度指数和不同长厚比值下无量纲固有频率与参考解的比较，数值结果与参考解吻合得很好。

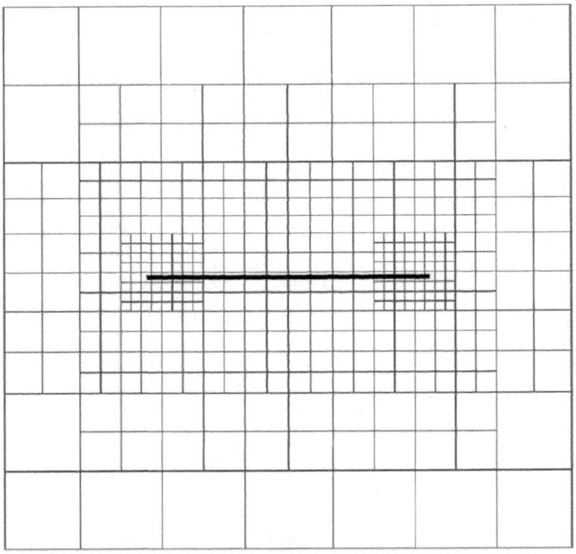

图 3.17 含中心裂纹的方板的 PHT 样条离散网格

表 3.8 含中心裂纹的方板的无量纲固有频率随材料梯度指数和长厚比的变化

a/h	方法	n		
		0	0.2	5
5	文献[31]	4.633	4.322	2.976
	文献[32]	4.7111	4.4151	3.0392
	本章	4.6187	4.3079	2.9684
50	文献[31]	5.318	4.937	3.496
	文献[32]	5.311	4.9404	3.4905
	本章	5.3139	4.9329	3.4926

2. 含中心裂纹的方板的热振动

考虑含中心裂纹的四边简支SUS304/Si_3N_4 方板的热振动问题，其边长和厚度比值表示为 $a/h=10$。裂纹长度与板边长的比值定义为 c/a。材料弹性模量和热膨胀系数随温度发生变化，其材料系数见表 3.5，计算公式见式(3.89)。温度场沿厚度方向呈非线性变化，其变化值由式(3.16)确定，其中 $T_m=300K$、$T_c=400K$ 和 $T_0=273.15K$。等效材料特性通过 Mori-Tanaka 均匀化技术计算。其他材料参数设置为，SUS304：$\rho_m=8166kg/m^3$ 和 $\kappa_m=12.04W/(m\cdot K)$；$Si_3N_4$：$\rho_c=2370kg/m^3$ 和 $\kappa_c=9.19W/(m\cdot K)$。泊松比均设定为 0.28。无量纲固有频率定义为 $\bar{\omega}=\omega a^2\sqrt{\rho_c h/D_c}$，其中 $D_c=E_c h^3\Big/\Big[12\big(1-\nu_c^2\big)\Big]$。裂纹长度和材料梯度指数对无量纲固有频率的影响如表 3.9 所示。数值结果与参考解吻合得很好。

表 3.9　裂纹长度和材料梯度指数对无量纲固有频率的影响

n	c/a (模态 1)				c/a (模态 2)			
	0.4		0.6		0.4		0.6	
	文献[19]	本章	文献[19]	本章	文献[19]	本章	文献[19]	本章
1	9.889	10.1081	9.337	9.5631	23.606	23.7742	18.820	19.5653
5	8.028	8.1186	7.584	7.6878	19.199	18.9647	15.285	15.6982
10	7.623	7.6886	7.204	7.2828	18.282	17.9633	14.556	14.9101

3. 含中心圆孔的方板的自由振动

考虑含中心圆孔的四边固支 Al/Al_2O_3 方板，其边长为 a=10m，长度与厚度比值为 $a/h=100$。圆孔半径与板边长的比值为 $r/a=0.1,0.2,0.3$。等效材料特性通过混合法则计算。图 3.18(a)和(b)分别展示了积分子胞元和自适应积分点沿圆孔边界

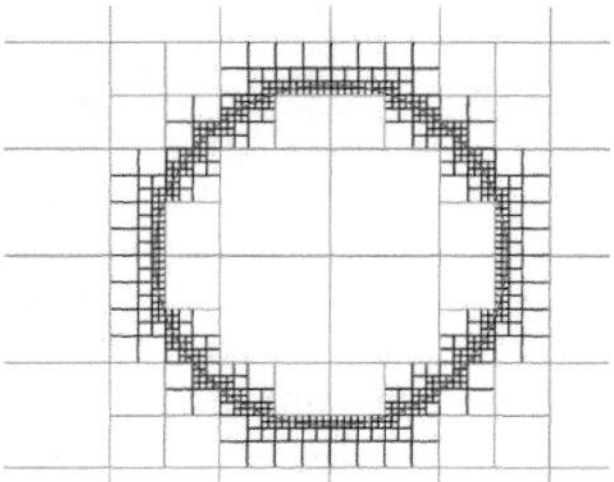

(a)

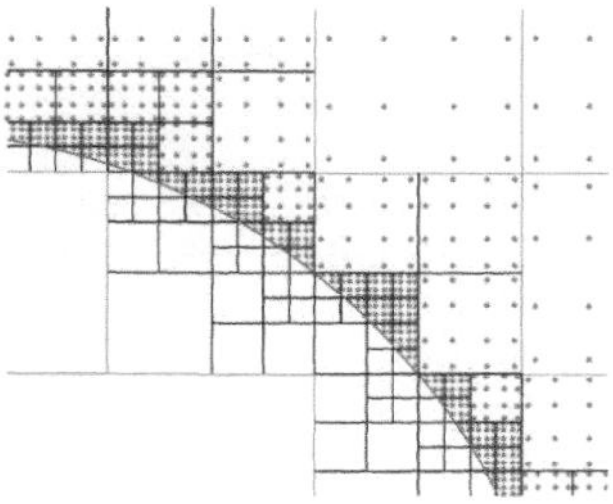

(b)

图 3.18　含中心圆孔的方板

(a) 求解网格(蓝线)和积分子胞元(黑线)；(b) 圆孔边界处积分点的分布

的分布。无量纲固有频率定义为 $\bar{\omega}=\left[\rho_{\mathrm{c}}h\omega^2\dfrac{a^4}{D_{\mathrm{c}}\left(1-\nu_{\mathrm{c}}^2\right)}\right]^{1/4}$，其中 $D_{\mathrm{c}}=\dfrac{E_{\mathrm{c}}h^3}{12\left(1-\nu_{\mathrm{c}}^2\right)}$。不同的圆孔半径和材料梯度指数对无量纲固有频率的影响如表 3.10 所示，可以观察到获得的数值解与参考解之间良好的一致性。

表 3.10　圆孔半径和材料梯度指数对无量纲固有频率的影响

r/a	n	方法	模态数				
			1	2	3	4	5
0.1	1	文献[30]	5.4026	7.5676	7.5676	9.1776	10.0905
		本章	5.3985	7.5493	7.5493	9.1772	10.0943
	2	文献[30]	5.1519	7.2160	7.2160	8.7513	9.6218
		本章	5.1473	7.1975	7.1975	8.7497	9.6238
	5	文献[30]	5.0164	7.0236	7.0236	8.5199	9.3668
		本章	5.0113	7.0046	7.0046	8.5164	9.3663
0.2	1	文献[30]	6.0234	7.3136	7.3136	8.9075	9.7506
		本章	6.0288	7.3206	7.3206	8.9129	9.7595
	2	文献[30]	5.7437	6.9739	6.9739	8.4937	9.2976
		本章	5.7482	6.9791	6.9791	8.4973	9.3041
	5	文献[30]	5.5927	6.7881	6.7881	8.2682	9.0497
		本章	5.5962	6.7905	6.7905	8.2690	9.0525
0.3	1	文献[30]	7.7452	8.2604	8.2604	9.0408	10.3112
		本章	7.7617	8.2765	8.2765	9.0538	10.3423
	2	文献[30]	7.3855	7.8767	7.8767	8.6207	9.8319
		本章	7.4003	7.8907	7.8907	8.6313	9.8593
	5	文献[30]	7.1909	7.6676	7.6676	8.3905	9.5683
		本章	7.2039	7.6790	7.6790	8.3976	9.5908

4. 含复杂孔洞形状的方板的自由振动

为了说明所提出方法对复杂孔洞结构的适用性，考虑了如图 3.19 所示的具有复杂孔洞的均质方板的自由振动。几何参数设置为：板长 $a=10\text{m}$，厚度 $h=0.05\text{m}$。材料参数设置为：弹性模量 $E=200\text{GPa}$，泊松比 $\nu=0.3$ 和密度 $\rho=8000\text{kg/m}^3$。PHT 样条离散网格和控制点分布如图 3.20(a)所示。图 3.20(b)给出了自适应积分子胞元的结构分布。无量纲固有频率定义为 $\bar{\omega}=\left(\dfrac{\rho h\omega^2 a^4}{D}\right)^{1/4}$，其中 $D=\dfrac{Eh^3}{12\left(1-\nu^2\right)}$。在简支和固支边界条件下，数值结果与文献[33]中参考解的比较如表 3.11 所示，可以观察到数值结果与参考解吻合得很好。值得注意的是，在文献[33]中该模型由八个 NURBS 片构成，并采用弯曲条法来实施片边界之间的 C^1 连续性，这显然增加了计算成本和复杂度。

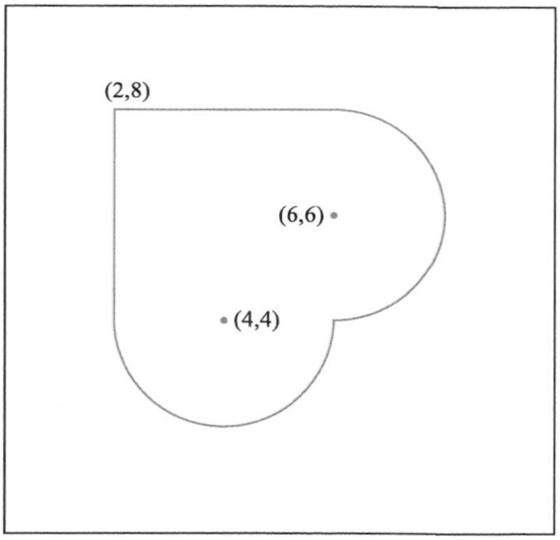

图 3.19　含复杂孔洞的方板的几何形状

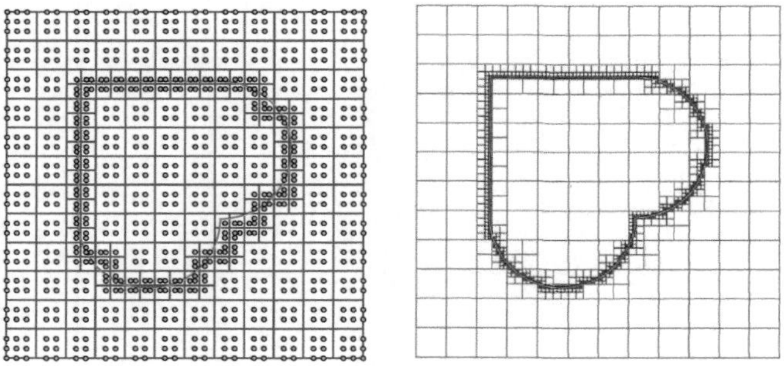

(a)　(b)

图 3.20　含复杂孔洞的方板

(a) PHT 样条网格及控制点；(b) 求解网格(蓝线)和积分子胞元(黑线)

表 3.11 含复杂孔洞的方板的无量纲固有频率

模态	SSSS			CCCC		
	IGA-FSDT[33]	IGA-S-FSDT[33]	本章	IGA-FSDT[33]	IGA-FSDT[33]	本章
1	4.914	5.098	4.917	7.453	7.431	7.450
2	6.390	6.608	6.385	9.825	9.880	9.818
3	6.762	6.929	6.759	9.845	9.992	9.831
4	8.568	8.644	8.557	10.964	11.077	10.935
5	8.982	9.031	8.972	11.165	11.254	11.153
6	10.683	10.591	10.659	12.381	12.424	12.340
7	10.934	10.946	10.883	12.953	12.862	12.825
8	11.694	11.800	11.626	13.721	13.678	13.469
9	12.852	12.517	12.821	14.511	14.227	14.451
10	13.229	13.001	13.165	14.792	14.613	14.714

3.6.3 含缺陷特性屈曲分析验证

1. 含中心裂纹的方板的热屈曲

考虑含中心裂纹的四边简支 Al/ZrO_2-2 板的热屈曲问题，假设温度沿厚度方向均匀变化，相关参数设置为：边长和厚度比值 $a/h=100$，裂纹长度与板边长的比值 $c/a=0.6$。等效材料特性通过混合法则计算。不同材料梯度指数下临界温度升高与文献[34]中参考解的比较如图 3.21 所示，数值结果与参考解吻合得很好。从图中可以看出，随着材料梯度指数的增加，临界温度的数值逐渐降低。

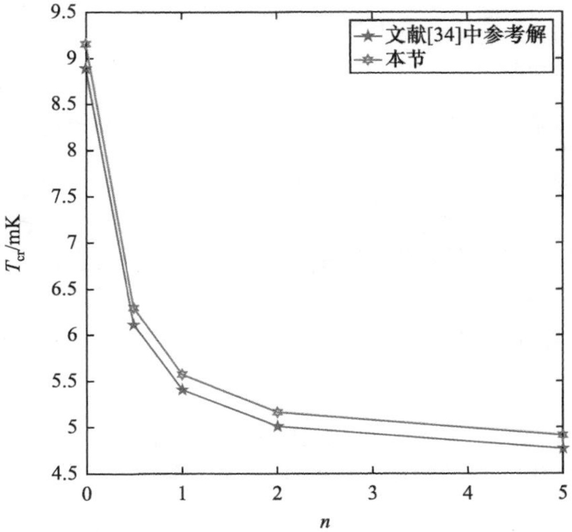

图 3.21 含中心裂纹的方板的热屈曲临界温度随材料梯度指数的变化

2. 含中心圆孔的方板的特征屈曲

考虑含中心圆孔的四边简支 Al/ZrO_2-2 板的单轴特征屈曲问题。几何参数设置为：长度与厚度比值 $a/h=100$，圆孔半径与板边长的比值 $r/a=0.1$。等效材料特性通过混合法则计算。临界屈曲载荷参数定义为：$\bar{N}_{cr}=\dfrac{N_{cr}a^2}{\pi^2 D_c}$，其中 $D_c=\dfrac{E_c h^3}{12\left(1-\nu_c^2\right)}$。表 3.12 列出了数值结果与参考解的比较。最大相对误差仅为 0.0985%。

表 3.12　不同材料梯度指数下，含中心圆孔的方板的临界屈曲载荷参数

方法	n		
	0	1	5
文献[30]	6.9999	4.8933	4.1625
本章	7.0048	4.8966	4.1656

3.6.4　含缺陷后屈曲分析验证

当由均质或对称复合材料组成的板受到面内压力载荷作用时，由于只产生薄膜力，板仍保持其初始的平面形状。当压力载荷达到临界值时，板会突然发生屈曲并产生横向变形，但在分岔点后板仍能承受相当大的附加压力载荷。因此，非线性特征值分析用于解决此类后屈曲问题。当外部载荷作用于 FGM 或非对称复合材料板时，由于拉弯耦合效应，板会随着压力载荷的增加而瞬时变形。在这种情况下，线性特征值分析无法准确地描述板的稳定状态。

在板的横截面方向施加微小的初始几何缺陷，用于初始化后屈曲路径。采用圆柱形弧长法和修正牛顿-拉弗森(Newton-Raphson)迭代法相结合的方式求解非线性平衡方程。任意形状孔洞的影响可通过 FCM 法中的自适应积分进行模拟。FCM 的理论已在第 2 章介绍过，此处不再赘述。随后，对后屈曲响应进行详细的参数研究，包括材料梯度指数、长厚比、边界条件、孔口尺寸等。各组分的材料性能参数列于表 3.13 中。

表 3.13　各组分的材料性能参数

	材料 1	材料 2	材料 3
弹性模量 E/GPa	205.1	70	151
泊松比 ν	0.3	0.3	0.3

无量纲的横向变形定义为

$$\bar{w} = \frac{w}{h} \tag{3.90}$$

式中，h代表板的厚度。

以特征屈曲分析中板的屈曲变形模态作为初始几何缺陷。由于 IGA 具有更高的连续性和精确的几何离散，则所用的初始几何缺陷是精确、光滑和连续的。无量纲缺陷系数定义为

$$\varDelta = \frac{\mu_{\mathrm{r}}}{h} \tag{3.91}$$

其中，μ_{r}是最大变形参考点处的模态横向位移，例如，一阶模态下均质方板的中心点处，或二阶模态下均质方板的四分之一点处。约定用字母 S、C 和 F 分别表示简支、固支和自由边界条件。除非另有说明，不同的边界条件定义如下所述。

简支边界条件(SSSS):

$$\begin{cases} v_0 = w_0 = \beta_y = 0, \ 在\ x = 0, a \\ u_0 = w_0 = \beta_x = 0, \ 在\ y = 0, b \end{cases} \tag{3.92}$$

固支边界条件(CCCC):

$$\begin{cases} v_0 = w_0 = \beta_y = w_{0,x} = 0, \ 在\ x = 0, a \\ u_0 = w_0 = \beta_x = w_{0,y} = 0, \ 在\ y = 0, b \end{cases} \tag{3.93}$$

式中，$w_{0,x}$ 和 $w_{0,y}$ 的实施方法与第 2 章所使用的方法一致。

由于在现有文献中没有含多孔的多向 FGM 板后屈曲分析的数值结果，为了说明现有方法的有效性和可靠性，利用本方法计算了各向同性板和单方向 FGM 板的后屈曲问题，并和文献中的参考结果进行了比较。

考虑各向同性板在单向压缩下的后屈曲响应。所涉及的无量纲参数和材料特性为：板的长宽比 $a/b=1$，长厚比 $a/h=120$，弹性模量 $E=3\times10^6$ Pa 和泊松比 $\nu=1/3$。在此算例中我们设置 $k_3=0$、$E_3=E$ 和 $\nu_3=\nu$。屈曲载荷参数用字母 λ 表示，λ_{cr} 代表线性屈曲分析下的临界屈曲载荷参数。相关的边界条件设置为

$$\begin{cases} v_0 = w_0 = \beta_y = w_{0,x} = 0, \ 在\ x = 0, a \\ w_0 = \beta_x = 0, \qquad 在\ y = 0, b \end{cases} \tag{3.94}$$

一阶模态形状伴随无量纲缺陷系数 $\varDelta = 0.001$ 用作初始几何缺陷。这里将三次 NURBS 网格下的无量纲载荷-挠度曲线与文献[35]中无初始缺陷下的参考结果进行比较，如图 3.22 所示。此外，图 3.23 还显示了不同幅度的初始几何缺陷下板的后屈曲载荷-挠度曲线之间的比较。可以观察到，后屈曲路径对初始缺陷十分敏

感。增加缺陷的大小会导致分岔曲线远离临界点。数值结果与参考解吻合得很好，可以观察到明显的收敛趋势。与参考解相比，12×12 的三次 NURBS 单元就足以获得很好的收敛结果。因此，该网格尺寸将应用于后续的数值算例分析。此外，在大变形阶段，数值结果与参考解的后屈曲路径之间的差异可能是由于 NURBS 具有更高的连续性(C^2 连续性)，从而产生了更精确的结果。

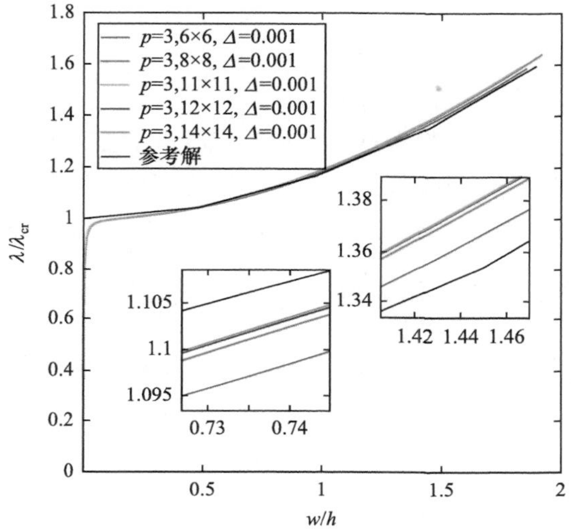

图 3.22 三次 NURBS 网格下的无量纲载荷-挠度曲线

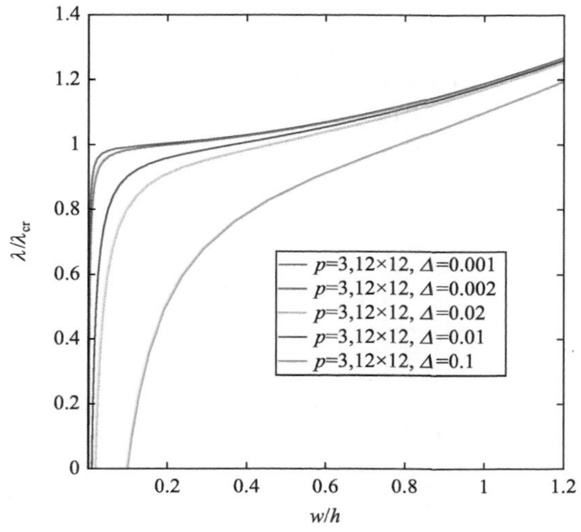

图 3.23 不同初始几何缺陷下板的后屈曲路径比较

3.7 数值算例

这里对同时含有裂纹和孔洞缺陷的 FGM 板在热和力载荷作用下的自由振动和特征屈曲进行分析，并考虑材料和几何方面等因素对固有频率和临界屈曲载荷的影响。

3.7.1 同时含多个裂纹和孔洞的矩形板的自由振动

考虑如图 3.24 所示的同时含有多个裂纹和孔洞的矩形板的自由振动问题，其长宽尺寸为 $b=2a=20\,\mathrm{m}$，且坐标原点位于板的左下角。两个边裂纹的垂直坐标分别为 6.4845m 和 13.5155m。三个圆孔的中心坐标分别为(2.5m,3.5m)、(2.5m,16.5m)和(5m,10m)。无量纲固有频率定义为：$\bar{\omega}=\omega a^2\sqrt{\dfrac{\rho_{\mathrm{c}}h}{D_{\mathrm{c}}}}$，其中 $D_{\mathrm{c}}=\dfrac{E_{\mathrm{c}}h^3}{12\left(1-\nu_{\mathrm{c}}^2\right)}$。接下来将从材料梯度指数、长厚比、边界条件和有效材料特性四个方面讨论对无量纲固有频率的影响。

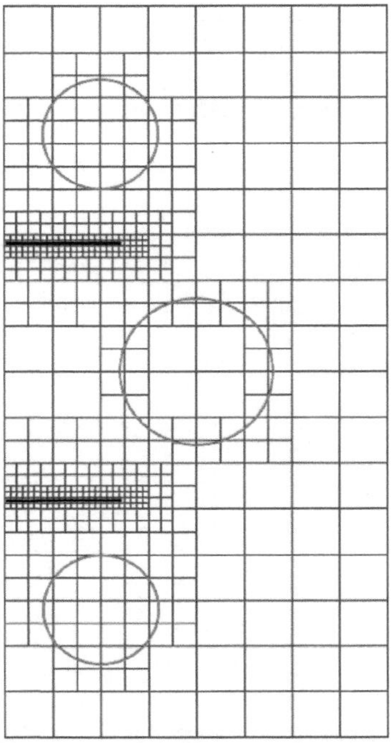

图 3.24 同时含多个裂纹和孔洞的矩形板的 PHT 样条离散网格

首先，研究长厚比对无量纲固有频率的影响，其变化趋势如图 3.25 所示。在本例中，使用混合规则计算材料的等效材料特性，并假定四边固支的边界条件。从图中可以看出，长厚比的增加导致无量纲固有频率的增加，而梯度指数的增加导致无量纲固有频率的降低。

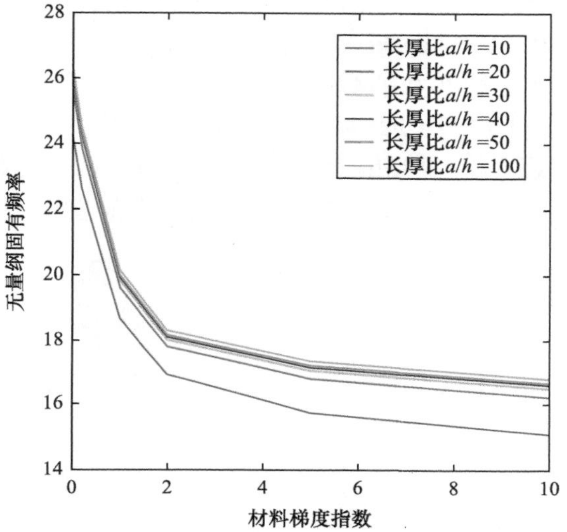

图 3.25　长厚比对无量纲固有频率的影响

其次，研究边界条件对无量纲固有频率的影响。在该算例中，材料的等效特性通过混合规则进行计算，且长厚比设置为 $a/h=100$ 。图 3.26 显示了在不同材料

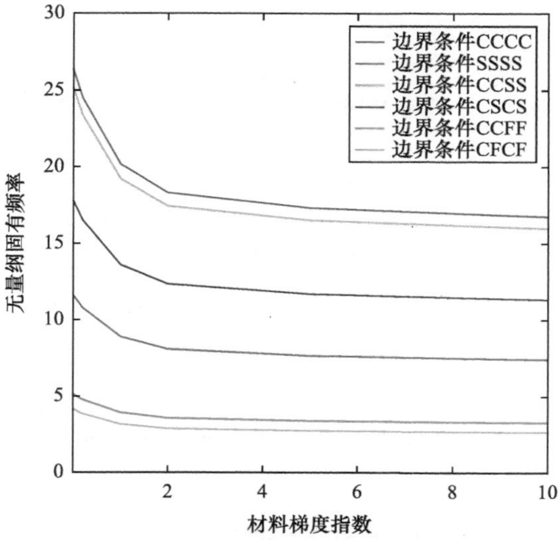

图 3.26　边界条件对无量纲固有频率的影响

梯度指数和边界条件下板无量纲固有频率的变化趋势。从图中可以看出，边界条件对 FGM 板的无量纲固有频率影响显著。四边固支边界条件下的无量纲固有频率幅值大于其他边界条件下的频率幅值，即加强约束能够提高板的固有频率。

最后，考虑等效材料特性对无量纲固有频率的影响。在该算例中，材料属性分别通过混合规则和 Mori-Tanaka 均匀化技术进行计算，并假定四边固支的边界条件。图 3.27 展示了无量纲固有频率随材料梯度指数、长厚比和等效材料性能计算规则的变化趋势。对于非均质材料，混合规则所得的频率参数比 Mori-Tanaka 均匀化技术求得的结果更高，这是因为混合规则条件下 FGM 板刚度的计算值更高。图 3.28 显示了在固支边界条件下，FGM 板的前六阶振型图($n=10$ ， $a/h=100$)。

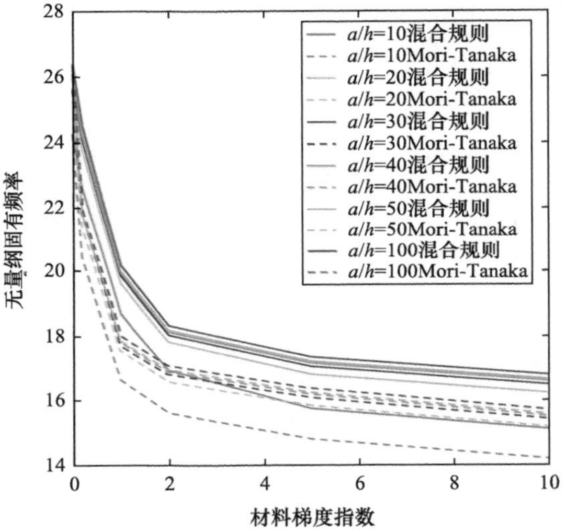

图 3.27 等效材料特性对无量纲固有频率的影响

3.7.2 同时含多个裂纹和孔洞的方板的热振动

考虑如图 3.29(a)所示的同时含有多个裂纹和孔洞的 Al/Al_2O_3 方板的热振动问题，其边长尺寸为 $a=20\,\text{m}$，且坐标原点位于板的中心。两个裂纹与边长的比值均为 $c/a=0.2$，且其垂直坐标分别为−1.1536m 和 1.1536m。孔洞的半径为 $r=1.5\text{m}$，且沿着圆心在坐标原点且半径为 7.5m 的圆周呈环形阵列。PHT 样条离散网格和自适应积分子胞元分别如图 3.29(a)和(b)所示。假设板顶面温度为 $T_c=300℃$ ，底面温度设置为 $T_m=20℃$ ，且无应力状态的参考温度为 $T_0=0℃$ 。无量纲固有频率定义为 $\bar{\omega}=\left[\rho_c h\omega^2\dfrac{a^4}{D_c\left(1-\nu_c^2\right)}\right]^{1/4}$ ，其中 $D_c=\dfrac{E_c h^3}{12\left(1-\nu_c^2\right)}$ 。分别考虑温度沿厚度方向

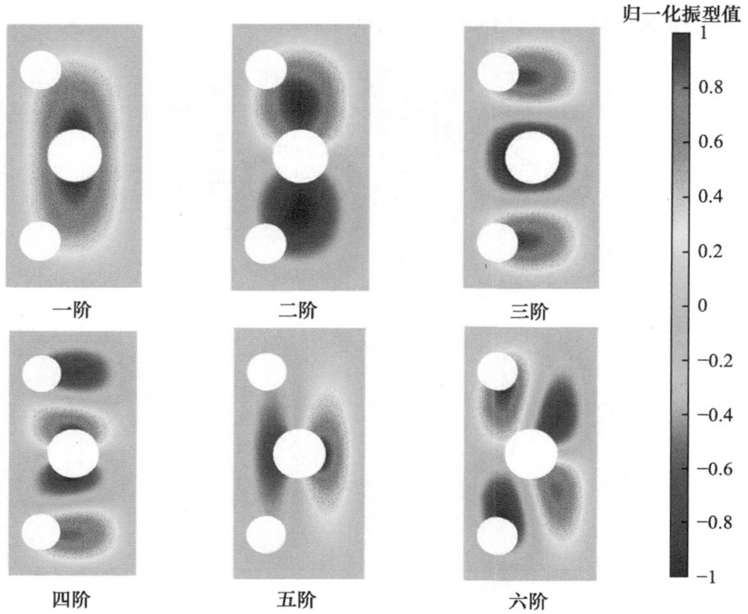

图 3.28　固支边界条件下，FGM 板的前六阶归一化振型图

呈线性和非线性变化，等效材料特性采用混合规则和 Mori-Tanaka 均匀化技术，以及材料梯度指数、长厚比、边界条件等因素对无量纲固有频率的影响。

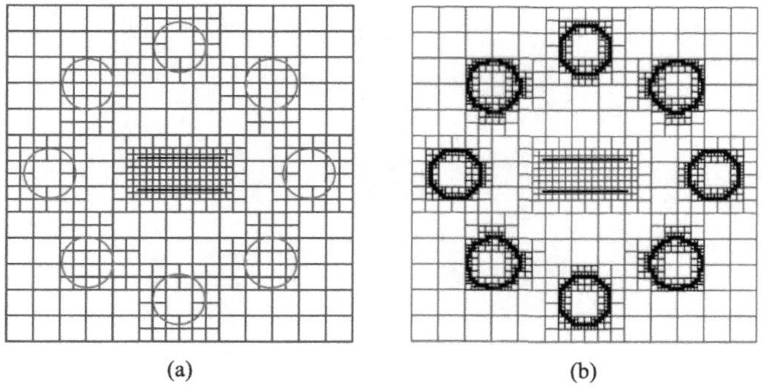

图 3.29　含多个裂纹和孔洞的方形板

(a) PHT 样条离散网格；(b) 求解网格(蓝线)和自适应积分子胞元(黑线)

四边固支和四边简支两种边界条件下的结果分别如图 3.30 和图 3.31 所示。数值结果表明，非线性温度变化下的无量纲固有频率值低于线性温度变化下的频率值。此外，对于非均质 FGM 板，热振动条件下，Mori-Tanaka 均匀化技术获得的无量纲固有频率值低于混合规则下的频率值。图 3.32 显示了在固支边界条件、

线性温度变化和混合规则下，FGM 板的前九阶振型图($n=2$，$a/h=50$)。

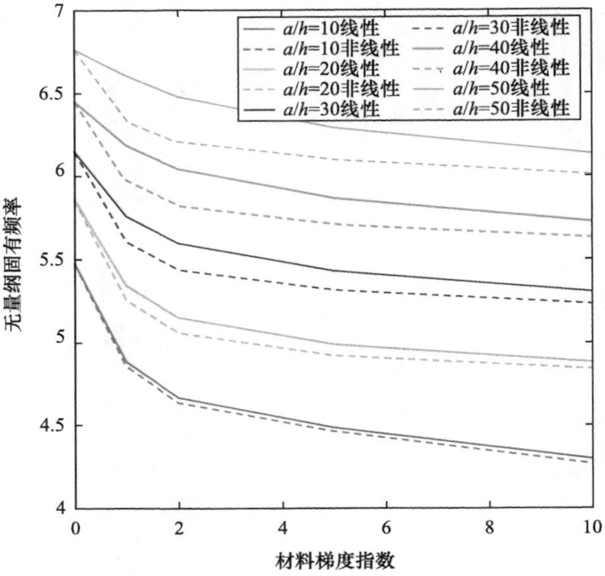

(a)

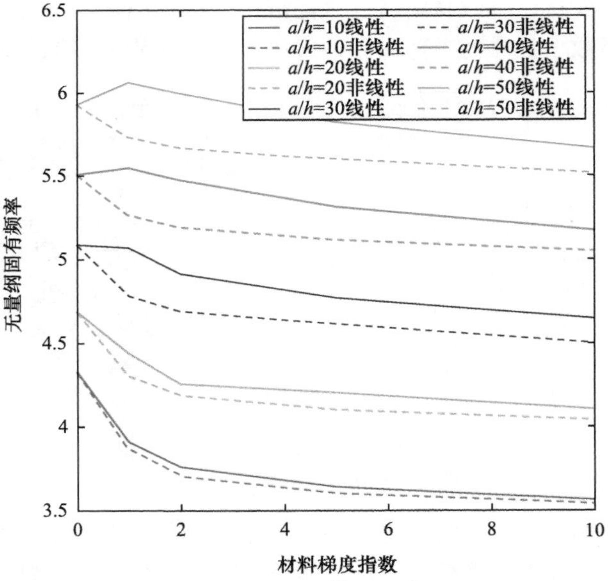

(b)

图 3.30 含多个缺陷的方形 FGM 板的无量纲固有频率变化(混合规则)

(a) CCCC；(b) SSSS

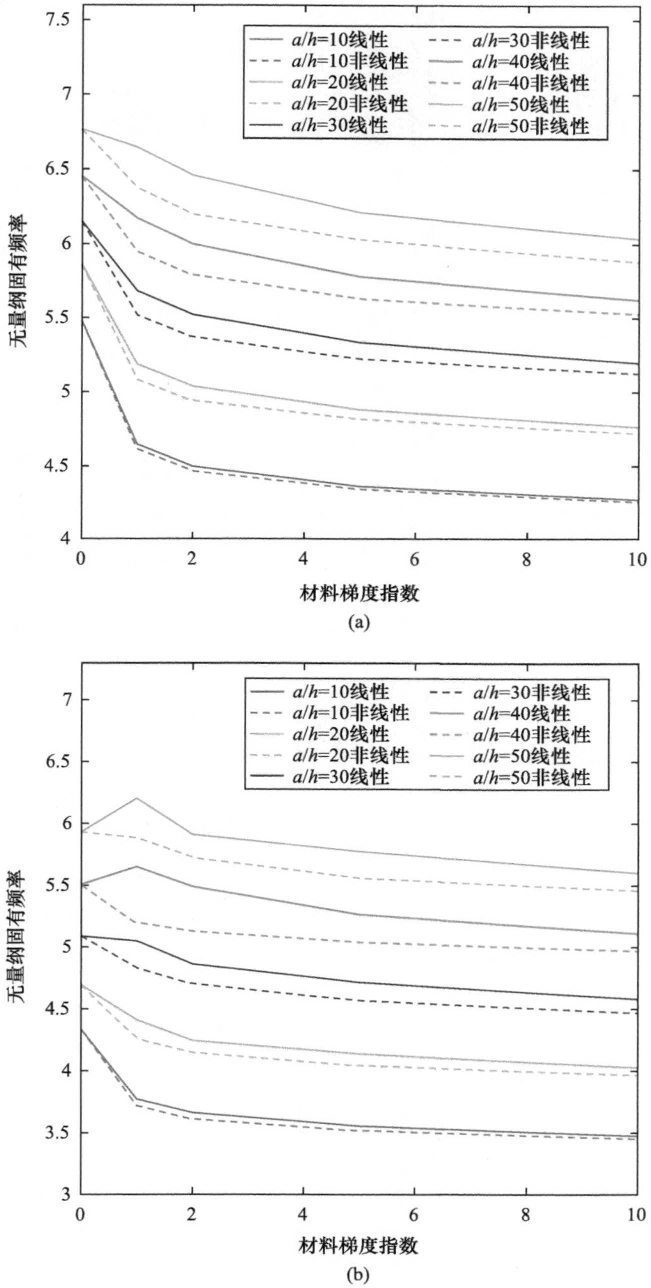

图 3.31　含多个缺陷的方形 FGM 板的无量纲固有频率变化(Mori-Tanaka)

(a) CCCC；(b) SSSS

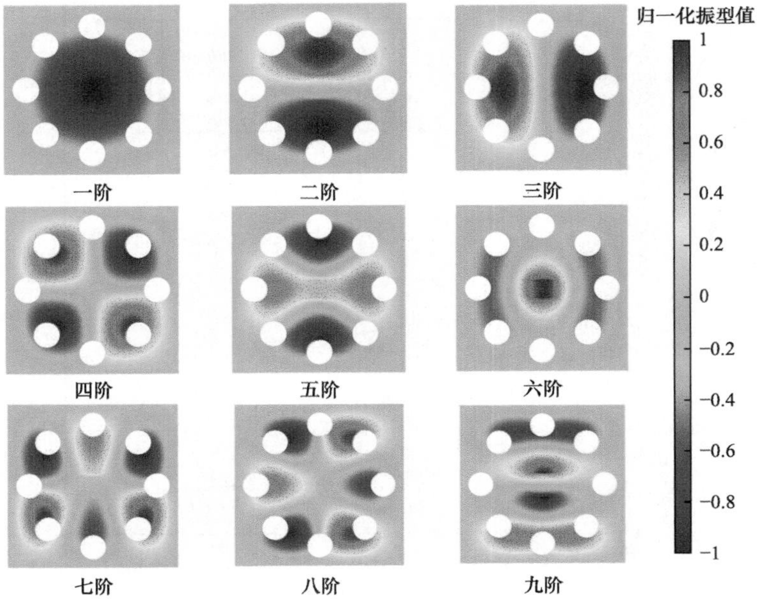

图 3.32 在固支边界条件、线性温度变化和混合规则下，FGM 板的前九阶归一化振型图

3.7.3 同时含裂纹和椭圆孔洞的方板的热力耦合屈曲

考虑如图 3.33(a)所示同时包含两个椭圆孔和一个中心裂纹的 Al/Al_2O_3 方板的热力耦合屈曲问题，其长度为 a=10m，坐标原点位于板的左下角。两个椭圆孔的中心坐标分别为(5,3)和(5,7)。考虑三种面内预屈曲载荷情况：水平方向单向压缩(记为 UniX)、垂直方向单向压缩(记为 UniY)和双向压缩(记为 Bi)。在该算例中，预屈曲热载荷和力载荷同时施加。力载荷和热载荷作用下的无量纲临界屈曲参数定义为 $N^*=\dfrac{N_{cr}a^2}{E_m h^2}$ 和 $T^*=\dfrac{\alpha_m T_{cr} a}{h}$，其中 N_{cr} 和 T_{cr} 分别为临界屈曲载荷和屈曲温度

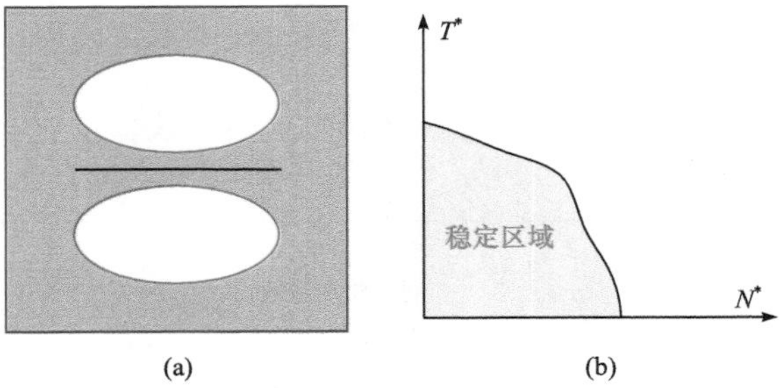

图 3.33 同时含裂纹和椭圆孔的方板
(a) 几何模型；(b) 稳定曲线

变化。以横轴为无量纲力屈曲参数 N^* 、纵轴为无量纲热屈曲参数 T^* ，绘制稳定曲线，如图 3.33(b)所示，曲线与坐标轴所围成的区域为稳定状态。研究材料梯度指数、裂纹长度、长厚比和边界条件因素对 FGM 板热力耦合屈曲响应的影响，并通过稳定曲线进行说明。

1. 材料梯度指数的影响

涉及的初始条件为：裂纹长度与板长度的比值 $c/a=0.5$ ，板长厚比 $a/h=100$ ，椭圆孔的长短轴比值 $E_a/E_b=2$ 。材料梯度指数的取值为 $n=0$ 、 $n=1$ 、 $n=2$ 、 $n=5$ 、 $n=10$ 和 $n=100$ 。图 3.34 显示了 CCCC 和 SSSS 边界条件下具有不同材料梯度指数的稳定性曲线。我们注意到，位于稳定曲线下方的点是稳定的，而位于每条曲线上方的点是不稳定的[36]。图中结果表明：①稳定区域随着材料梯度指数的增加而逐渐减小，并且对于纯陶瓷板，稳定区域达到最大，这是因为拥有较小热传导系数的陶瓷具有更好的抗热屈曲性能；②稳定区域随梯度指数从 0～1 的减小速率大于其随梯度指数从 1～100 的速率[36]；③在三种面内预屈曲压缩载荷情况下，同一材料梯度指数时，CCCC 边界条件获得的稳定区域均大于 SSSS 边界条件下的稳定区域；④稳定区域的面积大小顺序为 UniX 区域最大，UniY 区域次之，Bi 区域最小；⑤在大多数情况下，随着力载荷的增加，临界温度呈线性下降，而在 CCCC 边界条件和 UniX 情况下(图 3.34(a))，临界温度变化随力载荷的增加呈非线性趋势下降。

2. 板长厚比的影响

涉及的初始条件为：裂纹长度与板长度的比值 $c/a=0.5$ ，材料梯度指数为

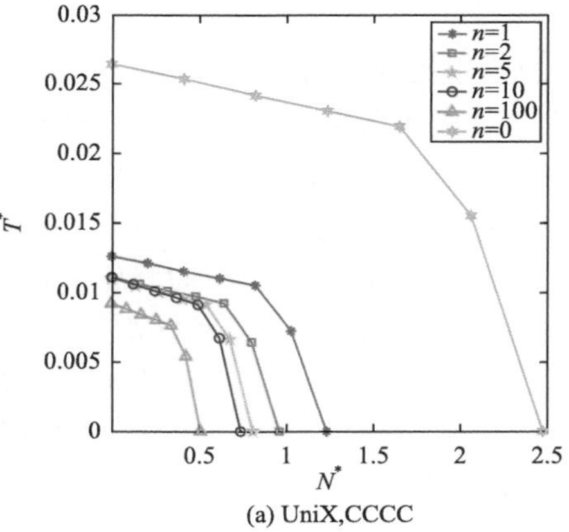

(a) UniX,CCCC

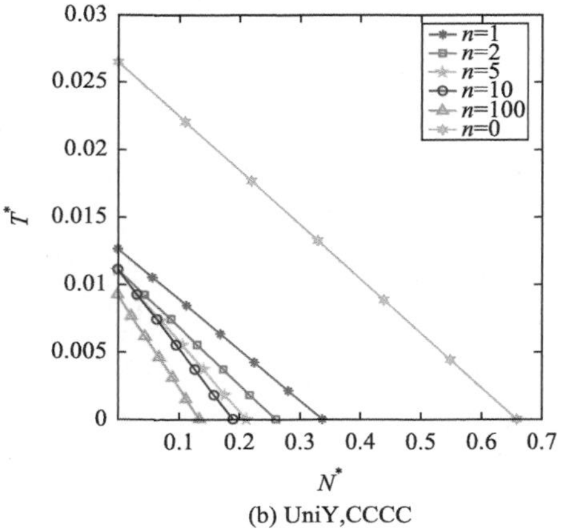

(b) UniY,CCCC

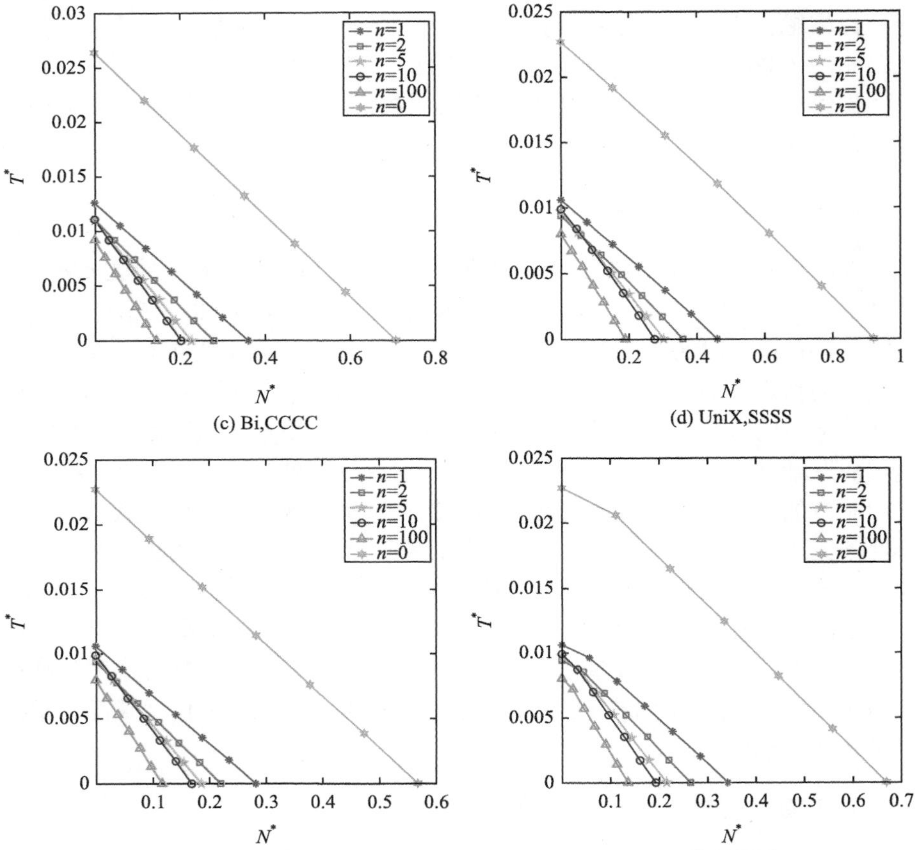

(c) Bi,CCCC
(d) UniX,SSSS
(e) UniY,SSSS
(f) Bi,SSSS

图 3.34　不同边界条件下材料梯度指数对稳定曲线的影响

$n=2$，椭圆孔的长短轴比值为 $E_a/E_b=2$。板长厚比的取值为 $a/h=10$、$a/h=20$、$a/h=30$、$a/h=40$、$a/h=50$ 和 $a/h=100$。图 3.35 显示了 CCCC 和 SSSS 边界条件下，具有不同长厚比的 FGM 板的稳定曲线。图中结果表明：①稳定区域随长厚比的增大而减小，表明长细板具有更好的抗热和力屈曲的能力；②在双向受压下，临界热屈曲载荷随力屈曲载荷的增加而线性降低，而在单向受压下，随力屈曲载荷的增加，稳定曲线呈现非线性下降趋势；③在相同长厚比情况下，CCCC 边界条件下板的稳定区域大于 SSSS 边界条件下的稳定区域，且在 UniX 下板的稳定区域最大，Bi 下板的稳定区域最小；④在纯力载荷的情况下，非均质和含缺陷板对压缩载荷类型的变化更为敏感。

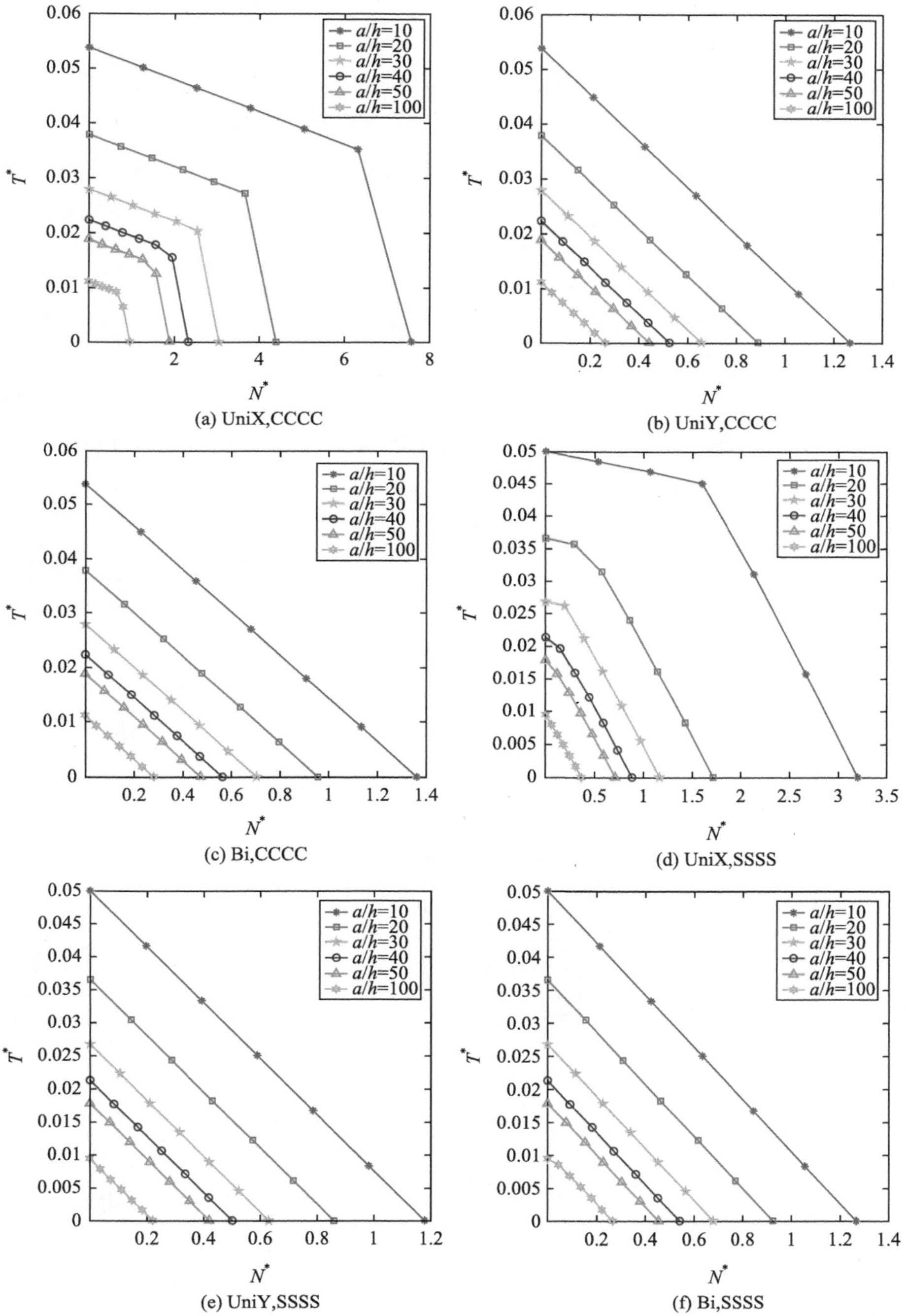

图 3.35 不同边界条件下板长厚比对稳定曲线的影响

3. 裂纹长度的影响

涉及的初始条件为：板长厚比的取值 $a/h=50$，材料梯度指数 $n=10$，椭圆孔的长短轴比值 $E_a/E_b=2$。裂纹长度与板长度的比值取为 $c/a=0.2$、$c/a=0.4$、$c/a=0.5$、$c/a=0.6$ 和 $c/a=0.8$。图 3.36 显示了 CCCC 和 SSSS 边界条件下，具有不同裂纹长度的 FGM 板的稳定曲线。图中结果表明：①稳定区域随裂纹长度的增大而减小；②在三种面内预屈曲压缩载荷情况下，CCCC 边界条件的稳定区域均大于 SSSS 边界条件的稳定区域，其中 UniX 时稳定区域最大，Bi 时稳定区域最小；③在纯热载荷的情况下，含缺陷的 FGM 板对压缩载荷类型的变化不敏感；④在 UniY 和 Bi 下，稳定曲线呈线性特征，而在 UniX 下，随着力载荷的增加，稳定曲线呈现非线性下降的趋势；⑤稳定区域随裂纹长度与板长度的比值为 0～0.5 的减小速率，小于其随裂纹长度与板长度的比值为 0.5～1 的速率。

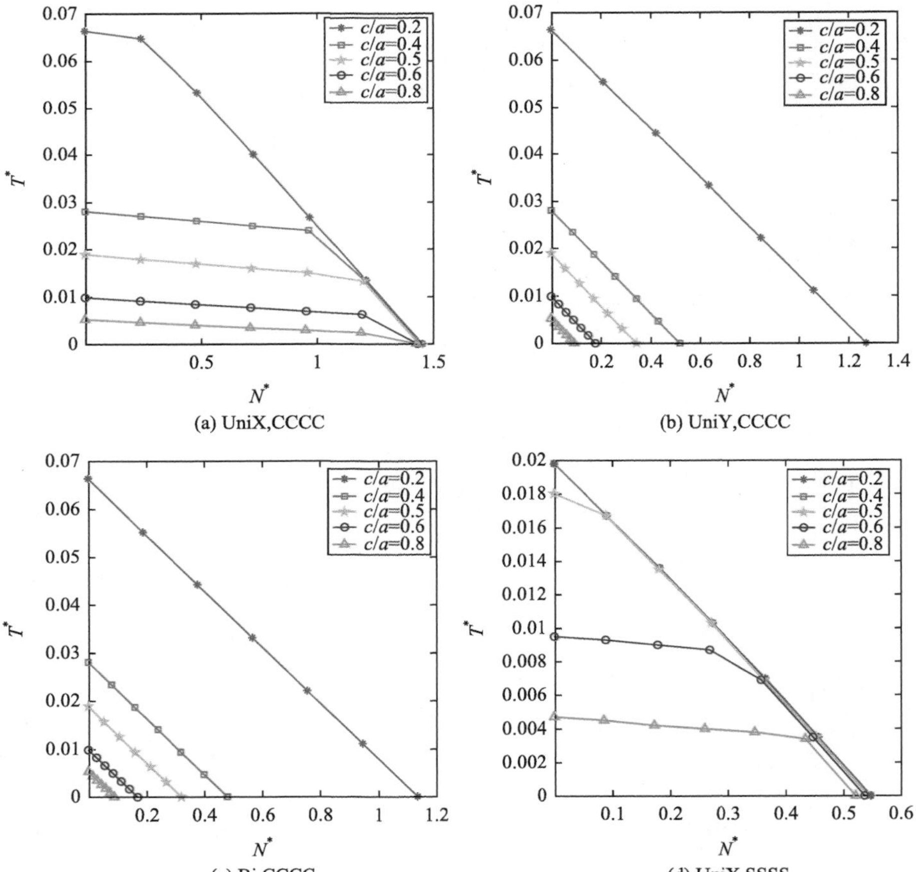

(a) UniX,CCCC

(b) UniY,CCCC

(c) Bi,CCCC

(d) UniX,SSSS

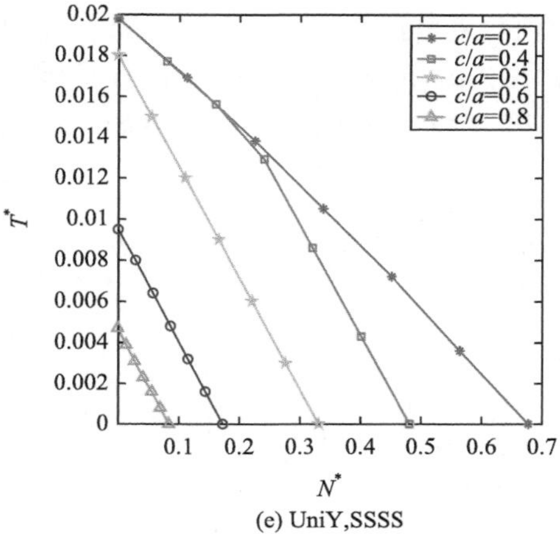

(e) UniY,SSSS

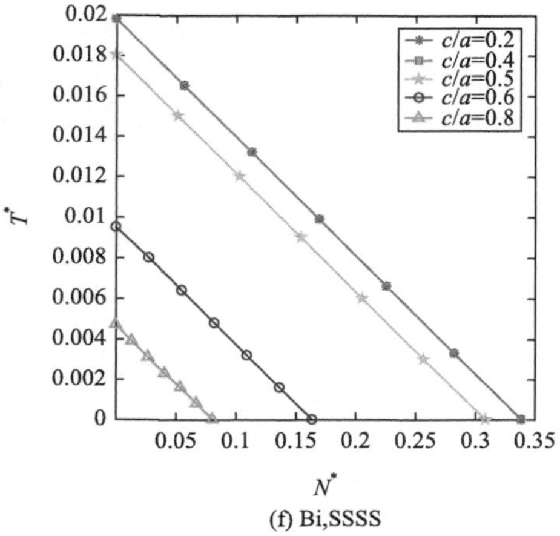

(f) Bi,SSSS

图 3.36　不同边界条件下裂纹长度对稳定曲线的影响

3.7.4　含多个孔洞的方板的后屈曲分析

在接下来的算例中，研究和讨论含多个孔洞的(包括圆形、椭圆形和复杂心形孔洞)多向 FGM 板在边缘压缩载荷作用下的后屈曲响应。除非另有说明，以下算例中均采用类型 A 情况下的体积分数规律。无量纲缺陷系数为 $\Delta=0.001$ 下板的一阶模态用作初始几何缺陷。

考虑如图 3.37 所示的含四个圆形孔洞的四边简支的多向 FGM 板在单轴压缩下的后屈曲响应。板的长宽尺寸为 $a=b=10\,\mathrm{m}$。四个圆孔的圆心坐标分别为(2.5m,2.5m)、(2.5m,7.5m)、(7.5m,2.5m)和(7.5m,7.5m)。无量纲载荷因子定义为 $\lambda=\dfrac{Pa^2}{\pi^2E_3h^3}$，其中 P 是压力载荷。

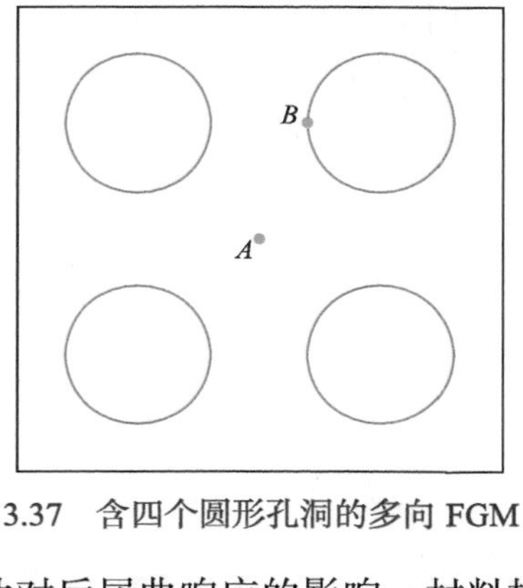

图 3.37　含四个圆形孔洞的多向 FGM 板

首先，考虑孔洞尺寸对后屈曲响应的影响。材料梯度指数和长厚比设置为

$k_1 = k_2 = k_3 = 2$ 和 $a/h = 50$ 。圆孔半径变化值为 0.5m、1m、1.5m 和 1.8m。需要强调的是，四个孔洞的半径同时变化。NURBS 单元网格和自适应积分子胞元如图 3.38 所示。图 3.39 给出了 A 点和 B 点(图 3.37)处不同开孔尺寸下的后屈曲载

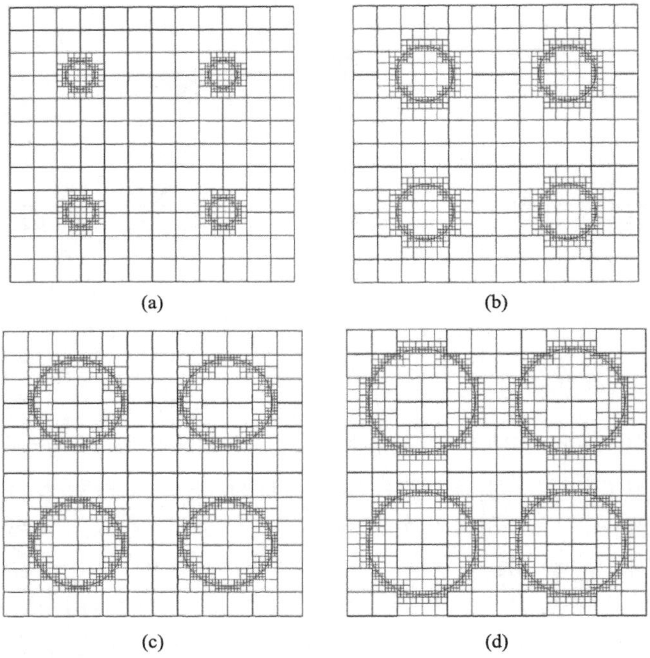

图 3.38　不同孔径下单元网格和自适应积分子胞元

(a) 半径 0.5m；(b) 半径 1m；(c) 半径 1.5m；(d) 半径 1.8m

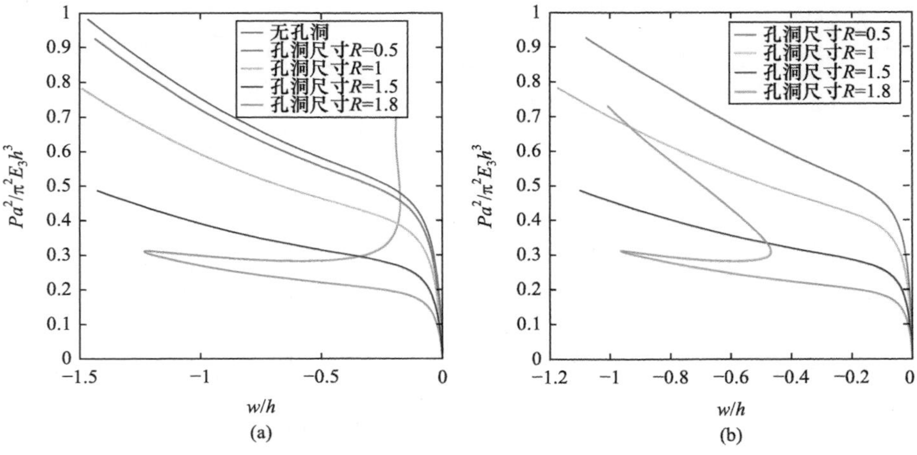

图 3.39　A 点和 B 点在不同孔洞尺寸下的后屈曲载荷-挠度曲线

(a) A 点；(b) B 点

荷-挠度曲线。图中还绘制了无孔洞的多向 FGM 板的后屈曲路径曲线。当孔洞尺寸变大时，突然翻转现象出现得更早，这是因为缺陷的积累加剧了 FGM 板的失稳倾向。增大孔洞尺寸会降低 FGM 板的刚度，导致载荷系数降低。不同孔洞尺寸下大变形阶段的变化趋势非常相似。

其次，研究长厚比对 FGM 板后屈曲行为的影响。孔洞半径设置为 1.5m。长厚比 a/h 取值为 10、20、50、100 和 200。其他参数的取值与上述算例相同。数值结果如图 3.40 所示。结果表明，随着板厚的减小，载荷系数略有增加。在大挠度后屈曲阶段，不同板厚情况下的后屈曲路径基本相同。

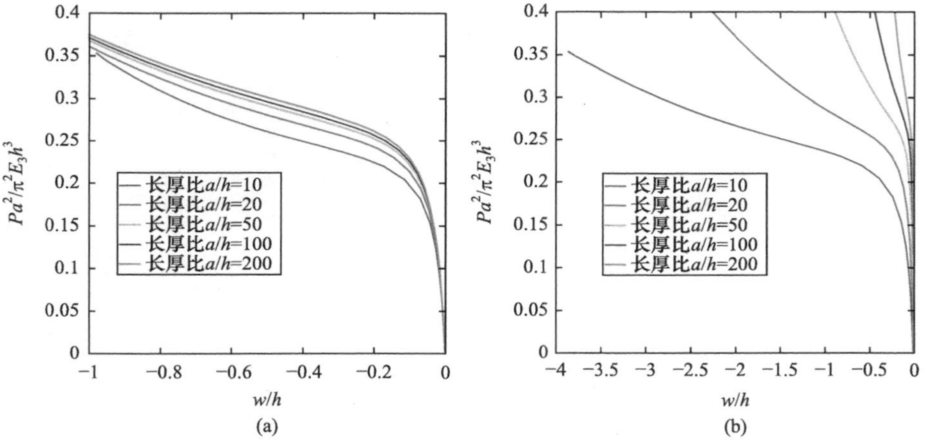

图 3.40 A 点和 B 点在不同长厚比下的后屈曲载荷-挠度曲线

(a) A 点；(b) B 点

随后，研究材料梯度指数对后屈曲响应的影响。梯度指数的变化对多向多孔 FGM 板的后屈曲行为有着重要的影响。长厚比设置为 $a/h=50$。除非特别说明，几何特性和载荷参数的选择与上述算例中的相同。图 3.41～图 3.43 分别展示了三种梯度指数变化情况下的 A 点和 B 点的载荷-挠度曲线。材料梯度指数在 0.1～10 变化，且当其中一个材料梯度指数发生变化时，其他梯度指数则设定为常数 2。

可以清楚地看到，与厚度方向的梯度指数变化相比，面内方向梯度指数的变化对后屈曲响应的影响更大，尤其是压力载荷作用的方向上 k_2 值的变化。图中数值结果表明：①从小变形阶段到大变形阶段，载荷系数随 k_1 值的增大先增大后减小；②当 k_2 的值从大于 1 变为小于 1 时，其对后屈曲响应的影响逐渐增强，甚至改变变形方向；③载荷系数随着 k_3 的值从 1～10 的变化而增加，但当 k_3 值小于 1 时，其变化趋势与随着 k_1 值的变化趋势一致。综上可知，在多个指定方向上性能发生变化的多向 FGM 板的力学响应比传统 FGM 板的响应表现得更为复杂。梯度指数 $k_2=0.5$ 下拐点 C 和 D(图 3.42(a))处的变形形状如图 3.44 所示。

此外，还探讨了边界条件对后屈曲响应的影响。板在两种边界条件(SSSS 和 CCCC)下承受单轴或双轴压力载荷作用。孔洞半径设置为 1.5m。所涉及的参数为：$a/b=1$ 、$a/h=50$ 和 $k_1=k_2=k_3=2$ 。数值结果如图 3.45 所示，从图中可以看出，CCCC 边界条件下板获得的载荷系数最大。增加约束可以提高板的极限承载能力。

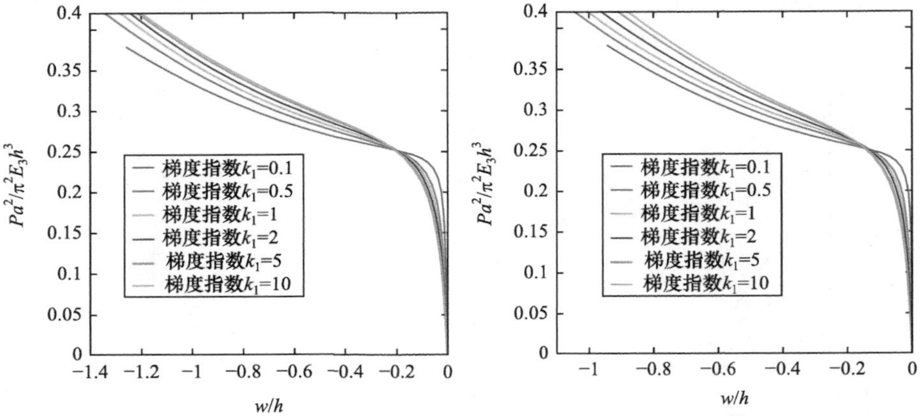

(a) (b)

图 3.41 A 点和 B 点在不同 k_1 值下的后屈曲载荷-挠度曲线

(a) A 点；(b) B 点

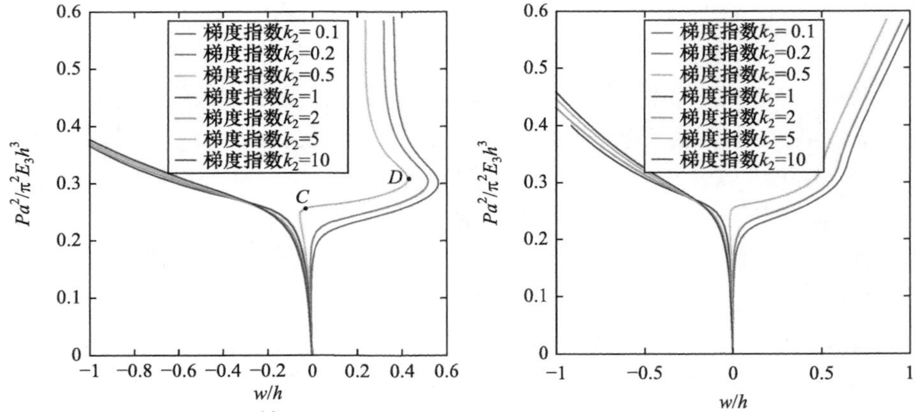

(a) (b)

图 3.42 A 点和 B 点在不同 k_2 值下的后屈曲载荷-挠度曲线

(a) A 点；(b) B 点

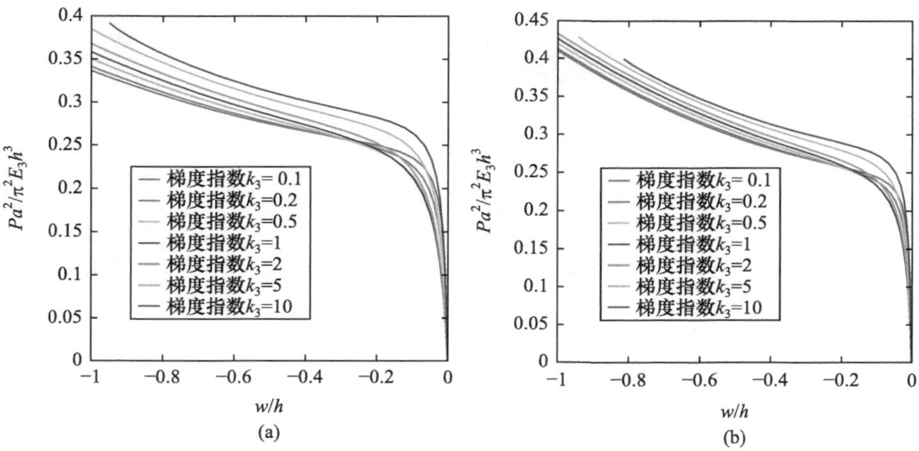

图 3.43　A 点和 B 点在不同 k_3 值下的后屈曲载荷-挠度曲线

(a) A 点；(b) B 点

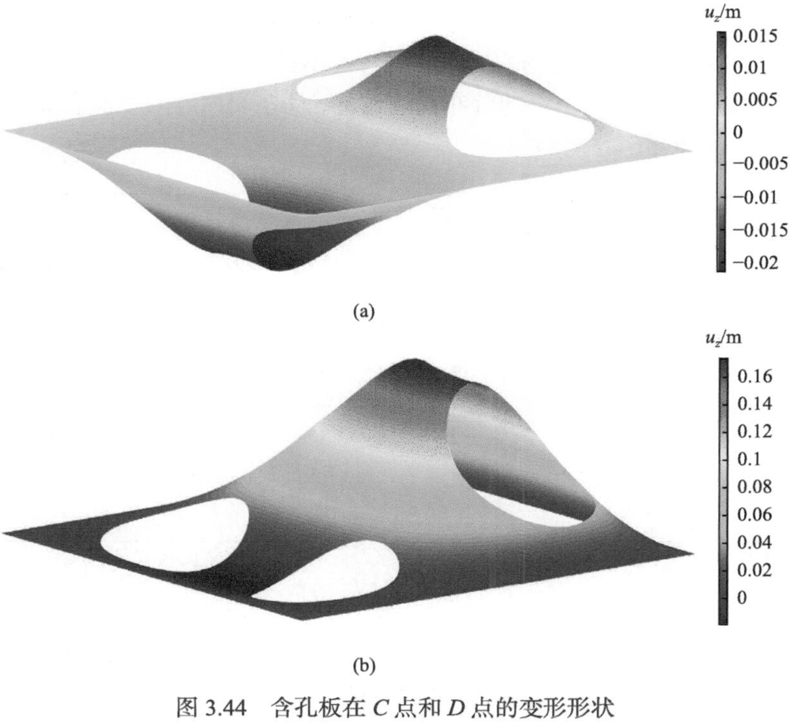

图 3.44　含孔板在 C 点和 D 点的变形形状

(a) C 点；(b) D 点

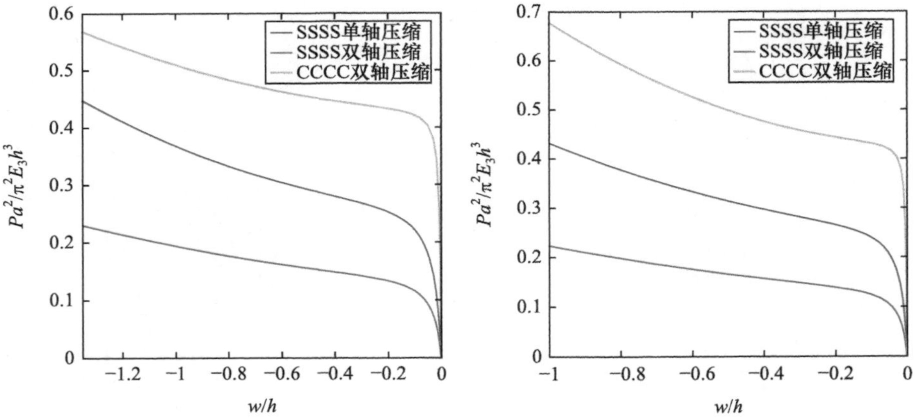

(a) (b)

图 3.45　*A* 点和 *B* 点在不同边界条件下的后屈曲载荷-挠度曲线

(a) *A* 点；(b) *B* 点

最后，比较不同面内体积分布函数对后屈曲路径的影响，结果如图 3.46 所示，可以观察到体积分布函数的影响比较轻微。A 型的载荷系数最大，而 C 型的载荷系数最小。这是由于 A 型材料的分布是连续的，且在面内方向上没有突然的变化，从而增加了板的抵抗变形能力。

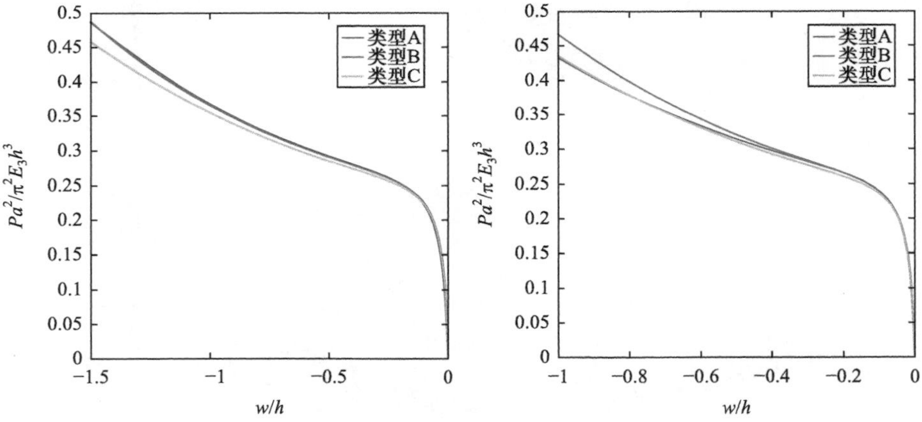

(a) (b)

图 3.46　*A* 点和 *B* 点在不同面内体积分布函数下的后屈曲载荷-挠度曲线

(a) *A* 点；(b) *B* 点

3.7.5　含多个椭圆形孔洞的方板的后屈曲分析

考虑如图 3.47 所示的具有三个椭圆形孔洞的多向 FGM 板在单轴压缩下的后屈曲响应。板的长宽尺寸为 $a=b=10\,\text{m}$，板的长厚比为 $a/h=50$。三个椭圆孔的中心坐标分别为(2.5m,2.5m)、(2.5m,7.5m)和(5m,7.5m)。椭圆长短半轴设置为 $E_a = 2\text{m}$ 和 $E_b = 1.5\text{m}$。椭圆的倾斜角度选取为 $\theta = 0°$、$\theta = 30°$、$\theta = 60°$、$\theta = 90°$、$\theta = 120°$ 和 $\theta = 150°$。材料梯度指数设置为 $k_1 = k_2 = k_3 = 2$。无量纲载荷因子定义为 $\lambda = \dfrac{Pa^2}{\pi^2 E_3 h^3}$，其中 P 是压力载荷。不同倾斜角度下，NURBS 离散网格和自适应积分子胞元如图 3.47 所示。

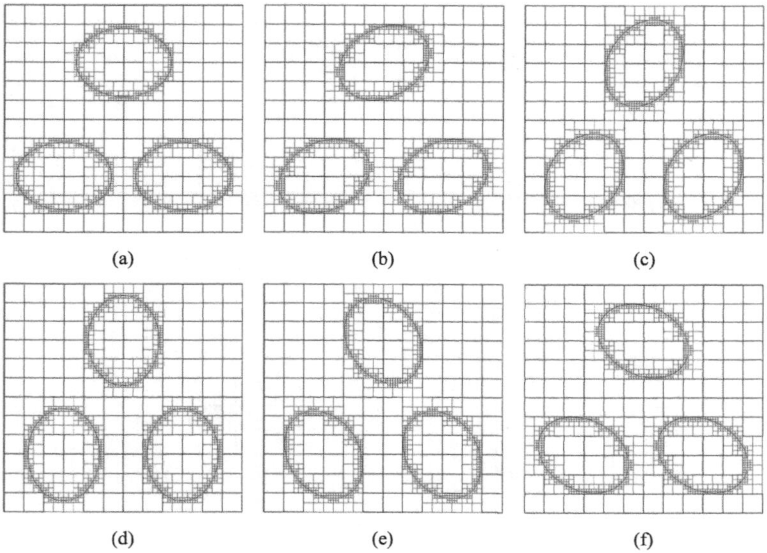

图 3.47 不同倾斜角度下，NURBS 离散网格和自适应积分子胞元

(a) 倾斜角 0°；(b) 倾斜角 30°；(c) 倾斜角 60°；(d) 倾斜角 90°；(e) 倾斜角 120°；(f) 倾斜角 150°

图 3.48(a)和(b)分别显示了 CCCC 和 SSSS 边界条件下不同椭圆孔倾斜角度下的载荷-挠度曲线。计算结果表明：①载荷系数随倾斜角度的增大而先增大后减小，当倾角达到 90°时达到最大值；②在 CCCC 边界条件下，互余角度下的载荷-挠度曲线基本一致；③在 SSSS 边界条件下，由于材料特性在这两个角度的面内方向上的分布是不对称的，因此互余角度下的载荷-挠度曲线之间存在差异。

其次，研究半长轴与半短轴比值对后屈曲行为的影响。在本例中，两个椭圆形孔洞的中心分别为(5m,2.5m)和(5m,7.5m)。几何参数为 $a = b = 10\text{m}$、$a/h = 50$ 和 $k_1 = k_2 = k_3 = 2$。在 $E_b = 1.5\text{m}$ 下，通过只改变 E_a 的值来控制半长轴与半短轴的比值，两个椭圆孔洞的比值同时改变。无量纲载荷因子定义为 $\lambda = \dfrac{Pa^2}{\pi^2 E_3 h^3}$，其中 P 是压力载荷。不同长短轴比值下，NURBS 离散网格和自适应积分子胞元如图 3.49 所示。

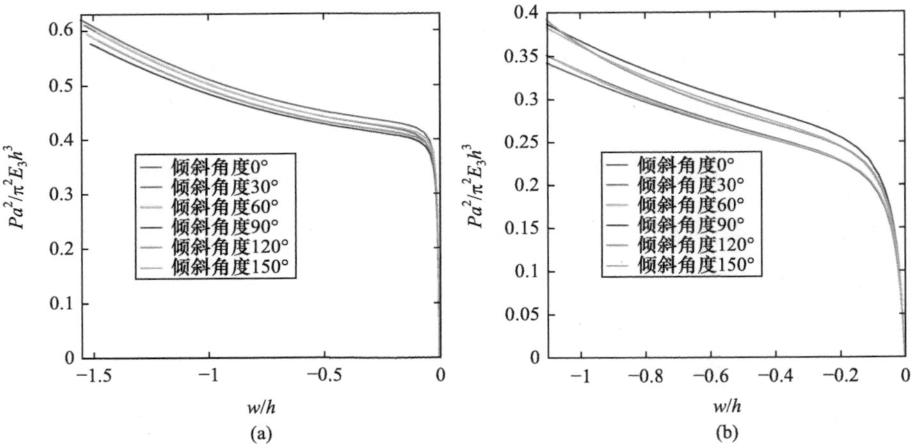

图 3.48　CCCC(a)和 SSSS(b)边界条件下，不同椭圆孔倾斜角度的后屈曲载荷-挠度曲线

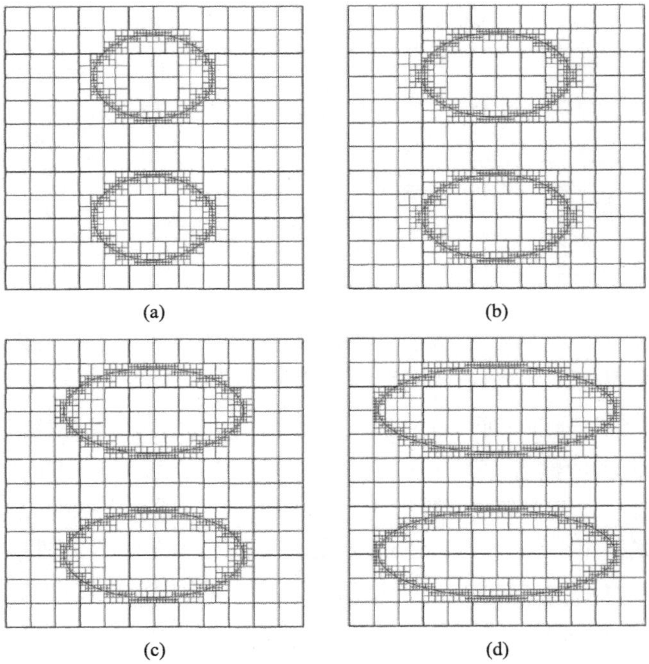

图 3.49　不同长短轴比值下，NURBS 离散网格和自适应积分子胞元

(a) 半长轴为 2m；(b) 半长轴为 2.5m；(c) 半长轴为 3m；(d) 半长轴为 4m

图 3.50(a)和(b)分别显示了在 CCCC 和 SSSS 边界条件下不同长短轴比值的后屈曲载荷-挠度曲线。数值结果表明：①随着半长轴与半短轴的比值的增加，板的刚度逐渐降低，板的承载能力也逐渐下降；②在其他参数相同的情况下，CCCC 边

界条件的载荷系数大于 SSSS 边界条件的载荷系数；③在 CCCC 边界条件下，当孔洞尺寸达到一定水平时，板的承载能力会突然下降。

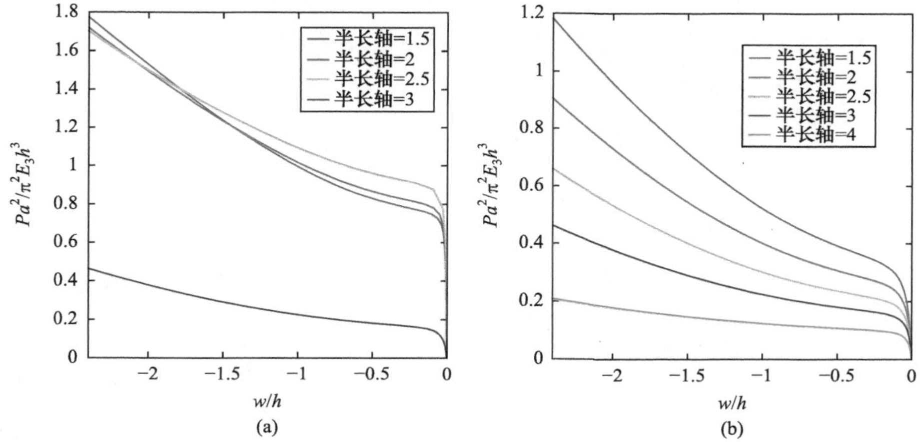

图 3.50　CCCC(a)和 SSSS(b)边界条件下，不同长短轴比值的后屈曲载荷-挠度曲线

3.7.6　含复杂孔洞的方板的后屈曲分析

这里研究含心形孔洞的多向 FGM 板的后屈曲响应，以展示 FCM 的灵活性和高效性。通常来说，传统方法分析此类结构需要使用多个 NURBS 片，而且还需处理片与片之间的连续性约束问题，这无疑增加了计算的成本和复杂度。除非另有说明，板的几何形状和材料特性的选择均与 3.7.4 节中算例的选择相同。

考虑如图 3.51(a)所示的具有心形孔洞的四边固支多向 FGM 板在单向压缩载荷作用下的后屈曲响应。无量纲载荷因子定义为 $\lambda = \dfrac{Pa^2}{\pi^2 E_3 h^3}$，其中 P 是压力载荷。

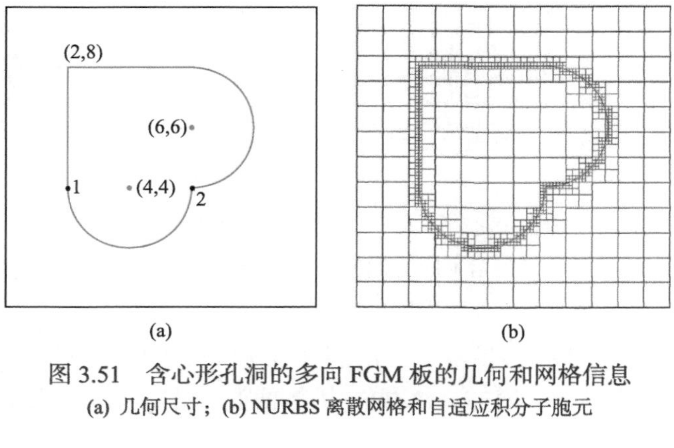

图 3.51　含心形孔洞的多向 FGM 板的几何和网格信息

(a) 几何尺寸；(b) NURBS 离散网格和自适应积分子胞元

NURBS 离散网格和自适应积分子胞元如图 3.51 (b)所示。

图 3.52～图 3.54 分别给出了所选点 1 和 2 处(图 3.51)随不同材料梯度指数变化的后屈曲载荷-挠度曲线，材料梯度指数的值均在 0.1～10 变化。当其中一个梯度指数发生变化时，其他梯度指数设定为常数 2。该模型在不同拐点 P 和 Q 处(图 3.53)的变形形状如图 3.55 所示。

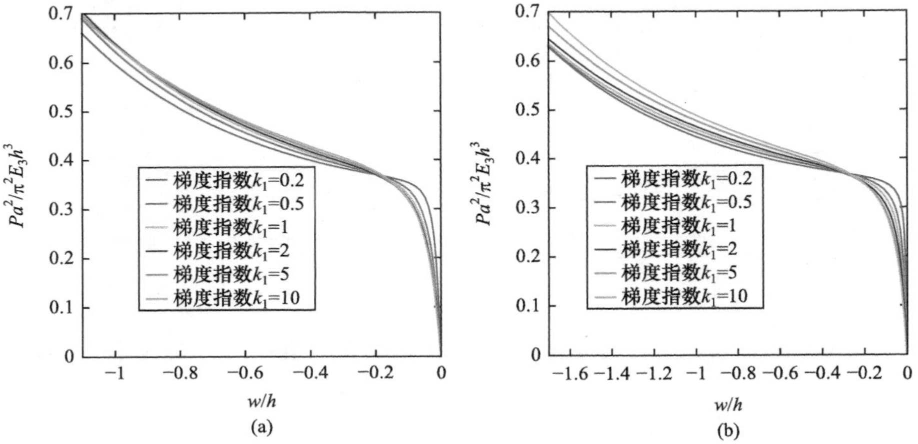

图 3.52　不同 k_1 值下的后屈曲载荷-挠度曲线

(a) 点 1；(b) 点 2

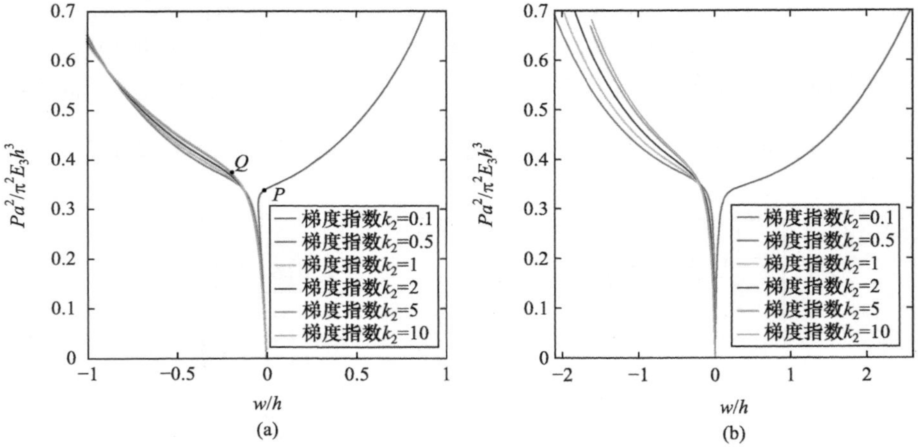

图 3.53　不同 k_2 值下的后屈曲载荷-挠度曲线

(a) 点 1；(b) 点 2

数值结果表明：①一般情况下，增加具有更大弹性模量的材料组分对应的梯度指数值时，板的刚度会增加，承载能力会增强，屈曲载荷系数也会随之增大；

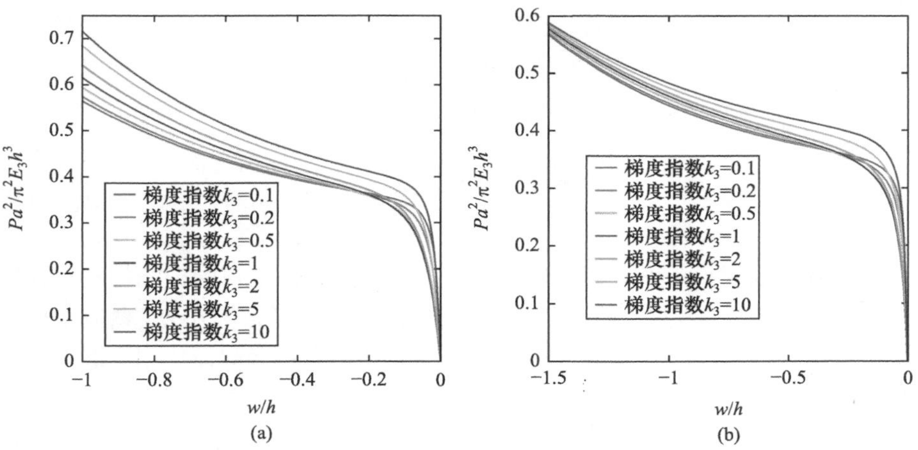

图 3.54　不同 k_3 值下的后屈曲载荷-挠度曲线

(a) 点 1；(b) 点 2

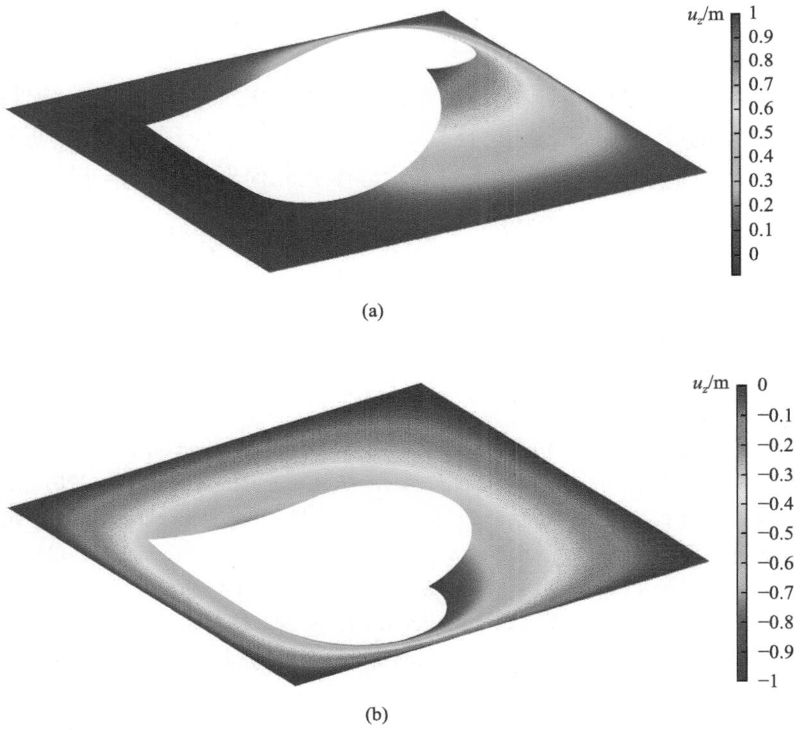

图 3.55　含心形孔洞的多向 FGM 板在 P 点和 Q 点处的变形形状

(a) P 点；(b) Q 点

②在其他因素相同的条件下，在面外方向梯度指数的改变对载荷系数的影响，大

于与面内方向相关梯度指数的改变所造成的影响。与前述算例放在一起可以总结出，孔洞形状对后屈曲路径有显著影响，因为孔洞的不同形状极大地改变了多向 FGM 板在面内方向上的材料特性分布。

3.8 小 结

本章将有限胞元法引入等几何分析中，模拟孔洞缺陷对薄壁结构动态性能的影响。首先，结合基于局部细分 PHT 样条的扩展等几何分析方法，研究了同时含裂纹和孔洞缺陷的功能梯度薄壁结构在热、力和热力耦合载荷作用下的自由振动和特征屈曲问题，并探究了材料参数、缺陷几何、结构几何和边界条件等因素对结构固有频率和临界屈曲载荷的影响；其次，结合考虑以特征屈曲分析中的无量纲一阶模态作为初始几何缺陷的弧长法公式，形成了能够有效分析含多孔多向功能梯度薄壁结构后屈曲行为的等几何算法。引入基于特征屈曲模态形状的初始几何缺陷来驱动后屈曲迭代过程，并研究了材料参数、缺陷形态和结构几何等因素对后屈曲路径的影响。

高阶连续样条函数可以自然地满足高阶板理论中对位移场 C^1 连续性的要求，并且在孔洞分割单元中采用基于分层四叉树分解的自适应积分法来逼近孔洞的几何形状，消除了需要使用多个样条片表示含多孔几何形状的复杂步骤，同时也能避免片与片之间位移场 C^1 连续性条件的复杂实施。

参 考 文 献

[1] Van Do T, Nguyen D K, Duc N D, et al. Analysis of bi-directional functionally graded plates by FEM and a new third-order shear deformation plate theory[J]. Thin-Walled Structures, 2017, 119: 687-699.

[2] Chu F, Wang L, Zhong Z, et al. Hermite radial basis collocation method for vibration of functionally graded plates with in-plane material inhomogeneity[J]. Computers & Structures, 2014, 142: 79-89.

[3] Chu F, He J, Wang L, et al. Buckling analysis of functionally graded thin plate with in-plane material inhomogeneity[J]. Engineering Analysis with Boundary Elements, 2016, 65: 112-125.

[4] Uymaz B, Aydogdu M, Filiz S. Vibration analyses of FGM plates with in-plane material inhomogeneity by Ritz method[J]. Composite Structures, 2012, 94(4): 1398-1405.

[5] Qian L F, Batra R C. Design of bidirectional functionally graded plate for optimal natural frequencies[J]. Journal of Sound and Vibration, 2005, 280(1): 415-424.

[6] Xiang T, Natarajan S, Man H, et al. Free vibration and mechanical buckling of plates with in-plane material inhomogeneity—a three-dimensional consistent approach[J]. Composite Structures, 2014, 118: 634-642.

[7] Adineh M, Kadkhodayan M. Three-dimensional thermo-elastic analysis and dynamic response of a multi-directional functionally graded skew plate on elastic foundation[J]. Composites Part B:

Engineering, 2017, 125: 227-240.

[8] Yin S H, Yu T T, Bui T Q, et al. In-plane material inhomogeneity of functionally graded plates: a higher-order shear deformation plate isogeometric analysis[J]. Composites Part B: Engineering, 2016, 106: 273-284.

[9] Lieu Q X, Lee S, Kang J, et al. Bending and free vibration analyses of in-plane bi-directional functionally graded plates with variable thickness using isogeometric analysis[J]. Composite Structures, 2018, 192: 434-451.

[10] Farzam A, Hassani B. Isogeometric analysis of in-plane functionally graded porous microplates using modified couple stress theory[J]. Aerospace Science and Technology, 2019, 91: 508-524.

[11] Yin S H, Yu T T, Bui T Q, et al. Rotation-free isogeometric analysis of functionally graded thin plates considering in-plane material inhomogeneity[J]. Thin-Walled Structures, 2017, 119: 385-395.

[12] Lieu Q X, Lee D, Kang J, et al. NURBS-based modeling and analysis for free vibration and buckling problems of in-plane bi-directional functionally graded plates[J]. Mechanics of Advanced Materials and Structures, 2019, 26(12): 1064-1080.

[13] Xue Y, Jin G, Ding H, et al. Free vibration analysis of in-plane functionally graded plates using a refined plate theory and isogeometric approach[J]. Composite Structures, 2018, 192: 193-205.

[14] Thai S, Thai H T, Vo T P, et al. Post-buckling of functionally graded microplates under mechanical and thermal loads using isogeomertic analysis[J]. Engineering Structures, 2017, 150: 905-917.

[15] Nguyen T N, Thai C H, Nguyen-Xuan H, et al. NURBS-based analyses of functionally graded carbon nanotube-reinforced composite shells[J]. Composite Structures, 2018, 203: 349-360.

[16] Nguyen T N, Thai C H, Luu A T, et al. NURBS-based postbuckling analysis of functionally graded carbon nanotube-reinforced composite shells[J]. Computer Methods in Applied Mechanics and Engineering, 2019, 347: 983-1003.

[17] Tran L V, Ly H A, Lee J, et al. Vibration analysis of cracked FGM plates using higher-order shear deformation theory and extended isogeometric approach[J]. International Journal of Mechanical Sciences, 2015, 96-97: 65-78.

[18] Ghatage P S, Kar V R, Sudhagar P E. On the numerical modelling and analysis of multi-directional functionally graded composite structures: a review[J]. Composite Structures, 2020, 236: 111837.

[19] Natarajan S, Baiz P M, Ganapathi M, et al. Linear free flexural vibration of cracked functionally graded plates in thermal environment[J]. Computers & Structures, 2011, 89(15): 1535-1546.

[20] Reddy J N. Analysis of functionally graded plates[J]. International Journal for Numerical Methods in Engineering, 2000, 47(1-3): 663-684.

[21] Valizadeh N, Natarajan S, Gonzalez-Estrada O A, et al. NURBS-based finite element analysis of functionally graded plates: static bending, vibration, buckling and flutter[J]. Composite Structures, 2013, 99: 309-326.

[22] Yang H S, Dong C Y, Qin X C, et al. Vibration and buckling analyses of FGM plates with multiple internal defects using XIGA-PHT and FCM under thermal and mechanical loads[J]. Applied Mathematical Modelling, 2020, 78: 433-481.

[23] Ruess M, Schillinger D, Özcan A I, et al. Weak coupling for isogeometric analysis of non-matching

and trimmed multi-patch geometries[J]. Computer Methods in Applied Mechanics and Engineering, 2014, 269: 46-71.

[24] Tran L V, Phung-Van P, Lee J, et al. Isogeometric analysis for nonlinear thermomechanical stability of functionally graded plates[J]. Composite Structures, 2016, 140: 655-667.

[25] Crisfield M A. A fast incremental/iterative solution procedure that handles "snap-through"[J]. Computers & Structures, 1981, 13(1): 55-62.

[26] De Borst R, Crisfield M A, Remmers J J C, et al. Non-linear Finite Element Analysis of Solids and Structures[M]. 2nd ed. New York: John Wiley & Sons, Inc, 2012.

[27] Huang C S, Mcgee O G, Chang M J. Vibrations of cracked rectangular FGM thick plates[J]. Composite Structures, 2011, 93(7): 1747-1764.

[28] Natarajan S, Baiz P M, Bordas S, et al. Natural frequencies of cracked functionally graded material plates by the extended finite element method[J]. Composite Structures, 2011, 93(11): 3082-3092.

[29] Liu S, Yu T, Van Lich L, et al. Size effect on cracked functional composite micro-plates by an XIGA-based effective approach[J]. Meccanica, 2018, 53(10): 2637-2658.

[30] Yin S H, Yu T T, Bui T Q, et al. Buckling and vibration extended isogeometric analysis of imperfect graded Reissner-Mindlin plates with internal defects using NURBS and level sets[J]. Computers & Structures, 2016, 177: 23-38.

[31] Huang C S, Yang P J, Chang M J. Three-dimensional vibration analyses of functionally graded material rectangular plates with through internal cracks[J]. Composite Structures, 2012, 94(9): 2764-2776.

[32] Tan P, Nguyen-Thanh N, Zhou K. Extended isogeometric analysis based on Bézier extraction for an FGM plate by using the two-variable refined plate theory[J]. Theoretical and Applied Fracture Mechanics, 2017, 89: 127-138.

[33] Yin S H, Hale J S, Yu T T, et al. Isogeometric locking-free plate element: a simple first order shear deformation theory for functionally graded plates[J]. Composite Structures, 2014, 118: 121-138.

[34] Yu T T, Bui T Q, Yin S H, et al. On the thermal buckling analysis of functionally graded plates with internal defects using extended isogeometric analysis[J]. Composite Structures, 2016, 136: 684-695.

[35] Yamaki N. Postbuckling behavior of rectangular plates with small initial curvature loaded in edge compression[J]. Journal of Applied Mechanics, 2021, 26(3): 407-414.

[36] Yu T T, Yin S H, Bui T Q, et al. Buckling isogeometric analysis of functionally graded plates under combined thermal and mechanical loads [J]. Composite Structures, 2017, 162: 54-69.

第4章

线性黏弹性材料的等几何有限元分析

4.1 引　言

实际工程中，有两类众所周知的材料元件：弹性固体和黏性流体。弹性固体具有确定的构型，在静载荷作用下发生形变且与时间无关，卸除外力后可以完全恢复原状。从能量的角度来说，外力在弹性体变形时所做的功全部以弹性势能的方式储存，且能在删除外载荷过程中释放。黏性流体没有确定的形状，或其形状由容器决定，外力作用下形变随时间而发展，产生不可逆的流动。实际工程材料通常同时具有弹性和黏性两种不同机理的形变，综合地体现黏性流体和弹性固体两者的特性，材料的这种性质称为黏弹性。

目前，分析黏弹性力学问题的方法主要有解析法、实验法和数值法。对于几何形状简单的线性问题，我们可以通过解析法进行分析。但是，解析法适用范围极其有限。此外，实验法由于受到试验技术水平限制，不仅难以模拟结构的真实应力状态，而且可测得的试验结果也往往非常有限，难以有效地解决复杂三维应力状态下的结构分析问题。随着计算机技术的不断发展，数值法的优势在非均质材料和复合结构中得到了较为广泛的应用，尤其是在求解非线性问题方面，已成为不可或缺的研究手段[1]。

FEM 是一种在计算机辅助工程领域发展较为成熟且有效的数值分析方法，可以用于模拟和分析各种材料的力学行为，其基本思想是将连续物体模型近似离散化为有限数量的单元,在每个单元内部通过变分原理建立数学模型以描述单元的物理行为(如应力、位移、温度等)，将偏微分方程的边界值问题转换为有限单元集合的代数方程组进行求解，并通过各单元之间的相互作用将各单元的数值解进行组合近似得到整个系统的全局性质。FEM 的离散化特点使得它在处理具有复杂几何结构和边界条件的非线性问题时有较高的精度和灵活性，已被广泛应用于结构力学[2]、传热分析[3]、电磁场[4]、流体力学[5]等多个学科领域，是现代工程中最重要的设计与分析工具之一。现在市场上的商业仿真软件 ABAQUS、ANSYS、COMSOL、NASTRAN 等均是基于 FEM 理论发展而来，它们可以帮助工程师模拟各种复杂的现象，分析不同的工程设计方案，从而优化设计并减少试验，节约成本并提高工程效率。

FEM 在黏弹性问题的计算过程中，以位移为未知量建立求解方程组，其中记忆应力可转换为位移的表达式，由单元刚度矩阵叠加获得整体刚度矩阵并结合等效节点载荷，求解出每一时刻域内所有单元节点的位移，然后在时间域内进行迭代求解，在计算过程中省略了位移-应变-应力的转换过程，最后再根据几何方程与本构关系分别得到应变与应力。FEM 在黏弹性问题的迭代求解过程中所需的变量较少且易于获得，同时 FEM 形成对称正定的带状稀疏系数矩阵，保证了计算的稳定性与高效率，且在施加体力、温度等载荷时操作简单，所以 FEM 在研究具有复杂材料结构和本构行为的黏弹性材料时具有特别显著的优势，已广泛应用于各类复杂的线性与非线性黏弹性力学问题。Srinatha 等[6]假设黏弹性材料是各向同性的、均匀的和线性的，其应力-应变规律以遗传积分的形式表示，并且认为黏弹性材料为热流变简单材料，然后推导出线性热黏性本构的 FEM 离散公式，并对线性热黏弹材料平面问题进行了应力分析。Shen 等[7]基于完全拉格朗日和更新拉格朗日方法推导出了非线性黏弹性大变形增量变分方程，并建立了三维问题的 FEM 离散公式和编写了求解程序。Zocher 等[8]推导出基于 FEM 的热黏弹性本构三维增量形式，并将其用于求解线性黏弹性介质经历热变形与机械变形时非耦合静态初始边值问题。Pavan 等[9]构造了一种适用于正交各向异性复合材料的连续损伤非线性黏弹性本构模型，并将其离散为 FEM 增量形式进行计算。Hinterhoelzl 等[10]考虑了黏弹性本构方程的时间依赖性，实现了具有损伤增长的颗粒复合材料三维黏弹性模型的 FEM 应用。Nguyen 等[11]采用 FEM 和拉普拉斯变换分析了线性黏弹性复合材料层压板在静态载荷和谐波载荷作用下的力学行为变化规律。

虽然 FEM 分析方法发展成熟且应用广泛，但是传统 FEM 在对力学问题进行模拟分析时仍存在一些亟待解决的难题。例如，FEM 构建分析模型时，通常采用网格近似几何模型，计算精度严重依赖离散化网格的质量，并且在离散复杂模型的过程中容易出现网格畸变和不连续问题，从而导致较大的离散化误差。另外，由于复杂模型细节信息增多，预处理和求解过程将耗费大量的计算资源和时间，因此在进行高精度的结构分析时，需要使用更加先进的数值计算方法来避免上述问题。2005 年，Hughes 及其合作者[12]首次提出的等几何分析(IGA)，有效解决了传统 FEM 在建立模型、网格划分与分析计算过程中存在的离散误差大、计算效率低等问题。

IGA 旨在将设计、分析和可视化进行集成，其根本思想是将能够精确描述几何形状特征的 NURBS 基函数作为数值计算的形函数，将用于模拟计算的分析模型建立在精确的几何模型上，从而提高计算精度。同时，统一的数学表达方式打破了传统数值分析中设计与分析分离的局面，实现了几何模型与分析模型的无缝集成，因此避免了不同几何表达之间数据的频繁交换和繁杂的网格生成而导致的计算成本的增加[13]。网格细分策略简单且不损失几何模型精度等特点有效地提高了 IGA 的计算效率与精度。正如绪论中所述，IGA 已经被广泛研究并成功应用于多个研究领域[14-19]。

本章首先介绍等几何分析的一些基本知识，然后给出线性黏弹性材料的等几何有限元法(IGAFEM)，最后通过几个数值算例验证本章给出的等几何有限元法的有效性。

4.2　等几何分析

B 样条与 NURBS 是构建几何造型的两种常见的曲线曲面插值方法。一个 NURBS 就是一个有理 B 样条，其能够以更加精确的形式表达自由且复杂的参数化曲线与曲面，具有更高的精度、稳定性和灵活性，在计算机图形领域与 CAD 系统中有着重要意义且应用广泛。IGA 无缝集成了基于 NURBS 基函数的几何模型和分析模型，通过设计和分析一体化的思想有效提高了计算精度和效率。

本章研究采用基于 IGA 框架的 FE 算法，其中 NURBS 基函数是 IGA 的重要理论基础。本节简要介绍与 B 样条和 NURBS 曲线曲面相关的理论基础知识与数学表达定义，更加详细的理论介绍与公式推导可参考文献[20]。

B 样条曲线本质上是分段多项式函数，由基函数和控制点的线性组合构成，这些基函数定义在给定参数区间上的节点向量 $\boldsymbol{\Phi}$ 上，该参数区间在参数空间中为单调非递减实数序列：

$$\boldsymbol{\Phi}=\left\{\phi_1,\phi_2,\cdots,\phi_{n+p+1}\right\},\quad \phi_i\in\mathbb{R}\text{和}\phi_1\leqslant\phi_2\leqslant\cdots\leqslant\phi_{n+p+1} \tag{4.1}$$

其中，下标 $i(=1,2,\cdots,n+p+1)$ 、n 和 p 分别表示节点编号、基函数个数和曲线多项式阶次。

B 样条基函数在节点处的连续性是可控的。如果某一个节点重复度为 m ，则该节点处的样条函数的连续性为 $p-m$ ，意味着样条函数在该处具有高达 $p-m$ 阶的连续导数。节点向量中任意相邻节点所组成的区间称为节点区间，如果节点向量中的首尾节点具有 $p+1$ 个重复度，即 $\phi_1=\cdots=\phi_{p+1}<\phi_{p+2}\leqslant\cdots\leqslant\phi_n<\phi_{n+1}=\cdots=\phi_{n+p+1}$ ，则该节点向量称为 p 阶开放节点向量。

第 i 个 p 阶 B 样条基函数 $N_i^p(\xi)$ 定义如下[12,13]：

$$\begin{aligned}
&p=0:\ N_i^0(\phi)=\begin{cases}1, & \phi_i\leqslant\phi<\phi_{i+1}\\ 0, & \text{其他}\end{cases}\\
&p>0:\ N_i^p(\phi)=\frac{\phi-\phi_i}{\phi_{i+p}-\phi_i}N_i^{p-1}(\phi)+\frac{\phi_{i+p+1}-\phi}{\phi_{i+p+1}-\phi_{i+1}}N_{i+1}^{p-1}(\phi)
\end{aligned} \tag{4.2}$$

上述公式称为 Cox-de Boor 递归公式，其关于参数坐标 ϕ 的一阶导数公式为

$$N_i'^{p}(\phi)=\frac{p}{\phi_{i+p}-\phi_i}N_i^{p-1}(\phi)+\frac{p}{\phi_{i+p+1}-\phi_{i+1}}N_{i+1}^{p-1}(\phi) \tag{4.3}$$

B 样条函数作为基函数具有以下理想特性：①非负性，$N_i^p(\phi)\geqslant 0$；②局部支撑性，$N_i^p(\phi)$存在于区间$[\phi_i,\phi_{i+p+1})$；③规范性，在区间$[\phi_i,\phi_{i+1}]$内$\sum\limits_{a=i-p}^{i}N_a^p(\phi)=1$；④兼备 Bezier 曲线几何不变性、仿射不变性等优点。

给定 B 样条基函数$N_i^p(\xi)$和相关的控制点$\boldsymbol{P}_i(i=0,1,\cdots,n)$后，分段多项式 B 样条曲线被定义为

$$\boldsymbol{C}(\phi)=\sum_{i=1}^{n}N_i^p(\phi)\boldsymbol{P}_i \tag{4.4}$$

这意味着，只需修改控制点$\boldsymbol{P}_i$的位置即可改变区间$[\phi_i,\phi_{i+p+1})$对应的曲线形状。NURBS 是 B 样条的一种拓展形式，在构建 B 样条曲线的要素中再引入与控制点一一对应的权重参数，相较于 B 样条能够更加灵活地控制曲线， NURBS 曲线的数学表达为

$$\boldsymbol{C}(\phi)=\sum_{i=1}^{n}R_i^p(\phi)\boldsymbol{P}_i \tag{4.5}$$

$$R_i^p(\phi)=\frac{N_i^p(\phi)w_i}{\sum\limits_{i'=1}^{n}N_{i'}^p(\phi)w_{i'}} \tag{4.6}$$

其中，$\boldsymbol{C}(\phi)$为 NURBS 曲线；$R_i^p(\phi)$为 NURBS 基函数；w_i是控制点$\boldsymbol{P}_i$的权值。

NURBS 实体是三元 NURBS 基函数$R_{i,j,k}^{p,q,r}(\phi,\eta,\zeta)$与控制点$\boldsymbol{P}_{i,j,k}$的线性组合，由式(4.6)类推可得到，三元变量 NURBS 基函数为

$$R_{i,j,k}^{p,q,r}(\phi,\eta,\zeta)=\frac{N_i^p(\phi)M_j^q(\eta)L_k^r(\zeta)w_{i,j,k}}{\sum\limits_{i'=1}^{n}\sum\limits_{j'=1}^{m}\sum\limits_{k'=1}^{l}N_{i'}^p(\phi)M_{j'}^q(\eta)L_{k'}^r(\zeta)w_{i',j',k'}} \tag{4.7}$$

其中，$w_{i,j,k}$是控制点$\boldsymbol{P}_{i,j,k}$的正权重；$N_i^p(\phi)$、$M_j^q(\eta)$和$L_k^r(\zeta)$分别为三个不同节点向量$\boldsymbol{\Phi}$、$\boldsymbol{\Psi}$和$\boldsymbol{\Xi}$上的单变量 B 样条基函数：

$$\boldsymbol{\Phi}=\left\{\underbrace{0,\cdots,0}_{p+1},\phi_{p+2},\cdots,\phi_n,\underbrace{1,\cdots,1}_{p+1}\right\} \tag{4.8}$$

$$\boldsymbol{\Psi}=\left\{\underbrace{0,\cdots,0}_{q+1},\eta_{q+2},\cdots,\eta_m,\underbrace{1,\cdots,1}_{q+1}\right\} \tag{4.9}$$

$$\boldsymbol{\Xi}=\left\{\underbrace{0,\cdots,0}_{r+1},\zeta_{r+2},\cdots,\zeta_l,\underbrace{1,\cdots,1}_{r+1}\right\} \tag{4.10}$$

则在三维空间中，NURBS 实体可定义为

$$\Omega(\phi,\eta,\zeta)=\sum_{i=1}^{n}\sum_{j=1}^{m}\sum_{k=1}^{l}R_{i,j,k}^{p,q,r}(\phi,\eta,\zeta)\boldsymbol{P}_{i,j,k} \tag{4.11}$$

其中，$\boldsymbol{P}_{i,j,k}$ 形成数量为 $n\times m\times l$ 的三维控制点网格；n、m 和 l 分别为对应 ϕ、η 和 ζ 三个参数方向上的 NURBS 基函数数量；p、q 和 r 分别为相应基函数多项式的阶次。

因此，三元 NURBS 基函数关于参数坐标 ϕ、η、ζ 的一阶导数分别表示为

$$\frac{\partial R_{i,j,k}^{p,q,r}(\phi,\eta,\zeta)}{\partial\phi}=\frac{N_i'^{p}(\phi)M_j^{q}(\eta)L_k^{r}(\zeta)w_{i,j,k}-R_{i,j,k}^{p,q,r}\left[\sum_{i'=1}^{n}\sum_{j'=1}^{m}\sum_{k'=1}^{l}N_{i'}'^{p}(\phi)M_{j'}^{q}(\eta)L_{k'}^{r}(\zeta)w_{i',j',k'}\right]}{\sum_{i'=1}^{n}\sum_{j'=1}^{m}\sum_{k'=1}^{l}N_{i'}^{p}(\phi)M_{j'}^{q}(\eta)L_{k'}^{r}(\zeta)w_{i',j',k'}} \tag{4.12}$$

$$\frac{\partial R_{i,j,k}^{p,q,r}(\phi,\eta,\zeta)}{\partial\eta}=\frac{N_i^{p}(\phi)M_j'^{q}(\eta)L_k^{r}(\zeta)w_{i,j,k}-R_{i,j,k}^{p,q,r}\left[\sum_{i'=1}^{n}\sum_{j'=1}^{m}\sum_{k'=1}^{l}N_{i'}^{p}(\phi)M_{j'}'^{q}(\eta)L_{k'}^{r}(\zeta)w_{i',j',k'}\right]}{\sum_{i'=1}^{n}\sum_{j'=1}^{m}\sum_{k'=1}^{l}N_{i'}^{p}(\phi)M_{j'}^{q}(\eta)L_{k'}^{r}(\zeta)w_{i',j',k'}} \tag{4.13}$$

$$\frac{\partial R_{i,j,k}^{p,q,r}(\phi,\eta,\zeta)}{\partial\zeta}=\frac{N_i^{p}(\phi)M_j^{q}(\eta)L_k'^{r}(\zeta)w_{i,j,k}-R_{i,j,k}^{p,q,r}\left[\sum_{i'=1}^{n}\sum_{j'=1}^{m}\sum_{k'=1}^{l}N_{i'}^{p}(\phi)M_{j'}^{q}(\eta)L_{k}'^{r}(\zeta)w_{i',j',k'}\right]}{\sum_{i'=1}^{n}\sum_{j'=1}^{m}\sum_{k'=1}^{l}N_{i'}^{p}(\phi)M_{j'}^{q}(\eta)L_{k'}^{r}(\zeta)w_{i',j',k'}} \tag{4.14}$$

其中，B 样条基函数一阶导数 $N_i'^{p}(\phi)$、$M_j'^{q}(\eta)$、$L_k'^{r}(\zeta)$ 由等式(4.3)计算得到。

与传统有限元方法不同，等几何分析方法采用等参变换思想来精确构建几何模型，而未知的物理场变量则用 NURBS 来描述，所研究的物理域与参数域为映射对应关系。以二维问题为例，根据参数空间到物理空间的映射关系可知，任意物理变量可以表示为

$$\boldsymbol{u}(\boldsymbol{x},\boldsymbol{y})=\sum_{i=1}^{n}R_i(\phi,\eta)\boldsymbol{u}_i \tag{4.15}$$

其中，$R_i(\phi,\eta)$ 为基函数；ϕ 与 η 为参数坐标；$\boldsymbol{u}_i$ 表示控制点处的变量；n 为单元的控制点个数。

物理空间与参数空间的映射过程需要满足一一对应关系，即需建立不同坐标系之间的转换关系，因此引入雅可比转换矩阵表示积分参数的转换。在参数空间 $(\phi,\eta)\in[\phi_a,\phi_{a+1}]\times[\eta_b,\eta_{b+1}]$，则有

$$\begin{Bmatrix} \dfrac{\partial}{\partial\phi} \\ \dfrac{\partial}{\partial\eta} \end{Bmatrix} = \begin{bmatrix} \dfrac{\partial x}{\partial\phi} & \dfrac{\partial y}{\partial\phi} \\ \dfrac{\partial x}{\partial\eta} & \dfrac{\partial y}{\partial\eta} \end{bmatrix} \begin{Bmatrix} \dfrac{\partial}{\partial x} \\ \dfrac{\partial}{\partial y} \end{Bmatrix} = \hat{\boldsymbol{J}} \begin{Bmatrix} \dfrac{\partial}{\partial x} \\ \dfrac{\partial}{\partial y} \end{Bmatrix} \tag{4.16}$$

其中，$\hat{\boldsymbol{J}}$ 为物理空间变换为参数空间的雅可比矩阵，定义为

$$\hat{\boldsymbol{J}} = \hat{\boldsymbol{J}}(\phi,\eta) = \begin{bmatrix} \dfrac{\partial x}{\partial\phi} & \dfrac{\partial y}{\partial\phi} \\ \dfrac{\partial x}{\partial\eta} & \dfrac{\partial y}{\partial\eta} \end{bmatrix} \tag{4.17}$$

由于等几何分析过程相较于传统有限元分析增加了参数空间的描述，因此在转换过程中存在物理空间与参数空间、参数空间与高斯空间的两次转换。

假设高斯空间为 $(\tilde{\phi},\tilde{\eta})\in[-1,1]\times[-1,1]$，高斯空间转换为参数空间的雅可比转换矩阵为

$$\left|\tilde{\boldsymbol{J}}\right| = \frac{(\phi_{a+1}-\phi_a)(\eta_{b+1}-\eta_b)}{4} \tag{4.18}$$

高斯空间与参数空间的映射关系满足下列等式：

$$\begin{aligned} \phi &= \frac{(\phi_{a+1}-\phi_a)\tilde{\phi}}{2} + \frac{(\phi_{a+1}+\phi_a)}{2} \\ \eta &= \frac{(\eta_{a+1}-\eta_a)\tilde{\eta}}{2} + \frac{(\eta_{a+1}+\eta_a)}{2} \end{aligned} \tag{4.19}$$

因此，函数 $f(x,y)$ 在物理空间的积分表达式可表示为

$$\begin{aligned} \int_\Omega f(x,y)\mathrm{d}\Gamma &= \sum_{e=1}^{N_e}\int_{\Omega_e} f(x,y)\mathrm{d}\Omega_e \\ &= \sum_{e=1}^{N_e}\int_{\hat{\Omega}_e} f[x(\phi,\eta),y(\phi,\eta)]\left|\hat{\boldsymbol{J}}(\phi,\eta)\right|\mathrm{d}\hat{\Omega}_e \\ &= \sum_{e=1}^{N_e}\int_{\tilde{\Omega}_e} f\left[x(\tilde{\phi},\tilde{\eta}),y(\tilde{\phi},\tilde{\eta})\right]\left|\hat{\boldsymbol{J}}(\phi,\eta)\right|\left|\hat{\boldsymbol{J}}(\tilde{\phi},\tilde{\eta})\right|\mathrm{d}\tilde{\Omega}_e \end{aligned} \tag{4.20}$$

其中，Ω 为积分物理域；Ω_e、$\hat{\Omega}_e$ 和 $\tilde{\Omega}_e$ 分别表示几何离散后的物理域、参数域

以及高斯域；N_e 为离散单元数量。

4.3　黏弹性问题的等几何有限元法

当不考虑温度时，三维各向同性线性黏弹性本构模型的积分形式可以写成如下 Stieltjes 卷积积分形式[21]：

$$\sigma_{il}(t)=\delta_{il}\int_{-\infty}^{t}\lambda(t-\tau)\frac{\partial\varepsilon_{nn}(\boldsymbol{x},\tau)}{\partial\tau}\mathrm{d}\tau+2\int_{-\infty}^{t}G(t-\tau)\frac{\partial\varepsilon_{il}(\boldsymbol{x},\tau)}{\partial\tau}\mathrm{d}\tau \tag{4.21}$$

其中，重复下标 n 遵循爱因斯坦求和约定；$\lambda=K-\dfrac{2}{3}G$，这里 G 为剪切松弛函数，K 为松弛函数体积。式(4.21)可以改写为

$$\begin{aligned}\sigma_{il}(t)=\delta_{il}\int_{-\infty}^{t}&\left[K(t-\tau)-\frac{2}{3}G(t-\tau)\right]\frac{\partial\varepsilon_{nn}(\boldsymbol{x},\tau)}{\partial\tau}\mathrm{d}\tau\\&+2\int_{-\infty}^{t}G(t-\tau)\frac{\partial\varepsilon_{il}(\boldsymbol{x},\tau)}{\partial\tau}\mathrm{d}\tau\end{aligned} \tag{4.22}$$

线性黏弹性材料的应力可分为弹性应力 σ_{il}^{e} 和记忆应力 σ_{il}^{m}：

$$\sigma_{il}=\sigma_{il}^{\mathrm{e}}-\sigma_{il}^{\mathrm{m}} \tag{4.23}$$

其中，

$$\sigma_{il}^{\mathrm{e}}(t)=\delta_{il}\left[K(0)-\frac{2}{3}G(0)\right]\varepsilon_{nn}(t)+2G(0)\varepsilon_{il}(t) \tag{4.24}$$

$$\begin{aligned}\sigma_{il}^{\mathrm{m}}(t)=2\int_{0}^{t}\frac{\partial G(t-\tau)}{\partial\tau}\varepsilon_{il}(\boldsymbol{x},\tau)\mathrm{d}\tau+\delta_{il}\int_{0}^{t}\frac{\partial K(t-\tau)}{\partial\tau}\varepsilon_{nn}(\boldsymbol{x},\tau)\mathrm{d}\tau\\-\delta_{il}\frac{2}{3}\int_{0}^{t}\frac{\partial G(t-\tau)}{\partial\tau}\varepsilon_{nn}(\boldsymbol{x},\tau)\mathrm{d}\tau\end{aligned} \tag{4.25}$$

其中，$K(0)$ 和 $G(0)$ 分别表示弹性状态下的体积模量和剪切模量。

式(4.21)可以写成如下矩阵形式：

$$\boldsymbol{\sigma}=\boldsymbol{D}_1\boldsymbol{\varepsilon}+\boldsymbol{L}_1\boldsymbol{\varepsilon}+\boldsymbol{L}_2\boldsymbol{\varepsilon} \tag{4.26}$$

其中，

$$\boldsymbol{\sigma}=\{\sigma_{11},\sigma_{22},\sigma_{33},\sigma_{12},\sigma_{23},\sigma_{31}\}^{\mathrm{T}} \tag{4.27}$$

$$\boldsymbol{\varepsilon}=\{\varepsilon_{11},\varepsilon_{22},\varepsilon_{33},\gamma_{12},\gamma_{23},\gamma_{31}\}^{\mathrm{T}} \tag{4.28}$$

此外，$\boldsymbol{D}_1$ 称为瞬态弹性矩阵；$\boldsymbol{L}_1$ 和 $\boldsymbol{L}_2$ 为积分算子矩阵。其具体形式如下：

$$
\boldsymbol{D}_1=\begin{bmatrix}
K(0)+\frac{4}{3}G(0) & K(0)-\frac{2}{3}G(0) & K(0)-\frac{2}{3}G(0) & 0 & 0 & 0\\
 & K(0)+\frac{4}{3}G(0) & K(0)-\frac{2}{3}G(0) & 0 & 0 & 0\\
 & & K(0)+\frac{4}{3}G(0) & 0 & 0 & 0\\
 & & & G(0) & 0 & 0\\
 & \text{对称} & & & G(0) & 0\\
 & & & & & G(0)
\end{bmatrix} \tag{4.29}
$$

$$
\boldsymbol{L}_1=\boldsymbol{D}_2\int_{0^+}^{t}\frac{\partial G(t-\tau)}{\partial \tau}\mathrm{d}\tau \tag{4.30}
$$

$$
\boldsymbol{L}_2=\boldsymbol{D}_3\int_{0^+}^{t}\frac{\partial K(t-\tau)}{\partial \tau}\mathrm{d}\tau \tag{4.31}
$$

其中，

$$
\boldsymbol{D}_2=\begin{bmatrix}
-\frac{4}{3} & \frac{2}{3} & \frac{2}{3} & 0 & 0 & 0\\
 & -\frac{4}{3} & \frac{2}{3} & 0 & 0 & 0\\
 & & -\frac{4}{3} & 0 & 0 & 0\\
 & \text{对称} & & -1 & 0 & 0\\
 & & & & -1 & 0\\
 & & & & & -1
\end{bmatrix} \tag{4.32}
$$

$$
\boldsymbol{D}_3=\begin{bmatrix}
-1 & -1 & -1 & 0 & 0 & 0\\
 & -1 & -1 & 0 & 0 & 0\\
 & & -1 & 0 & 0 & 0\\
 & \text{对称} & & 0 & 0 & 0\\
 & & & & 0 & 0\\
 & & & & & 0
\end{bmatrix} \tag{4.33}
$$

式(4.33)以矩阵形式给出了线性黏弹性问题的本构方程。我们利用虚功原理推导平衡方程和几何方程的等效积分弱形式。变形体的虚功原理可以描述为：变形体中任意满足平衡的力系在任意满足协调条件的变形状态上做的虚功等于零，即体系外的虚功和内力的虚功之和等于零。

虚功原理是虚位移原理和虚应力原理的统称。它们都可以被认为是与某些控制方程相等效的积分“弱”形式。虚位移原理是平衡方程和力边界条件的等效积分“弱”形式；虚应力原理是几何方程和位移边界条件的等效积分“弱”形式。

为了方便起见，我们将导出张量形式的结果，并将其写成矩阵形式[22]。

平衡方程和边界条件可以写成如下等效的积分形式：

$$\int_{\Omega}\delta u_i\sigma_{il,l}\mathrm{d}\Omega+\int_{\Omega}\delta u_i b_i\mathrm{d}\Omega-\int_{S}\delta u_i\sigma_{il}n_l\mathrm{d}S+\int_{S}\delta u_i\hat{f}_i\mathrm{d}S=0 \tag{4.34}$$

式中，S 是物体域 Ω 的边界；δu_i 是真实位移的变分，同时在给定位移的边界 S_u 上 $\delta u_i=0$。对式(4.34)的第一项体积分进行分部积分可得

$$\begin{aligned}\int_{\Omega}\delta u_i\sigma_{il,l}\mathrm{d}\Omega&=\int_{\Omega}(\delta u_i\sigma_{il})_{,l}\mathrm{d}\Omega-\int_{\Omega}\frac{1}{2}(\delta u_{i,l}+\delta u_{l,i})\sigma_{il}\mathrm{d}\Omega\\&=\int_{S}\delta u_i\sigma_{il}n_l\mathrm{d}S-\int_{\Omega}\frac{1}{2}(\delta u_{i,l}+\delta u_{l,i})\sigma_{il}\mathrm{d}\Omega\end{aligned} \tag{4.35}$$

由几何方程可得，式(4.35)中的 $(\delta u_{i,l}+\delta u_{l,i})/2$ 可以用 $\delta\varepsilon_{il}$ 代替。将 $\delta\varepsilon_{il}$ 代入式(4.35)，然后再将式(4.35)代入式(4.34)可得

$$\int_{\Omega}(-\delta\varepsilon_{il}\sigma_{il}+\delta u_i b_i)\mathrm{d}\Omega+\int_{S}\delta u_i\hat{f}\mathrm{d}S=0 \tag{4.36}$$

式(4.36)体积分中的第一项是变形体内的应力在虚应变上所做功的负值，即内力的虚功；体积分中的第二项和面积分分别为体力和面力在虚位移上所做的功，即外力的虚功。外力的虚功和内力的虚功之和等于零，这就是虚功原理。现在的虚功是外力和内力分别在虚位移和与之相对应的虚应变上所做的功，所以得到的是虚功原理中的虚位移原理。这是平衡方程和力边界条件的积分弱形式，其矩阵形式可写成

$$\int_{\Omega}\delta\boldsymbol{\varepsilon}^{\mathrm{T}}\boldsymbol{\sigma}\mathrm{d}\Omega-\int_{\Omega}\delta\boldsymbol{u}^{\mathrm{T}}\boldsymbol{b}\mathrm{d}\Omega-\int_{S}\delta\boldsymbol{u}^{\mathrm{T}}\hat{\boldsymbol{f}}\mathrm{d}S=0 \tag{4.37}$$

虚位移原理的力学意义是：如果力系(包括内力 $\boldsymbol{\sigma}$、体力 $\boldsymbol{b}$ 和面力 $\hat{\boldsymbol{f}}$)是平衡的(即在内部满足平衡方程，在给定外力边界上满足力边界条件)，则它在虚位移(在给定位移边界上满足 $\delta u_i=0$)和虚应变(与虚位移对应，满足几何方程)上所做之功的和等于零。反之，如果力系在虚位移(及虚应变)上所做之功的和等于零，则它们一定是满足平衡的。所以虚位移原理表述了力系平衡的充分必要条件。

值得注意的是，在导出虚位移原理的过程中，我们没有涉及本构方程(应力-应变关系)，因此虚位移原理不仅可以用于线弹性问题，还可以用于黏弹性、超弹性和弹塑性等非线性问题。等几何分析基于等参有限元概念，将用于描述几何模型的 NURBS 基函数用作位移场的插值函数，所以近似位移场可表示为

$$\boldsymbol{u}(\phi,\eta,\zeta)=\sum_{i=1}^{n}\sum_{j=1}^{m}\sum_{k=1}^{l}R_{i,j,k}(\phi,\eta,\zeta)\boldsymbol{u}^{e} \tag{4.38}$$

其中，$\boldsymbol{u}^{e}$ 是单元控制点的位移列阵。

由几何方程可得单元应变向量 $\boldsymbol{\varepsilon}$ 为

$$\boldsymbol{\varepsilon} = \boldsymbol{L}\boldsymbol{u} = \boldsymbol{L}\boldsymbol{R}\boldsymbol{u}^e = \boldsymbol{B}\boldsymbol{u}^e \tag{4.39}$$

其中，$\boldsymbol{R}$ 为插值函数矩阵；$\boldsymbol{L}$ 为微分算子，

$$\boldsymbol{L} = \begin{bmatrix} \frac{\partial}{\partial x} & 0 & 0 & \frac{\partial}{\partial y} & 0 & \frac{\partial}{\partial z} \\ 0 & \frac{\partial}{\partial y} & 0 & \frac{\partial}{\partial x} & \frac{\partial}{\partial z} & 0 \\ 0 & 0 & \frac{\partial}{\partial z} & 0 & \frac{\partial}{\partial y} & \frac{\partial}{\partial x} \end{bmatrix}^{\mathrm{T}} \tag{4.40}$$

$\boldsymbol{B}$ 为应变-位移转换矩阵，即

$$\boldsymbol{B}=\begin{bmatrix} R_{1,x} & 0 & 0 & \cdots & R_{(p+1)(q+1)(r+1),x} & 0 & 0 \\ 0 & R_{1,y} & 0 & \cdots & 0 & R_{(p+1)(q+1)(r+1),y} & 0 \\ 0 & 0 & R_{1,z} & \cdots & 0 & 0 & R_{(p+1)(q+1)(r+1),z} \\ R_{1,y} & R_{1,x} & 0 & \cdots & R_{(p+1)(q+1)(r+1),y} & R_{(p+1)(q+1)(r+1),x} & 0 \\ 0 & R_{1,z} & R_{1,y} & \cdots & 0 & R_{(p+1)(q+1)(r+1),z} & R_{(p+1)(q+1)(r+1),y} \\ R_{1,z} & 0 & R_{1,x} & \cdots & R_{(p+1)(q+1)(r+1),z} & 0 & R_{(p+1)(q+1)(r+1),x} \end{bmatrix} \tag{4.41}$$

式中，$R_{(p+1)(q+1)(r+1),x}$、$R_{(p+1)(q+1)(r+1),y}$ 和 $R_{(p+1)(q+1)(r+1),z}$ 分别是序号为 $(p+1)(q+1)(r+1)$ 的 NURBS 基函数对坐标 x、y 和 z 的一阶导数。

我们对计算域 Ω 进行单元离散，在每个单元上建立近似函数，每个单元上积分满足式(4.37)，即

$$\int_{\Omega_e} \delta\boldsymbol{\varepsilon}^{\mathrm{T}}\boldsymbol{\sigma}\mathrm{d}\Omega - \int_{\Omega_e} \delta\boldsymbol{u}^{\mathrm{T}}\boldsymbol{b}\mathrm{d}\Omega - \int_{S_e} \delta\boldsymbol{u}^{\mathrm{T}}\hat{\boldsymbol{f}}\mathrm{d}S = 0 \tag{4.42}$$

式中，S_e 是单元域 Ω_e 的边界；$\delta\boldsymbol{u}$ 是单元中任意一点处的位移变分，其对应的应变变分为 $\delta\boldsymbol{\varepsilon} = \boldsymbol{B}\delta\boldsymbol{u}^e$。

将位移变分和应变变分代入式(4.42)中，并消除每个积分中相同项 $\delta\boldsymbol{u}^e$ 后可得

$$\int_{\Omega_e} \boldsymbol{B}^{\mathrm{T}}\boldsymbol{\sigma}\mathrm{d}\Omega - \int_{\Omega_e} \boldsymbol{R}^{\mathrm{T}}\boldsymbol{b}\mathrm{d}\Omega - \int_{S_e} \boldsymbol{R}^{\mathrm{T}}\hat{\boldsymbol{f}}\mathrm{d}S = 0 \tag{4.43}$$

将本构方程的矩阵形式，即式(4.26)，代入式(4.43)，可得有限元的求解方程组：

$$\boldsymbol{K}^e\boldsymbol{u}^e + \int_{\tilde{\Omega}_e} \boldsymbol{B}^{\mathrm{T}}\boldsymbol{L}_1\boldsymbol{B}\boldsymbol{u}^e \left|\hat{\boldsymbol{J}}\right|\left|\tilde{\boldsymbol{J}}\right|\mathrm{d}\tilde{\Omega} + \int_{\tilde{\Omega}_e} \boldsymbol{B}^{\mathrm{T}}\boldsymbol{L}_2\boldsymbol{B}\boldsymbol{u}^e \left|\hat{\boldsymbol{J}}\right|\left|\tilde{\boldsymbol{J}}\right|\mathrm{d}\tilde{\Omega} = \boldsymbol{F}_f^e + \boldsymbol{F}_b^e \tag{4.44}$$

其中，$\boldsymbol{K}^e$ 表示单元弹性刚度矩阵，即

$$\boldsymbol{K}^e = \int_{\tilde{\Omega}_e} \boldsymbol{B}^{\mathrm{T}} \boldsymbol{D}_1 \boldsymbol{B} \left|\hat{\boldsymbol{J}}\right| \left|\tilde{\boldsymbol{J}}\right| \mathrm{d}\tilde{\Omega} \tag{4.45}$$

$\boldsymbol{F}_f^e$ 和 $\boldsymbol{F}_b^e$ 分别表示单元在外部载荷和体力作用下的等效节点力向量，具体表达式如下：

$$\boldsymbol{F}_f^e = \int_{\tilde{S}_e} \boldsymbol{R}^{\mathrm{T}} \hat{\boldsymbol{f}} \left|\hat{\boldsymbol{J}}\right| \left|\tilde{\boldsymbol{J}}\right| \mathrm{d}\tilde{S} \tag{4.46}$$

$$\boldsymbol{F}_b^e = \int_{\tilde{\Omega}_e} \boldsymbol{R}^{\mathrm{T}} \boldsymbol{b} \left|\hat{\boldsymbol{J}}\right| \left|\tilde{\boldsymbol{J}}\right| \mathrm{d}\tilde{\Omega} \tag{4.47}$$

式(4.44)中的两个积分是与时间相关的，我们需要对它们分别进行处理。对于第一项积分，其可以写成如下形式：

$$\int_{\tilde{\Omega}_e} \boldsymbol{B}^{\mathrm{T}} \boldsymbol{L}_1 \boldsymbol{B} \boldsymbol{u}^e \left|\hat{\boldsymbol{J}}\right| \left|\tilde{\boldsymbol{J}}\right| \mathrm{d}\tilde{\Omega} = \int_{\tilde{\Omega}_e} \boldsymbol{B}^{\mathrm{T}} \boldsymbol{D}_2 \boldsymbol{B} \left[\int_{0^+}^{t} \frac{\partial G(t-\tau)}{\partial \tau} \boldsymbol{u}^e(\tau) \mathrm{d}\tau \right] \left|\hat{\boldsymbol{J}}\right| \left|\tilde{\boldsymbol{J}}\right| \mathrm{d}\tilde{\Omega} \tag{4.48}$$

利用梯形公式[23-26]求解式(4.48)中的遗传积分，将 $[0,t]$ 分为 k 个时间段，即 $[0,t_1],\cdots,[t_{k-1},t_k]$，可得

$$\begin{aligned}\int_{0^+}^{t} \frac{\partial G(t-\tau)}{\partial \tau} \boldsymbol{u}^e(\tau) \mathrm{d}\tau &= \sum_{i=1}^{k-1} \int_{t_i}^{t_{i+1}} \frac{\partial G(t-\tau)}{\partial \tau} \boldsymbol{u}^e(\tau) d\tau \\ &= \sum_{i=1}^{k-1} \left[\left.\frac{\partial G(t-\tau)}{\partial \tau}\right|_{\tau=t_i} \boldsymbol{u}^e(t_i) + \left.\frac{\partial G(t-\tau)}{\partial \tau}\right|_{\tau=t_{i+1}} \boldsymbol{u}^e(t_{i+1}) \right] \frac{\Delta t}{2}\end{aligned} \tag{4.49}$$

其中，$\Delta t = t_{i+1} - t_i$。利用向前差分近似有

$$\left.\frac{\partial G(\gamma-\gamma')}{\partial \gamma'}\right|_{\gamma'=t_i} = \frac{1}{\Delta t}[G(\gamma_k - \gamma_{i+1}) - G(\gamma_k - \gamma_i)] \tag{4.50}$$

利用后差分公式可得

$$\left.\frac{\partial G(\tau-\tau')}{\partial \tau'}\right|_{\tau'=t_{i+1}} = \frac{1}{\Delta t}[G(\tau_k - \tau_{i+1}) - G(\tau_k - \tau_i)] \tag{4.51}$$

其中，$\gamma_i = \gamma(t_i)$，将式(4.50)和式(4.51)代入式(4.49)中，可得

$$\begin{aligned}\int_{0^+}^{t} \frac{\partial G(t-\tau)}{\partial \tau} \boldsymbol{u}^e(\tau) \mathrm{d}\tau &= \sum_{i=1}^{k-1} \int_{t_i}^{t_{i+1}} \frac{\partial G(t-\tau)}{\partial \tau} \boldsymbol{u}^e(\tau) \mathrm{d}\gamma' \\ &= \sum_{i=1}^{k-1} \left\{ [G(t_k - t_{i+1}) - G(t_k - t_i)] \frac{\boldsymbol{u}^e(t_i) + \boldsymbol{u}^e(t_{i+1})}{2} \right\} \\ &= \sum_{i=1}^{k-2} \left\{ [G(t_k - t_{i+1}) - G(t_k - t_i)] \frac{\boldsymbol{u}^e(t_i) + \boldsymbol{u}^e(t_{i+1})}{2} \right\} \\ &\quad + [G(0) - G(t_k - t_{k-1})] \frac{\boldsymbol{u}^e(t_{k-1})}{2} + [G(0) - G(t_k - t_{k-1})] \frac{\boldsymbol{u}^e(t_k)}{2}\end{aligned} \tag{4.52}$$

式(4.52)中的前 k–1 步的位移 $\boldsymbol{u}^e(t_{k-1})$ 是已知的，因此将式(4.52)代入式(4.48)中，可得

$$\int_{\tilde{\Omega}^e} \boldsymbol{B}^{\mathrm{T}} \boldsymbol{L}_1 \boldsymbol{B} \boldsymbol{u}^e \left|\hat{\boldsymbol{J}}\right|\left|\tilde{\boldsymbol{J}}\right| \mathrm{d}\tilde{\Omega} = \boldsymbol{F}_1^e + \boldsymbol{K}_1^e \boldsymbol{u}^e(t_k) \tag{4.53}$$

其中，

$$\boldsymbol{K}_1^e = \frac{1}{2}\int_{\tilde{\Omega}_e} \boldsymbol{B}^{\mathrm{T}} \boldsymbol{D}_2 \boldsymbol{B}\left[G(0) - G(t_k - t_{k-1})\right]\left|\hat{\boldsymbol{J}}\right|\left|\tilde{\boldsymbol{J}}\right| \mathrm{d}\tilde{\Omega} \tag{4.54}$$

$$\begin{aligned}\boldsymbol{F}_1^e = \int_{\tilde{\Omega}_e} \boldsymbol{B}^{\mathrm{T}} \boldsymbol{D}_2 \boldsymbol{B} \Bigg(&\sum_{i=1}^{k-2}\left\{\left[G(t_k - t_{i+1}) - G(t_k - t_i)\right]\frac{\boldsymbol{u}^e(t_i) + \boldsymbol{u}^e(t_{i+1})}{2}\right\} \\ &+ \left[G(0) - G(t_k - t_{k-1})\right]\frac{\boldsymbol{u}^e(t_{k-1})}{2}\Bigg)\left|\hat{\boldsymbol{J}}\right|\left|\tilde{\boldsymbol{J}}\right| \mathrm{d}\tilde{\Omega}\end{aligned} \tag{4.55}$$

由式(4.55)可知，在计算 $\boldsymbol{F}_1^e$ 时，需要存储 t_k 前每一步的位移。为了避免存储，我们利用 Dirichlet-Prony 级数将剪切模量 $G(t)$ 和体积模量 $K(t)$ 展开成如下形式：

$$G(t) = G_0 + \sum_{w=1}^{m} G_w \exp(-t/\tau_w^G) \tag{4.56}$$

$$K(t) = K_0 + \sum_{w=1}^{m} K_w \exp(-t/\tau_w^K) \tag{4.57}$$

其中，τ_w^G 和 τ_w^K 表示剪切响应和体积响应对应的松弛时间；G_0、K_0、G_w、K_w 分别表示剪切模量和体积模量的展开系数；m 表示展开的项数。将式(4.56)代入式(4.55)中的累加部分可得

$$\begin{aligned}\sum_{w=1}^{m} G_w \boldsymbol{q}_{w,k} &= \sum_{i=1}^{k-2}\left[G(t_k - t_{i+1}) - G(t_k - t_i)\right]\frac{\boldsymbol{u}^e(t_i) + \boldsymbol{u}^e(t_{i+1})}{2} \\ &= \sum_{w=1}^{m} G_w \sum_{i=1}^{k-2}\left\{\exp\left[(-(t_k - t_{i+1})/\tau_w^G\right] - \exp\left[-(t_k - t_i)/\tau_w^G\right]\right\}\frac{\boldsymbol{u}^e(t_i) + \boldsymbol{u}^e(t_{i+1})}{2} \\ &= \sum_{w=1}^{m} G_w \sum_{i=1}^{k-2}\left[\exp(-t_k/\tau_w^G)(\exp(t_{i+1}/\tau_w^G) - \exp(t_i/\tau_w^G))\right]\frac{\boldsymbol{u}^e(t_i) + \boldsymbol{u}^e(t_{i+1})}{2}\end{aligned} \tag{4.58}$$

最终可以得到如下递推格式：

$$\boldsymbol{q}_{w,k} = \mathrm{e}^{-\frac{t_k - t_{k-1}}{\tau_w^G}}\left[\left(1 - \mathrm{e}^{-\frac{t_{k-1} - t_{k-2}}{\tau_w^G}}\right)\frac{\boldsymbol{u}^e(t_{k-2}) + \boldsymbol{u}^e(t_{k-1})}{2} + \boldsymbol{q}_{w,k-1}\right] \tag{4.59}$$

将式(4.58)代入式(4.55)可得

$$\boldsymbol{F}_1^e = \int_{\tilde{\Omega}_e} \boldsymbol{B}^{\mathrm{T}} \boldsymbol{D}_2 \boldsymbol{B} \left\{ \sum_{w=1}^{m} G_w \boldsymbol{q}_{w,k} + \frac{1}{2} \left[G(0) - G(t_k - t_{k-1}) \right] \boldsymbol{u}^e(t_{k-1}) \right\} \left| \hat{\boldsymbol{J}} \right| \left| \tilde{\boldsymbol{J}} \right| \mathrm{d}\tilde{\Omega} \tag{4.60}$$

对于式(4.44)中的第二项积分，其可以写成如下形式：

$$\begin{aligned}\int_{\tilde{\Omega}_e} \boldsymbol{B}^{\mathrm{T}} \boldsymbol{L}_2 \boldsymbol{B} \boldsymbol{u}^e \left| \hat{\boldsymbol{J}} \right| \left| \tilde{\boldsymbol{J}} \right| \mathrm{d}\tilde{\Omega} &= \int_{\tilde{\Omega}_e} \boldsymbol{B}^{\mathrm{T}} \boldsymbol{D}_3 \boldsymbol{B} \left[\int_{0^+}^{t} \frac{\partial K(t-\tau)}{\partial \tau} \boldsymbol{u}^e(\tau) \mathrm{d}\tau \right] \left| \hat{\boldsymbol{J}} \right| \left| \tilde{\boldsymbol{J}} \right| \mathrm{d}\tilde{\Omega} \\ &= \boldsymbol{F}_2^e + \boldsymbol{K}_2^e \boldsymbol{u}^e(t_k)\end{aligned} \tag{4.61}$$

其中，

$$\boldsymbol{K}_2^e = \frac{1}{2} \int_{\tilde{\Omega}_e} \boldsymbol{B}^{\mathrm{T}} \boldsymbol{D}_3 \boldsymbol{B} \left[K(0) - K(t_k - t_{k-1}) \right] \left| \hat{\boldsymbol{J}} \right| \left| \tilde{\boldsymbol{J}} \right| \mathrm{d}\tilde{\Omega} \tag{4.62}$$

$$\boldsymbol{F}_2^e = \int_{\tilde{\Omega}_e} \boldsymbol{B}^{\mathrm{T}} \boldsymbol{D}_2 \boldsymbol{B} \left\{ \sum_{w=1}^{m} K_w \boldsymbol{q}_{w,k} + \frac{1}{2} \left[K(0) - K(t_k - t_{k-1}) \right] \boldsymbol{u}^e(t_{k-1}) \right\} \left| \hat{\boldsymbol{J}} \right| \left| \tilde{\boldsymbol{J}} \right| \mathrm{d}\tilde{\Omega} \tag{4.63}$$

基于式(4.53)和式(4.61)，式(4.44)可以写成如下形式：

$$\boldsymbol{K}^e \boldsymbol{u}^e(t_k) = \boldsymbol{F}_f^e + \boldsymbol{F}_b^e + \boldsymbol{F}_1^e + \boldsymbol{F}_2^e - (\boldsymbol{K}_1^e + \boldsymbol{K}_2^e) \boldsymbol{u}^e(t_k) \tag{4.64}$$

对于时刻 t_k 的第 $i+1$ 步迭代，我们有如下公式：

$$\boldsymbol{K}^e \boldsymbol{u}_{i+1}^e(t_k) = \boldsymbol{F}_f^e + \boldsymbol{F}_b^e + \boldsymbol{F}_1^e + \boldsymbol{F}_2^e - (\boldsymbol{K}_1^e + \boldsymbol{K}_2^e) \boldsymbol{u}_i^e(t_k) \tag{4.65}$$

对于每一时刻 t_k 的第一步迭代，我们有 $\boldsymbol{u}^e(t_k) = \boldsymbol{u}^e(t_{k-1})$。迭代循环结束的判断准则为

$$\left\| \frac{\boldsymbol{u}_{i+1}^e(t_k) - \boldsymbol{u}_i^e(t_k)}{\boldsymbol{u}_{i+1}^e(t_k)} \right\| \leqslant \mathrm{tol} \tag{4.66}$$

其中，tol 表示给定的迭代收敛的阈值。

4.4 数值算例

以下算例中弹性材料杨氏模量为 $E = 2\times10^5$ Pa，泊松比为 $\nu = 0.3$，推进剂药柱黏弹性材料参数采用 Prony 级数表示[27]：

$$\begin{cases} G(t) = G_0 + G_1 \exp(-t/\gamma_1) \\ K(t) = K(0) = 20000\mathrm{Pa} \end{cases} \tag{4.67}$$

其中，剪切模量系数 G_0 和 G_1 分别为 100Pa 与 9900Pa；松弛时间 $\gamma_1 = 0.4170\mathrm{s}$。

4.4.1 受内压的二维线性黏弹性厚壁圆筒

通过分析承受均布内压载荷的二维厚壁圆筒验证 IGAFEM 求解黏弹性力学

问题的收敛性与正确性，几何模型如图 4.1 所示。为了缩减建模成本并减少计算工作量，可利用轴对称特性将轴对称结构的二维厚壁圆筒进行简化分析，简化后的 1/4 等效二维模型如图 4.2 所示。厚壁圆筒内径 R_1 与外径 R_2 分别为 1m 和 2m，在厚壁圆筒内壁施加均布内压 $\hat{P}=10\,\text{Pa}$，并在对称边界施加轴向约束，取内径监测点为 A_1 (1m，0)和外径监测点为 A_2 (2m，0)。

图 4.1 厚壁圆筒

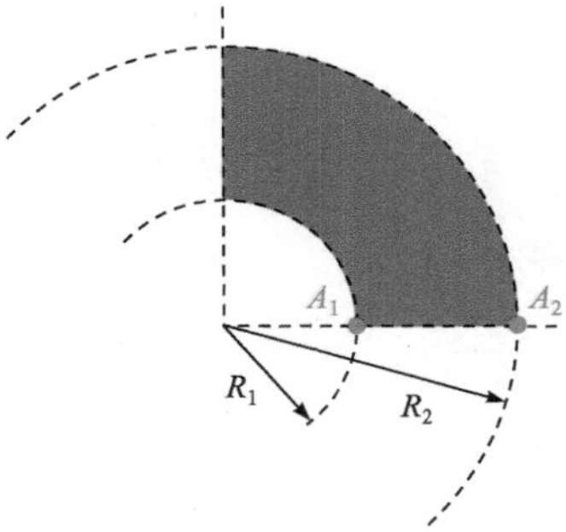

图 4.2 厚壁圆筒的 1/4 等效二维模型

在这个例子中，通过 IGAFEM 计算得到的数值结果与平面应变问题的解析解进行对比。在极坐标形式下，承受均布内压 $\hat{p}$ 作用的黏弹性材料厚壁圆筒在 t 时刻的径向位移解析解为[28,29]

$$u(r)=\frac{\hat{P}{R_b}^2}{(R_a^2-R_b^2)}\left\{3r\left[\frac{1}{6K+q_0^G}+\left(\frac{p_1^G}{6Kp_1^G+q_1^G}-\frac{1}{6K+q_0^G}\right)\mathrm{e}^{-\frac{\left(6K+q_0^G\right)t}{\left(6Kp_1^G+q_1^G\right)}}\right]\right.$$
$$\left.+\frac{a^2}{q_0^G r}\left[1+\left(\frac{q_0^G p_1^G}{q_1^G}-1\right)\mathrm{e}^{-\frac{q_0^G t}{q_1^G}}\right]\right\} \tag{4.68}$$

其中，R_a 和 R_b 分别表示厚壁圆筒的内径和外径；r 是极坐标系中的半径；K 为体积模量。此外，与 Prony 级数相关的其他参数可以表示为

$$p_1^G=\gamma_1,\quad q_0^G=2G_0,\quad q_1^G=2\gamma_1(G_0+G_1) \tag{4.69}$$

IGAFEM 结果与解析解的相对误差 E_r 定义为

$$E_\mathrm{r}=(u_\text{numerical}-u_\text{analytical})/u_\text{analytical}\times 100\% \tag{4.70}$$

其中，$u_\text{numerical}$ 与 $u_\text{analytical}$ 分别为径向位移的 IGAFEM 解与解析解。

本算例的厚壁圆筒采用阶次 p 为 2 的 NURBS 基函数，在两个不同参数方向的初始节点向量在表 4.1 中给出。为了讨论细化次数对数值结果准确性与收敛性的影响，分别对厚壁圆筒进行 1 次细化、2 次细化、4 次细化、8 次细化，不同细化次数的 NURBS 网格情况如图 4.3 所示。

表 4.1　1/4 厚壁圆筒不同参数方向的初始节点向量

参数方向	节点向量
ϕ	$\boldsymbol{\Psi}=\{0,0,0,1,1,1\}$
η	$\boldsymbol{\Phi}=\{0,0,0,1,1,1\}$

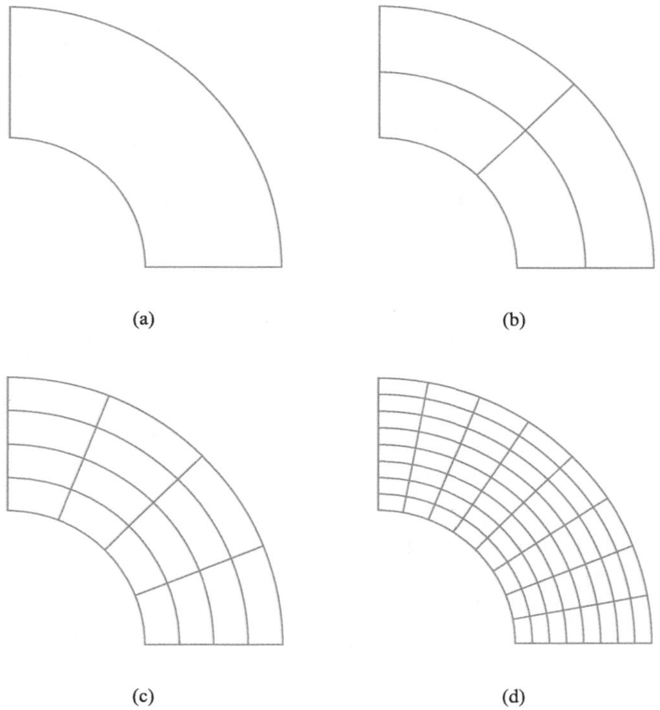

图 4.3　不同细化次数的 NURBS 网格模型

(a) 1 次细化；(b) 2 次细化；(c) 4 次细化；(d) 8 次细化

图 4.4 分别显示了在不同细化次数下，从时间 $t=0\text{s}$ 到 $t=300\text{s}$ 时监测点 A_1 的径向位移随时间的变化，计算时取固定步长为 $\Delta t=0.05\text{s}$。本算例中黏弹性计算迭代收敛的预期精度 tol 为 1×10^{-5}。可以看出，细化次数对计算结果的影响是比较明显的，数值结果随着细化次数的增加而逐渐收敛。原因是在 IGA 分析中，少数控制点可以准确地构建几何模型，但不能准确地描述物理场。当细化次数越多时控制点越多，IGAFEM 描述的物理量就越准确。

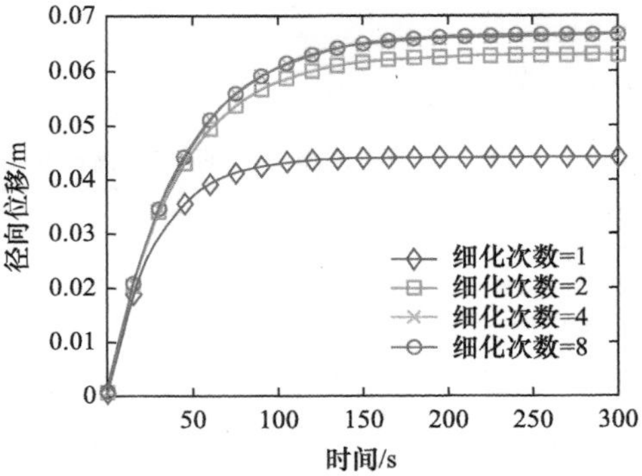

图 4.4　不同细化次数下监测点 A_1 的径向位移对比曲线

如图 4.5(a)所示，当细化次数为 8 时，IGAFEM 计算得到监测点 A_1 和 A_2 的数值结果与解析解吻合较好。图 4.5(b)展示的是 IGAFEM 解与解析解的相对误差曲线。可以看到，当细化次数为 8 时，监测点 A_1 和 A_2 在迭代最终时刻的相对误差 E_r 分别为 0.0112%与 0.04164%，验证了 IGAFEM 求解黏弹性问题的正确性与收敛性。计算模型的径向位移 IGAFEM 解与解析解的对比云图如图 4.6 所示。

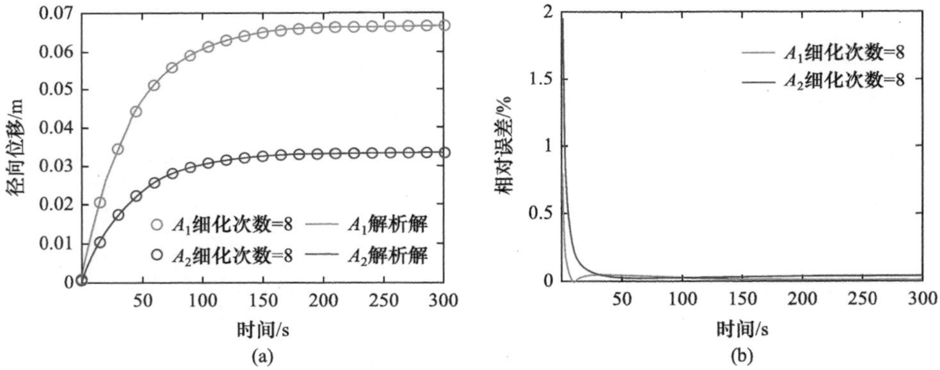

图 4.5　径向位移 IGAFEM 解与解析解对比

(a) 随时间变化曲线；(b) 相对误差曲线

黏弹性材料力学响应与时间密切相关，本章采用梯形差分形式对积分型黏弹性本构时间域进行离散，因此时间步长 Δt 对计算结果精度的影响不可忽略。接下来本算例将研究分析时间步长大小对数值结果的准确性与收敛性的影响。图 4.7 展示了细化次数为 8 的计算模型，在时间步长 Δt 分别为 3s、1s、0.5s、0.1s 和 0.05s 时的计算结果。图 4.8 为不同时间步长的数值结果与解析解的相对误差曲线。可以看出，时间步长的选取对计算结果影响非常明显，时间步长越小则计算精度越高。

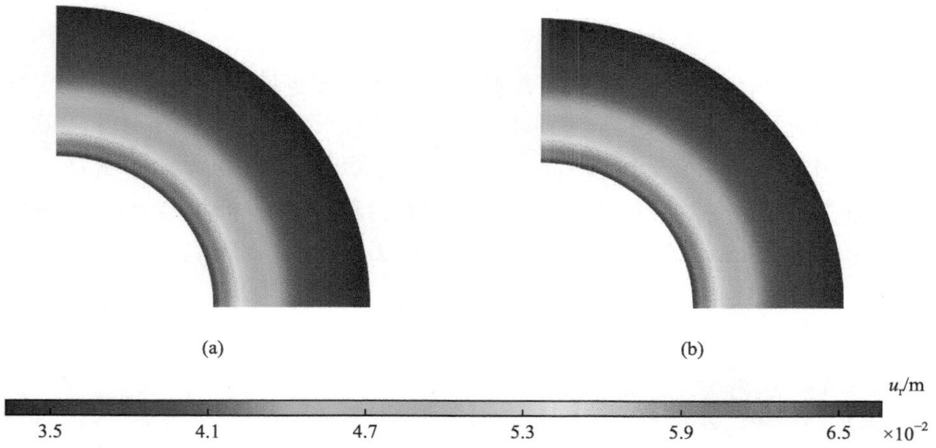

图 4.6　厚壁圆筒径向位移分布云图

(a) 解析解；(b) IGAFEM 解

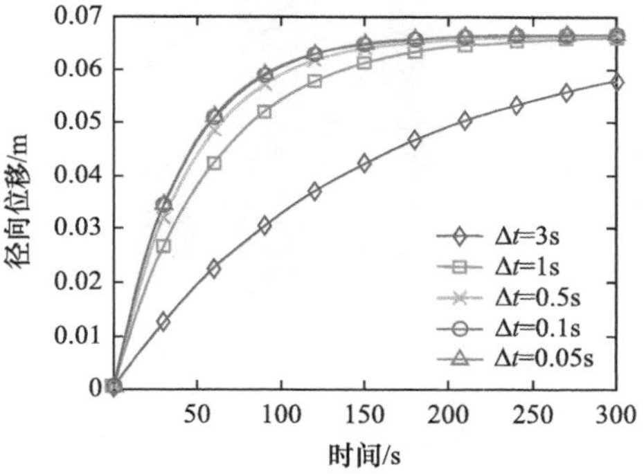

图 4.7　不同时间步长时 A_1 点径向位移对比曲线

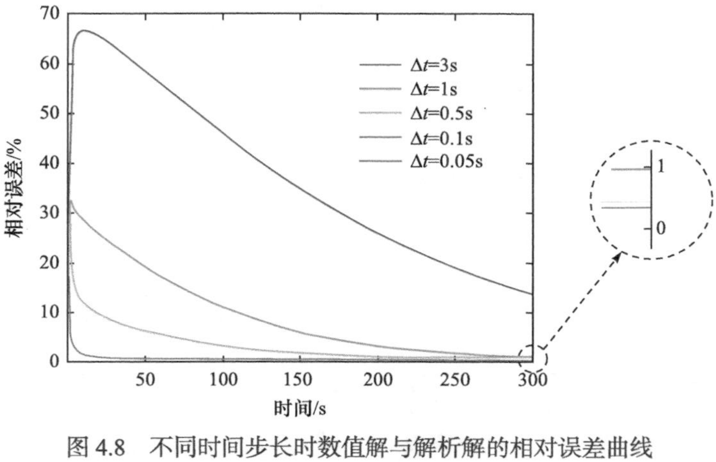

图 4.8　不同时间步长时数值解与解析解的相对误差曲线

4.4.2 受内压的三维线性黏弹性厚壁圆筒

几何模型如图 4.9 所示。为了缩减建模成本并减少计算工作量，可利用轴对称特性将轴对称结构的三维厚壁圆筒模型进行简化分析，简化后的 1/8 等效三维模型如图 4.10 所示。厚壁圆筒内径 R_1 与外径 R_2 分别为 1m 和 2m，高度 H 为 5m，在厚壁圆筒内壁施加均匀内压 $\hat{P}=10\,\text{Pa}$，并在对称边界施加轴向约束，取内径监测点为 A_1 (1m，0m，2.5m)和外径监测点为 A_2 (2m，0m，2.5m)。

图 4.9 厚壁圆筒整体模型示意图

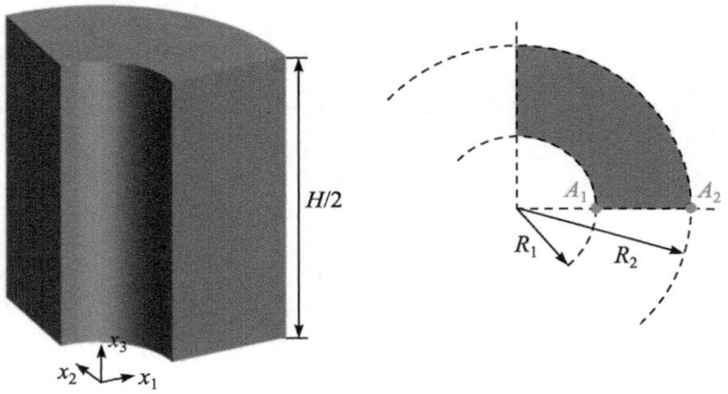

图 4.10 1/8 等效模型示意图

本算例中将 IGAFEM 解与解析解进行对比。在极坐标形式下，承受均匀内压 $\hat{p}$ 作用的黏弹性材料厚壁圆筒在 t 时刻的应力与径向位移解析解为[28,29]

$$
\begin{cases}
\sigma_{rr}=\dfrac{R_a^2\hat{P}}{R_b^2-R_a^2}\left(1-\dfrac{R_b^2}{r^2}\right) \\
\sigma_{\theta\theta}=\dfrac{R_a^2\hat{P}}{R_b^2-R_a^2}\left(1+\dfrac{R_b^2}{r^2}\right) \\
u(r)=\dfrac{\left[R_b{}^2\hat{P}\left(9KR_a{}^2+3Kr^2+2q_0^G r^2\right)\right]}{9Kq_0^G r(R_a+R_b)(R_a-R_b)} \\
\qquad -\dfrac{\left[R_b{}^2\hat{P}\mathrm{e}^{-\frac{q_0^G t}{q_1^G}}\left(3R_a{}^2q_1^G+q_1^G r^2-3R_a{}^2p_0^G q_1^G-p_0^G q_0^G r^2\right)\right]}{3q_0^G q_1^G r(R_a+R_b)(R_a-R_b)}
\end{cases}
\tag{4.71}
$$

与 Prony 级数相关的其他参数见式(4.67)。如图 4.11(a)所示，当细化次数为 8

时，IGAFEM 计算得到监测点 A_1 和 A_2 的数值结果与解析解吻合较好。图 4.11 (b) 展示的是 IGAFEM 解与解析解的相对误差曲线。可以看到，当细化次数为 8 时，相对误差大部分在 0.5%以内，其中监测点 A_1 和 A_2 在迭代初始时刻的相对误差最大，分别为 0.4211%与 0.3915%，迭代最终时刻的相对误差分别为 0.02236%与 0.01684%。这些结果验证了 IGAFEM 求解黏弹性问题的正确性与收敛性，计算模型的径向位移 IGAFEM 解与解析解的对比云图如图 4.12 所示。

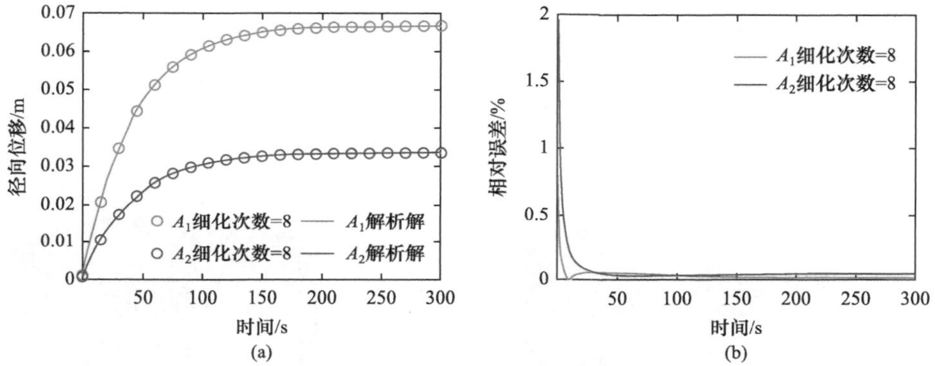

图 4.11　径向位移 IGAFEM 解与解析解的对比

(a) 随时间变化曲线；(b) 相对误差曲线

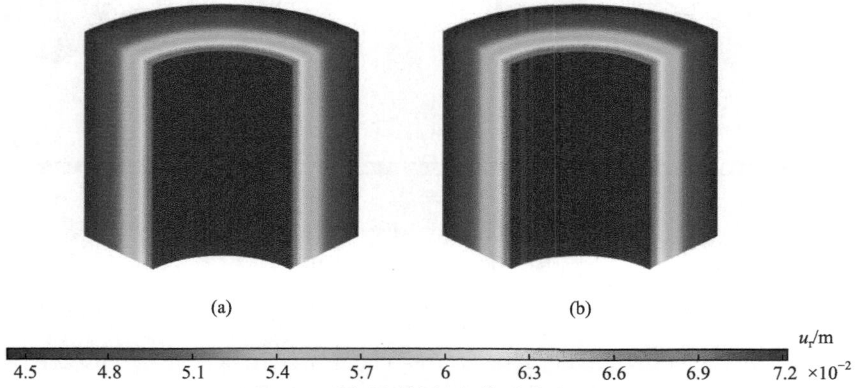

图 4.12　厚壁圆筒径向位移分布云图

(a) 解析解；(b) IGAFEM 解

图 4.13 分别为监测点 A_1 与 A_2 的 von Mises 应力随时间变化的曲线和径向路径 von Mises 应力曲线，并将 IGAFEM 解与解析解进行了对比，可以看到对比曲线非常吻合。厚壁圆筒模型的 von Mises 应力与解析解的对比云图如图 4.14 所示，其中 von Mises 应力计算公式如下：

$$\sigma_{\text{Mises}} = \sqrt{\frac{(\sigma_x - \sigma_y)^2 + (\sigma_y - \sigma_z)^2 + (\sigma_z - \sigma_x)^2 + 6(\sigma_{xy}^2 + \sigma_{yz}^2 + \sigma_{zx}^2)}{2}} \tag{4.72}$$

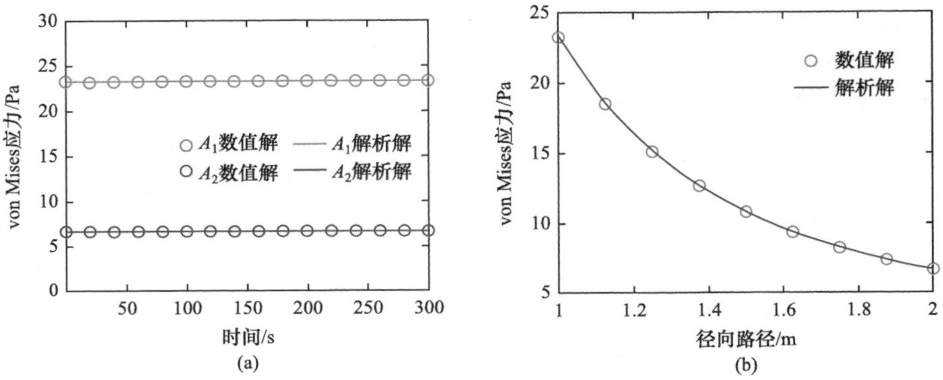

图 4.13 von Mises 应力 IGAFEM 解与解析解对比

(a) 随时间变化曲线；(b) 径向路径变化曲线

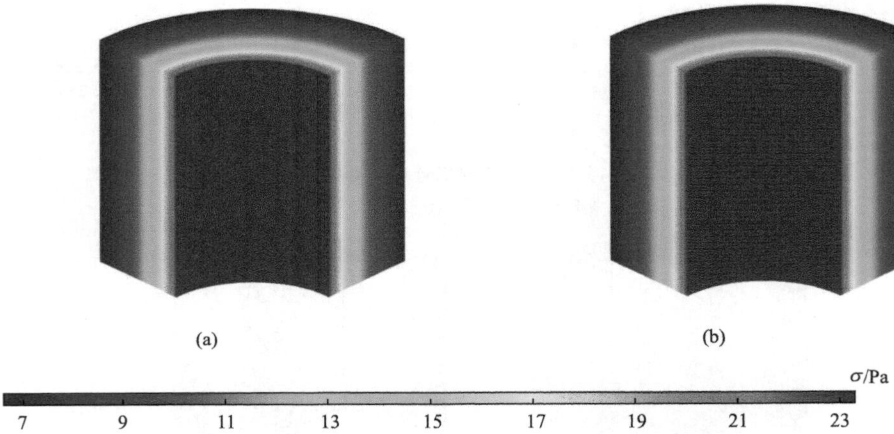

图 4.14 厚壁圆筒 von Mises 应力分布云图

(a) 解析解；(b) IGAFEM 解

黏弹性材料力学响应与时间密切相关，本章采用梯形差分形式对积分型黏弹性本构时间域进行离散，因此时间步长 Δt 对计算结果精度的影响不可忽略。接下来，本算例将分析时间步长大小对数值结果准确性与收敛性的影响。将细化次数为 8 的计算模型在 Δt 分别为 3s、1s、0.5s、0.1s 和 0.05s 的时间步长下的计算结果进行对比，如图 4.15 所示。

图 4.16 显示了不同时间步长的 IGAFEM 解与解析解的相对误差曲线。可以看出，时间步长的选取对计算结果有着非常明显的影响。时间步长越小，计算精度就越高。在不同的时间步长下，最大相对误差与最终时刻的相对误差的具体数值如表 4.2 所示。

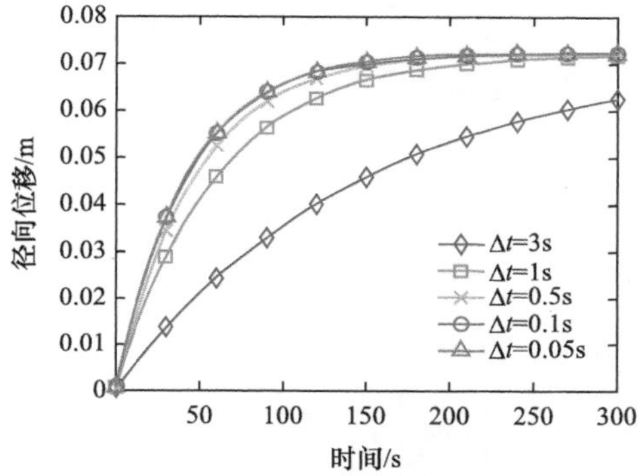

图 4.15　不同时间步长时 A_1 点径向位移对比曲线

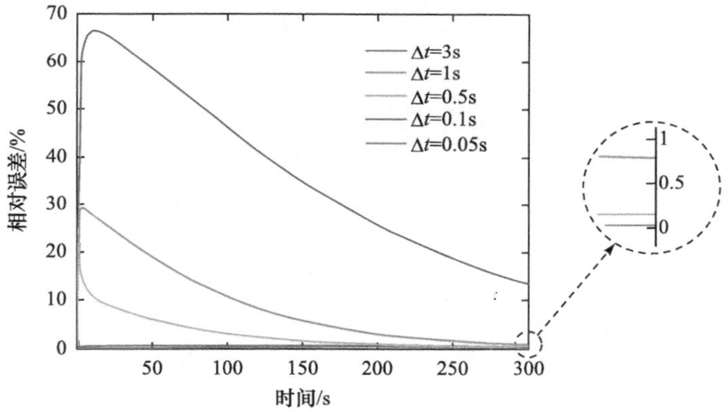

图 4.16　IGAFEM 解与解析解的相对误差

表 4.2　不同时间步长(Δt)条件下的最大相对误差与最终相对误差

	3s	1s	0.5s	0.1s	0.05s
最大相对误差/%	66.4848	29.1997	21.4958	1.6734	0.4211
最终相对误差/%	13.3999	0.6445	0.1258	0.0268	0.0223

4.4.3　复杂几何推进剂药柱

通常情况下，为了满足不同类型的发动机与飞行器的战略任务，需要依据需求不断调整药柱结构增加燃烧面积以提高发动机性能。因此，在装药结构设计中较为常见的有八星孔型药柱模型。本算例将对含八星孔型药柱发动机燃烧室模型进行等几何有限元分析，如图 4.17 所示。根据其对称性采用 1/4 等效简化模型进行分析，在对称界面以及上下两端施加轴向约束，在内表面施加均布载荷 10Pa，图 4.18 中八星孔型药柱的内表面过渡圆弧半径 r_1 与星孔半径 r_2 均为 0.4m，R_1 为 3.2m，R_2 为 4.5m，模型的高度为 8m，分别取星孔内表面处 P_1(2.2627m，2.2627m，

0m)、星孔对称轴线耦合界面处 P_2 (3.1820m，3.1820m，0m)以及药柱内径最小位置处 P_3 (1.07m，2.58m，0m)为模型监测点。

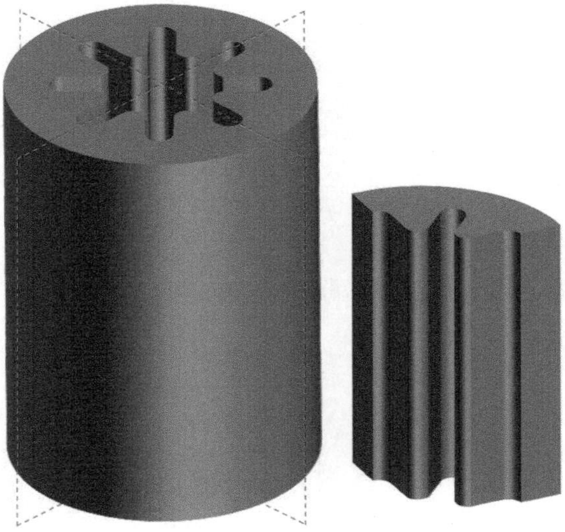

图 4.17 八星孔型药柱发动机燃烧室整体模型示意图

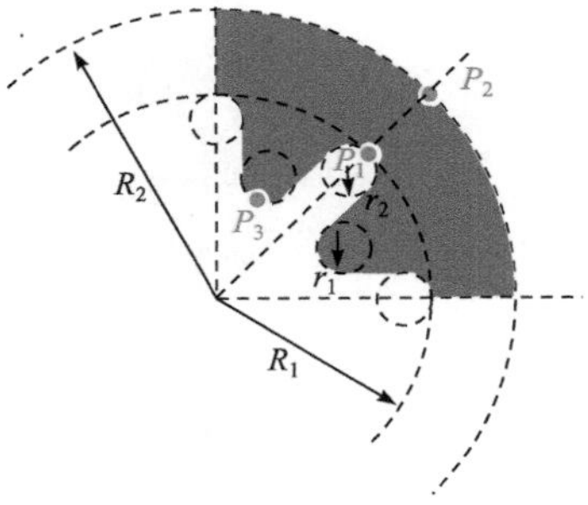

图 4.18 1/4 八星孔型药柱发动机燃烧室等效模型

构建的八星孔型药柱模型在三个参数方向的初始单元数分别为 10、4 和 4，初始节点向量在表 4.3 中给出。经过两次细化后，药柱模型有 1280 个单元和 5239 个控制点，NURBS 网格如图 4.19 所示。本算例将数值结果与 ABAQUS 计算结果进行对比，由 ABAQUS 八节点六面体单元构建的模型有限元网格如图 4.19 所示，共有 78240 个单元与 84912 个节点。

表 4.3 1/4 八星孔型药柱发动机燃烧室模型初始节点向量

参数方向	节点向量
ϕ^{FE}	$\boldsymbol{\Phi}^{\mathrm{FE}}=\left\{0,0,0,\frac{1}{10},\frac{1}{10},\frac{2}{10},\frac{2}{10},\ldots,\frac{8}{10},\frac{8}{10},\frac{9}{10},\frac{9}{10},1,1,1\right\}$
η^{FE}	$\boldsymbol{\Psi}^{\mathrm{FE}}=\left\{0,0,0,\frac{1}{4},\frac{1}{4},\frac{2}{4},\frac{2}{4},\frac{3}{4},\frac{3}{4},1,1,1\right\}$
ζ^{FE}	$\boldsymbol{\Xi}^{\mathrm{FE}}=\left\{0,0,0,\frac{1}{4},\frac{1}{4},\frac{2}{4},\frac{2}{4},\frac{3}{4},\frac{3}{4},1,1,1\right\}$

八星孔型药柱各监测点径向位移随时间的变化曲线如图 4.20 所示，并与 ABAQUS 计算结果进行了比较，可以看到曲线吻合情况比较理想，图 4.21 为八星孔型药柱整体模型的径向位移云图。图 4.22 为八星孔型药柱 von Mises 应力云

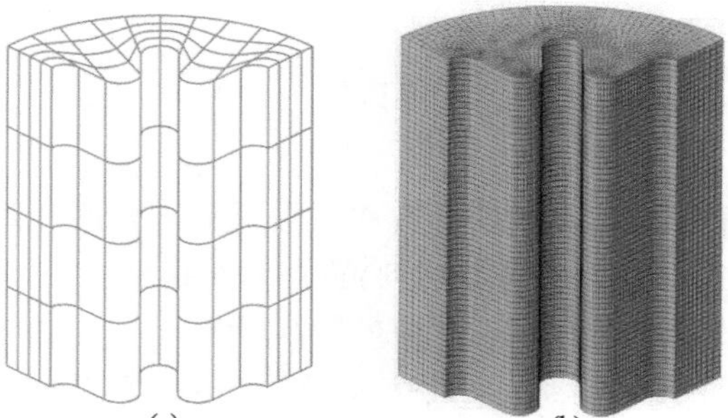

图 4.19　八星孔型药柱模型网格划分
(a) NURBS 网格；(b) ABAQUS 网格

图，可以直观地观察到八星孔型药柱在承受均布内压载荷的作用下，内表面在星孔深度最大位置处应力值最大，在内表面内径最小位置处应力最小，因此星孔凹槽部分为应力危险区域。

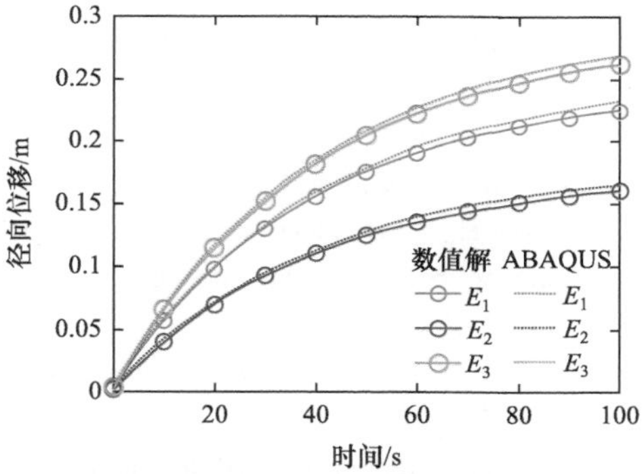

图 4.20　八星孔型药柱各监测点径向位移随时间的变化曲线

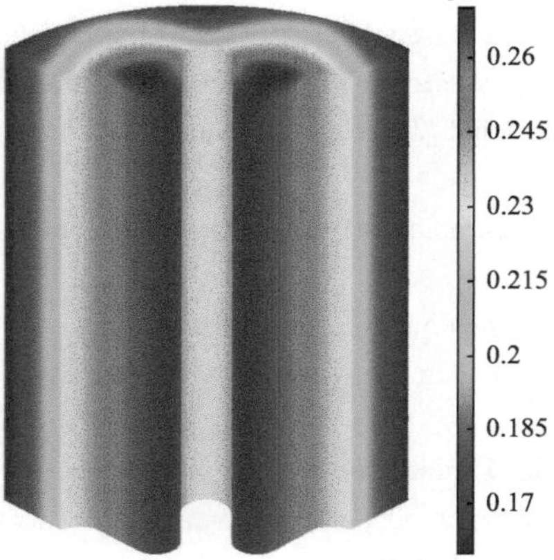

图 4.21　八星孔型药柱整体模型的径向位移云图

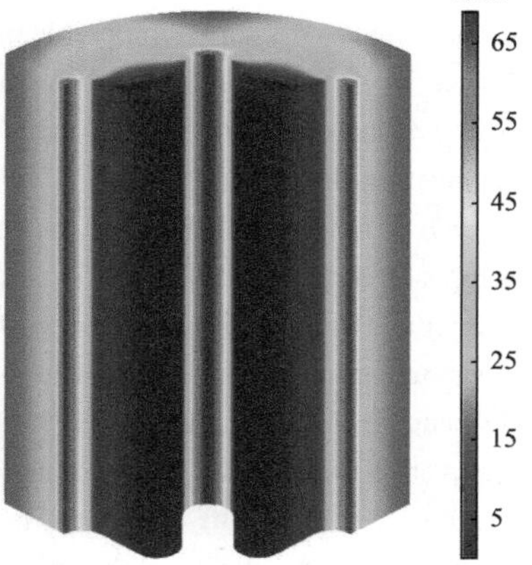

图 4.22　八星孔型药柱 von Mises 应力云图

4.5 小 结

本章给出了求解黏弹性问题的等几何有限元分析方法。该算法可用于分析黏弹性材料(如固体火箭推进剂和橡胶材料)的力学性能。

针对承受内压载荷的二维、三维厚壁圆筒模型，分别讨论了细化次数以及时间步长对计算结果精度的影响。从 IGAFEM 数值结果与解析解的位移对比曲线可以发现，模型细化次数越多，时间步长越小，则计算精度越高，且当细化次数为 8、固定步长为 $\Delta t = 0.05\text{s}$ 时，相对误差可达到 $O(10^{-4})$ 数量级。本章验证了 IGAFEM 求解黏弹性问题的正确性与收敛性，在计算不同力学问题时具有较高的计算精度，且达到同样计算精度时，IGAFEM 所需单元数量及节点数量相比传统有限元少很多。

参 考 文 献

[1] 王家林. 固体火箭推进剂有限元理论分析与计算[D]. 哈尔滨：哈尔滨工程大学, 2003.

[2] 赵经文, 王宏钰. 结构有限元分析[M]. 北京：科学出版社, 2001.

[3] Sardar Bilal D, Khan N, Fatima I. Mixed convective heat transfer in a power-law fluid in a square enclosure: Higher order finite element solutions[J]. Frontiers in Physics, 2023, 10: 1079641.

[4] Albertos-Cabanas M, Lopez-Pascual D, Valiente-Blanco I. A novel ultra-low power consumption electromagnetic actuator based on potential magnetic energy: theoretical and finite element analysis[J]. Actuators, 2023, 12: 87-115.

[5] Abali B E. An accurate finite element method for the numerical solution of isothermal and incompressible flow of viscous fluid[J]. Fluids, 2019, 4(1): 5-24.

[6] Srinatha H R, Lewis R W. A finite element method for thermoviscoelastic analysis of plane problems[J]. Computer Methods in Applied Mechanics and Engineering, 1981, 25(1): 21-33.

[7] Shen Y P, Hasebe N, Lee L X. The finite element method of three-dimensional nonlinear viscoelastic large deformation problems[J]. Computers & Structures, 1995, 55(4): 659-666.

[8] Zocher M A, Groves S E, Allen D H. A three-dimensional finite element formulation for thermoviscoelastic orthotropic media[J]. International Journal for Numerical Methods in Engineering, 1997, 40(12): 2267-2288.

[9] Pavan R C, Oliveira B F, Maghous S. A model for anisotropic viscoelastic damage in composites[J]. Composite Structures, 2010, 92(5): 1223-1228.

[10] Hinterhoelzl R M, Schapery R A. FEM implementation of a three-dimensional viscoelastic constitutive model for particulate composites with damage growth[J]. Mechanics of Time-Dependent Materials, 2004, 8(1): 65-94.

[11] Nguyen S N, Lee J, Cho M. Application of the Laplace transformation for the analysis of viscoelastic composite laminates based on equivalent single-layer theories[J]. International Journal of Aeronautical and Space Sciences, 2012, 13: 458-467.

[12] Hughes T J R, Cottrell J A, Bazilevs Y. Isogeometric analysis: CAD, finite elements, NURBS,

exact geometry and mesh refinement[J]. Computer Methods in Applied Mechanics and Engineering, 2005, 194 (39): 4135-4195.

[13] Cottrell J A, Hughes T J R, Bazilevs Y. Isogeometric Analysis: Toward Integration of CAD and FEA[M]. New York: John Wiley & Sons, Inc., 2009.

[14] Yin S, Yu T, Bui T Q, et al. In-plane material inhomogeneity of functionally graded plates: a higherorder shear deformation plate isogeometric analysis[J]. Composites Part B: Engineering, 2016, 106: 273-284.

[15] Farzam A, Hassani B. Isogeometric analysis of in-plane functionally graded porous microplates using modified couple stress theory[J]. Aerospace Science and Technology, 2019, 91: 508-524.

[16] Simpson R, Scott M, Taus M, et al. Acoustic isogeometric boundary element analysis[J]. Computer Methods in Applied Mechanics and Engineering, 2014, 269: 265-290.

[17] Sun D Y, Dong C Y. Shape optimization of heterogeneous materials based on isogeometric boundary element method[J]. Computer Methods in Applied Mechanics and Engineering, 2020, 370: 113279.

[18] Benson D J, Bazilevs Y, Hsu M C. Isogeometric shell analysis: the Reissner-Mindlin shell[J]. Computer Methods in Applied Mechanics and Engineering, 2010, 199(5): 276-289.

[19] Sun F L, Dong C Y, Yang H S. Isogeometric boundary element method for crack propagation based on Bézier extraction of NURBS[J]. Engineering Analysis with Boundary Elements, 2019, 99: 76-88.

[20] Piegl L, Tiller W. 非均匀有理 B 样条[M]. 2 版. 赵罡, 穆国旺, 王拉柱, 译. 北京: 清华大学出版社, 2010.

[21] Muki R, Sternberg E. On transient thermal stresses in viscoelastic materials with temperature-dependent properties[J]. Journal of Applied Mechanics, 1961, 28: 193-207.

[22] 王勖成. 有限单元法. 北京：清华大学出版社，2002.

[23] Lee E H, Rogers T G. Solution of viscoelastic stress analysis problems using measured creep or relaxation functions[J]. Journal of Applied Mechanics, 1963, 30: 127-133.

[24] Zhan Y S, Xu C, Yang H S, et al. Isogeometric FE-BE method with non-conforming coupling interface for solving elasto-thermoviscoelastic problems[J]. Engineering Analysis with Boundary Elements, 2022, 141:199-221.

[25] Xu C, Zhan Y S, Dai R, et al. RI-IGAEM for 3D viscoelastic problems with body force[J]. Computer Methods in Applied Mechanics and Engineering, 2022, 394: 114911.

[26] Xu C, Yang H S, Zhan Y S, et al. Non-conforming coupling RI-IGABEM for solving multidimensional and multiscale thermoelastic-viscoelastic thermoelastic-viscoelastic problems[J]. Computer Methods in Applied Mechanics and Engineering, 2023, 403: 115725.

[27] 王本华. 固体推进剂的热黏弹性有限元分析[J]. 推进技术, 1985, (3): 18-26.

[28] Marques S P C, Creus G J. Computational Viscoelasticity[M]. New York: Springer, 2012.

[29] Timoshenko S P, Goodier J N. Theory of Elasticity[M]. New York: McGraw-Hill, 1969.

第5章

瞬态热传导问题的等几何边界元分析

5.1 引　言

在工程应用中，材料的选择对于确保设备在极端环境下的安全稳定运行至关重要。普通金属材料难以满足工程上的需求，功能梯度材料(FGM)[1,2]由于其良好的性能，被广泛应用于冶金行业[3]、航空航天[4]和制造业[5]等。因此，近几十年来，大量学者对 FGM 的瞬态热传导问题进行了研究，主要采用了无网格法(meshless method)[6]、FEM[7,8]和 BEM[9,10]等数值方法。

到目前为止，求解 FGM 瞬态热传导问题的传统数值方法已经比较成熟。然而，如何减小模型离散化过程中的几何误差仍然是一个具有挑战性的课题。众所周知，网格划分过程将花费大量时间，网格划分不当将严重影响最终结果。为了 CAD 与 CAE 之间的无缝连接，Hughes 等[11]采用 NURBS 基函数作为 FEM 的形函数，并首次将这种方法命名为等几何分析(IGA)。IGA 的优点是能精确描述几何形状，无需网格离散，基函数具有非负高阶次，单元具有高连续性。

众所周知，BEM 的降维和半解析的优点使其成为一种强大的数值工具[12]。此外，由于 BEM 只对边界进行离散，而 CAD 也只提供表面数据，因此 IGA 可以更自然地应用于 BEM。Simpson 等[13]将 IGA 和 BEM 结合起来求解二维弹性问题，首次提出 IGABEM 的概念。与传统 BEM 相比，IGABEM 不仅保留了半解析性和仅边界离散化的特点，而且具有基函数的非负高阶次、单元具有高连续性和网格细化灵活等诸多优点[14]。近年来，IGABEM 已被应用于求解势问题[15,16]、弹性问题[17,18]、声学问题[19,20]等。本章将利用 IGABEM 求解 FGM 瞬态热传导问题。

如何有效处理域积分是 IGABEM 需要面对的问题。由于采用势问题的基本解来推导 FGM 瞬态热传导问题的边界积分方程，因此，随坐标变化的热传导系数、热源和初始温度[21]都会导致积分方程中存在域积分。求解域积分常用的方法是将积分域划分为单元[22]，但这种方法会使边界元失去只对边界离散的优势。因此，将域积分转化为等效的边界积分是十分必要的。近几十年来，学者们提出了很多方法将域积分转换为等效的边界积分。其中，最常用的将域积分转换为等效边界积分的方法是 Nardini 和 Brebbia[23]提出的双重互易法(DRM)。DRM 的核心思想是在控制方程的算子上求解域内函数的特解，然后用格林公式将域积分转化为边界积分。由于域内函数的特解难以求出，在 DRM 中通常采用径向基函数(RBF)对域内函数进行插值，从而求得基函数的特解。DRM 已被广泛用于解决泊松(Poisson)方程、亥姆霍兹(Helmholtz)方程和纳维(Navier)方程等许多问题[24,25]。DRM 的缺点是，如果积分方程中域积分的核函数与对应问题的基本解不同[26]，则该方法难以实现。此外，该方法需要基函数的特解，这限制了其在复杂问题中的应用[27]。本章我们将径向积分法(RIM)应用到 IGABEM 中求解二维及三维问题 FGM 瞬态热传导问题。RIM 的一个重要特点是它可以将任何复杂的域积分转化为等效的边界积分，而不需要使用拉普拉斯算子和问题的特解。此外，它还可以去除域积分的一阶奇异性[28,29]。同时，RIM 可以很自然地用于求解多边界问题的域积分[30]。对于核函数已知的情况，我们可以直接利用 RIM 将域积分转换为等效的边界积分。对于核函数未知的情况，我们首先利用四阶样条紧支径向基函数[31]对未知核函数进行展开，然后再利用 RIM 将其转换为等效的边界积分。

一般来说，求解瞬态热传导问题的方法可分为两大类：频域法[32]和时域法[33]。频域法可以得到更精确的结果。然而，该方法的计算精度取决于变换参数。此外，其逆变换也不适用于大规模计算。对于时域法，一般采用逐步积分法，而逐步积分法又可进一步分为显式方法和隐式方法。显式方法不需要迭代，且每个周期的计算效率高。然而，为了保证计算结果的准确性和稳定性，时间步长必须足够小，这往往会导致大量的计算成本。隐式方法通过选择合适的参数来保证计算结果的稳定性。为了保证结果的收敛性，每一步都需要迭代，所以适当增加时间步长仍然可以保证得到较好的结果。然而，由于每一步都需要迭代，计算效率较低，因此，传统的逐步积分法难以同时兼顾计算精度、稳定性和效率。本章将 PIM 引入 IGABEM 中分析了瞬态热传导问题。PIM 由钟万勰 [34]于 1994 年提出，已被广泛应用于求解动力学 [35]、瞬态热传导[36]等问题。该方法的关键思想是计算指数矩阵，并将每个时间步分成 2^N (N 为给定的正整数)段。因此，时间步长对计算结果影响不大[37]。

和传统 BEM 类似，当源点和场点在一个单元内，且源点趋近于场点时会出现不同阶次的奇异积分，如何正确求解不同阶次的奇异积分是保证 IGABEM 求解精度的关键。Simpson 等[13]使用奇异值提取技术(SST)[38]来计算 IGABEM 中二维弹性问题的强奇异积分。Gong 和 Dong[39,40]采用幂级数展开法[41]求解 IGABEM 中的三维强奇异积分。众所周知，与 SST 和幂级数展开法相比，常位势法[42]是传统 BEM 中求解强奇异积分最简单的方法。但是，由于 NURBS 基函数的特殊性质，常位势法不能直接用于 IGABEM 中。在本章中，通过控制点到配点的变换，将常位势法应用于 IGABEM 中的强奇异积分的求解。

5.2 问题描述

功能梯度材料瞬态热传导问题的控制方程可以写成如下形式：

$$\frac{\partial}{\partial x_i}\left[\lambda(\boldsymbol{x})\frac{\partial T(\boldsymbol{x},t)}{\partial x_i}\right]+Q(\boldsymbol{x},t)=\rho\tilde{c}\frac{\partial T(\boldsymbol{x},t)}{\partial t},\quad \boldsymbol{x}\in\Omega \tag{5.1}$$

其中，下标 $i=1,2$ (二维问题)或 1,2,3(三维问题)；$\boldsymbol{x}$ 表示计算域 Ω 中的点(内点和边界点)；$T(\boldsymbol{x},t)$ 和 $Q(\boldsymbol{x},t)$ 分别表示 t 时刻点 $\boldsymbol{x}$ 处的温度和热源强度；ρ 表示材料密度；$\tilde{c}$ 表示材料的比热容；$\lambda(\boldsymbol{x})$ 表示随坐标变化的热传导系数。温度和热流的边界条件如下：

$$\begin{cases}T(\boldsymbol{x},t)=\overline{T}(\boldsymbol{x},t), & \boldsymbol{x}\in\Gamma_T\\ -\lambda(\boldsymbol{x})\dfrac{\partial T(\boldsymbol{x},t)}{\partial n}=\overline{q}(\boldsymbol{x},t), & \boldsymbol{x}\in\Gamma_q\end{cases} \tag{5.2}$$

其中，$\overline{T}(\boldsymbol{x},t)$ 表示边界 Γ_T 上给定的温度值；$\overline{q}(\boldsymbol{x},t)$ 表示边界 Γ_q 上给定的热流值。此外，边界 Γ_T 和 Γ_q 满足：$\Gamma_T\cap\Gamma_q=\varnothing$ 和 $\Gamma_T\cup\Gamma_q=\Gamma$。初始条件可以用如下形式表示：

$$T(\boldsymbol{x},t_0)=\overline{T}_0(\boldsymbol{x},t_0) \tag{5.3}$$

其中，$\overline{T}_0$ 表示初始温度。

5.3 边界域积分方程

由于本章采用拉普拉斯方程的基本解推导边界域积分方程，因此，瞬态热传导问题中的随坐标变化的热源、热传导系数以及初始温度会导致边界积分方程中存在域积分。在本节中，我们将详细描述如何将 PIM 和 IGABEM 结合起来，以解决含有热源的 FGM 的瞬态热传导问题。

5.3.1 规则化边界域积分方程

对于控制方程(5.1)，可以采用伽辽金加权余量法和高斯散度定理来获得以下形式的边界域积分方程：

$$\begin{aligned}\tilde{T}(\boldsymbol{p},t)=&-\int_\Gamma T^*(\tilde{\boldsymbol{Q}},\boldsymbol{p})q(\tilde{\boldsymbol{Q}},t)\mathrm{d}\Gamma(\tilde{\boldsymbol{Q}})-\int_\Gamma q^*(\tilde{\boldsymbol{Q}},\boldsymbol{p})\tilde{T}(\tilde{\boldsymbol{Q}},t)\mathrm{d}\Gamma(\tilde{\boldsymbol{Q}})\\&+\int_\Omega T^*(\tilde{\boldsymbol{q}},\boldsymbol{p})Q(\tilde{\boldsymbol{q}},t)\mathrm{d}\Omega(\tilde{\boldsymbol{q}})+\int_\Omega V(\boldsymbol{p},\tilde{\boldsymbol{q}})\tilde{T}(\tilde{\boldsymbol{q}},t)\mathrm{d}\Omega(\tilde{\boldsymbol{q}})\\&-\int_\Omega T^*(\tilde{\boldsymbol{q}},\boldsymbol{p})\tilde{\rho}(\tilde{\boldsymbol{q}})\dot{\tilde{T}}(\tilde{\boldsymbol{q}},t)\mathrm{d}\Omega(\tilde{\boldsymbol{q}})\end{aligned} \tag{5.4}$$

其中，$\boldsymbol{p}$ 表示内部源点；$\tilde{\boldsymbol{Q}}$ 表示边界场点；$\tilde{\boldsymbol{q}}$ 表示内部场点；$\dot{\tilde{T}}$ 表示规则化温度 $\tilde{T}$ 对时间的导数。此外，式(5.4)中的其他参数在下面给出：

$$q^*(\tilde{\boldsymbol{Q}},\boldsymbol{p})=\frac{\partial T^*(\tilde{\boldsymbol{Q}},\boldsymbol{p})}{\partial n}=\frac{-r_{,i}}{2\pi\mu r^{\mu}}n_i \tag{5.5}$$

$$\tilde{\rho}(\tilde{\boldsymbol{q}})=\rho(\tilde{\boldsymbol{q}})\tilde{c}(\tilde{\boldsymbol{q}})/\lambda(\tilde{\boldsymbol{q}}) \tag{5.6}$$

$$V(\boldsymbol{p},\tilde{\boldsymbol{q}})=\frac{\partial T^*(\boldsymbol{p},\tilde{\boldsymbol{q}})}{\partial x_i}\frac{\partial\tilde{\lambda}(\tilde{\boldsymbol{q}})}{\partial x_i}=T^*_{,i}\tilde{\lambda}_{,i} \tag{5.7}$$

$$\tilde{T}(\tilde{\boldsymbol{q}},t)=\lambda(\tilde{\boldsymbol{q}})T(\tilde{\boldsymbol{q}},t) \tag{5.8}$$

$$\tilde{\lambda}(\tilde{\boldsymbol{q}})=\ln\lambda(\tilde{\boldsymbol{q}}) \tag{5.9}$$

$$T^*_{,i}=\frac{\partial T^*(\tilde{\boldsymbol{q}},\boldsymbol{p})}{\partial x_i}=\frac{-r_{,i}}{2\pi\mu r^{\mu}} \tag{5.10}$$

其中，$\mu=1$ (二维问题)或 2(三维问题)；格林函数 T^* 可以写成如下形式：

$$\begin{cases} T^*(\tilde{\boldsymbol{q}},\boldsymbol{p})=\dfrac{1}{2\pi}\ln\left(\dfrac{1}{r}\right), & \text{二维} \\ T^*(\tilde{\boldsymbol{q}},\boldsymbol{p})=\dfrac{1}{4\pi r}, & \text{三维} \end{cases} \tag{5.11}$$

式中，r 表示源点 $\boldsymbol{p}$ 到场点 $\tilde{\boldsymbol{q}}$ 的距离，即 $r=\|\boldsymbol{p}-\tilde{\boldsymbol{q}}\|$。此外，$r_{,i}$ 表示 r 对 x_i 求偏导，即 $r_{,i}=\partial r/\partial x_i=(x_i^{\tilde{Q}}-x_i^{p})/r(\tilde{\boldsymbol{Q}},\boldsymbol{p})$。式(5.4)为内点的规则化边界域积分方程。边界节点的规则化边界域积分方程可以写成如下形式：

$$\begin{aligned} c(\boldsymbol{P})\tilde{T}(\boldsymbol{P},t)=&-\int_{\Gamma}T^*(\tilde{\boldsymbol{Q}},\boldsymbol{P})q(\tilde{\boldsymbol{Q}},t)\mathrm{d}\Gamma(\tilde{\boldsymbol{Q}})-\int_{\Gamma}q^*(\tilde{\boldsymbol{Q}},\boldsymbol{P})\tilde{T}(\tilde{\boldsymbol{Q}},t)\mathrm{d}\Gamma(\tilde{\boldsymbol{Q}}) \\ &+\int_{\Omega}T^*(\tilde{\boldsymbol{q}},\boldsymbol{P})Q(\tilde{\boldsymbol{q}},t)\mathrm{d}\Omega(\tilde{\boldsymbol{q}})+\int_{\Omega}V(\tilde{\boldsymbol{q}},\boldsymbol{P})\tilde{T}(\tilde{\boldsymbol{q}},t)\mathrm{d}\Omega(\tilde{\boldsymbol{q}}) \\ &-\int_{\Omega}T^*(\tilde{\boldsymbol{q}},\boldsymbol{P})\tilde{\rho}(\tilde{\boldsymbol{q}})\dot{\tilde{T}}(\tilde{\boldsymbol{q}},t)\mathrm{d}\Omega(\tilde{\boldsymbol{q}}) \end{aligned} \tag{5.12}$$

其中，$\boldsymbol{P}$ 表示边界点；$c(\boldsymbol{P})$ 是与边界点 $\boldsymbol{P}$ 相关的常数。

5.3.2　利用径向积分法将域积分转换为边界积分

从边界域积分方程式(5.4)和式(5.12)中可以清楚地发现，随坐标变化的热源、热传导系数和初始温度将导致积分方程中存在域积分，即 $\int_{\Omega}T^*(\tilde{\boldsymbol{q}},\boldsymbol{p})Q(\tilde{\boldsymbol{q}},t)\mathrm{d}\Omega(\tilde{\boldsymbol{q}})$、$\int_{\Omega}V(\boldsymbol{p},\tilde{\boldsymbol{q}})\tilde{T}(\tilde{\boldsymbol{q}},t)\mathrm{d}\Omega(\tilde{\boldsymbol{q}})$ 和 $\int_{\Omega}T^*(\tilde{\boldsymbol{q}},\boldsymbol{p})\tilde{\rho}(\tilde{\boldsymbol{q}})\dot{\tilde{T}}(\tilde{\boldsymbol{q}},t)\mathrm{d}\Omega(\tilde{\boldsymbol{q}})$。第一个域积分是由热源项产生的，它一般是一个常数或是坐标和时间的函数，可以应用径向积分法将该域

积分精确地转化为等效的边界积分。第二个域积分和第三个域积分分别由随坐标变化的热传导系数和初始温度产生。由于被积函数为未知的，我们需要用已知基函数表示未知值，然后利用径向积分法将域积分转化为边界积分。

1. 已知核函数

第一个域积分是由热源引起的。因为热源一般是一个常数或坐标和时间的函数，所以被积函数为已知的。因此，可以利用径向积分法直接将其转化为如下形式的等效边界积分：

$$\int_{\Omega} T^*(\tilde{\boldsymbol{q}},\boldsymbol{p})Q(\tilde{\boldsymbol{q}},t)\mathrm{d}\Omega(\tilde{\boldsymbol{q}}) = \int_{\Gamma} \frac{1}{r^{\mu}(\tilde{\boldsymbol{Q}},\boldsymbol{p})}\frac{\partial r}{\partial n}F(\tilde{\boldsymbol{Q}},\boldsymbol{p},t)\mathrm{d}\Gamma(\tilde{\boldsymbol{Q}}) \tag{5.13}$$

其中，$r(\tilde{\boldsymbol{Q}},\boldsymbol{p})$ 表示源点 $\boldsymbol{p}$ 到边界场点 $\tilde{\boldsymbol{Q}}$ 的距离。径向积分 $F(\tilde{\boldsymbol{Q}},\boldsymbol{p},t)$ 可以写成如下形式：

$$F(\tilde{\boldsymbol{Q}},\boldsymbol{p},t) = \int_0^{r(\tilde{\boldsymbol{Q}},\boldsymbol{p})} T^*(\tilde{\boldsymbol{q}},\boldsymbol{p})Q(\tilde{\boldsymbol{q}},t)r^{\mu}\mathrm{d}r(\tilde{\boldsymbol{q}}) \tag{5.14}$$

式(5.14)中的被积函数形式简单，可直接用高斯积分求解。

2. 未知核函数

域积分 $\int_{\Omega} V(\boldsymbol{p},\tilde{\boldsymbol{q}})\tilde{T}(\tilde{\boldsymbol{q}},t)\mathrm{d}\Omega(\tilde{\boldsymbol{q}},t)$ 和 $\int_{\Omega} T^*(\tilde{\boldsymbol{q}},\boldsymbol{p})\tilde{\rho}(\tilde{\boldsymbol{q}})\dot{\tilde{T}}(\tilde{\boldsymbol{q}},t)\mathrm{d}\Omega(\tilde{\boldsymbol{q}})$ 含有未知被积函数 $\tilde{T}(\tilde{\boldsymbol{q}},t)$ 和 $\dot{\tilde{T}}(\tilde{\boldsymbol{q}},t)$，因此不能直接应用径向积分法。在进行转换之前，未知被积函数需要利用基函数进行展开。我们采用增强的紧支径向基函数将 $\tilde{T}(\tilde{\boldsymbol{q}},t)$ 和 $\dot{\tilde{T}}(\tilde{\boldsymbol{q}},t)$ 展开为如下形式：

$$\tilde{T} = \sum_{A=1}^{N} \alpha_A \phi^A(R) + a_0 + \sum_{i=1}^{k} a_i x_i^{\tilde{q}} \tag{5.15}$$

$$\dot{\tilde{T}} = \sum_{A=1}^{N} \beta_A \phi^A(R) + b_0 + \sum_{i=1}^{k} b_i x_i^{\tilde{q}} \tag{5.16}$$

其中，参数 α_A 和 β_A 应满足下列条件：

$$\sum_{A=1}^{N} \alpha_A = \sum_{A=1}^{N} \alpha_A x_i^A = 0,\quad i=1,2\ (\text{二维问题});\ i=1,2,3\ (\text{三维问题}) \tag{5.17}$$

$$\sum_{A=1}^{N} \beta_A = \sum_{A=1}^{N} \beta_A x_i^A = 0,\quad i=1,2\ (\text{二维问题});\ i=1,2,3\ (\text{三维问题}) \tag{5.18}$$

式中，A 表示应用点，由所有边界配点和内点组成；N 表示应用点的数目，

$N = N_{\mathrm{b}} + N_{\mathrm{I}}$，这里，$N_{\mathrm{b}}$ 表示边界配点的个数，N_{I} 表示内点的个数；$k=2$ 或 3；α_A、β_A、a_0、b_0、a_i 和 b_i 为待定系数；$\phi^A(R)$ 为径向基函数。为了平衡算法的精度和计算效率，我们采用四阶样条紧支径向基函数。式(5.15)～式(5.18)可以写成如下矩阵形式：

$$\begin{Bmatrix} \tilde{\boldsymbol{T}}_{N\times 1} \\ \boldsymbol{0}_{(k+1)\times 1} \end{Bmatrix} = \boldsymbol{\Phi}_{(N+k+1)\times(N+k+1)} \begin{Bmatrix} \boldsymbol{\alpha}_{N\times 1} \\ \boldsymbol{a}_{(k+1)\times 1} \end{Bmatrix} \tag{5.19}$$

$$\begin{Bmatrix} \dot{\tilde{\boldsymbol{T}}}_{N\times 1} \\ \boldsymbol{0}_{(k+1)\times 1} \end{Bmatrix} = \boldsymbol{\Phi}_{(N+k+1)\times(N+k+1)} \begin{Bmatrix} \boldsymbol{\beta}_{N\times 1} \\ \boldsymbol{b}_{(k+1)\times 1} \end{Bmatrix} \tag{5.20}$$

将式(5.15)代入域积分 $\int_{\Omega} V(\boldsymbol{p},\tilde{\boldsymbol{q}})\tilde{T}(\tilde{\boldsymbol{q}},t)\mathrm{d}\Omega(\tilde{\boldsymbol{q}})$ 中，即可利用径向积分法将其转化为如下形式的等效边界积分：

$$\begin{aligned} \int_{\Omega} V(\tilde{\boldsymbol{q}},\boldsymbol{p})\tilde{T}(\tilde{\boldsymbol{q}},t)\mathrm{d}\Omega(\tilde{\boldsymbol{q}}) = {} & \alpha_A \int_{\Gamma} \frac{1}{r^{\mu}(\tilde{\boldsymbol{Q}},\boldsymbol{p})} \frac{\partial r}{\partial n} F_A^{\alpha}(\tilde{\boldsymbol{q}},\boldsymbol{p})\mathrm{d}\Gamma(\tilde{\boldsymbol{Q}}) \\ & + a_0 \int_{\Gamma} \frac{1}{r^{\mu}(\tilde{\boldsymbol{Q}},\boldsymbol{p})} \frac{\partial r}{\partial n} F_0^{\alpha}(\tilde{\boldsymbol{q}},\boldsymbol{p})\mathrm{d}\Gamma(\tilde{\boldsymbol{Q}}) \\ & + a_i \int_{\Gamma} \frac{1}{r^{\mu}(\tilde{\boldsymbol{Q}},\boldsymbol{p})} \frac{\partial r}{\partial n} F_1^{\alpha}(\tilde{\boldsymbol{q}},\boldsymbol{p})\mathrm{d}\Gamma(\tilde{\boldsymbol{Q}}) \end{aligned} \tag{5.21}$$

其中，

$$F_A^{\alpha}(\tilde{\boldsymbol{q}},\boldsymbol{p}) = \int_0^{r(\boldsymbol{p},\tilde{\boldsymbol{Q}})} V(\tilde{\boldsymbol{q}},\boldsymbol{p})\phi^A(R) r^{\mu}(\tilde{\boldsymbol{q}},\boldsymbol{p})\mathrm{d}r(\tilde{\boldsymbol{q}}) \tag{5.22}$$

$$F_0^{\alpha}(\tilde{\boldsymbol{q}},\boldsymbol{p}) = \int_0^{r(\boldsymbol{p},\tilde{\boldsymbol{Q}})} V(\tilde{\boldsymbol{q}},\boldsymbol{p}) r^{\mu}(\tilde{\boldsymbol{q}},\boldsymbol{p})\mathrm{d}r(\tilde{\boldsymbol{q}}) \tag{5.23}$$

$$F_1^{\alpha}(\tilde{\boldsymbol{q}},\boldsymbol{p}) = \int_0^{r(\boldsymbol{p},\tilde{\boldsymbol{Q}})} V(\tilde{\boldsymbol{q}},\boldsymbol{p})(x_i^p + r_{,i} r) r^{\mu}(\tilde{\boldsymbol{q}},\boldsymbol{p})\mathrm{d}r(\tilde{\boldsymbol{q}}) \tag{5.24}$$

同理，将式(5.16)代入域积分 $\int_{\Omega} T^*(\tilde{\boldsymbol{q}},\boldsymbol{p})\tilde{\rho}(\tilde{\boldsymbol{q}})\dot{\tilde{T}}(\tilde{\boldsymbol{q}},t)\mathrm{d}\Omega(\tilde{\boldsymbol{q}})$ 中，由径向积分法将其转换为如下的等效边界积分：

$$\begin{aligned} \int_{\Omega} T^*(\tilde{\boldsymbol{q}},\boldsymbol{p})\tilde{\rho}(\tilde{\boldsymbol{q}})\dot{\tilde{T}}(\tilde{\boldsymbol{q}},t)\mathrm{d}\Omega(\tilde{\boldsymbol{q}}) = {} & \beta_A \int_{\Gamma} \frac{1}{r^{\mu}(\tilde{\boldsymbol{Q}},\boldsymbol{p})} \frac{\partial r}{\partial n} F_A^{\beta}(\tilde{\boldsymbol{q}},\boldsymbol{p})\mathrm{d}\Gamma(\tilde{\boldsymbol{Q}}) \\ & + b_0 \int_{\Gamma} \frac{1}{r^{\mu}(\tilde{\boldsymbol{Q}},\boldsymbol{p})} \frac{\partial r}{\partial n} F_0^{\beta}(\tilde{\boldsymbol{q}},\boldsymbol{p})\mathrm{d}\Gamma(\tilde{\boldsymbol{Q}}) \\ & + b_i \int_{\Gamma} \frac{1}{r^{\mu}(\tilde{\boldsymbol{Q}},\boldsymbol{p})} \frac{\partial r}{\partial n} F_1^{\beta}(\tilde{\boldsymbol{q}},\boldsymbol{p})\mathrm{d}\Gamma(\tilde{\boldsymbol{Q}}) \end{aligned} \tag{5.25}$$

其中，

$$F_A^{\beta}(\tilde{\boldsymbol{q}},\boldsymbol{p})=\int_0^{r(\boldsymbol{p},\tilde{\boldsymbol{Q}})}T^*(\tilde{\boldsymbol{q}},\boldsymbol{p})\tilde{\rho}(\tilde{\boldsymbol{q}})\phi^A(R)r^{\mu}(\tilde{\boldsymbol{q}},\boldsymbol{p})\mathrm{d}r(\tilde{\boldsymbol{q}}) \tag{5.26}$$

$$F_0^{\beta}(\tilde{\boldsymbol{q}},\boldsymbol{p})=\int_0^{r(\boldsymbol{p},\tilde{\boldsymbol{Q}})}T^*(\tilde{\boldsymbol{q}},\boldsymbol{p})\tilde{\rho}(\tilde{\boldsymbol{q}})r^{\mu}(\tilde{\boldsymbol{q}},\boldsymbol{p})\mathrm{d}r(\tilde{\boldsymbol{q}}) \tag{5.27}$$

$$F_1^{\beta}(\tilde{\boldsymbol{q}},\boldsymbol{p})=\int_0^{r(\boldsymbol{p},\tilde{\boldsymbol{Q}})}T^*(\tilde{\boldsymbol{q}},\boldsymbol{p})\tilde{\rho}(\tilde{\boldsymbol{q}})(x_i^p+r_{,i}r)r^{\mu}(\tilde{\boldsymbol{q}},\boldsymbol{p})\mathrm{d}r(\tilde{\boldsymbol{q}}) \tag{5.28}$$

式中，R 表示场点 $\tilde{\boldsymbol{q}}$ 到应用点 A 的距离。为了便于使用径向积分法，我们需要将 R 转换为 r 的函数。该方法不仅适用于单连通问题，也适用于多连通问题。

5.4 边界积分方程的等几何分析

与传统 BEM 类似，IGABEM 也具有高精度和降维的优点。同时，IGABEM 可以精确描述模型的几何形状。本章采用 NURBS 来描述模型的几何形状和近似物理量。关于 NURBS 曲线和曲面的更多的内容可参考文献[43]。

5.4.1 边界积分方程的 NURBS 离散

与传统 BEM 不同，IGABEM 可以精确描述几何形状。边界配点 $\boldsymbol{x}(\xi)$ 、规则化温度 $\tilde{T}(\xi,t)$ 和热流 $q(\xi,t)$ 可以由控制点的物理量系数插值成如下形式：

$$\begin{cases}\boldsymbol{x}(\xi)=\sum\limits_{i=1}^{\hat{p}+1}R_{i,\hat{p}}(\xi)\bar{\boldsymbol{P}}_i\\ \tilde{T}(\xi,t)=\sum\limits_{i=1}^{\hat{p}+1}R_{i,\hat{p}}(\xi)\tilde{T}_i(t)\\ q(\xi,t)=\sum\limits_{i=1}^{\hat{p}+1}R_{i,\hat{p}}(\xi)q_i(t)\end{cases} \tag{5.29}$$

其中，$\bar{\boldsymbol{P}}_i$ 、$\tilde{T}_i(t)$ 和 $q_i(t)$ 分别表示 t 时刻控制点的坐标、规则化的温度和热流系数；$\hat{p}$ 是 NURBS 基函数 R 的阶次。此外，二维问题的积分方程可以离散为以下形式：

$$\begin{aligned}\tilde{T}(\boldsymbol{p},t)=&-\sum_{e=1}^{N_e}\sum_{l=1}^{\hat{p}+1}\left\{\int_{-1}^{1}q^*\left[\tilde{\boldsymbol{Q}}(\hat{\xi}),\boldsymbol{p}\right]R_{l,p}^e(\hat{\xi})J^e(\hat{\xi})\mathrm{d}\hat{\xi}\right\}\tilde{T}_l^e(t)\\ &-\sum_{e=1}^{N_e}\sum_{l=1}^{\hat{p}+1}\left\{\int_{-1}^{1}T^*\left[\tilde{\boldsymbol{Q}}(\hat{\xi}),\boldsymbol{p}\right]R_{l,p}^e(\hat{\xi})J^e(\hat{\xi})\mathrm{d}\hat{\xi}\right\}q_l^e(t)\\ &+\sum_{e=1}^{N_e}\sum_{l=1}^{\hat{p}+1}\left\{\int_{-1}^{1}\frac{1}{r\left[\tilde{\boldsymbol{Q}}(\hat{\xi}),\boldsymbol{p}\right]}\frac{\partial r}{\partial n}F\left[\tilde{\boldsymbol{Q}}(\hat{\xi}),\boldsymbol{p},t\right]J^e(\hat{\xi})\mathrm{d}\hat{\xi}\right\}+\boldsymbol{V}(\boldsymbol{p})\tilde{\boldsymbol{T}}(t)-\boldsymbol{C}(\boldsymbol{p})\dot{\tilde{\boldsymbol{T}}}(t)\end{aligned} \tag{5.30}$$

$$c(\boldsymbol{P})\sum_{l=1}^{\hat{p}+1}R_{l,p}^{e'}(\hat{\xi}')\tilde{T}_l^{e'}(t)+\sum_{e=1}^{N_e}\sum_{l=1}^{\hat{p}+1}\left\{\int_{-1}^{1}q^*\left[\tilde{\boldsymbol{Q}}(\hat{\xi}),\boldsymbol{P}\right]R_{l,p}^{e}(\hat{\xi})J^e(\hat{\xi})\mathrm{d}\hat{\xi}\right\}\tilde{T}_l^e(t)$$
$$=\sum_{e=1}^{N_e}\sum_{l=1}^{\hat{p}+1}\left\{\int_{-1}^{1}T^*\left[\tilde{\boldsymbol{Q}}(\hat{\xi}),\boldsymbol{P}\right]R_{l,p}^{e}(\hat{\xi})J^e(\hat{\xi})\mathrm{d}\hat{\xi}\right\}q_l^e(t)$$
$$+\sum_{e=1}^{N_e}\sum_{l=1}^{\hat{p}+1}\left\{\int_{-1}^{1}\frac{1}{r\left[\tilde{\boldsymbol{Q}}(\hat{\xi}),\boldsymbol{P}\right]}\frac{\partial r}{\partial n}F\left[\tilde{\boldsymbol{Q}}(\hat{\xi}),\boldsymbol{P},t\right]J^e(\hat{\xi})\mathrm{d}\hat{\xi}\right\}+\boldsymbol{V}(\boldsymbol{P})\tilde{\boldsymbol{T}}(t)-\boldsymbol{C}(\boldsymbol{P})\dot{\tilde{\boldsymbol{T}}}(t)$$

(5.31)

其中，局部坐标 $\hat{\xi}$ 为−1 到 1(图 5.1 (a))；$\hat{\xi}'$ 是源点所在单元 e' 的局部坐标；N_e 表示单元的个数。此外，雅可比行列式可以写为如下形式：

$$J^e(\hat{\xi})=\frac{\mathrm{d}\Gamma}{\mathrm{d}\xi}\frac{\mathrm{d}\xi}{\mathrm{d}\hat{\xi}}=\sqrt{\left(\frac{\mathrm{d}x_1}{\mathrm{d}\xi}\right)^2+\left(\frac{\mathrm{d}x_2}{\mathrm{d}\xi}\right)^2}\cdot\frac{\xi_2-\xi_1}{2} \tag{5.32}$$

其中，ξ_1 和 ξ_2 表示参数空间中单元两端的值。此外，行向量 $\boldsymbol{V}$ 和 $\boldsymbol{C}$ 可以写为如下形式：

$$\left\{\boldsymbol{V}_{1\times N}\right\}=\left\{\bar{\boldsymbol{V}}_{1\times N}\right\}\begin{bmatrix}\boldsymbol{\Psi}_{N_\mathrm{b}\times N_\mathrm{b}} & \boldsymbol{0}_{N_\mathrm{b}\times N_\mathrm{I}}\\ \boldsymbol{0}_{N_\mathrm{I}\times N_\mathrm{b}} & \boldsymbol{I}_{N_\mathrm{I}\times N_\mathrm{I}}\end{bmatrix}_{N\times N} \tag{5.33}$$

$$\left\{\boldsymbol{C}_{1\times N}\right\}=\left\{\bar{\boldsymbol{C}}_{1\times N}\right\}\begin{bmatrix}\boldsymbol{\Psi}_{N_\mathrm{b}\times N_\mathrm{b}} & \boldsymbol{0}_{N_\mathrm{b}\times N_\mathrm{I}}\\ \boldsymbol{0}_{N_\mathrm{I}\times N_\mathrm{b}} & \boldsymbol{I}_{N_\mathrm{I}\times N_\mathrm{I}}\end{bmatrix}_{N\times N} \tag{5.34}$$

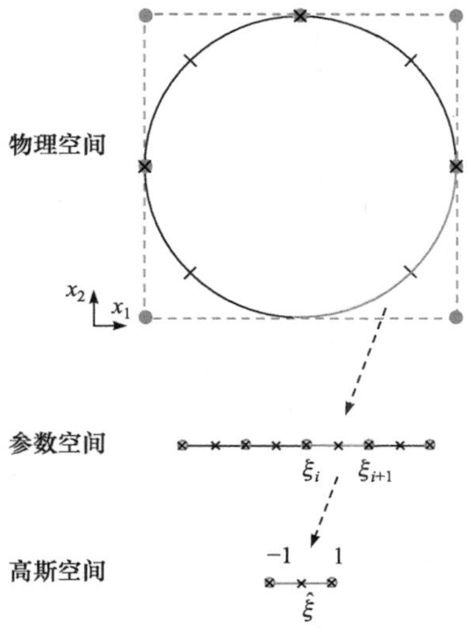

(a)

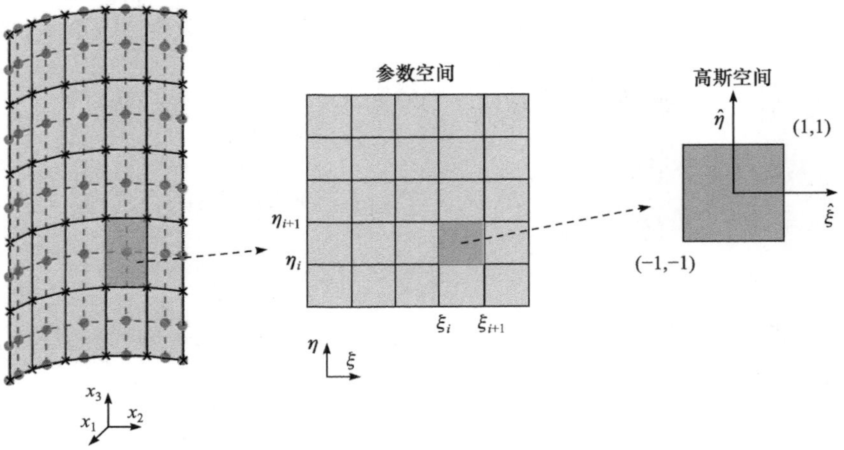

(b)

图 5.1 等几何边界元法中的积分区域定义

(a) 一维问题；(b) 二维问题

其中，$\boldsymbol{I}$ 为单位矩阵；$\boldsymbol{\Psi}$ 表示控制点到配点的转换矩阵；$\bar{\boldsymbol{V}}$ 和 $\bar{\boldsymbol{C}}$ 为 N 维行向量，这里第 j 个元素 $\bar{V}_j$ 和 $\bar{C}_j$ 可以写为如下形式：

$$\begin{aligned}\bar{V}_j=&\left\{\varPhi_{1j}^{-1}\int_\varGamma\frac{1}{r}\frac{\partial r}{\partial n}F_1^\alpha\mathrm{d}\varGamma+\varPhi_{2j}^{-1}\int_\varGamma\frac{1}{r}\frac{\partial r}{\partial n}F_2^\alpha\mathrm{d}\varGamma+\cdots+\varPhi_{Nj}^{-1}\int_\varGamma\frac{1}{r}\frac{\partial r}{\partial n}F_N^\alpha\mathrm{d}\varGamma\right.\\&+\varPhi_{(N+1)j}^{-1}\int_\varGamma\frac{1}{r}\frac{\partial r}{\partial n}F_0^\alpha\mathrm{d}\varGamma+\varPhi_{(N+2)j}^{-1}\left[(x_1^p+r_{,1}r)\int_\varGamma\frac{1}{r}\frac{\partial r}{\partial n}F_1^\alpha\mathrm{d}\varGamma\right]\\&\left.+\varPhi_{(N+3)j}^{-1}\left[(x_2^p+r_{,2}r)\int_\varGamma\frac{1}{r}\frac{\partial r}{\partial n}F_1^\alpha\mathrm{d}\varGamma\right]\right\}\end{aligned}\tag{5.35}$$

$$\begin{aligned}\bar{C}_j=&\left\{\varPhi_{1j}^{-1}\int_\varGamma\frac{1}{r}\frac{\partial r}{\partial n}F_1^\beta\mathrm{d}\varGamma+\varPhi_{2j}^{-1}\int_\varGamma\frac{1}{r}\frac{\partial r}{\partial n}F_2^\beta\mathrm{d}\varGamma+\ldots+\varPhi_{Nj}^{-1}\int_\varGamma\frac{1}{r}\frac{\partial r}{\partial n}F_N^\beta\mathrm{d}\varGamma\right.\\&+\varPhi_{(N+1)j}^{-1}\int_\varGamma\frac{1}{r}\frac{\partial r}{\partial n}F_0^\beta\mathrm{d}\varGamma+\varPhi_{(N+2)j}^{-1}\left[(x_1^p+r_{,1}r)\int_\varGamma\frac{1}{r}\frac{\partial r}{\partial n}F_1^\beta\mathrm{d}\varGamma\right]\\&\left.+\varPhi_{(N+3)j}^{-1}\left[(x_2^p+r_{,2}r)\int_\varGamma\frac{1}{r}\frac{\partial r}{\partial n}F_1^\beta\mathrm{d}\varGamma\right]\right\}\end{aligned}\tag{5.36}$$

对于三维问题，我们需要一张 NURBS 曲面来描述模型的几何形状。为了构造一张 NURBS 曲面，需要两个方向上的节点矢量 $\boldsymbol{U}$ 和 $\boldsymbol{V}$，$m\times n$ 个控制点 $\{\bar{P}_{i,j}\}$，以及每个方向的基函数阶次 $\hat{p}$ 和 $\hat{q}$。类似于二维问题，配点坐标 $\boldsymbol{x}(\xi,\eta)$、规则化温度 $\tilde{T}(\xi,\eta,t)$ 和热流 $q(\xi,\eta,t)$ 可以插值为如下形式：

$$\begin{cases} \boldsymbol{x}(\xi,\eta)=\sum_{i=1}^{(\hat{p}+1)}\sum_{j=1}^{(\hat{q}+1)}R_{i,j}^{\hat{p},\hat{q}}(\xi,\eta)\overline{\boldsymbol{P}}_{i,j} \\ \tilde{T}(\xi,\eta,t)=\sum_{i=1}^{(\hat{p}+1)}\sum_{j=1}^{(\hat{q}+1)}R_{i,j}^{\hat{p},\hat{q}}(\xi,\eta)\tilde{T}_{i,j}(t) \\ q(\xi,\eta,t)=\sum_{i=1}^{(\hat{p}+1)}\sum_{j=1}^{(\hat{q}+1)}R_{i,j}^{\hat{p},\hat{q}}(\xi,\eta)q_{i,j}(t) \end{cases} \tag{5.37}$$

三维规则化的边界域积分方程可以离散为如下形式：

$$\begin{aligned} \tilde{T}(\boldsymbol{p},t)= & -\sum_{e=1}^{N_e}\sum_{l=1}^{(\hat{p}+1)(\hat{q}+1)}\left\{\int_{-1}^{1}\int_{-1}^{1}q^*\left[\tilde{\boldsymbol{Q}}(\hat{\xi},\hat{\eta}),\boldsymbol{p}\right]R_l^e(\hat{\xi},\hat{\eta})J^e(\hat{\xi},\hat{\eta})\mathrm{d}\hat{\xi}\mathrm{d}\hat{\eta}\right\}\tilde{T}_l^e(t) \\ & -\sum_{e=1}^{N_e}\sum_{l=1}^{(\hat{p}+1)(\hat{q}+1)}\left\{\int_{-1}^{1}\int_{-1}^{1}T^*\left[\tilde{\boldsymbol{Q}}(\hat{\xi},\hat{\eta}),\boldsymbol{p}\right]R_l^e(\hat{\xi},\hat{\eta})J^e(\hat{\xi},\hat{\eta})\mathrm{d}\hat{\xi}\mathrm{d}\hat{\eta}\right\}q_l^e(t) \\ & +\sum_{e=1}^{N_e}\sum_{l=1}^{(\hat{p}+1)(\hat{q}+1)}\left\{\int_{-1}^{1}\int_{-1}^{1}\frac{1}{r\left[\tilde{\boldsymbol{Q}}(\hat{\xi},\hat{\eta}),\boldsymbol{p}\right]}\frac{\partial r}{\partial n}F\left[\tilde{\boldsymbol{Q}}(\hat{\xi},\hat{\eta}),\boldsymbol{p},t\right]J^e(\hat{\xi},\hat{\eta})\mathrm{d}\hat{\xi}\mathrm{d}\hat{\eta}\right\} \\ & +\boldsymbol{V}(\boldsymbol{p})\tilde{\boldsymbol{T}}(t)-\boldsymbol{C}(\boldsymbol{p})\dot{\tilde{\boldsymbol{T}}}(t) \end{aligned} \tag{5.38}$$

$$\begin{aligned} & c(\boldsymbol{P})\sum_{l=1}^{(\hat{p}+1)(\hat{q}+1)}R_l^{e'}(\hat{\xi}',\hat{\eta}')\tilde{T}_l^{e'}(t)+\sum_{e=1}^{N_e}\sum_{l=1}^{(\hat{p}+1)(\hat{q}+1)}\left\{\int_{-1}^{1}\int_{-1}^{1}q^*\left[\tilde{\boldsymbol{Q}}(\hat{\xi},\hat{\eta}),\boldsymbol{P}\right]R_l^e(\hat{\xi},\hat{\eta})J^e(\hat{\xi},\hat{\eta})\mathrm{d}\hat{\xi}\right\}\tilde{T}_l^e(t) \\ = & \sum_{e=1}^{N_e}\sum_{l=1}^{(\hat{p}+1)(\hat{q}+1)}\left\{\int_{-1}^{1}\int_{-1}^{1}T^*\left[\tilde{\boldsymbol{Q}}(\hat{\xi},\hat{\eta}),\boldsymbol{P}\right]R_{l,p}^e(\hat{\xi},\hat{\eta})J^e(\hat{\xi},\hat{\eta})\mathrm{d}\hat{\xi}\mathrm{d}\hat{\eta}\right\}q_l^e(t) \\ & +\sum_{e=1}^{N_e}\sum_{l=1}^{(\hat{p}+1)(\hat{q}+1)}\left\{\int_{-1}^{1}\int_{-1}^{1}\frac{1}{r\left[\tilde{\boldsymbol{Q}}(\hat{\xi}),\boldsymbol{P}\right]}\frac{\partial r}{\partial n}F\left[\tilde{\boldsymbol{Q}}(\hat{\xi},\hat{\eta}),\boldsymbol{P},t\right]J^e(\hat{\xi},\hat{\eta})\mathrm{d}\hat{\xi}\mathrm{d}\hat{\eta}\right\} \\ & +\boldsymbol{V}(\boldsymbol{P})\tilde{\boldsymbol{T}}(t)-\boldsymbol{C}(\boldsymbol{P})\dot{\tilde{\boldsymbol{T}}}(t) \end{aligned} \tag{5.39}$$

其中，局部坐标$\hat{\xi}$和$\hat{\eta}$皆为–1到1(图5.1(b))，雅可比行列式可以写为如下形式：

$$J^e(\hat{\xi},\hat{\eta})=\frac{\mathrm{d}^2\varGamma}{\mathrm{d}\hat{\xi}\mathrm{d}\hat{\eta}}=\frac{\mathrm{d}^2\varGamma}{\mathrm{d}\xi\mathrm{d}\eta}\frac{\mathrm{d}\xi}{\mathrm{d}\hat{\xi}}\frac{\mathrm{d}\eta}{\mathrm{d}\hat{\eta}} \tag{5.40}$$

类似于式(5.35)和式(5.36)，三维问题的$\overline{V}_j$和$\overline{C}_j$可以写为如下形式：

$$\begin{aligned} \overline{V}_j=\Bigg\{ & \varPhi_{1j}^{-1}\int_\varGamma\frac{1}{r}\frac{\partial r}{\partial n}F_1^\alpha\mathrm{d}\varGamma+\varPhi_{2j}^{-1}\int_\varGamma\frac{1}{r}\frac{\partial r}{\partial n}F_2^\alpha\mathrm{d}\varGamma+\cdots+\varPhi_{Nj}^{-1}\int_\varGamma\frac{1}{r}\frac{\partial r}{\partial n}F_N^\alpha\mathrm{d}\varGamma \\ & +\varPhi_{(N+1)j}^{-1}\int_\varGamma\frac{1}{r}\frac{\partial r}{\partial n}F_0^\alpha\mathrm{d}\varGamma+\varPhi_{(N+2)j}^{-1}\left[(x_1^p+r_{,1}r)\int_\varGamma\frac{1}{r}\frac{\partial r}{\partial n}F_1^\alpha\mathrm{d}\varGamma\right] \end{aligned}$$

$$+\varPhi_{(N+3)j}^{-1}\left[(x_2^p+r_{,2}r)\int_{\varGamma}\frac{1}{r}\frac{\partial r}{\partial n}F_1^{\alpha}\mathrm{d}\varGamma\right]+\varPhi_{(N+4)j}^{-1}\left[(x_3^p+r_{,3}r)\int_{\varGamma}\frac{1}{r}\frac{\partial r}{\partial n}F_1^{\alpha}\mathrm{d}\varGamma\right]\Bigg\} \tag{5.41}$$

$$\begin{aligned}\bar{C}_j=&\Bigg\{\varPhi_{1j}^{-1}\int_{\varGamma}\frac{1}{r}\frac{\partial r}{\partial n}F_1^{\beta}\mathrm{d}\varGamma+\varPhi_{2j}^{-1}\int_{\varGamma}\frac{1}{r}\frac{\partial r}{\partial n}F_2^{\beta}\mathrm{d}\varGamma+\cdots+\varPhi_{Nj}^{-1}\int_{\varGamma}\frac{1}{r}\frac{\partial r}{\partial n}F_N^{\beta}\mathrm{d}\varGamma\\&+\varPhi_{(N+1)j}^{-1}\int_{\varGamma}\frac{1}{r}\frac{\partial r}{\partial n}F_0^{\beta}\mathrm{d}\varGamma+\varPhi_{(N+2)j}^{-1}\left[(x_1^p+r_{,1}r)\int_{\varGamma}\frac{1}{r}\frac{\partial r}{\partial n}F_1^{\beta}\mathrm{d}\varGamma\right]\\&+\varPhi_{(N+3)j}^{-1}\left[(x_2^p+r_{,2}r)\int_{\varGamma}\frac{1}{r}\frac{\partial r}{\partial n}F_1^{\beta}\mathrm{d}\varGamma\right]+\varPhi_{(N+4)j}^{-1}\left[(x_3^p+r_{,3}r)\int_{\varGamma}\frac{1}{r}\frac{\partial r}{\partial n}F_1^{\beta}\mathrm{d}\varGamma\right]\Bigg\}\end{aligned} \tag{5.42}$$

5.4.2 利用精细积分法求解时域问题代数方程组

二维和三维问题的离散边界域积分方程可以写为如下矩阵形式：

$$\tilde{\boldsymbol{T}}_{\mathrm{I}}=\boldsymbol{G}_{\mathrm{I}}\boldsymbol{q}_{\mathrm{b}}-\boldsymbol{H}_{\mathrm{I}}\tilde{\boldsymbol{T}}_{\mathrm{b}}+\boldsymbol{Q}_{\mathrm{I}}+\boldsymbol{V}_{\mathrm{I}}\tilde{\boldsymbol{T}}-\boldsymbol{C}_{\mathrm{I}}\dot{\tilde{\boldsymbol{T}}} \tag{5.43}$$

$$\boldsymbol{H}_{\mathrm{b}}\tilde{\boldsymbol{T}}_{\mathrm{b}}=\boldsymbol{G}_{\mathrm{b}}\boldsymbol{q}_{\mathrm{b}}+\boldsymbol{Q}_{\mathrm{b}}+\boldsymbol{V}_{\mathrm{b}}\tilde{\boldsymbol{T}}-\boldsymbol{C}_{\mathrm{b}}\dot{\tilde{\boldsymbol{T}}} \tag{5.44}$$

上面两个方程组需要联立求解。本章考虑两类边界条件，即单一温度边界条件和温度-热流混合边界条件。为了利用精细积分法(PIM)解决上述二阶常微分方程组，我们需要对这些矩阵和向量进行分块和重组。下标 I 和 b 分别代表内点和边界。

1. 温度边界条件

只考虑温度边界条件时，式(5.43)和式(5.44)可以进行如下形式的分块：

$$\tilde{\boldsymbol{T}}_{\mathrm{I}}=\boldsymbol{G}_{\mathrm{I}}\boldsymbol{q}_{\mathrm{b}}-\boldsymbol{H}_{\mathrm{I}}\tilde{\boldsymbol{T}}_{\mathrm{b}}+\boldsymbol{Q}_{\mathrm{I}}+[\boldsymbol{V}_{\mathrm{Ib}}\ \ \boldsymbol{V}_{\mathrm{II}}]\begin{Bmatrix}\tilde{\boldsymbol{T}}_{\mathrm{b}}\\\tilde{\boldsymbol{T}}_{\mathrm{I}}\end{Bmatrix}-[\boldsymbol{C}_{\mathrm{Ib}}\ \ \boldsymbol{C}_{\mathrm{II}}]\begin{Bmatrix}\dot{\tilde{\boldsymbol{T}}}_{\mathrm{b}}\\\dot{\tilde{\boldsymbol{T}}}_{\mathrm{I}}\end{Bmatrix} \tag{5.45}$$

$$\boldsymbol{H}_{\mathrm{b}}\tilde{\boldsymbol{T}}_{\mathrm{b}}=\boldsymbol{G}_{\mathrm{b}}\boldsymbol{q}_{\mathrm{b}}+\boldsymbol{Q}_{\mathrm{b}}+[\boldsymbol{V}_{\mathrm{bb}}\ \ \boldsymbol{V}_{\mathrm{bI}}]\begin{Bmatrix}\tilde{\boldsymbol{T}}_{\mathrm{b}}\\\tilde{\boldsymbol{T}}_{\mathrm{I}}\end{Bmatrix}-[\boldsymbol{C}_{\mathrm{bb}}\ \ \boldsymbol{C}_{\mathrm{bI}}]\begin{Bmatrix}\dot{\tilde{\boldsymbol{T}}}_{\mathrm{b}}\\\dot{\tilde{\boldsymbol{T}}}_{\mathrm{I}}\end{Bmatrix} \tag{5.46}$$

式(5.45)和式(5.46)需要同时求解，可写为如下形式：

$$\begin{bmatrix}\boldsymbol{A}_{11}&\boldsymbol{A}_{12}\\\boldsymbol{A}_{21}&\boldsymbol{A}_{22}\end{bmatrix}\begin{Bmatrix}\boldsymbol{q}_{\mathrm{b}}\\\tilde{\boldsymbol{T}}_{\mathrm{I}}\end{Bmatrix}=\begin{Bmatrix}\boldsymbol{Y}_{\mathrm{b}}\\\boldsymbol{Y}_{\mathrm{I}}\end{Bmatrix}-\begin{bmatrix}\boldsymbol{C}_{\mathrm{bb}}&\boldsymbol{C}_{\mathrm{bI}}\\\boldsymbol{C}_{\mathrm{Ib}}&\boldsymbol{C}_{\mathrm{II}}\end{bmatrix}\begin{Bmatrix}\dot{\tilde{\boldsymbol{T}}}_{\mathrm{b}}\\\dot{\tilde{\boldsymbol{T}}}_{\mathrm{I}}\end{Bmatrix} \tag{5.47}$$

其中，

$$\begin{cases} \boldsymbol{A}_{11} = -\boldsymbol{G}_{\mathrm{b}} \\ \boldsymbol{A}_{12} = -\boldsymbol{V}_{\mathrm{bI}} \\ \boldsymbol{A}_{21} = -\boldsymbol{G}_{\mathrm{I}} \\ \boldsymbol{A}_{22} = \boldsymbol{I} - \boldsymbol{V}_{\mathrm{II}} \end{cases} \tag{5.48}$$

$$\begin{Bmatrix} \boldsymbol{Y}_{\mathrm{b}} \\ \boldsymbol{Y}_{\mathrm{I}} \end{Bmatrix} = \begin{bmatrix} -\boldsymbol{H}_{\mathrm{b}} \\ -\boldsymbol{H}_{\mathrm{I}} \end{bmatrix} \tilde{\boldsymbol{T}}_{\mathrm{b}} + \begin{Bmatrix} \boldsymbol{Q}_{\mathrm{b}} \\ \boldsymbol{Q}_{\mathrm{I}} \end{Bmatrix} \tag{5.49}$$

边界处的热流 $\boldsymbol{q}_{\mathrm{b}}(t_{k+1})$ 可由式(5.47)得到

$$\boldsymbol{q}_{\mathrm{b}}(t_{k+1}) = \boldsymbol{A}_{11}^{-1}\left[-\boldsymbol{A}_{12}\tilde{\boldsymbol{T}}_{\mathrm{I}}(t_{k+1}) - \boldsymbol{C}_{\mathrm{bI}}\dot{\tilde{\boldsymbol{T}}}_{\mathrm{I}}(t_{k+1}) + \boldsymbol{Y}_{\mathrm{b}} - \boldsymbol{C}_{\mathrm{bb}}\dot{\tilde{\boldsymbol{T}}}_{\mathrm{b}}(t_{k+1})\right] \tag{5.50}$$

将式(5.50)代入式(5.47)，可以得到

$$\dot{\tilde{\boldsymbol{T}}}_{\mathrm{I}}(t_{k+1}) = \boldsymbol{B}_{\mathrm{I}}\tilde{\boldsymbol{T}}_{\mathrm{I}}(t_{k+1}) + \hat{\boldsymbol{F}}_{\mathrm{I}}(t_{k+1}) \tag{5.51}$$

其中，

$$\boldsymbol{B}_{\mathrm{I}} = (\boldsymbol{C}_{\mathrm{II}} - \boldsymbol{A}_{21}\boldsymbol{A}_{11}^{-1}\boldsymbol{C}_{\mathrm{bI}})^{-1}(\boldsymbol{A}_{21}\boldsymbol{A}_{11}^{-1}\boldsymbol{A}_{12} - \boldsymbol{A}_{22}) \tag{5.52}$$

$$\begin{aligned} \hat{\boldsymbol{F}}_{\mathrm{I}}(t_{k+1}) = (\boldsymbol{C}_{\mathrm{II}} - \boldsymbol{A}_{21}\boldsymbol{A}_{11}^{-1}\boldsymbol{C}_{\mathrm{bb}})^{-1}\Big[&\boldsymbol{Y}_{\mathrm{I}}(t_{k+1}) - \boldsymbol{A}_{21}\boldsymbol{A}_{11}^{-1}\boldsymbol{Y}_{\mathrm{b}}(t_{k+1}) \\ &+(\boldsymbol{A}_{21}\boldsymbol{A}_{11}^{-1}\boldsymbol{C}_{\mathrm{bb}} - \boldsymbol{C}_{\mathrm{Ib}})\dot{\tilde{\boldsymbol{T}}}_{\mathrm{b1}}(t_{k+1})\Big] \end{aligned} \tag{5.53}$$

2. 混合边界条件

当考虑温度和热流两类边界条件时，式(5.43)和式(5.44)可以写为如下形式：

$$\begin{aligned} \tilde{\boldsymbol{T}}_{\mathrm{I}} = [\boldsymbol{G}_{\mathrm{I1}} \quad \boldsymbol{G}_{\mathrm{I2}}]\begin{Bmatrix} \boldsymbol{q}_{\mathrm{b1}} \\ \boldsymbol{q}_{\mathrm{b2}} \end{Bmatrix} - [\boldsymbol{H}_{\mathrm{I1}} \quad \boldsymbol{H}_{\mathrm{I2}}]\begin{Bmatrix} \tilde{\boldsymbol{T}}_{\mathrm{b1}} \\ \tilde{\boldsymbol{T}}_{\mathrm{b2}} \end{Bmatrix} + \{\boldsymbol{Q}_{\mathrm{I}}\} \\ + [\boldsymbol{V}_{\mathrm{I1}} \quad \boldsymbol{V}_{\mathrm{I2}} \quad \boldsymbol{V}_{\mathrm{II}}]\begin{Bmatrix} \tilde{\boldsymbol{T}}_{\mathrm{b1}} \\ \tilde{\boldsymbol{T}}_{\mathrm{b2}} \\ \tilde{\boldsymbol{T}}_{\mathrm{I}} \end{Bmatrix} - [\boldsymbol{C}_{\mathrm{I1}} \quad \boldsymbol{C}_{\mathrm{I2}} \quad \boldsymbol{C}_{\mathrm{II}}]\begin{Bmatrix} \dot{\tilde{\boldsymbol{T}}}_{\mathrm{b1}} \\ \dot{\tilde{\boldsymbol{T}}}_{\mathrm{b2}} \\ \dot{\tilde{\boldsymbol{T}}}_{\mathrm{I}} \end{Bmatrix} \end{aligned} \tag{5.54}$$

$$\begin{aligned} \begin{bmatrix} \boldsymbol{H}_{\mathrm{b11}} & \boldsymbol{H}_{\mathrm{b12}} \\ \boldsymbol{H}_{\mathrm{b21}} & \boldsymbol{H}_{\mathrm{b22}} \end{bmatrix}\begin{Bmatrix} \tilde{\boldsymbol{T}}_{\mathrm{b1}} \\ \tilde{\boldsymbol{T}}_{\mathrm{b2}} \end{Bmatrix} = \begin{bmatrix} \boldsymbol{G}_{\mathrm{b11}} & \boldsymbol{G}_{\mathrm{b12}} \\ \boldsymbol{G}_{\mathrm{b21}} & \boldsymbol{G}_{\mathrm{b11}} \end{bmatrix}\begin{Bmatrix} \boldsymbol{q}_{\mathrm{b1}} \\ \boldsymbol{q}_{\mathrm{b2}} \end{Bmatrix} + \begin{Bmatrix} \boldsymbol{Q}_{\mathrm{b1}} \\ \boldsymbol{Q}_{\mathrm{b2}} \end{Bmatrix} \\ + \begin{bmatrix} \boldsymbol{V}_{\mathrm{b11}} & \boldsymbol{V}_{\mathrm{b12}} & \boldsymbol{V}_{\mathrm{b1I}} \\ \boldsymbol{V}_{\mathrm{b21}} & \boldsymbol{V}_{\mathrm{b22}} & \boldsymbol{V}_{\mathrm{b2I}} \end{bmatrix}\begin{Bmatrix} \tilde{\boldsymbol{T}}_{\mathrm{b1}} \\ \tilde{\boldsymbol{T}}_{\mathrm{b2}} \\ \tilde{\boldsymbol{T}}_{\mathrm{I}} \end{Bmatrix} - \begin{bmatrix} \boldsymbol{C}_{\mathrm{b11}} & \boldsymbol{C}_{\mathrm{b12}} & \boldsymbol{C}_{\mathrm{b1I}} \\ \boldsymbol{C}_{\mathrm{b21}} & \boldsymbol{C}_{\mathrm{b22}} & \boldsymbol{C}_{\mathrm{b2I}} \end{bmatrix}\begin{Bmatrix} \dot{\tilde{\boldsymbol{T}}}_{\mathrm{b1}} \\ \dot{\tilde{\boldsymbol{T}}}_{\mathrm{b2}} \\ \dot{\tilde{\boldsymbol{T}}}_{\mathrm{I}} \end{Bmatrix} \end{aligned} \tag{5.55}$$

其中，上式中的下标 b1 、b2 和 I 分别表示温度边界点、热流边界点和内点的编号。

类似于式(5.47)，式(5.54)和式(5.55)可以联立为如下形式：

$$\begin{bmatrix} \boldsymbol{A}_{11} & \boldsymbol{A}_{12} \\ \boldsymbol{A}_{21} & \boldsymbol{A}_{22} \end{bmatrix} \begin{Bmatrix} \boldsymbol{q}_{\mathrm{b1}} \\ \tilde{\boldsymbol{T}}_{\mathrm{u}} \end{Bmatrix} = \begin{Bmatrix} \boldsymbol{Y}_{\mathrm{b1}} \\ \boldsymbol{Y}_{\mathrm{u}} \end{Bmatrix} - \begin{bmatrix} \boldsymbol{C}_{11} & \boldsymbol{C}_{12} \\ \boldsymbol{C}_{21} & \boldsymbol{C}_{22} \end{bmatrix} \begin{Bmatrix} \dot{\tilde{\boldsymbol{T}}}_{\mathrm{b1}} \\ \dot{\tilde{\boldsymbol{T}}}_{\mathrm{u}} \end{Bmatrix} \tag{5.56}$$

其中，

$$\begin{aligned} &\boldsymbol{A}_{11} = -\boldsymbol{G}_{\mathrm{b11}} \\ &\boldsymbol{A}_{12} = \begin{bmatrix} \boldsymbol{H}_{\mathrm{b12}} - \boldsymbol{V}_{\mathrm{b12}} & -\boldsymbol{V}_{\mathrm{b1I}} \end{bmatrix} \\ &\boldsymbol{A}_{21} = \begin{bmatrix} -\boldsymbol{G}_{\mathrm{b21}} \\ -\boldsymbol{G}_{\mathrm{I1}} \end{bmatrix} \\ &\boldsymbol{A}_{22} = \begin{bmatrix} \boldsymbol{H}_{\mathrm{b22}} - \boldsymbol{V}_{\mathrm{b22}} & -\boldsymbol{V}_{\mathrm{b2I}} \\ \boldsymbol{H}_{\mathrm{I2}} - \boldsymbol{V}_{\mathrm{I2}} & \boldsymbol{I} - \boldsymbol{V}_{\mathrm{II}} \end{bmatrix} \end{aligned} \tag{5.57}$$

$$\begin{cases} \tilde{\boldsymbol{T}}_{\mathrm{u}} = \{\{\tilde{\boldsymbol{T}}_{\mathrm{b2}}\}^{\mathrm{T}}, \{\tilde{\boldsymbol{T}}_{\mathrm{I}}\}^{\mathrm{T}}\}^{\mathrm{T}} \\ \dot{\tilde{\boldsymbol{T}}}_{\mathrm{u}} = \{\{\dot{\tilde{\boldsymbol{T}}}_{\mathrm{b2}}\}^{\mathrm{T}}, \{\dot{\tilde{\boldsymbol{T}}}_{\mathrm{I}}\}^{\mathrm{T}}\}^{\mathrm{T}} \end{cases} \tag{5.58}$$

$$\begin{Bmatrix} \boldsymbol{Y}_{\mathrm{b1}} \\ \boldsymbol{Y}_{\mathrm{u}} \end{Bmatrix} = \begin{bmatrix} \boldsymbol{V}_{\mathrm{b12}} - \boldsymbol{H}_{\mathrm{b12}} & \boldsymbol{G}_{\mathrm{b12}} \\ \boldsymbol{V}_{\mathrm{b22}} - \boldsymbol{H}_{\mathrm{b22}} & \boldsymbol{G}_{\mathrm{b22}} \\ \boldsymbol{V}_{\mathrm{I2}} - \boldsymbol{H}_{\mathrm{I2}} & \boldsymbol{G}_{\mathrm{I2}} \end{bmatrix} \begin{Bmatrix} \tilde{\boldsymbol{T}}_{\mathrm{b1}} \\ \boldsymbol{q}_{\mathrm{b2}} \end{Bmatrix} + \begin{Bmatrix} \boldsymbol{Q}_{\mathrm{b1}} \\ \boldsymbol{Q}_{\mathrm{b2}} \\ \boldsymbol{Q}_{\mathrm{I}} \end{Bmatrix} \tag{5.59}$$

$$\begin{aligned} &\boldsymbol{C}_{11} = \boldsymbol{C}_{\mathrm{b11}} \\ &\boldsymbol{C}_{12} = \begin{bmatrix} \boldsymbol{C}_{\mathrm{b12}} & \boldsymbol{C}_{\mathrm{b1I}} \end{bmatrix} \\ &\boldsymbol{C}_{21} = \begin{bmatrix} \boldsymbol{C}_{\mathrm{b21}} \\ \boldsymbol{C}_{\mathrm{I1}} \end{bmatrix} \\ &\boldsymbol{C}_{22} = \begin{bmatrix} \boldsymbol{C}_{\mathrm{b22}} & \boldsymbol{C}_{\mathrm{b2I}} \\ \boldsymbol{C}_{\mathrm{I2}} & \boldsymbol{C}_{\mathrm{II}} \end{bmatrix} \end{aligned} \tag{5.60}$$

由式(5.56)可以得到如下的热流表达形式：

$$\boldsymbol{q}_{\mathrm{b1}}(t_{k+1}) = \boldsymbol{A}_{11}^{-1}\left[-\boldsymbol{A}_{12}\tilde{\boldsymbol{T}}_{\mathrm{u}}(t_{k+1}) - \boldsymbol{C}_{12}\dot{\tilde{\boldsymbol{T}}}_{\mathrm{u}}(t_{k+1}) + \boldsymbol{Y}_{\mathrm{b1}} - \boldsymbol{C}_{11}\dot{\tilde{\boldsymbol{T}}}_{\mathrm{b1}}(t_{k+1})\right] \tag{5.61}$$

将式(5.61)代入式(5.56)，可得

$$\dot{\tilde{\boldsymbol{T}}}_{\mathrm{u}}(t_{k+1}) = \boldsymbol{B}_{\mathrm{u}}\tilde{\boldsymbol{T}}_{\mathrm{u}}(t_{k+1}) + \hat{\boldsymbol{F}}_{\mathrm{u}}(t_{k+1}) \tag{5.62}$$

其中，

$$\boldsymbol{B}_{\mathrm{u}} = (\boldsymbol{C}_{22} - \boldsymbol{A}_{21}\boldsymbol{A}_{11}^{-1}\boldsymbol{C}_{12})^{-1}(\boldsymbol{A}_{21}\boldsymbol{A}_{11}^{-1}\boldsymbol{A}_{12} - \boldsymbol{A}_{22}) \tag{5.63}$$

$$\hat{\boldsymbol{F}}_{\mathrm{u}}(t_{k+1}) = (\boldsymbol{C}_{22} - \boldsymbol{A}_{21}\boldsymbol{A}_{11}^{-1}\boldsymbol{C}_{12})^{-1}\left[\boldsymbol{Y}_{\mathrm{u}}(t_{k+1}) - \boldsymbol{A}_{21}\boldsymbol{A}_{11}^{-1}\boldsymbol{Y}_{\mathrm{b1}}(t_{k+1})\right.$$

$$-(\boldsymbol{C}_{21}+\boldsymbol{A}_{21}\boldsymbol{A}_{11}^{-1}\boldsymbol{C}_{11})\dot{\tilde{\boldsymbol{T}}}_{\mathrm{b1}}(t_{k+1})\Big] \tag{5.64}$$

3. 利用精细积分法求解二阶常微分方程组

对于温度边界条件问题，内点任一时段$[t_k,t_{k+1}]$的规则化温度$\tilde{\boldsymbol{T}}_{\mathrm{I}}(t_{k+1})$可以写成如下形式：

$$\tilde{\boldsymbol{T}}_{\mathrm{I}}(t_{k+1})=\boldsymbol{E}_{\mathrm{I}}\tilde{\boldsymbol{T}}_{\mathrm{I}}(t_k)+\int_0^{\Delta t}\exp\left[\boldsymbol{B}_{\mathrm{I}}(\Delta t-\xi)\right]\hat{\boldsymbol{F}}_{\mathrm{I}}(t_k+\xi)\mathrm{d}\xi \tag{5.65}$$

其中，$\boldsymbol{E}_{\mathrm{I}}=\exp(\boldsymbol{B}_{\mathrm{I}}\Delta t)$为指数矩阵；$\Delta t$为时间步长；$\boldsymbol{B}_{\mathrm{I}}$表示矩阵$\boldsymbol{B}$对应的内点部分。这里的$\hat{\boldsymbol{F}}_{\mathrm{I}}(t_k+\xi)$在时间间隔$[t_k,t_{k+1}]$内采用的是线性插值。值得注意的是，指数矩阵$\boldsymbol{E}_{\mathrm{I}}$可以写成$\boldsymbol{E}_{\mathrm{I}}=[\exp(\boldsymbol{B}_{\mathrm{I}}\Delta t/\overline{m})]^{\overline{m}}$，其中$\overline{m}=2^M$，$M=20$。假设$\overline{\eta}=\Delta t/\overline{m}$，指数矩阵可以写成如下形式：

$$\boldsymbol{E}_{\mathrm{I}}=\exp(\boldsymbol{B}_{\mathrm{I}}\overline{\eta})\approx\boldsymbol{I}+\boldsymbol{E}_{\mathrm{a}} \tag{5.66}$$

其中，

$$\boldsymbol{E}_{\mathrm{a}}=\boldsymbol{I}+(\boldsymbol{B}_{\mathrm{I}}\overline{\eta})^2/2!+(\boldsymbol{B}_{\mathrm{I}}\overline{\eta})^3/3!+(\boldsymbol{B}_{\mathrm{I}}\overline{\eta})^4/4! \tag{5.67}$$

其中，$\boldsymbol{E}_{\mathrm{a}}$是一个小量矩阵。为了保证计算精度，式(5.66)可以写成如下形式：

$$\boldsymbol{E}_{\mathrm{I}}=\boldsymbol{I}+[\mathrm{for}(\mathrm{iter}=1;\mathrm{iter}<M+1;\mathrm{iter}++)\boldsymbol{E}_{\mathrm{a}}=2\boldsymbol{E}_{\mathrm{a}}+\boldsymbol{E}_{\mathrm{a}}\times\boldsymbol{E}_{\mathrm{a}}] \tag{5.68}$$

最终，式(5.65)可以写成如下形式：

$$\tilde{\boldsymbol{T}}_{\mathrm{I}}(t_{k+1})=\boldsymbol{E}_{\mathrm{I}}\left[\tilde{\boldsymbol{T}}_{\mathrm{I}}(t_k)+\boldsymbol{B}_{\mathrm{I}}^{-1}(\hat{\boldsymbol{R}}_0+\boldsymbol{B}_{\mathrm{I}}^{-1}\hat{\boldsymbol{R}}_1)\right]-\boldsymbol{B}_{\mathrm{I}}^{-1}(\hat{\boldsymbol{R}}_0+\boldsymbol{B}_{\mathrm{I}}^{-1}\hat{\boldsymbol{R}}_1+\Delta t\hat{\boldsymbol{R}}_1) \tag{5.69}$$

其中，$\hat{\boldsymbol{R}}_0=\hat{\boldsymbol{F}}_{\mathrm{I}}(t_k)$，$\hat{\boldsymbol{R}}_1=\left[\hat{\boldsymbol{F}}_{\mathrm{u}}(t_{k+1})-\hat{\boldsymbol{F}}_{\mathrm{u}}(t_k)\right]/\Delta t$。

类似于式(5.65)～式(5.69)，对于温度和热流两类边界条件问题，每一时刻未知规则化温度可以写成如下形式：

$$\tilde{\boldsymbol{T}}_{\mathrm{u}}(t_{k+1})=\boldsymbol{E}_{\mathrm{u}}\left[\tilde{\boldsymbol{T}}_{\mathrm{u}}(t_k)+\boldsymbol{B}_{\mathrm{u}}^{-1}(\hat{\boldsymbol{R}}_0+\boldsymbol{B}_{\mathrm{u}}^{-1}\hat{\boldsymbol{R}}_1)\right]-\boldsymbol{B}_{\mathrm{u}}^{-1}(\hat{\boldsymbol{R}}_0+\boldsymbol{B}_{\mathrm{u}}^{-1}\hat{\boldsymbol{R}}_1+\Delta t\hat{\boldsymbol{R}}_1) \tag{5.70}$$

其中，$\hat{\boldsymbol{R}}_0=\hat{\boldsymbol{F}}_{\mathrm{u}}(t_k)$和$\hat{\boldsymbol{R}}_1=\left[\hat{\boldsymbol{F}}_{\mathrm{u}}(t_{k+1})-\hat{\boldsymbol{F}}_{\mathrm{u}}(t_k)\right]/\Delta t$。

为了更好地说明本章的框架，图 5.2 给出了基于精细积分法的等几何边界元法的流程图。

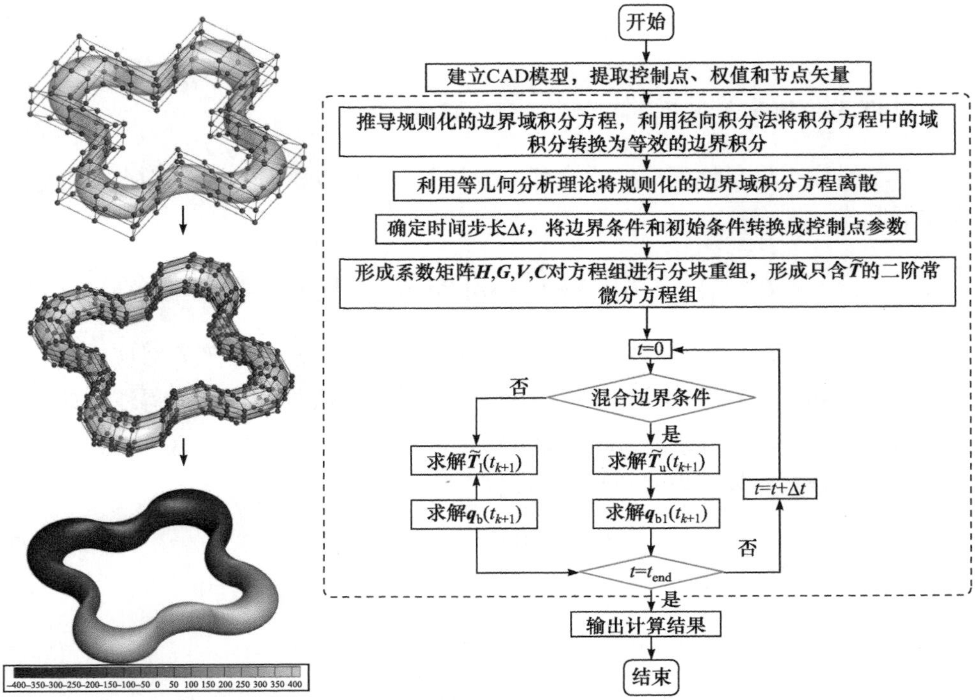

图 5.2　基于精细积分法的等几何边界元法流程图

5.5　数值算例

在本节中，我们将检验几个二维和三维数值例子，以表明基于径向积分法和精细积分法的等几何边界元法在求解含热源的FGM瞬态热传导问题上的正确性。采用精确解验证该模型的计算结果，并使用绝对误差(ABSERR)和均方根误差(RMSERR)来检验所提方法的收敛性。

$$\mathrm{ABSERR}=\left|T_{\mathrm{numerical},i}(\boldsymbol{x},t)-T_{\mathrm{exact},i}(\boldsymbol{x},t)\right| \tag{5.71}$$

$$\mathrm{RMSERR}=\sqrt{\sum_{i=1}^{N}\left[T_{\mathrm{numerical},i}(\boldsymbol{x},t)-T_{\mathrm{exact},i}(\boldsymbol{x},t)\right]^{2}\Big/\sum_{i=1}^{N}T_{\mathrm{exact},i}^{2}(\boldsymbol{x},t)} \tag{5.72}$$

其中，$T_{\mathrm{numerical},i}$和$T_{\mathrm{exact},i}$分别代表第i个节点的数值解和解析解；N是节点的数目。

5.5.1　二维厚壁圆筒的瞬态热传导问题

图 5.3(a)所示的是一个二维厚壁圆筒。材料的热传导系数为 1 W/(m·℃)。圆

筒外边界半径为 1m，外边界施加如下的温度边界条件：

$$T_{b1}(\boldsymbol{x},t)=(x_1^2+x_2^2+100)\exp(t) \tag{5.73}$$

圆筒的内边界半径为 0.5m，内边界施加如下的热流边界条件：

$$q_{b2}(\boldsymbol{x},t)=-2\sqrt{x_1^2+x_2^2}\exp(t) \tag{5.74}$$

热源强度和初始条件如下：

$$Q(\boldsymbol{x},t)=(x_1^2+x_2^2+96)\exp(t) \tag{5.75}$$

$$T_0(\boldsymbol{x},0)=x_1^2+x_2^2+100 \tag{5.76}$$

我们采用一条 2 阶 NURBS 曲线描述厚壁圆筒，外边界 Γ_{b1} 和内边界 Γ_{b2} 采用相同的节点矢量如下：

$$\Xi_{1,2}=\left\{0,0,0,\frac{1}{4},\frac{1}{4},\frac{1}{2},\frac{1}{2},\frac{3}{4},\frac{3}{4},1,1,1\right\} \tag{5.77}$$

多连通模型控制点和配点的初始分布如图 5.3 (b)所示。由于圆筒几何形状简单，我们只对其进行一次细化。首先讨论内部应用点的数量对结果的影响。图 5.4 给出了不同数目的内点(即 N_{I} 等于 4、8、16 和 32)的分布。本算例的时间步长选取 $\Delta t=0.1\,\mathrm{s}$。

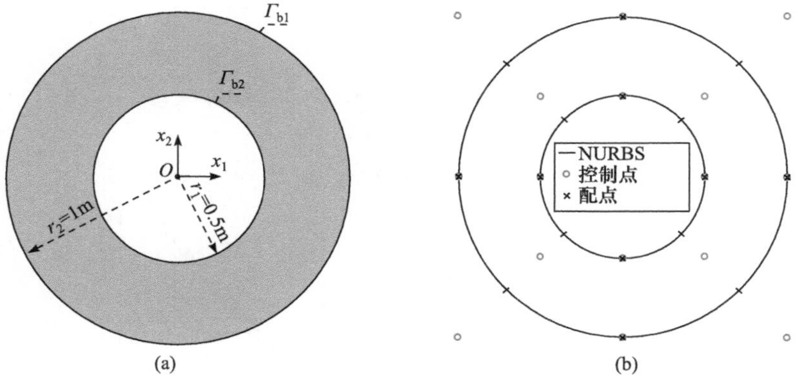

图 5.3 计算模型及 IGA 网格

(a) 模型的几何形状和尺寸；(b) 模型控制点和配点的分布

图 5.5 给出了 $t=0\mathrm{s}$ 到 $t=2\mathrm{s}$ 时间段内每一时刻 A 点(在不同数目内点的情况下)的等几何边界元解和解析解。不难看出，等几何边界元在求解二维简单模型瞬态热传导问题时，即使只采用初始的 NURBS 样条(即细化次数为 1)和较少的内部应用点，也都可以获得较为准确的等几何边界元解。此外，图 5.6 给出了 $t=2\mathrm{s}$ 时刻的解析解、等几何边界元解的温度云图以及绝对误差云图。

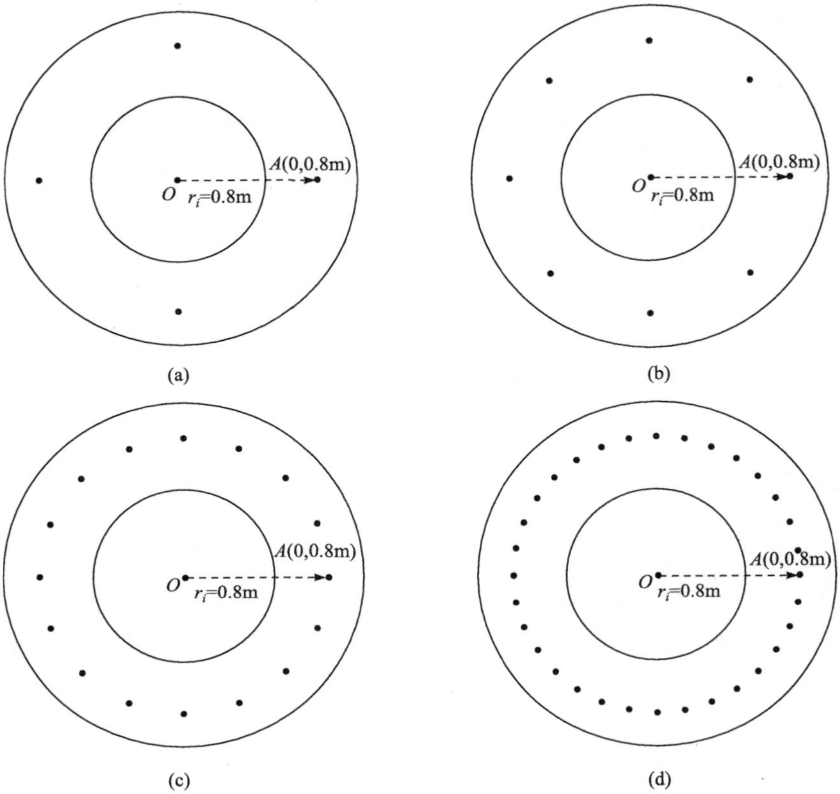

图 5.4 计算模型不同数目的内点分布

(a) $N_I=4$ ；(b) $N_I=8$ ；(c) $N_I=16$ ；(d) $N_I=32$

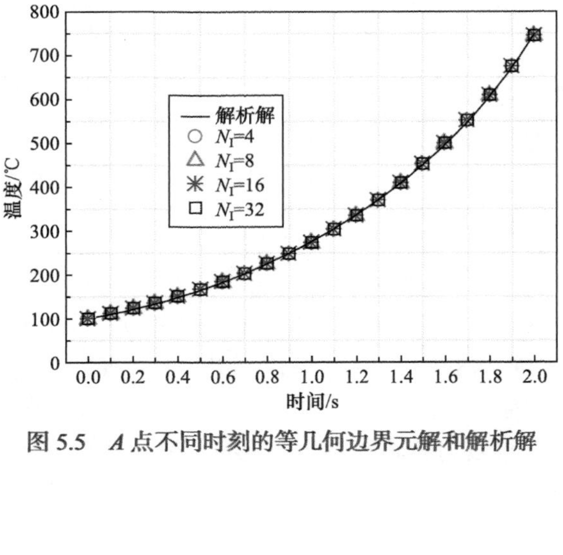

图 5.5 A 点不同时刻的等几何边界元解和解析解

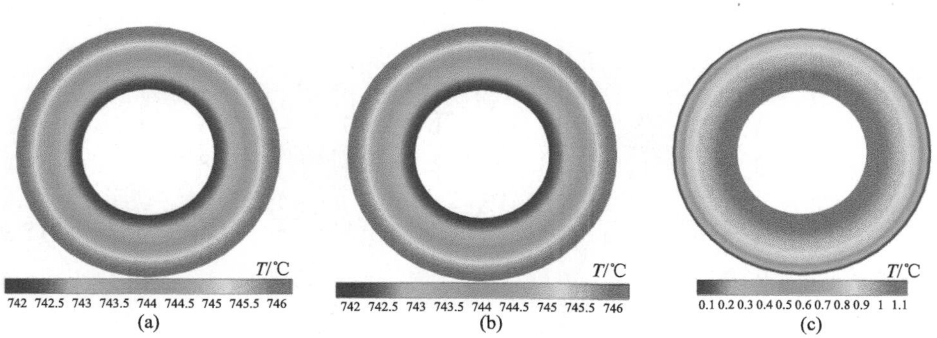

图 5.6　在 $t=2\,\mathrm{s}$ 时刻的温度云图和绝对误差云图

(a) $t=2\,\mathrm{s}$ 时刻解析解温度云图；(b) $t=2\,\mathrm{s}$ 时刻等几何边界元解温度云图；(c) $t=2\,\mathrm{s}$ 时刻绝对误差云图

此外，从图 5.7 中可以看出，即使不同情况下数值解的误差都很小，但是随着内点数的增加，等几何边界元解的精度也会继续增加。

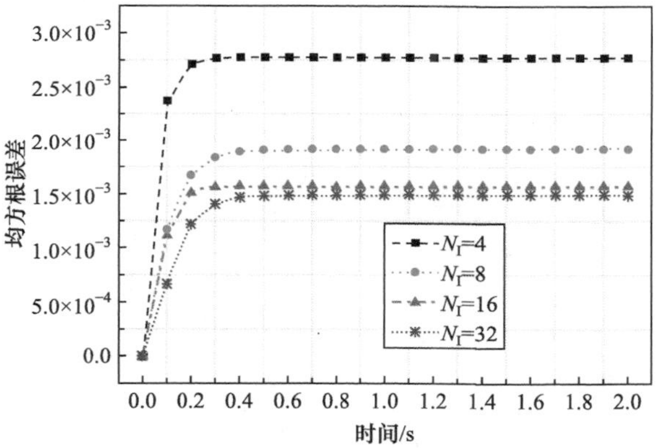

图 5.7　等几何边界元解的均方根误差

5.5.2　二维复杂几何模型的瞬态热传导问题

图 5.8 所示的是一个复杂的功能梯度材料二维模型。材料的热传导系数随坐标变化，即 $\lambda(\boldsymbol{x})=x_1+x_2\,(\mathrm{W/(m\cdot{}^\circ C)})$，全域的热源强度可以表示为如下形式：

$$Q(\boldsymbol{x},t)=[5(x_1+x_2)+100]\cos t-10\sin t\quad(\mathrm{W/m^3}) \tag{5.78}$$

该模型的温度解析解为如下形式：

$$T(\boldsymbol{x},t)=[5(x_1+x_2)+100]\sin t\quad(℃) \tag{5.79}$$

为了检验 IGABEM 的收敛性，我们考虑了四种不同的细化模型，即 1 次细化、2 次细化、5 次细化和 10 次细化(图 5.9)。NURBS 基函数的阶次为 2。几何模型的初始节点矢量如下：

$$\Xi = \left\{0,0,0,\frac{1}{24},\frac{1}{24},\frac{1}{12},\frac{1}{12},\frac{1}{8},\frac{1}{8},\frac{1}{6},\frac{1}{6},\frac{5}{24},\frac{5}{24},\frac{1}{4},\frac{1}{4},\frac{7}{24},\frac{7}{24},\right.$$
$$\frac{1}{3},\frac{1}{3},\frac{3}{8},\frac{3}{8},\frac{5}{12},\frac{5}{12},\frac{11}{24},\frac{11}{24},\frac{1}{2},\frac{1}{2},\frac{13}{24},\frac{13}{24},\frac{7}{12},\frac{7}{12},\frac{5}{8},\frac{5}{8},$$
$$\left.\frac{2}{3},\frac{2}{3},\frac{17}{24},\frac{17}{24},\frac{3}{4},\frac{3}{4},\frac{19}{24},\frac{19}{24},\frac{5}{6},\frac{5}{6},\frac{7}{8},\frac{7}{8},\frac{11}{12},\frac{11}{12},\frac{23}{24},\frac{23}{24},1,1,1\right\} \tag{5.80}$$

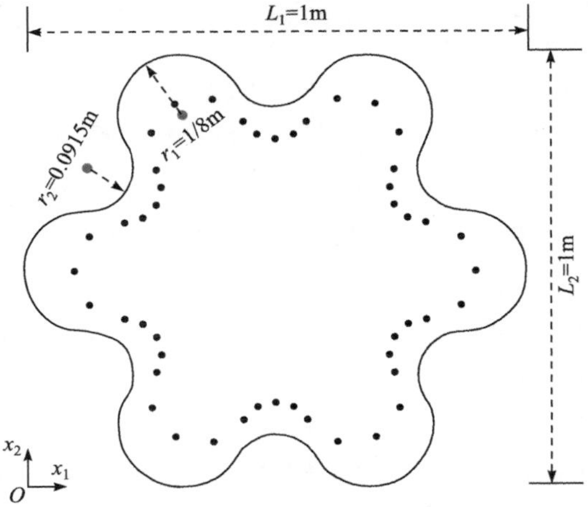

图 5.8　模型的几何、尺寸和内点的分布

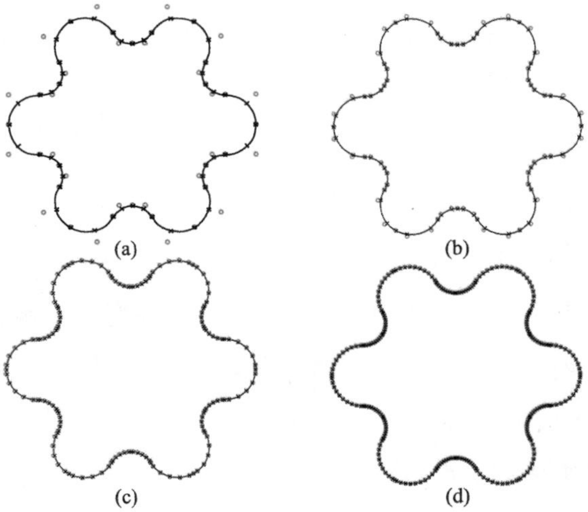

图 5.9　模型不同细化次数下控制点和配点的分布

(a) 1 次细化；(b) 2 次细化；(c) 5 次细化；(d) 10 次细化

此算例的时间步长选取为 $\Delta t = 0.1\text{s}$。图 5.10 给出了内点 $t = 0\text{s}$ 到 $t = 2\text{s}$ 期间每一时刻的解析解。从图 5.11 和图 5.12 可以看出，对于复杂几何形状，为了提高

数值结果的精度，需要对模型进行细化。

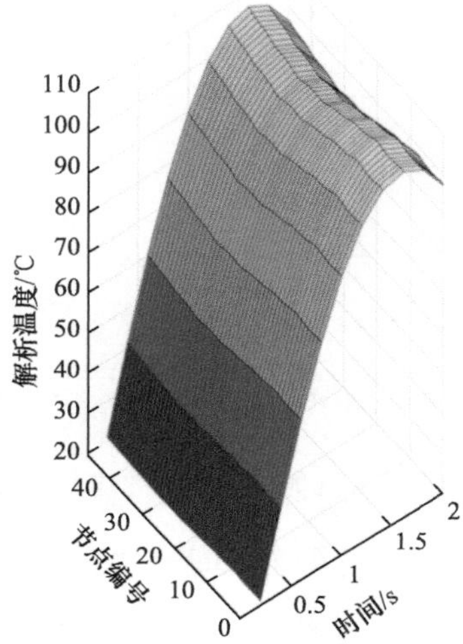

图 5.10　不同时刻 48 个内点的解析解

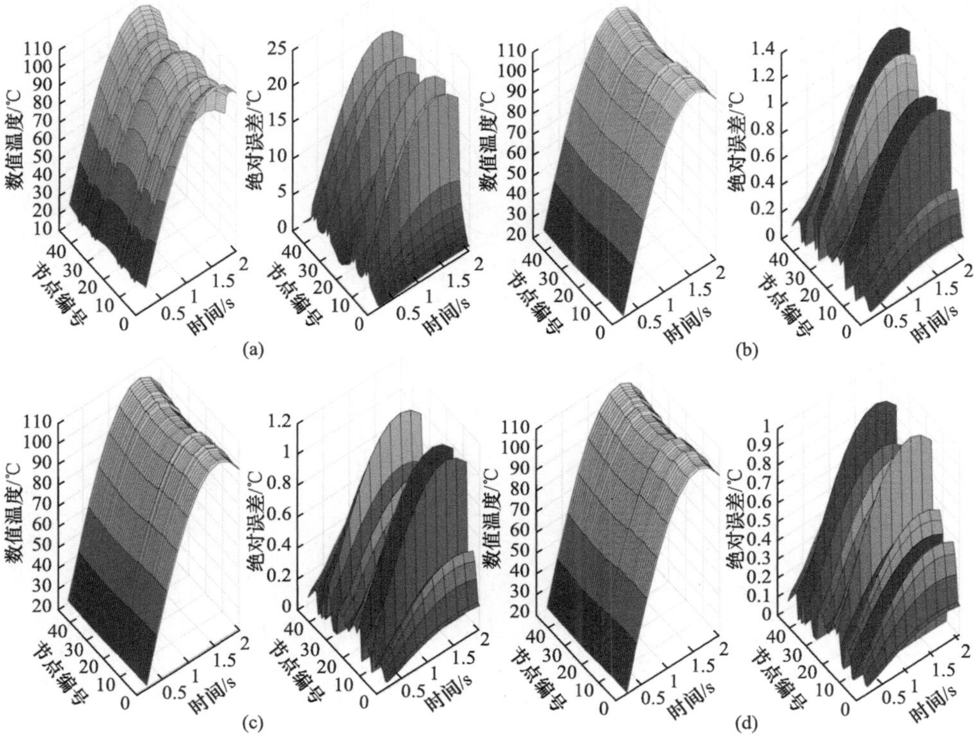

图 5.11　不同时刻的等几何边界元解分布及绝对误差分布

(a) 1 次细化；(b) 2 次细化；(c) 5 次细化；(d) 10 次细化

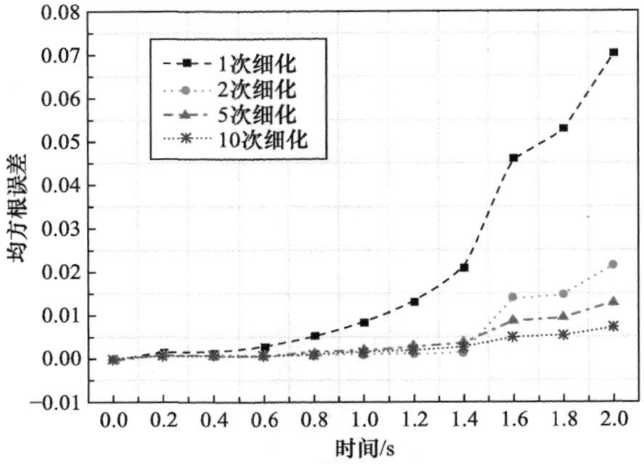

图 5.12 等几何边界元解的均方根误差

5.5.3 二维多连通复杂几何模型的瞬态热传导问题

图 5.13 所示的是一个多连通模型。其中，内边界为一个半径为 0.2m 的圆，外边界为一个复杂的几何形状。不同阶次的 NURBS 网格在图 5.14 中给出，用于描述外边界和内边界的初始节点矢量 $\Xi_1(=\Xi)$ 和 $\Xi_2(=\Xi_{1,2})$ 分别在式(5.80)和式(5.77)中给出。材料的热传导系数为 $\lambda(\boldsymbol{x})=x_1+x_2$ (W/(m·℃))，全域的热源强度可以写成如下形式：

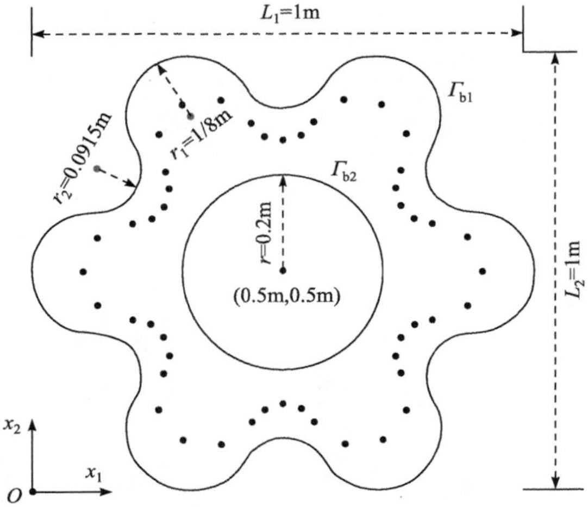

图 5.13 计算模型的几何尺寸和内部应用点分布

$$Q(\boldsymbol{x},t)=[5(x_1+x_2)+90]\exp(t)\quad (\mathrm{W/m^3}) \tag{5.81}$$

该问题的解析解可以写成

$$T(\boldsymbol{x},t)=[5(x_1+x_2)+100]\exp(t)\quad(℃) \tag{5.82}$$

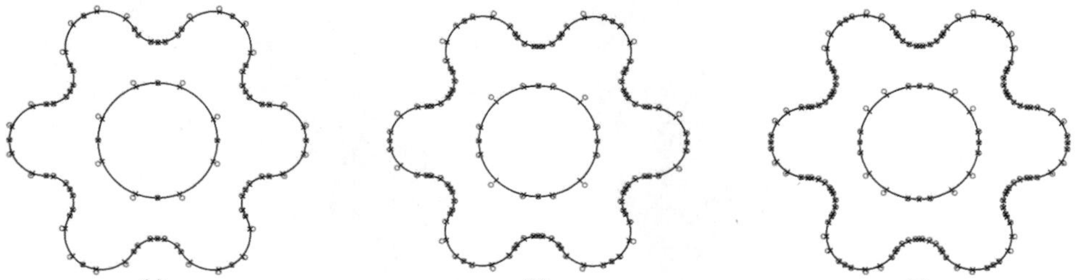

(a)　(b)　(c)

图 5.14　不同阶次基函数 2 次细化时模型的控制点和配点分布

(a) $\hat{p}=2$; (b) $\hat{p}=3$; (c) $\hat{p}=4$

解析解的分布如图 5.15 所示。此算例中，我们选取的时间步长为 $\Delta t=0.2\text{s}$ 。从图 5.16 中可以清楚地看到，对于不同的 NURBS 基函数阶次，在 $t=0\text{s}$ 到 $t=2\text{s}$ 期间每一时刻的 IGABEM 解与解析解吻合很好。图 5.17 给出了不同的 NURBS 基函数阶次情况下 IGABEM 解的均方根误差。从图 5.16 和图 5.17 中可以看出，对于一个相对复杂的模型，提高基函数阶次可以在一定程度上提高计算结果的精度。

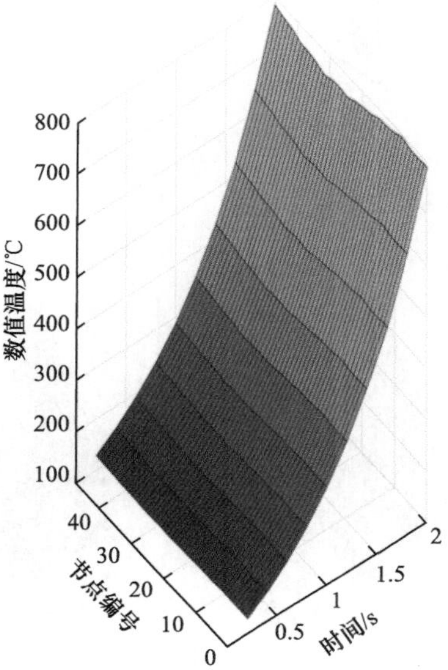

图 5.15　不同时刻 48 个内点的解析解

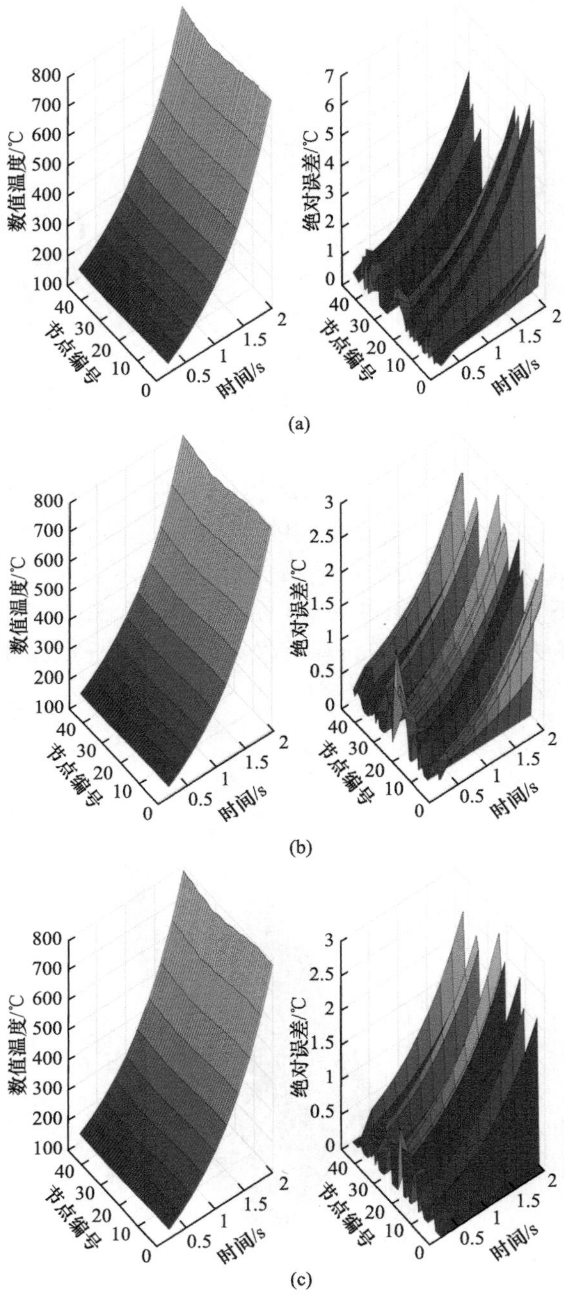

图 5.16 不同时刻的等几何边界元解分布及绝对误差分布

(a) $\hat{p}=2$；(b) $\hat{p}=3$；(c) $\hat{p}=4$

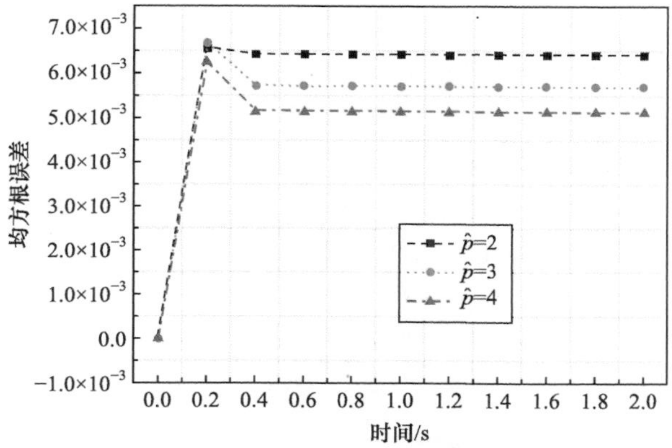

图 5.17　等几何边界元解的均方根误差

5.5.4　三维复杂模型的瞬态热传导问题

一个复杂模型的几何形状如图 5.18 (a)所示。模型的横截面尺寸在图 5.18 (b)中给出。我们用一张 NURBS 曲面描述这个几何形状，该曲面两个参数方向的阶次及节点矢量在表 5.1 中给出，相应的网格及控制点分布在图 5.18 (c)中给出。

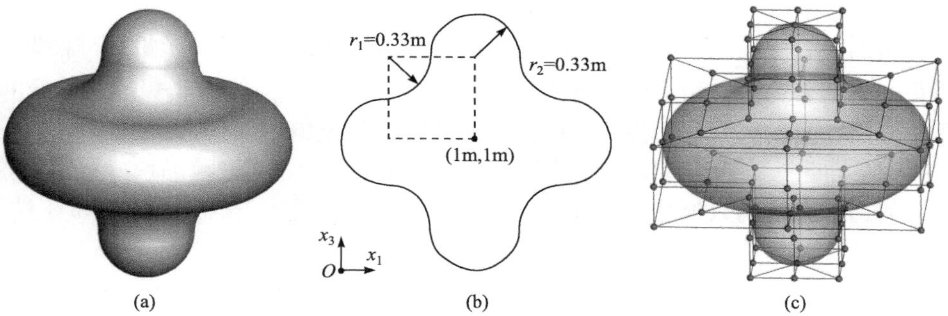

图 5.18　计算模型及 NURBS 网格
(a) 模型的几何形状；(b) 横截面尺寸；(c) 控制点的分布

表 5.1　NURBS 阶次和节点向量

参数方向	阶次	节点矢量
ξ	$\hat{p}=2$	$\boldsymbol{U}=\{0,0,0,1,1,2,2,3,3,4,4,4\}$
η	$\hat{q}=2$	$\boldsymbol{V}=\{0,0,0,1,1,2,2,3,3,4,4,5,5,6,6,6\}$

与坐标相关的热传导系数为$\lambda(\boldsymbol{x})=x_1+x_2+x_3$ (W/(m · °C))。计算域的热源强度为

$$Q(\boldsymbol{x},t)=[10(x_1+x_2+x_3)+70]\exp(t)\quad (\mathrm{W/m^3}) \tag{5.83}$$

该问题的解析解为

$$T(\boldsymbol{x},t)=[10(x_1+x_2+x_3)+100]\exp(t) \quad (℃) \tag{5.84}$$

为了验证算法的正确性，我们选取 90 个内点分布在曲面 S_1 上(图 5.19)。本算例用于讨论时间步长对计算结果的影响。选取三个时间步长 Δt ，分别为 0.4s 、0.05s 和 0.02s 。为了验证算法的有效性，我们还给出了基于有限差分(finite difference, FD)的径向积分等几何边界元解(RI-IGABEM)。关于有限差分的细节可以参考文献[44]。本算例的欧拉因子取为 1。

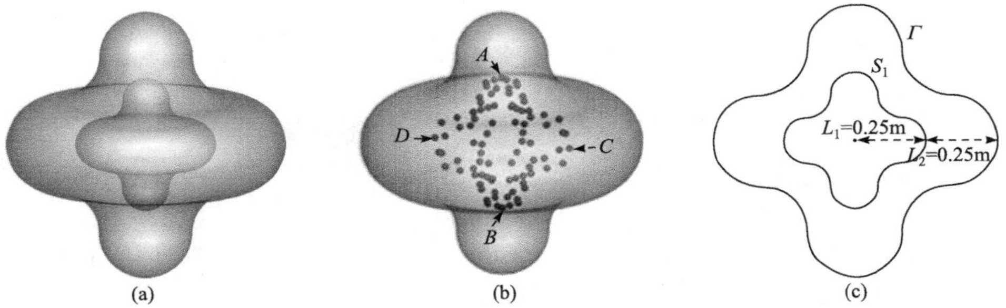

图 5.19 计算模型内点分布示意图

(a) 面 S_1 的位置；(b) 90 个内点的分布；(c) 横截面尺寸

由图 5.20 可以看出，节点 A(1m,1m,1.5m)的数值结果与各时刻的精确解吻合得很好。此外，数值结果的均方根误差可以达到10^{-3} 量级。由图 5.21 可以看出，基于有限差分的等几何边界元法的数值结果趋于稳定，精度随着时间步长的减小而提高。同时，与基于有限差分的等几何边界元法相比，基于精细积分法(PIM)的等几何边界元法的数值结果更加稳定。由表 5.2 可以看出，时间步长对基于精细积分法的等几何边界元法的计算结果影响不大。

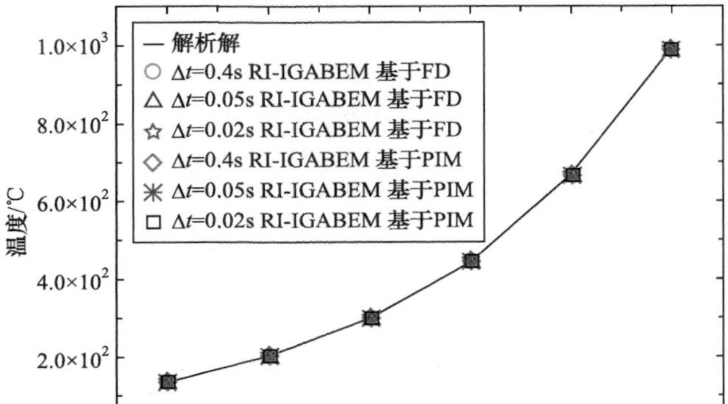

图 5.20 节点 A(1m,1m,1.5m)的解析解和等几何边界元解

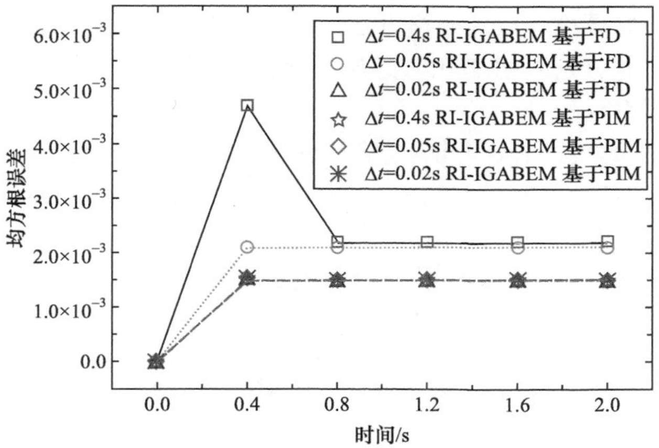

图 5.21 等几何边界元解的均方根误差

表 5.2 $t=2s$ 时刻不同内点的温度

节点坐标/m	PIM			FD		
	$\Delta t=0.4s$	$\Delta t=0.05s$	$\Delta t=0.02s$	$\Delta t=0.4s$	$\Delta t=0.05s$	$\Delta t=0.02s$
$A(1,1,1.5)$	992.5251	992.5255	992.5259	991.7322	992.7568	992.5656
$B(1,1,0.5)$	912.7607	912.7602	912.7604	911.7194	913.0856	912.7604
$C(1.5,1,1)$	991.3366	991.3369	991.3371	990.1955	991.6357	991.3371
$D(0.5,1,1)$	908.6833	908.6829	908.6831	906.5123	909.1052	908.7757

5.6 小 结

本章将精细积分法和等几何边界元法相结合，对含热源的功能梯度材料的二维和三维瞬态热传导问题进行了分析。基于径向积分的等几何边界元法是对等几何边界元法的扩展，可应用于冶金、航空航天和制造等领域的许多实际瞬态热传导工程问题。

我们推导了规则化的边界域积分方程，并将径向积分应用于等几何边界元法处理瞬态热传导问题中存在的域积分，这些域积分是由热源、随坐标变化的热传导系数和初始温度引起的。

与传统边界元法相比，等几何边界元法不仅保留了半解析性质和仅进行边界离散化的优点，而且体现了一定的优势。首先，等几何边界元法采用与 CAD 软件相同的基函数，可以精确地描述模型的几何形状，没有离散误差。其次，无需划分网格，节省了大量的计算成本。最后，NURBS 基函数具有非负高阶次的特点。

本章通过控制点到配点的变换方法，将常位势法应用于等几何边界元法强奇异积分的求解，并分别采用 Telles 变换法和单元子分方法求解二维及三维等几何边界元法中的弱奇异积分。

本章将精细积分法引入等几何边界元法中求解时域问题。与一般时域方法相比，精细积分法对时间步长不敏感。采用较大的时间步长依然能保证计算结果的稳定性和准确性，节约了计算成本。

在本章所有的例子中，我们将等几何边界元解和解析解进行了比较，表明了基于精细积分法的等几何边界元法可以用于分析功能梯度材料瞬态热传导问题。

参考文献

[1] Comi C, Mariani S. Extended finite element simulation of quasi-brittle fracture in functionally graded materials[J]. Computer Methods in Applied Mechanics and Engineering, 2007, 196: 4013-4026.

[2] Shen H S, Xiang Y. Postbuckling behavior of functionally graded graphene-reinforced composite laminated cylindrical shells under axial compression in thermal environments[J]. Computer Methods in Applied Mechanics and Engineering, 2018, 330: 64-82.

[3] Yu B, Xu C, Zhou H L, et al. A novel non-iterative method for estimating boundary conditions and geometry of furnace inner wall made of FGMs[J]. Applied Thermal Engineering, 2019, 147 : 251-271.

[4] Wakamatsu Y, Saito T, Ueda S, et al. Development of a thermal shock evaluation device for functionally gradient materials for aerospace applications [M]//Schneider G A, Petzow G. Thermal Shock and Thermal Fatigue Behavior of Advanced Ceramics. NATO ASI Series, vol 241. Dordrecht: Springer, 1993, 555-565.

[5] Jang J, Lee C. Fabrication and mechanical properties of glass fibre-carbon fibre polypropylene functionally gradient materials[J]. Journal of Materials Science, 1998, 33: 5445-5450.

[6] Hosseini A, Mahmoud S. Application of a hybrid meshless technique for natural frequencies analysis in functionally graded thick hollow cylinder subjected to suddenly thermal loading[J]. Applied Mathematical Modelling, 2014, 38: 425-436.

[7] Nie C B, Yu B. Inversing heat flux boundary conditions based on precise integration FEM without iteration and estimation of thermal stress in FGMs[J]. International Journal of Thermal Sciences, 2019, 140: 201-224.

[8] Xu C, Yu B. A novel domain propulsion and adaptive modified inversion method for the inverse geometry heat conduction analysis of FGMs[J]. Numerical Heat Transfer Part A—Applications, 2020, 78 : 1-31.

[9] Feng W Z, Gao X W. An interface integral equation method for solving transient heat conduction in multi-medium materials with variable thermal properties[J]. International Journal of Heat and Mass Transfer, 2016, 98: 227-239.

[10] Yang K, Jiang G H, Peng H F, et al. A new modified Levenberg-Marquardt algorithm for identifying the temperature-dependent conductivity of solids based on the radial integration

boundary element method[J]. International Journal of Heat and Mass Transfer, 2019, 114 : 118-141.

[11] Hughes T J R, Cottrell J A, Bazilevs Y. Isogeometric analysis: CAD, finite elements, NURBS, exact geometry and mesh refinement[J]. Computer Methods in Applied Mechanics and Engineering, 2005, 194 : 4135-4195.

[12] Guo S P, Li X J, Zhang J M, et al. A triple reciprocity method in Laplace transform boundary element method for three-dimensional transient heat conduction problems[J]. International Journal of Heat and Mass Transfer, 2017, 114 : 258-267.

[13] Simpson R N, Bordas S P A, Trevelyan J, et al. A two-dimensional isogeometric boundary element method for elastostatic analysis[J]. Computer Methods in Applied Mechanics and Engineering, 2012, 209 : 87-100.

[14] Cordeiro S G F, Leonel E D. Mechanical modelling of three-dimensional cracked structural components using the isogeometric dual boundary element method[J]. Applied Mathematical Modelling, 2018, 63: 415-444.

[15] Gu J L, Zhang J M, Li G Y. Isogeometric analysis in BIE for 3-D potential problem[J]. Engineering Analysis with Boundary Elements, 2012, 36: 858-865.

[16] Gong Y P, Trevelyan J, Hattori G, et al. Hybrid nearly singular integration for three-dimensional isogeometric boundary element analysis of coatings and other thin structures[J]. Computer Methods in Applied Mechanics and Engineering, 2019, 367: 642-673.

[17] Nguyen B H, Zhuang X, Wriggers P, et al. Isogeometric symmetric Galerkin boundary element method for three-dimensional elasticity problems[J]. Computer Methods in Applied Mechanics and Engineering, 2017, 323: 132-150.

[18] Taus M, Rodin G J, Hughes T J R, et al. Isogeometric boundary element methods and patch tests for linear elastic problems: formulation, numerical integration, and applications[J]. Computer Methods in Applied Mechanics and Engineering, 2019, 357: 112591.

[19] Keuchel K, Hagelstein N C, Zaleski O, et al. Evaluation of hypersingular and nearly singular integrals in the isogeometric boundary element method for acoustics[J]. Computer Methods in Applied Mechanics and Engineering, 2017, 325 : 488-504.

[20] Chen L L, Zhang Y, Lian H, et al. Seamless integration of computer-aided geometric modeling and acoustic simulation: isogeometric boundary element methods based on Catmull-Clark subdivision surfaces[J]. Advances in Engineering Software, 2020, 149 : 102879.

[21] Cui M, Xu B B, Feng W Z, et al. A radial integration boundary element method for solving transient heat conduction problems with heat sources and variable thermal conductivity[J]. Numerical Heat Transfer Part B—Fundamentals, 2018, 73: 1-18.

[22] Gao X W , Davies T G. Boundary Element Programming in Mechanics[M]. Cambridge: Cambridge University Press, 2002.

[23] Nardini D, Brebbia C A. A New Approach for Free Vibration Analysis Using Boundary Elements[M]. Berlin: Springer, 1982.

[24] Partridge P W, Brebbia C A, Wrobel L C. The Dual Reciprocity Boundary Element Method[M]. Southampton: Computational Mechanics Publications, 1992.

[25] Cheng A H D, Young D L, Tsai C C. Solution of Poisson's equation by iterative DRBEM using compactly supported, positive definite radial basis function[J]. Engineering Analysis with Boundary Elements, 2000, 24 : 549-557.

[26] Ochiai Y. Meshless thermo-elastoplastic analysis by triple-reciprocity boundary element method[J]. International Journal for Numerical Methods in Engineering, 2010, 81 : 1609-1634.

[27]Yang K, Gao X W, Liu Y F. Using analytical expressions in radial integration BEM for variable coefficient heat conduction problems[J]. Engineering Analysis with Boundary Elements, 2011, 35 : 1085-1089.

[28] Feng W Z, Gao L F, Du J M, et al. A meshless interface integral BEM for solving heat conduction in multi-non-homogeneous media with multiple heat sources[J]. International Communications in Heat and Mass Transfer, 2019,104 : 70-82.

[29] Feng W Z, Li H Y, Gao L F, et al. Hypersingular flux interface integral equation for multi-medium heat transfer analysis[J]. International Journal of Heat and Mass Transfer, 2019, 138: 852-865.

[30] Dong C Y, Lo S H, Cheung Y K. Numerical solution for elastic inclusion problems by domain integral equation with integration by means of radial basis functions[J]. Engineering Analysis with Boundary Elements, 2004, 28 : 623-632.

[31] Peng H F, Cui M, Yang K, et al. Radial integration BEM for steady convection-conduction problem with spatially variable velocity and thermal conductivity[J]. International Journal of Heat and Mass Transfer, 2018, 126: 1150-1161.

[32] Guo S P, Zhang J M, Li G Y, et al. Three dimensional transient heat conduction analysis by Laplace transformation and multiple reciprocity boundary face method[J]. Engineering Analysis with Boundary Elements, 2013, 37 : 15-22.

[33] Wang C H, Grigoriev M M, Dargush G F. A fast multi-level convolution boundary element method for transient diffusion problems[J]. Engineering Analysis with Boundary Elements, 2005, 62: 1895-1926.

[34] Zhong W X, Williams F W. A precise time step integration method[J]. Proceedings of the Institution of Mechanical Engineers Part C—Journal of Mechanical Engineering Sci., 1994, 208: 427-430.

[35] Fung T C. A precise time-step integration method by step-response and impulsive-response matrices for dynamic problems[J]. International Journal for Numerical Methods in Engineering, 2015, 40: 4501-4527.

[36] Li Q H, Chen S S, Kou G X. Transient heat conduction analysis using the MLPG method and modified precise time step integration method[J]. Journal of Computational Physics, 2011, 230 : 2736-2750.

[37] Zhong W X, Cai Z Q. Precise integration method for LQG optimal measurement feedback control problem[J]. Applied Mathematics and Mechanics—English Edition, 2000, 21: 1417-1422.

[38] Guiggiani M, Casalini P. Direct computation of Cauchy principal value integralsin advanced boundary elements[J]. International Journal for Numerical Methods in Engineering, 1987, 24 : 1711-1720.

[39] Gong Y P, Dong C Y, Qin X C. An isogeometric boundary element method for three dimensional

potential problems[J]. Journal of Computational and Applied Mathematics, 2017, 313 : 454-468.
[40] Gong Y P, Dong C Y. An isogeometric boundary element method using adaptive integral method for 3D potential problems[J]. Journal of Computational and Applied Mathematics, 2017,319: 141-158.
[41] Gao X W. An effective method for numerical evaluation of general 2D and 3D high order singular boundary integrals[J]. Computer Methods in Applied Mechanics and Engineering, 2010, 199: 2856-2864.
[42] Rizzo F J, Shippy D J. An advanced boundary integral equation method for three-dimensional thermoelasticity[J]. International Journal for Numerical Methods in Engineering, 1977, 11 : 1753-1768.
[43] 李晟泽. 三维弹性问题等几何边界元快速计算方法研究[D]. 长沙：国防科技大学, 2018.
[44] Yang K, Gao X W. Radial integration BEM for transient heat conduction problems[J]. International Journal for Numerical Methods in Engineering, 2010, 34 : 557-563.

第6章

三维黏弹性材料的等几何边界元分析

6.1 引　　言

在实际工程中，橡胶、推进剂药柱和混凝土等材料都是黏弹性材料。黏弹性材料模型是由遵循胡克定律的弹性元件和遵循牛顿黏性定律的黏壶有机组合而成。推进剂药柱是固体火箭发动机(solid rocket motor, SRM)的动力来源和主要受力部分，同时也是最薄弱的环节。因此对推进剂药柱进行结构完整性分析非常重要。推进剂药柱在固化降温、长期储存、运输和发射过程中，会受到温度载荷、内压和重力等多重载荷的作用。近些年来，学者利用 FEM[1,2]、BEM[3,4]等算法来研究黏弹性力学问题。众所周知，FEM 是研究黏弹性问题最常用的方法[5]，因为它理论简单，易于编程实现。与 FEM 相比，关于黏弹性力学问题的 BEM 文献较少，通常仅限于二维问题[6,7]。但是，与 FEM 相比，BEM 具有降维和高精度的优点，在研究线性黏弹性力学问题时具有独特的优势[8,9]。本章我们研究用于分析线性黏弹性力学问题的 IGABEM。

当使用传统的数值方法分析问题时，通常需要首先在 CAD 软件中建模，采用的是 NURBS 数学表达，然后对模型进行离散，采用的是拉格朗日基函数数学表达，并且，在计算完成后的后处理阶段采用的是双线性插值函数。此外，设计、分析和可视化阶段的分离，以及数学表达的不统一，导致运算时频繁的数据传输和交换，极大地增加了分析成本。整个几何建模和网格划分过程占据了近 80%的分析时间。拉格朗日多项式插值还会引起几何离散误差，其拉格朗日单元之间仅为 C^0 连续。2005 年，美国得克萨斯州立大学的 Hughes 等[10]提出了 IGA 的概念。IGAFEM 的核心思想是，利用 CAD 软件中的 NURBS 作为 FEM 的形函数直接对模型进行数值分析。IGA 的优点是可以精确描述几何形状，无离散误差，基函数为非负高阶次，单元具有高阶连续性[11,12]。和 FEM 不同，BEM 只需要对边界进行离散，而 CAD 软件也只提供模型表面的信息。因此，IGABEM 真正实现了 CAD 到 CAE 的无缝连接。2009 年，IGABEM 的萌芽思想被提出[13]。2012 年，Simpson 等[14,15]首次提出了 IGABEM 的概念，并将其用于求解弹性力学问题。近十年来，IGABEM 被广泛应用于求解弹性动力学[16]、形状优化 [17,18]等问题。2019 年，经过十年的发展和沉淀，Beer 等[19]出版了第一本关于 IGABEM 的专著 *The Isogeometric Boundary Element Method*。该书总结了 IGABEM 在过去十年中的发展和应用，肯定了其潜力和优势。该书的出版标志着 IGABEM 逐渐走向成熟。但是，IGABEM 在黏弹性材料中的研究相对较少。与 IGAFEM 不同，IGABEM 在分析黏弹性问题时需要在迭代过程中求解应力和应变。因此，准确计算各时刻的应力和应变，对该方法计算结果的准确性和稳定性至关重要。本章采用应力/应变积分方程求解内部节点的应力和应变。为了避免超奇异积分，我们用黏弹性力学的面力恢复法(TRM) [20]来计算边界处的应力和应变。由于 IGABEM 可以精确地描述几何形状，可以在每个配点精确地获得与底层几何相关的数值，没有任何误差[21]，因此，与传统 BEM 相比，IGABEM 在每次迭代中都能更好地求解复杂模型的应力和应变，并且数值结果更稳定。

与传统 BEM 类似，如何将域积分转换为等效边界积分，以及如何准确求解奇异积分，是利用 IGABEM 分析塑性、黏弹性等非线性问题的关键。在本章中我们使用由弹性静力学 Kelvin 基本解导出的位移、应变边界域积分方程，其中的域积分是由体力和记忆应力产生的。众所周知，将计算域划分为单元是求解域积分最有效和最简洁的方法，但该方法使 IGABEM 失去了仅离散边界的优势[22]。近些年来，通常采用两种方案来消除积分方程中的域积分[23]。第一种方法是找到相关问题的基本解，但对于复杂问题，很难找到其基本解。因此，这种方法有一定的局限性，难以编写通用代码[24]。第二种方法是将域积分转换为等效边界积分。1982 年之前，学者提出了多种将域积分转换为等效边界积分的方法[25-27]，但这些方法一般只能解决特殊问题，不具有普适性。1982 年，Nardini 和 Brebbia[28]提出了具有较强通用性的 DRM。近几十年来，DRM 已被广泛应用于求解各类问题[29-31]。DRM 的核心思想是对域内函数进行插值，从而求得基函数的特解。然而，对于一些复杂的三维基函数，很难获得其特解。此外，即使对于已知的体力问题，该方法仍然需要近似函数。因此，计算结果不如其他方法准确。本章将径向积分法引入 IGABEM，将域积分转换为等效边界积分。径向积分法的优点是，它能够将任何复杂的域积分转换为边界积分，而不需要使用拉普拉斯算子和问题的特解。它还可以降低各种奇异积分[32,33]的奇异性阶数，也可以自然地用于求解多边界问题的域积分[34]。由于被积核函数是$1/r$ (r 为源点到场点的距离)的函数，因此，当源点和场点在同一单元内，且源点逐渐接近场点时，会出现不同阶次的奇异积分[35,36]。IGABEM 的精度在很大程度上取决于求解各阶奇异积分的精度。为了兼顾编程的准确性、效率和方便性，我们使用不同的方法来求解不同阶次奇异积分。对于弱奇异积分，我们采用幂级数展开法[37]进行求解。通过一个简单的变换，可用刚体位移法求解强奇异积分。

由于位移积分方程中含有记忆应力，因此求解记忆应力需要每一时刻的应变。我们使用正则化的应变积分方程[38]来求解内点的应变。对于边界节点的应变，也可以使用应变积分方程求解，但必须处理超奇异积分[39]。为了避免上述问题，我们提出了一种三维黏弹性力学的面力恢复法[40,41]，它可以利用边界节点的位移和面力来求解边界节点的应变。

6.2 节将介绍三维黏弹性力学问题的一些基本公式和 IGABEM 的数值实施过程，6.3 节给出四个例子来验证 IGABEM 在求解黏弹性问题中的有效性。

6.2 三维黏弹性力学问题的等几何边界元法

6.2.1 黏弹性力学问题的本构方程及记忆应力

线性黏弹性材料的本构方程可以用 Stieltjes 积分[42]表示为

$$\sigma_{il}(\boldsymbol{x},t)=\delta_{il}\int_{-\infty}^{t}\left[K(t-t')-\frac{2}{3}G(t-t')\right]\frac{\partial\varepsilon_{nn}(\boldsymbol{x},t')}{\partial t'}\mathrm{d}t' +2\int_{-\infty}^{t}G(t-t')\frac{\partial\varepsilon_{il}(\boldsymbol{x},t')}{\partial t'}\mathrm{d}t' \tag{6.1}$$

其中，K 和 G 分别表示体积模量和剪切模量；σ_{il} 和 ε_{il} $(i, l=1, 2, 3)$分别表示应力和应变分量；$\varepsilon_{nn}=\varepsilon_{11}+\varepsilon_{22}+\varepsilon_{33}$；$\delta_{il}$ 表示 Kronecker-delta 函数；t' 表示时间变量；t 表示加载时间。线性黏弹性材料的应力可分为弹性应力 σ_{il}^{e} 和记忆应力 σ_{il}^{m}：

$$\sigma_{il}=\sigma_{il}^{\mathrm{e}}-\sigma_{il}^{\mathrm{m}} \tag{6.2}$$

其中，

$$\sigma_{il}^{\mathrm{e}}(\boldsymbol{x},t)=\delta_{il}\left[K(0)-\frac{2}{3}G(0)\right]\varepsilon_{nn}(\boldsymbol{x},t)+2G(0)\varepsilon_{il}(\boldsymbol{x},t) \tag{6.3}$$

$$\sigma_{il}^{\mathrm{m}}=2\int_{0}^{t}\frac{\partial G(t-t')}{\partial t'}\varepsilon_{il}(\boldsymbol{x},t')\mathrm{d}t'+\delta_{il}\int_{0}^{t}\frac{\partial K(t-t')}{\partial t'}\varepsilon_{nn}(\boldsymbol{x},t')\mathrm{d}t' -\delta_{il}\frac{2}{3}\int_{0}^{t}\frac{\partial G(t-t')}{\partial t'}\varepsilon_{nn}(\boldsymbol{x},t')\mathrm{d}t' \tag{6.4}$$

其中，$K(0)$ 和 $G(0)$ 分别表示弹性状态下的体积模量和剪切模量。

如式(6.3)所示，当获得应变 ε_{il} 时，可以计算弹性应力 σ_{il}^{e}。然而，如果要计算记忆应力，就需要计算遗传积分。为了简化计算，节省计算机内存，$G(t)$ 和 $K(t)$ 可以展开成如下形式：

$$G(t)=G_0+\sum_{w=1}^{\bar{m}}G_w\exp(-t/\tau_w^G) \tag{6.5}$$

$$K(t)=K_0+\sum_{w=1}^{\bar{n}}K_w\exp(-t/\tau_w^K) \tag{6.6}$$

其中，τ_w^G 和 τ_w^K 分别表示剪切响应和体积响应的弛豫时间；$\bar{m}$ 和 $\bar{n}$ 分别表示 Prony 级数展开的项数；G_0、G_w、K_0 和 K_w 分别表示与材料有关的常数。在本章中，假设体积模量为常数，因此，$\int_0^t \frac{\partial K(t-t')}{\partial t'}\varepsilon_{nn}(\boldsymbol{x},t')\mathrm{d}t'=0$，式(6.4)可以写成如下形式：

$$\sigma_{il}^{\mathrm{m}}=\int_0^t \frac{\partial G(t-t')}{\partial t'}\hat{\varepsilon}_{il}(\boldsymbol{x},t')\mathrm{d}t' \tag{6.7}$$

其中，

$$\begin{Bmatrix}\hat{\varepsilon}_{11}\\ \hat{\varepsilon}_{22}\\ \hat{\varepsilon}_{33}\\ \hat{\varepsilon}_{12}\\ \hat{\varepsilon}_{13}\\ \hat{\varepsilon}_{21}\\ \hat{\varepsilon}_{23}\\ \hat{\varepsilon}_{31}\\ \hat{\varepsilon}_{32}\end{Bmatrix}=\begin{bmatrix}\frac{4}{3} & -\frac{2}{3} & -\frac{2}{3} & 0 & 0 & 0 & 0 & 0 & 0\\ -\frac{2}{3} & \frac{4}{3} & -\frac{2}{3} & 0 & 0 & 0 & 0 & 0 & 0\\ -\frac{2}{3} & -\frac{2}{3} & \frac{4}{3} & 0 & 0 & 0 & 0 & 0 & 0\\ 0 & 0 & 0 & 2 & 0 & 0 & 0 & 0 & 0\\ 0 & 0 & 0 & 0 & 2 & 0 & 0 & 0 & 0\\ 0 & 0 & 0 & 0 & 0 & 2 & 0 & 0 & 0\\ 0 & 0 & 0 & 0 & 0 & 0 & 2 & 0 & 0\\ 0 & 0 & 0 & 0 & 0 & 0 & 0 & 2 & 0\\ 0 & 0 & 0 & 0 & 0 & 0 & 0 & 0 & 2\end{bmatrix}\begin{Bmatrix}\varepsilon_{11}\\ \varepsilon_{22}\\ \varepsilon_{33}\\ \varepsilon_{12}\\ \varepsilon_{13}\\ \varepsilon_{21}\\ \varepsilon_{23}\\ \varepsilon_{31}\\ \varepsilon_{32}\end{Bmatrix} \tag{6.8}$$

在时间段$[0,t_S]$，遗传积分可以用梯形积分写成如下形式：

$$\begin{aligned}&\int_0^{t_S}\frac{\partial G(t_S-t')}{\partial t'}\hat{\varepsilon}_{il}(\boldsymbol{x},t')\mathrm{d}t'\\ &=\sum_{j=1}^{S-1}\left[\frac{\partial G(t_S-t'_j)}{\partial t'_j}\hat{\varepsilon}_{il}(\boldsymbol{x},t'_j)+\frac{\partial G(t_S-t'_{j+1})}{\partial t'_{j+1}}\hat{\varepsilon}_{il}(\boldsymbol{x},t'_{j+1})\right]\frac{1}{2}(t'_{j+1}-t'_j)\end{aligned} \tag{6.9}$$

在时间段$[t'_j,t'_{j+1}]$，我们作出如下假设：

$$\frac{\partial G(t_S-t'_j)}{\partial t'_j}=\frac{\partial G(t_S-t'_{j+1})}{\partial t'_j}=\frac{G(t_S-t'_{j+1})-G(t_S-t'_j)}{t'_{j+1}-t'_j} \tag{6.10}$$

将式(6.10)代入式(6.9)，式(6.9)可以写成如下形式：

$$\begin{aligned}\int_0^{t_S}\frac{\partial G(t_S-t')}{\partial t'}\hat{\varepsilon}_{il}(\boldsymbol{x},t')\mathrm{d}t' &= \sum_{j=1}^{S-1}[G(t_S-t'_{j+1})-G(t_S-t'_j)]\hat{\varepsilon}^*_{il}(\boldsymbol{x},t'_{j+1}) \\ &= \sum_{j=1}^{S-2}[G(t_S-t'_{j+1})-G(t_S-t'_j)]\hat{\varepsilon}^*_{il}(\boldsymbol{x},t'_{j+1}) \\ &\quad+\frac{1}{2}[G(0)-G(t_S-t'_{S-1})]\hat{\varepsilon}_{il}(\boldsymbol{x},t'_{S-1}) \\ &\quad+\frac{1}{2}[G(0)-G(t_S-t'_{S-1})]\hat{\varepsilon}_{il}(\boldsymbol{x},t'_S)\end{aligned} \tag{6.11}$$

将式(6.5)代入式(6.11)，可以得到如下形式的递归关系：

$$\begin{aligned}\int_0^{t_S}\frac{\partial G(t_S-t')}{\partial t'}\hat{\varepsilon}_{il}(\boldsymbol{x},t')\mathrm{d}t' &= \sum_{w=1}^{\bar{m}}G_w\exp\left(-t_S/\tau_w^G\right)\sum_{i=1}^{S-2}\exp\left(t'_{j+1}/\tau_w^G\right) \\ &\quad\times\left\{1-\exp\left[-(t'_{j+1}-t'_j)/\tau_w^G\right]\right\}\hat{\varepsilon}^*_{il}(\boldsymbol{x},t'_{j+1}) \\ &\quad+\frac{1}{2}[G(t_S-t'_{S-1})-G(0)]\hat{\varepsilon}_{il}(\boldsymbol{x},t'_{S-1}) \\ &\quad+\frac{1}{2}[G(t_S-t'_{S-1})-G(0)]\hat{\varepsilon}_{il}(\boldsymbol{x},t'_S)\end{aligned} \tag{6.12}$$

因此，在时间段$[0,t_S]$，我们可以通过如下递归关系计算记忆应力：

$$\begin{aligned}\boldsymbol{\sigma}^{\mathrm{m}} &= \int_0^{t_S}\frac{\partial G(t_S-t')}{\partial t'}\hat{\boldsymbol{\varepsilon}}(\boldsymbol{x},t')\mathrm{d}t' \\ &= \sum_{w=1}^{\bar{m}}G_w\boldsymbol{q}_{w,S}+\frac{1}{2}[G(0)-G(t_S-t'_{S-1})]\hat{\boldsymbol{\varepsilon}}(\boldsymbol{x},t'_{S-1}) \\ &\quad+\frac{1}{2}[G(0)-G(t_S-t'_{S-1})]\hat{\boldsymbol{\varepsilon}}(\boldsymbol{x},t'_S)\end{aligned} \tag{6.13}$$

其中，

$$\boldsymbol{q}_{w,S}=\exp\left[-(t_S-t'_{S-1})/\tau_w^G\right]\left(\left\{1-\exp\left[-(t'_{S-1}-t'_{S-2})/\tau_w^G\right]\right\}\hat{\boldsymbol{\varepsilon}}^*(\boldsymbol{x},t'_{S-1})+\boldsymbol{q}_{w,S-1}\right) \tag{6.14}$$

式中，$\hat{\boldsymbol{\varepsilon}}^*(\boldsymbol{x},t'_{j+1})=\frac{1}{2}\left[\hat{\boldsymbol{\varepsilon}}(\boldsymbol{x},t'_j)+\hat{\boldsymbol{\varepsilon}}(\boldsymbol{x},t'_{j+1})\right]$。此外，$\boldsymbol{q}_{w,1}$和$\boldsymbol{q}_{w,2}$为$\mathbf{0}$向量。

6.2.2 位移边界域积分方程

黏弹性力学的控制方程为

$$\sigma_{kj,k}+b_j=0,\quad x_j\in\Omega \tag{6.15}$$

边界条件如下：

$$u_j = \bar{u}_j, \quad x_j \in \Gamma_{\mathrm{b1}} \tag{6.16}$$

$$f_j = \bar{f}_j, \quad x_j \in \Gamma_{\mathrm{b2}} \tag{6.17}$$

其中，Ω是求解的区域；$\Gamma_{\mathrm{b1}} \cup \Gamma_{\mathrm{b2}} = \Gamma$ 是Ω域的边界；b_j、u_j、f_j和x_j分别表示体力、位移、面力和坐标分量，$\bar{u}_j$和$\bar{f}_j$分别是边界Γ_{b1}和Γ_{b2}上位移和面力的给定值，这里j=1,2,3。由加权余量法，可以获得下式：

$$\int_\Omega \sigma_{kj,k} U_j^* \mathrm{d}\Omega + \int_\Omega b_j U_j^* \mathrm{d}\Omega = -\int_{\Gamma_{\mathrm{b1}}} (u_j - \bar{u}_j) T_j^* \mathrm{d}\Gamma + \int_{\Gamma_{\mathrm{b2}}} (f_j - \bar{f}_j) U_j^* \mathrm{d}\Gamma \tag{6.18}$$

其中，

$$\begin{cases} U_j^*(\boldsymbol{q}) = U_{ij}^*(\boldsymbol{q}, \boldsymbol{p}) e_i^*(\boldsymbol{p}) \\ T_j^*(\boldsymbol{q}) = T_{ij}^*(\boldsymbol{q}, \boldsymbol{p}) e_i^*(\boldsymbol{p}) \end{cases} \tag{6.19}$$

其中，$U_j^*(\boldsymbol{q})$和$T_j^*(\boldsymbol{q})$分别表示在$\boldsymbol{p}$处x_i方向上作用于无限域上单位集中力$e_i^*(\boldsymbol{p})$所引起的点$\boldsymbol{q}$处x_j方向上的位移和面力。式(6.18)中的第一项可以由分部积分写成如下形式：

$$\int_\Omega \sigma_{kj,k} U_j^* \mathrm{d}\Omega = \int_\Omega (\sigma_{ij} U_j^*)_{,k} \mathrm{d}\Omega - \int_\Omega \sigma_{kj} U_{j,k}^* \mathrm{d}\Omega \tag{6.20}$$

式(6.20)中的$\int_\Omega (\sigma_{kj} U_j^*)_{,k} \mathrm{d}\Omega$可以由高斯积分写成如下形式：

$$\int_\Omega (\sigma_{kj} U_j^*)_{,k} \mathrm{d}\Omega = \int_\Gamma \sigma_{kj} U_j^* n_k \mathrm{d}\Gamma = \int_\Gamma (\sigma_{kj} n_k) U_j^* \mathrm{d}\Gamma = \int_\Gamma f_j U_j^* \mathrm{d}\Gamma \tag{6.21}$$

因此，式(6.20)可以写成

$$\int_\Omega \sigma_{kj,k} U_j^* \mathrm{d}\Omega = \int_\Gamma f_j U_j^* \mathrm{d}\Gamma - \int_\Omega \sigma_{kj} U_{j,k}^* \mathrm{d}\Omega \tag{6.22}$$

将式(6.22)和式(6.2)代入式(6.18)，可以得到

$$\begin{aligned} & -\int_\Omega \sigma_{kj}^{\mathrm{e}} U_{j,k}^* \mathrm{d}\Omega + \int_\Omega \sigma_{kj}^{\mathrm{m}} U_{j,k}^* \mathrm{d}\Omega + \int_\Omega b_j U_j^* \mathrm{d}\Omega \\ & = -\int_{\Gamma_{\mathrm{b1}}} \left(u_j - \bar{u}_j\right) T_j^* \mathrm{d}\Gamma - \int_{\Gamma_{\mathrm{b2}}} \bar{f}_j U_j^* \mathrm{d}\Gamma - \int_{\Gamma_{\mathrm{b1}}} f_j U_j^* \mathrm{d}\Gamma \end{aligned} \tag{6.23}$$

对于式(6.23)中的第一项$\int_\Omega \sigma_{kj}^{\mathrm{e}} U_{j,k}^* \mathrm{d}\Omega$，由贝蒂(Betti)互易定理和高斯公式可得

$$\begin{aligned} \int_\Omega \sigma_{kj}^{\mathrm{e}} U_{j,k}^* \mathrm{d}\Omega &= \int_\Omega \sigma_{kj}^{\mathrm{e}*} u_{j,k} \mathrm{d}\Omega \\ &= \int_\Omega \left(\sigma_{kj}^{\mathrm{e}*} u_j\right)_{,k} \mathrm{d}\Omega - \int_\Omega \sigma_{kj,k}^{\mathrm{e}*} u_j \mathrm{d}\Omega \\ &= \int_\Gamma \sigma_{kj}^{\mathrm{e}*} u_j n_k \mathrm{d}\Gamma - \int_\Omega \sigma_{kj,k}^{\mathrm{e}*} u_j \mathrm{d}\Omega \\ &= \int_\Gamma T_j^* u_j \mathrm{d}\Gamma - \int_\Omega \sigma_{kj,k}^{\mathrm{e}*} u_j \mathrm{d}\Omega \end{aligned} \tag{6.24}$$

式(6.23)中的第二项 $\int_{\Omega}\sigma_{kj}^{\mathrm{m}}U_{j,k}^{*}\mathrm{d}\Omega$ 可以写成如下等价形式：

$$\int_{\Omega}\sigma_{kj}^{\mathrm{m}}U_{j,k}^{*}\mathrm{d}\Omega=\int_{\Omega}\sigma_{kj}^{\mathrm{m}}\frac{1}{2}\left(U_{j,k}^{*}+U_{k,j}^{*}\right)\mathrm{d}\Omega=\int_{\Omega}\sigma_{jk}^{\mathrm{m}}\varepsilon_{jk}^{*}\mathrm{d}\Omega \tag{6.25}$$

将式(6.24)和式(6.25)代入式(6.23)，可得如下形式：

$$\int_{\Omega}\sigma_{kj,k}^{\mathrm{e}*}u_{j}\mathrm{d}\Omega+\int_{\Omega}\sigma_{jk}^{\mathrm{m}}\varepsilon_{jk}^{*}\mathrm{d}\Omega+\int_{\Omega}b_{j}U_{j}^{*}\mathrm{d}\Omega=\int_{\Gamma}u_{j}T_{j}^{*}\mathrm{d}\Gamma-\int_{\Gamma}f_{j}U_{j}^{*}\mathrm{d}\Gamma \tag{6.26}$$

对于式(6.26)中的第一项，由基本解的性质可得如下形式：

$$\int_{\Omega}\sigma_{kj,k}^{\mathrm{e}*}u_{j}\mathrm{d}\Omega=-\int_{\Omega}\delta(\boldsymbol{q},\boldsymbol{p})\delta_{ij}e_{i}^{*}(\boldsymbol{p})u_{j}(\boldsymbol{q})\mathrm{d}\Omega \tag{6.27}$$

将式(6.19)和式(6.27)代入式(6.26)，可得

$$\begin{aligned}&-\int_{\Omega}\delta(\boldsymbol{q},\boldsymbol{p})\delta_{ij}e_{i}^{*}(\boldsymbol{p})u_{j}(\boldsymbol{q})\mathrm{d}\Omega+\int_{\Omega}\sigma_{jk}^{\mathrm{m}}(\boldsymbol{q})\varepsilon_{ijk}^{*}(\boldsymbol{q},\boldsymbol{p})e_{i}^{*}(\boldsymbol{p})\mathrm{d}\Omega\\&+\int_{\Omega}b_{j}(\boldsymbol{q})U_{ij}^{*}(\boldsymbol{q},\boldsymbol{p})e_{i}^{*}(\boldsymbol{p})\mathrm{d}\Omega\\&=\int_{\Gamma}u_{j}(\boldsymbol{Q})T_{ij}^{*}(\boldsymbol{Q},\boldsymbol{p})e_{i}^{*}(\boldsymbol{p})\mathrm{d}\Gamma-\int_{\Gamma}f_{j}(\boldsymbol{Q})U_{j}^{*}(\boldsymbol{Q},\boldsymbol{p})e_{i}^{*}(\boldsymbol{p})\mathrm{d}\Gamma\end{aligned} \tag{6.28}$$

式中的单位向量在所有积分中都是一样的，而且每个分量都是相互独立的。此外，$-\int_{\Omega}\delta(\boldsymbol{q},\boldsymbol{p})\delta_{ij}u_{j}(\boldsymbol{q})\mathrm{d}\Omega=-u_{i}(\boldsymbol{p})$，因此式(6.28)最终简化为如下形式：

$$\begin{aligned}u_{i}(\boldsymbol{p})=&\int_{\Gamma}f_{j}(\boldsymbol{Q})U_{ij}^{*}(\boldsymbol{Q},\boldsymbol{p})\mathrm{d}\Gamma(\boldsymbol{Q})-\int_{\Gamma}u_{j}(\boldsymbol{Q})T_{ij}^{*}(\boldsymbol{Q},\boldsymbol{p})\mathrm{d}\Gamma(\boldsymbol{Q})\\&+\int_{\Omega}b_{j}(\boldsymbol{q})U_{ij}^{*}(\boldsymbol{q},\boldsymbol{p})\mathrm{d}\Omega(\boldsymbol{q})+\int_{\Omega}\sigma_{jk}^{\mathrm{m}}(\boldsymbol{q})\varepsilon_{ijk}^{*}(\boldsymbol{q},\boldsymbol{p})\mathrm{d}\Omega(\boldsymbol{q})\end{aligned} \tag{6.29}$$

其中，

$$\begin{cases}U_{ij}^{*}=\dfrac{1}{16\pi G(1-\nu)r}[(3-4\nu)\delta_{ij}+r_{,i}r_{,j}]\\T_{ij}^{*}=-\dfrac{1}{8\pi(1-\nu)r^{2}}\left\{\dfrac{\partial r}{\partial n}[(1-2\nu)\delta_{ij}+3r_{,i}r_{,j}]+(1-2\nu)\left(-r_{,j}n_{i}+r_{,i}n_{j}\right)\right\}\\\varepsilon_{ijk}^{*}=-\dfrac{1}{16\pi(1-\nu)Gr^{2}}[(1-2\nu)\left(\delta_{ik}r_{,j}+\delta_{ij}r_{,k}\right)-\delta_{jk}r_{,i}+3r_{,i}r_{,j}r_{,k}]\end{cases} \tag{6.30}$$

其中，$\boldsymbol{Q}$ 表示边界场点；$\boldsymbol{q}$ 为域内部场点；$\boldsymbol{p}$ 表示域内部源点；r 表示源点 $\boldsymbol{p}$ 到场点 $\boldsymbol{Q}$ 的距离。此外，$r_{,i}=\partial r/\partial x_{i}=\left(x_{i}^{Q}-x_{i}^{p}\right)\big/r(\boldsymbol{Q},\boldsymbol{p})$ 表示 r 对坐标 x_{i} 的偏导；ν 表示泊松比。边界点的边界域积分方程可写成如下形式：

$$
\begin{aligned}
c_{ij}(\boldsymbol{P})u_j(\boldsymbol{P}) = & \int_\Gamma f_j(\boldsymbol{Q})U_{ij}^*(\boldsymbol{Q},\boldsymbol{P})\mathrm{d}\Gamma(\boldsymbol{Q}) - \int_\Gamma u_j(\boldsymbol{Q})T_{ij}^*(\boldsymbol{Q},\boldsymbol{P})\mathrm{d}\Gamma(\boldsymbol{Q}) \\
& + \int_\Omega b_j(\boldsymbol{q})U_{ij}^*(\boldsymbol{q},\boldsymbol{p})\mathrm{d}\Omega(\boldsymbol{q}) + \int_\Omega \sigma_{jk}^{\mathrm{m}}(\boldsymbol{q})\varepsilon_{ijk}^*(\boldsymbol{q},\boldsymbol{p})\mathrm{d}\Omega(\boldsymbol{q})
\end{aligned} \tag{6.31}
$$

其中，$c_{ij}(\boldsymbol{P})$ 与边界上源点 $\boldsymbol{P}$ 的位置有关。

6.2.3 应力和应变边界积分方程

通过几何方程，可以得到内点的应变积分方程：

$$
\begin{aligned}
\varepsilon_{il}(\boldsymbol{p}) = & \int_\Gamma f_j(\boldsymbol{Q})U_{ilj}^{\varepsilon*}(\boldsymbol{Q},\boldsymbol{p})\mathrm{d}\Gamma(\boldsymbol{Q}) - \int_\Gamma u_j(\boldsymbol{Q})T_{ilj}^{\varepsilon*}(\boldsymbol{Q},\boldsymbol{p})\mathrm{d}\Gamma(\boldsymbol{Q}) \\
& + \int_\Omega b_j(\boldsymbol{q})U_{ilj}^{\varepsilon*}(\boldsymbol{q},\boldsymbol{p})\mathrm{d}\Omega(\boldsymbol{q}) + \int_{\Omega-\Omega_\zeta} \sigma_{jk}^{\mathrm{m}}(\boldsymbol{q})\varepsilon_{iljk}^*(\boldsymbol{q},\boldsymbol{p})\mathrm{d}\Omega(\boldsymbol{q}) \\
& + \int_{\Omega_\zeta} \sigma_{jk}^{\mathrm{m}}(\boldsymbol{q})\varepsilon_{iljk}^*(\boldsymbol{q},\boldsymbol{p})\mathrm{d}\Omega(\boldsymbol{q})
\end{aligned} \tag{6.32}
$$

其中，

$$
\begin{cases}
U_{ilj}^{\varepsilon*} = \dfrac{1}{16\pi G(1-\nu)r^2}[(1-2\nu)(\delta_{jl}r_{,i} + \delta_{ij}r_{,l}) - \delta_{il}r_{,j} + 3r_{,i}r_{,j}r_{,l}] \\
T_{ilj}^{\varepsilon*} = \dfrac{-1}{8\pi(1-\nu)r^3}\{(1-2\nu)n_j(\delta_{il} - 3r_{,i}r_{,l}) \\
\qquad - n_i[(1-2\nu)\delta_{jl} + 3\nu r_{,j}r_{,l}] - n_l[(1-2\nu)\delta_{ij} + 3\nu r_{,i}r_{,j}] \\
\qquad - 3\dfrac{\partial r}{\partial n}[\nu\delta_{ij}r_{,l} + \nu\delta_{jl}r_{,i} + \delta_{il}r_{,j} - 5r_{,i}r_{,j}r_{,l}]\} \\
\varepsilon_{iljk}^* = \dfrac{1}{16\pi G(1-\nu)r^3}\Big[(1-2\nu)\big(\delta_{ij}\delta_{kl} + \delta_{ik}\delta_{jl}\big) - \delta_{il}\delta_{jk} \\
\qquad + 3\nu\big(\delta_{ij}r_{,k}r_{,l} + \delta_{ik}r_{,j}r_{,l} + \delta_{jl}r_{,i}r_{,k} + \delta_{kl}r_{,i}r_{,j}\big) \\
\qquad + 3\big(\delta_{jk}r_{,i}r_{,l} + \delta_{il}r_{,j}r_{,k}\big) - 15r_{,i}r_{,j}r_{,k}r_{,l}\Big]
\end{cases} \tag{6.33}
$$

对于柯西主值积分 $\int_{\Omega-\Omega_\zeta} \sigma_{jk}^{\mathrm{m}}(\boldsymbol{q})\varepsilon_{iljk}^*(\boldsymbol{q},\boldsymbol{p})\mathrm{d}\Omega(\boldsymbol{q})$，如图 6.1 所示，为了降低其积分的奇异性，我们需要将其正则化为如下形式：

$$
\begin{aligned}
\int_{\Omega-\Omega_\zeta} \sigma_{jk}^{\mathrm{m}}(\boldsymbol{q})\varepsilon_{iljk}^*(\boldsymbol{q},\boldsymbol{p})\mathrm{d}\Omega(\boldsymbol{q}) = & \int_{\Omega-\Omega_\zeta}\left[\sigma_{jk}^{\mathrm{m}}(\boldsymbol{q}) - \sigma_{jk}^{\mathrm{m}}(\boldsymbol{p})\right]\varepsilon_{iljk}^*(\boldsymbol{q},\boldsymbol{p})\mathrm{d}\Omega(\boldsymbol{q}) \\
& + \int_{\Omega-\Omega_\zeta} \sigma_{jk}^{\mathrm{m}}(\boldsymbol{p})\varepsilon_{iljk}^*(\boldsymbol{q},\boldsymbol{p})\mathrm{d}\Omega(\boldsymbol{q})
\end{aligned} \tag{6.34}
$$

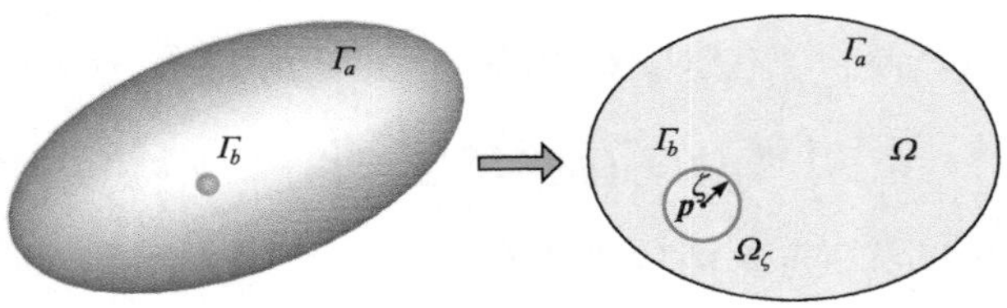

图 6.1 在奇异点附近进行无穷小区域分割

在域 Ω_ζ 内，我们假设 $\sigma^{\mathrm{m}}(\boldsymbol{q})=\sigma^{\mathrm{m}}(\boldsymbol{p})$。因此，式(6.32)的最后两项域积分可以写成

$$
\begin{aligned}
&\int_{\Omega-\Omega_\zeta}\sigma_{jk}^{\mathrm{m}}(\boldsymbol{q})\varepsilon_{iljk}^{*}(\boldsymbol{q},\boldsymbol{p})\mathrm{d}\Omega(\boldsymbol{q})+\int_{\Omega_\zeta}\sigma_{jk}^{\mathrm{m}}(\boldsymbol{q})\varepsilon_{iljk}^{*}(\boldsymbol{q},\boldsymbol{p})\mathrm{d}\Omega(\boldsymbol{q})\\
&=\int_{\Omega-\Omega_\zeta}\left[\sigma_{jk}^{\mathrm{m}}(\boldsymbol{q})-\sigma_{jk}^{\mathrm{m}}(\boldsymbol{p})\right]\varepsilon_{iljk}^{*}(\boldsymbol{q},\boldsymbol{p})\mathrm{d}\Omega(\boldsymbol{q})+\sigma_{jk}^{\mathrm{m}}(\boldsymbol{p})\int_{\Omega-\Omega_\zeta}\varepsilon_{iljk}^{*}(\boldsymbol{q},\boldsymbol{p})\mathrm{d}\Omega(\boldsymbol{q})\\
&\quad+\sigma_{jk}^{\mathrm{m}}(\boldsymbol{q})\int_{\Omega-\Omega_\zeta}\varepsilon_{iljk}^{*}(\boldsymbol{q},\boldsymbol{p})\mathrm{d}\Omega(\boldsymbol{q})\\
&=\int_{\Omega-\Omega_\zeta}\left[\sigma_{jk}^{\mathrm{m}}(\boldsymbol{q})-\sigma_{jk}^{\mathrm{m}}(\boldsymbol{p})\right]\varepsilon_{iljk}^{*}(\boldsymbol{q},\boldsymbol{p})\mathrm{d}\Omega(\boldsymbol{q})+\sigma_{jk}^{\mathrm{m}}(\boldsymbol{p})\int_{\Omega}\varepsilon_{iljk}^{*}(\boldsymbol{q},\boldsymbol{p})\mathrm{d}\Omega(\boldsymbol{q})
\end{aligned}\tag{6.35}
$$

式中的域积分 $\int_{\Omega}\varepsilon_{iljk}^{*}(\boldsymbol{q},\boldsymbol{p})\mathrm{d}\Omega(\boldsymbol{q})$ 可以转换为如下形式：

$$
\begin{aligned}
\int_{\Omega}\varepsilon_{iljk}^{*}(\boldsymbol{q},\boldsymbol{p})\mathrm{d}\Omega(\boldsymbol{q})&=\int_{\Omega}\frac{1}{2}\left[\frac{\partial\varepsilon_{ijk}^{*}(\boldsymbol{q},\boldsymbol{p})}{\partial x_l^{p}}+\frac{\partial\varepsilon_{ljk}^{*}(\boldsymbol{q},\boldsymbol{p})}{\partial x_i^{p}}\right]\mathrm{d}\Omega(\boldsymbol{q})\\
&=\int_{\Omega}\frac{1}{2}\left[\frac{\partial\varepsilon_{ilj}^{*}(\boldsymbol{q},\boldsymbol{p})}{\partial x_k^{q}}+\frac{\partial\varepsilon_{ilk}^{*}(\boldsymbol{q},\boldsymbol{p})}{\partial x_j^{q}}\right]\mathrm{d}\Omega(\boldsymbol{q})\\
&=\int_{\Gamma}\frac{1}{2}\left[\varepsilon_{ilj}^{*}(\boldsymbol{Q},\boldsymbol{p})n_k+\varepsilon_{ilk}^{*}(\boldsymbol{Q},\boldsymbol{p})n_j\right]\mathrm{d}\Gamma(\boldsymbol{Q})
\end{aligned}\tag{6.36}
$$

将式(6.36)代入式(6.32)，可得正则化的应变积分方程：

$$
\begin{aligned}
\varepsilon_{il}(\boldsymbol{p})=&\int_{\Gamma}f_j(\boldsymbol{Q})U_{ilj}^{\varepsilon*}(\boldsymbol{Q},\boldsymbol{p})\mathrm{d}\Gamma(\boldsymbol{Q})-\int_{\Gamma}u_j(\boldsymbol{Q})T_{ilj}^{\varepsilon*}(\boldsymbol{Q},\boldsymbol{p})\mathrm{d}\Gamma(\boldsymbol{Q})\\
&+\int_{\Omega}b_j(\boldsymbol{q})U_{ilj}^{\varepsilon*}(\boldsymbol{q},\boldsymbol{p})\mathrm{d}\Omega(\boldsymbol{q})+\int_{\Omega_A-\Omega_\zeta}[\sigma_{jk}^{\mathrm{m}}(\boldsymbol{q})-\sigma_{jk}^{\mathrm{m}}(\boldsymbol{p})]\varepsilon_{iljk}^{*}(\boldsymbol{q},\boldsymbol{p})\mathrm{d}\Omega(\boldsymbol{q})\\
&+\frac{1}{2}\sigma_{jk}^{\mathrm{m}}(\boldsymbol{p})\int_{\Gamma}[\varepsilon_{ilj}^{*}(\boldsymbol{Q},\boldsymbol{p})n_k+\varepsilon_{ilk}^{*}(\boldsymbol{Q},\boldsymbol{p})n_j]\mathrm{d}\Gamma(\boldsymbol{Q})
\end{aligned}\tag{6.37}
$$

其中，n_k 和 n_j 表示边界点 $\boldsymbol{Q}$ 处的单位法向量分量。与应变积分方程类似，可得内点应力积分方程为

$$\begin{aligned}\sigma_{il}^{\mathrm{e}}(\boldsymbol{p})=&\int_{\Gamma} f_j(\boldsymbol{Q})D_{ilj}^{\sigma*}(\boldsymbol{p},\boldsymbol{Q})\mathrm{d}\Gamma(\boldsymbol{Q})-\int_{\Gamma} u_j(\boldsymbol{Q})V_{ilj}^{\sigma*}(\boldsymbol{p},\boldsymbol{Q})\mathrm{d}\Gamma(\boldsymbol{Q})\\&+\int_{\Omega} b_j(\boldsymbol{q})D_{ilj}^{\sigma*}(\boldsymbol{p},\boldsymbol{Q})\mathrm{d}\Omega(\boldsymbol{q})\\&-\int_{\Omega_A-\Omega_\zeta} W_{iljk}^{\sigma*}(\boldsymbol{p},\boldsymbol{q})\sigma_{jk}^{\mathrm{m}}(\boldsymbol{q})\mathrm{d}\Omega(\boldsymbol{q})+\bar{F}_{il}(\boldsymbol{p})\end{aligned}\tag{6.38}$$

其中，

$$\bar{F}_{il}(\boldsymbol{p})=-\frac{1}{8(1-\nu)}[(6-8\nu)\sigma_{il}^{m}(\boldsymbol{p})-(1-4\nu)\sigma_{nn}^{m}(\boldsymbol{p})\delta_{il}]\tag{6.39}$$

$$\begin{cases}D_{ilj}^{\sigma*}=\dfrac{1}{8\pi(1-\nu)r^2}[(1-2\nu)(\delta_{jl}r_{,i}+\delta_{ij}r_{,l}-\delta_{il}r_{,j})+3r_{,i}r_{,j}r_{,l}]\\V_{ilj}^{\sigma*}=\dfrac{G(0)}{4\pi(1-\nu)r^3}\left\{3\dfrac{\partial r}{\partial n}[(1-2\nu)\delta_{il}r_{,j}+\nu(\delta_{ij}r_{,l}+\delta_{lj}r_{,i})-5r_{,i}r_{,j}r_{,l}]\right.\\\qquad\left.+3\nu(n_ir_{,l}r_{,j}+n_lr_{,i}r_{,j})+(1-2\nu)(3n_jr_{,i}r_{,l}+n_l\delta_{ij}+n_i\delta_{lj})-(1-4\nu)n_j\delta_{il}\right\}\\W_{iljk}^{\sigma*}=2G(0)\varepsilon_{iljk}^{*}+\delta_{il}\left[K(0)-\dfrac{2}{3}G(0)\right]\varepsilon_{iljk}^{*}\end{cases}\tag{6.40}$$

6.2.4　三维黏弹性力学的面力恢复法

利用边界积分方程计算边界点的应力和应变时，需要求解其中的超奇异积分。为了避免求解超奇异积分，我们将使用黏弹性面力恢复法来求解边界点的应力和应变。面力恢复法的关键是通过对边界元的形函数求导而得到表面的局部应变，然后利用广义胡克定律和面力确定边界点的应力场。最后，利用本构方程得到边界点处的应变。

边界源点的位移和面力可以通过下式获得：

$$\begin{cases}\boldsymbol{u}(\boldsymbol{P})=\boldsymbol{\Phi}\boldsymbol{u}(\boldsymbol{Q})\\\boldsymbol{f}(\boldsymbol{P})=\boldsymbol{\Phi}\boldsymbol{f}(\boldsymbol{Q})\end{cases}\tag{6.41}$$

其中，$\boldsymbol{\Phi}$ 表示控制点 $\boldsymbol{Q}$ 到配点 $\boldsymbol{P}$ 的转换矩阵。如图 6.2(a)所示，局部单位切向量 $\hat{\boldsymbol{e}}_i(i=1,2)$ 和局部单位法向量 $\hat{\boldsymbol{e}}_3$ 可以由下式获得：

$$\begin{cases}\hat{\boldsymbol{e}}_1=\boldsymbol{m}_1/|\boldsymbol{m}_1|\\\hat{\boldsymbol{e}}_2=\hat{\boldsymbol{e}}_1\times\hat{\boldsymbol{e}}_3\\\hat{\boldsymbol{e}}_3=\boldsymbol{n}/|\boldsymbol{n}|\end{cases}\tag{6.42}$$

其中，$\boldsymbol{n}$ 表示法向量；$\boldsymbol{m}$ 表示切向量，可由下式获得：

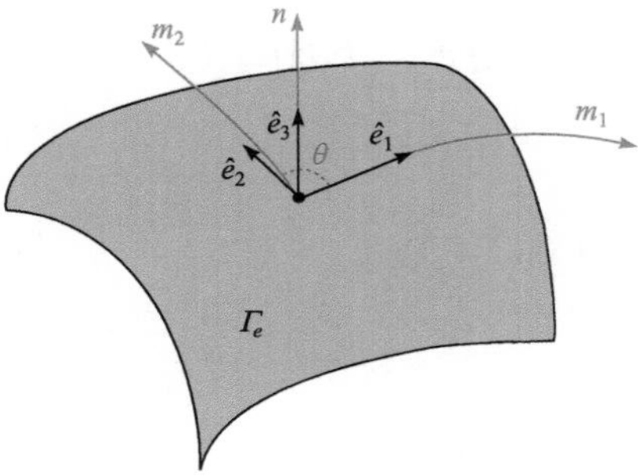

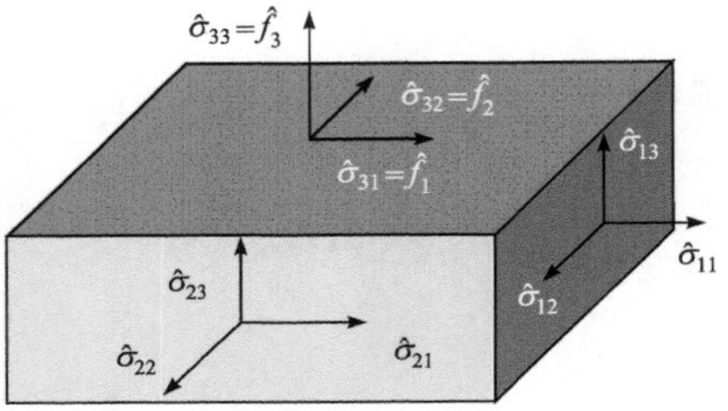

图 6.2 面力恢复法示意图

(a) 局部坐标系；(b) 边界点的应力和面力

$$\begin{cases}\boldsymbol{m}_1(\hat{\xi}_1,\hat{\xi}_2)=\partial\boldsymbol{r}/\partial\xi_1\\ \boldsymbol{m}_2(\hat{\xi}_1,\hat{\xi}_2)=\partial\boldsymbol{r}/\partial\xi_2\\ \boldsymbol{n}(\hat{\xi}_1,\hat{\xi}_2)=\boldsymbol{m}_1(\hat{\xi}_1,\hat{\xi}_2)\times\boldsymbol{m}_2(\hat{\xi}_1,\hat{\xi}_2)\Big/\left|\boldsymbol{m}_1(\hat{\xi}_1,\hat{\xi}_2)\times\boldsymbol{m}_2(\hat{\xi}_1,\hat{\xi}_2)\right|\end{cases}\tag{6.43}$$

因此，我们可以得到整体坐标系到局部坐标系的转换矩阵如下：

$$\boldsymbol{A}=\begin{Bmatrix}\hat{\boldsymbol{e}}_1\\ \hat{\boldsymbol{e}}_2\\ \hat{\boldsymbol{e}}_3\end{Bmatrix}\tag{6.44}$$

局部黏弹性应变可表示为

$$\hat{\varepsilon}_{il}=\frac{\partial\hat{u}_i}{\partial\hat{x}_l}=\frac{\partial\hat{u}_i}{\partial\xi_j}\frac{\partial\xi_j}{\partial\hat{x}_l}\tag{6.45}$$

其中，$\partial\hat{u}_i/\partial\xi_j=A_{il}\,\partial u_l/\partial\xi_j$。局部坐标系下的边界点的应力 $\hat{\sigma}_{ij}$ 和面力 $\hat{f}_i\,(i,j=1,2,3)$如图 6.2(b)所示，弹性应力可由如下公式获得：

$$\begin{cases}\hat{\sigma}_{13}^{\mathrm{e}}=\hat{\sigma}_{13}+\hat{\sigma}_{13}^{\mathrm{m}}=\hat{f}_1+\hat{\sigma}_{13}^{\mathrm{m}}\\ \hat{\sigma}_{23}^{\mathrm{e}}=\hat{\sigma}_{23}+\hat{\sigma}_{23}^{\mathrm{m}}=\hat{f}_2+\hat{\sigma}_{23}^{\mathrm{m}}\\ \hat{\sigma}_{33}^{\mathrm{e}}=\hat{\sigma}_{33}+\hat{\sigma}_{33}^{\mathrm{m}}=\hat{f}_3+\hat{\sigma}_{33}^{\mathrm{m}}\\ \hat{\sigma}_{11}^{\mathrm{e}}=\dfrac{E}{1-\nu^2}\left(\hat{\varepsilon}_{11}+\nu\hat{\varepsilon}_{22}\right)+\dfrac{\nu}{1-\nu}\hat{\sigma}_{33}^{\mathrm{e}}\\ \hat{\sigma}_{12}^{\mathrm{e}}=2G(0)\hat{\varepsilon}_{12}\\ \hat{\sigma}_{22}^{\mathrm{e}}=\dfrac{E}{1-\nu^2}\left(\hat{\varepsilon}_{22}+\nu\hat{\varepsilon}_{11}\right)+\dfrac{\nu}{1-\nu}\hat{\sigma}_{33}^{\mathrm{e}}\end{cases}\tag{6.46}$$

其中，

$$\begin{bmatrix} \hat{\boldsymbol{\sigma}}_{11}^{\mathrm{m}} & \hat{\boldsymbol{\sigma}}_{12}^{\mathrm{m}} & \hat{\boldsymbol{\sigma}}_{13}^{\mathrm{m}} \\ \hat{\boldsymbol{\sigma}}_{21}^{\mathrm{m}} & \hat{\boldsymbol{\sigma}}_{22}^{\mathrm{m}} & \hat{\boldsymbol{\sigma}}_{23}^{\mathrm{m}} \\ \hat{\boldsymbol{\sigma}}_{31}^{\mathrm{m}} & \hat{\boldsymbol{\sigma}}_{32}^{\mathrm{m}} & \hat{\boldsymbol{\sigma}}_{33}^{\mathrm{m}} \end{bmatrix} = \boldsymbol{A} \begin{bmatrix} \boldsymbol{\sigma}_{11}^{\mathrm{m}} & \boldsymbol{\sigma}_{12}^{\mathrm{m}} & \boldsymbol{\sigma}_{13}^{\mathrm{m}} \\ \boldsymbol{\sigma}_{21}^{\mathrm{m}} & \boldsymbol{\sigma}_{22}^{\mathrm{m}} & \boldsymbol{\sigma}_{23}^{\mathrm{m}} \\ \boldsymbol{\sigma}_{31}^{\mathrm{m}} & \boldsymbol{\sigma}_{32}^{\mathrm{m}} & \boldsymbol{\sigma}_{33}^{\mathrm{m}} \end{bmatrix} \tag{6.47}$$

基于上式，并通过 $\sigma_{il}^{\mathrm{e}} = A_{ji}A_{kl}\hat{\sigma}_{jk}^{\mathrm{e}}$ 来计算弹性应力分量。此外，还可以通过下式得到黏弹性应变和应力：

$$\begin{cases} \varepsilon_{il} = \dfrac{1}{2G(0)}\sigma_{il}^{\mathrm{e}} - \delta_{il}\left\{\dfrac{1}{2G_0}\left[K(0) - \dfrac{2}{3}G(0)\right]\right\}\varepsilon_{nn} \\ \sigma_{il} = \sigma_{il}^{\mathrm{e}} - \sigma_{il}^{\mathrm{m}} \end{cases} \tag{6.48}$$

6.2.5　利用径向积分法将域积分转换为边界积分

由于使用 Kelvin 基本解来推导积分方程，因此积分方程中必然存在域积分。为了保证等几何边界元法只具有边界离散的优点，利用径向积分法将域积分转换为等效边界积分。

1.已知核函数

式(6.29)、式(6.37)和式(6.38)中含有由体力项 $b_j(\boldsymbol{q})$ 引起的域积分 $\int_\Omega b_j(\boldsymbol{q})U_{ij}^*(\boldsymbol{p},\boldsymbol{q})\mathrm{d}\Omega(\boldsymbol{q})$、$\int_\Omega b_j(\boldsymbol{q})U_{ilj}^{\varepsilon*}(\boldsymbol{q},\boldsymbol{p})\mathrm{d}\Omega(\boldsymbol{q})$ 和 $\int_\Omega b_j(\boldsymbol{q})D_{ilj}^{\sigma*}(\boldsymbol{p},\boldsymbol{Q})\mathrm{d}\Omega(\boldsymbol{q})$。这三个域积分的核函数都是已知的，因此，我们可以通过径向积分法将其直接转换为等效边界积分：

$$\begin{cases} \displaystyle\int_\Omega b_j(\boldsymbol{q})U_{ij}^*(\boldsymbol{p},\boldsymbol{q})\mathrm{d}\Omega(\boldsymbol{q}) = \int_\Gamma \frac{1}{r^2(\boldsymbol{p},\boldsymbol{Q})}\frac{\partial r}{\partial n}F_{\mathrm{b1}}(\boldsymbol{p},\boldsymbol{Q})\mathrm{d}\Gamma(\boldsymbol{Q}) \\ \displaystyle\int_\Omega b_j(\boldsymbol{q})U_{ilj}^{\varepsilon*}(\boldsymbol{q},\boldsymbol{p})\mathrm{d}\Omega(\boldsymbol{q}) = \int_\Gamma \frac{1}{r^2(\boldsymbol{p},\boldsymbol{Q})}\frac{\partial r}{\partial n}F_{\mathrm{b2}}(\boldsymbol{p},\boldsymbol{Q})\mathrm{d}\Gamma(\boldsymbol{Q}) \\ \displaystyle\int_\Omega b_j(\boldsymbol{q})D_{ilj}^{\sigma*}(\boldsymbol{p},\boldsymbol{Q})\mathrm{d}\Omega(\boldsymbol{q}) = \int_\Gamma \frac{1}{r^2(\boldsymbol{p},\boldsymbol{Q})}\frac{\partial r}{\partial n}F_{\mathrm{b3}}(\boldsymbol{p},\boldsymbol{Q})\mathrm{d}\Gamma(\boldsymbol{Q}) \end{cases} \tag{6.49}$$

其中，

$$\begin{cases} F_{\mathrm{b1}} = \displaystyle\int_0^r U_{ij}^*\rho g\delta_{j3}r^2\mathrm{d}r \\ F_{\mathrm{b2}} = \displaystyle\int_0^r U_{ilj}^{\varepsilon*}\rho g\delta_{j3}r^2\mathrm{d}r \\ F_{\mathrm{b3}} = \displaystyle\int_0^r D_{ilj}^{\sigma*}\rho g\delta_{j3}r^2\mathrm{d}r \end{cases} \tag{6.50}$$

其中，ρ 表示材料的密度；g 表示重力加速度。

2. 未知核函数

式(6.29)中的第二个域积分 $\int_{\Omega}\sigma_{jk}^{\mathrm{m}}(\boldsymbol{q})\varepsilon_{ijk}^{*}(\boldsymbol{q},\boldsymbol{p})\mathrm{d}\Omega(\boldsymbol{q})$ 含有未知的记忆应力，径向积分法无法直接应用。因此，我们需要将未知核函数 $\sigma_{jk}^{\mathrm{m}}(\boldsymbol{q})$ 近似为一系列径向基函数，其形式如下：

$$\sigma_{jk}^{\mathrm{m}}(\boldsymbol{q})=\sum_{A=1}^{N_A}\alpha_A^{jk}\phi_A\left(R_{Aq}\right)+a_0+\sum_{i=1}^{3}a_i x_i^q,\quad j,k=1,2,3 \tag{6.51}$$

为了使问题可解，系数 α 还需要满足如下关系：

$$\sum_{A=1}^{N_A}\alpha_A^{jk}=\sum_{A=1}^{N_A}\alpha_A^{jk}x_i^A=0,\quad i=1,2,3 \tag{6.52}$$

其中，A 表示应用点；$N_A=N_{\mathrm{b}}+N_{\mathrm{I}}$ 表示应用点的数目，这里 N_{b} 表示边界配点的数目；N_{I} 表示内点的数目；$R=\|\boldsymbol{q}-\boldsymbol{A}\|$ 表示场点到应用点的距离(图 6.3)；α_A 和 a_i 是待定系数。采用四阶样条紧支径向基函数对未知核函数进行展开，其具体形式如下：

$$\phi(R/d_A)=\begin{cases}1-6\left(\dfrac{R}{d_A}\right)^2+8\left(\dfrac{R}{d_A}\right)^3-3\left(\dfrac{R}{d_A}\right)^4, & 0\leqslant R\leqslant d_A\\ 0, & R>d_A\end{cases} \tag{6.53}$$

其中，d_A 表示应用点 A 的紧支域的大小；R、r 和 $\bar{R}$ 的关系如下(图 6.3)：

$$\begin{cases}R=\sqrt{r^2+2sr+\bar{R}^2}\\ \bar{R}=\sqrt{\bar{R}_i\bar{R}_i}\\ \bar{R}_i=x_i^p-x_i^A\\ s=r_{,i}\bar{R}_i\end{cases} \tag{6.54}$$

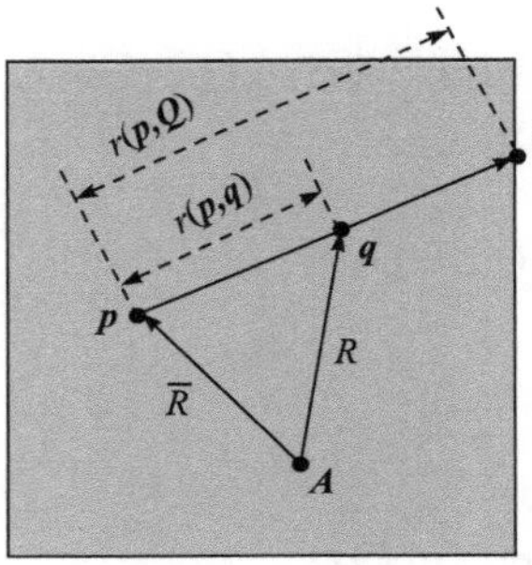

图 6.3 r、$\bar{R}$ 和 R 的关系

将式(6.51)代入域积分 $\int_{\Omega}\sigma_{jk}^{\mathrm{m}}(\boldsymbol{q})\varepsilon_{ijk}^{*}(\boldsymbol{q},\boldsymbol{p})\mathrm{d}\Omega(\boldsymbol{q})$，可得

$$\begin{aligned}\int_{\Omega}\sigma_{jk}^{\mathrm{m}}(\boldsymbol{q})\varepsilon_{ijk}^{*}(\boldsymbol{q},\boldsymbol{p})\mathrm{d}\Omega(\boldsymbol{q})=&\sum_{A=1}^{N_A}\alpha_A^{jk}\int_{\Gamma}\frac{1}{r^2(\boldsymbol{Q},\boldsymbol{p})}\frac{\partial r}{\partial n}F_A(\boldsymbol{q},\boldsymbol{p})\mathrm{d}\Gamma(\boldsymbol{Q})\\&+a_0^{jk}\int_{\Gamma}\frac{1}{r^2(\boldsymbol{Q},\boldsymbol{p})}\frac{\partial r}{\partial n}F_0(\boldsymbol{q},\boldsymbol{p})\mathrm{d}\Gamma(\boldsymbol{Q})\\&+\sum_{s=1}^{3}a_s^{jk}\int_{\Gamma}\frac{1}{r^2(\boldsymbol{Q},\boldsymbol{p})}\frac{\partial r}{\partial n}F_i(\boldsymbol{q},\boldsymbol{p})\mathrm{d}\Gamma(\boldsymbol{Q})\end{aligned} \tag{6.55}$$

其中，

$$\begin{cases} F_A = \int_0^{r(\boldsymbol{p},Q)} \varepsilon_{ijk}^*(\boldsymbol{q},\boldsymbol{p})\phi^A(R)r^2(\boldsymbol{q},\boldsymbol{p})\mathrm{d}r(\boldsymbol{q}) \\ F_0 = \int_0^{r(\boldsymbol{p},Q)} \varepsilon_{ijk}^*(\boldsymbol{q},\boldsymbol{p})r^2(\boldsymbol{q},\boldsymbol{p})\mathrm{d}r(\boldsymbol{q}) \\ F_s = \int_0^{r(\boldsymbol{p},Q)} \varepsilon_{ijk}^*(\boldsymbol{q},\boldsymbol{p})(x_s^p + r_{,s}r)r^2(\boldsymbol{q},\boldsymbol{p})\mathrm{d}r(\boldsymbol{q}) \end{cases} \tag{6.56}$$

与式(6.51)～式(6.56)类似，可以用相同的方法来处理域积分 $\int_{\Omega-\zeta} W_{iljk}^{\sigma^*}(\boldsymbol{p},\boldsymbol{q})\sigma_{jk}^{\mathrm{m}}(\boldsymbol{q})\mathrm{d}\Omega(\boldsymbol{q})$ 和 $\int_{\Omega-\zeta}[\sigma_{jk}^{\mathrm{m}}(\boldsymbol{q})-\sigma_{jk}^{\mathrm{m}}(\boldsymbol{p})]\varepsilon_{iljk}^*(\boldsymbol{q},\boldsymbol{p})\mathrm{d}\Omega(\boldsymbol{q})$。

6.2.6 边界积分方程的等几何分析

与传统边界元法不同，等几何边界元法中的形函数是 NURBS，其可以精确地描述几何形状。本节只介绍 NURBS 的一些结论，其详细介绍可在参考文献[19]中找到。

1. NURBS 曲面

NURBS 曲面可以写成如下形式：

$$S(\xi,\eta) = \sum_{i=1}^{n}\sum_{j=1}^{m} R_{i,j}^{\hat{p},\hat{q}} \bar{\boldsymbol{P}}_{i,j} \tag{6.57}$$

其中，$R_{i,j}^{\hat{p},\hat{q}}$ 表示 NURBS 基函数；$\bar{\boldsymbol{P}}_{i,j}$ 表示网格控制点，它是一个 $n\times m$ 的矩阵(i=1,2,⋯,n, j=1,2,⋯,m)；$\hat{p}$ 和 $\hat{q}$ 分别表示 NURBS 曲面两个方向 ξ 和 η 上的阶次。一般来说，一张 NURBS 曲面很难描述一个三维实体，因此，我们将使用多张 NURBS 片对模型进行描述(图 6.4)。

2. NURBS 离散

模型的几何形状和物理量采用相同的基函数描述如下：

$$\begin{cases} \boldsymbol{x}(\xi,\eta) = \sum_{i=1}^{(\hat{p}+1)}\sum_{j=1}^{(\hat{q}+1)} R_{i,j}^{\hat{p},\hat{q}}(\xi,\eta)\bar{\boldsymbol{P}}_{i,j} \\ \boldsymbol{u}(\xi,\eta,t) = \sum_{i=1}^{(\hat{p}+1)}\sum_{j=1}^{(\hat{q}+1)} R_{i,j}^{\hat{p},\hat{q}}(\xi,\eta)\boldsymbol{u}_{i,j}(t) \\ \boldsymbol{f}(\xi,\eta,t) = \sum_{i=1}^{(\hat{p}+1)}\sum_{j=1}^{(\hat{q}+1)} R_{i,j}^{\hat{p},\hat{q}}(\xi,\eta)\boldsymbol{f}_{i,j}(t) \end{cases} \tag{6.58}$$

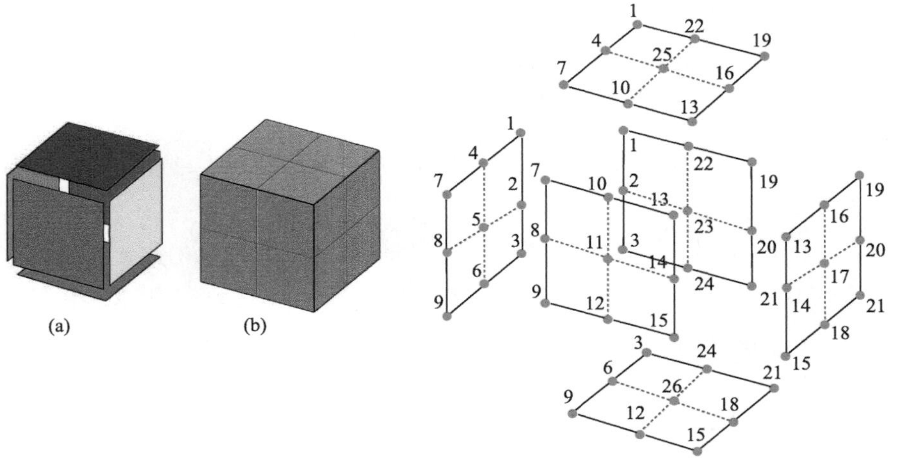

图 6.4 立方体边界的多个 NURBS 片($\hat{p}=2$ 和 $\hat{q}=2$)
(a) 6 个 NURBS 片；(b) NURBS 网格；(c) NURBS 片的整体编号

其中，$\boldsymbol{x}(\xi,\eta)$ 表示边界配点的坐标；$\boldsymbol{u}(\xi,\eta,t)$ 和 $\boldsymbol{f}(\xi,\eta,t)$ 分别表示位移和面力。对于力不连续问题，采用了多个角点，角点处位移连续。未知力的数量与已知位移的数量相同，因此，问题是可解的。

内点和边界点的位移边界域积分方程可以离散为如下形式：

$$\begin{aligned}\boldsymbol{u}(\boldsymbol{p})=&-\sum_{e=1}^{N_e}\sum_{l=1}^{(\hat{p}+1)(\hat{q}+1)}\left\{\int_{-1}^{1}\int_{-1}^{1}T^*\left[\boldsymbol{Q}(\hat{\xi},\hat{\eta}),\boldsymbol{p}\right]R_l^e(\hat{\xi},\hat{\eta})J^e(\hat{\xi},\hat{\eta})\mathrm{d}\hat{\xi}\mathrm{d}\hat{\eta}\right\}\boldsymbol{u}_l^e\\&+\sum_{e=1}^{N_e}\sum_{l=1}^{(\hat{p}+1)(\hat{q}+1)}\left\{\int_{-1}^{1}\int_{-1}^{1}U^*\left[\boldsymbol{Q}(\hat{\xi},\hat{\eta}),\boldsymbol{p}\right]R_l^e(\hat{\xi},\hat{\eta})J^e(\hat{\xi},\hat{\eta})\mathrm{d}\hat{\xi}\mathrm{d}\hat{\eta}\right\}\boldsymbol{f}_l^e\\&+\boldsymbol{f}(\boldsymbol{p})+\boldsymbol{B}(\boldsymbol{p})\boldsymbol{\sigma}^{\mathrm{m}}\end{aligned}\tag{6.59}$$

$$\begin{aligned}&c(\boldsymbol{P})\sum_{l=1}^{(\hat{p}+1)(\hat{q}+1)}R_l^{e'}(\hat{\xi}',\hat{\eta}')\boldsymbol{u}_l^{e'}+\sum_{e=1}^{N_e}\sum_{l=1}^{(\hat{p}+1)(\hat{q}+1)}\left\{\int_{-1}^{1}\int_{-1}^{1}T^*\left[\boldsymbol{Q}(\hat{\xi},\hat{\eta}),\boldsymbol{P}\right]R_l^e(\hat{\xi},\hat{\eta})J^e(\hat{\xi},\hat{\eta})\mathrm{d}\hat{\xi}\right\}\boldsymbol{u}_l^e\\&=\sum_{e=1}^{N_e}\sum_{l=1}^{(\hat{p}+1)(\hat{q}+1)}\left\{\int_{-1}^{1}\int_{-1}^{1}U^*\left[\boldsymbol{Q}(\hat{\xi},\hat{\eta}),\boldsymbol{P}\right]R_{l,p}^e(\hat{\xi},\hat{\eta})J^e(\hat{\xi},\hat{\eta})\mathrm{d}\hat{\xi}\mathrm{d}\hat{\eta}\right\}\boldsymbol{f}_l^e+\boldsymbol{f}(\boldsymbol{P})+\boldsymbol{B}(\boldsymbol{P})\boldsymbol{\sigma}^{\mathrm{m}}\end{aligned}\tag{6.60}$$

其中，局部坐标 $(\hat{\xi},\hat{\eta})$ 见图 6.5；$(\hat{\xi}',\hat{\eta}')$ 表示单元 e' 的局部坐标；N_e 表示单元的数目。雅可比转换行列式 $J^e(\hat{\xi},\hat{\eta})$ 可以写成如下形式：

$$J^e(\hat{\xi},\hat{\eta})=\frac{\mathrm{d}^2\Gamma}{\mathrm{d}\hat{\xi}\mathrm{d}\hat{\eta}}=\frac{\mathrm{d}^2\Gamma}{\mathrm{d}\xi\mathrm{d}\eta}\frac{\mathrm{d}\xi}{\mathrm{d}\hat{\xi}}\frac{\mathrm{d}\eta}{\mathrm{d}\hat{\eta}}\tag{6.61}$$

其中，ξ 和 η 表示参数空间的坐标(图 6.5)。此外，$\boldsymbol{B}$ 是一个行向量，其第 j 个值为

$$
\begin{aligned}
B_j=&\left\{\Phi_{1j}^{-1}\int_\Gamma\frac{1}{r}\frac{\partial r}{\partial n}F_1^\alpha\mathrm{d}\Gamma+\Phi_{2j}^{-1}\int_\Gamma\frac{1}{r}\frac{\partial r}{\partial n}F_2^\alpha\mathrm{d}\Gamma+\cdots+\Phi_{Nj}^{-1}\int_\Gamma\frac{1}{r}\frac{\partial r}{\partial n}F_{N_A}^\alpha\mathrm{d}\Gamma\right.\\
&+\Phi_{(N+1)j}^{-1}\int_\Gamma\frac{1}{r}\frac{\partial r}{\partial n}F_0^\alpha\mathrm{d}\Gamma+\Phi_{(N+2)j}^{-1}\left[\int_\Gamma\frac{1}{r}\frac{\partial r}{\partial n}F_1^\alpha(x_1^p+r_{,1}r)\mathrm{d}\Gamma\right]\\
&\left.+\Phi_{(N+3)j}^{-1}\left[\int_\Gamma\frac{1}{r}\frac{\partial r}{\partial n}F_1^\alpha(x_2^p+r_{,2}r)\mathrm{d}\Gamma\right]+\Phi_{(N+4)j}^{-1}\left[\int_\Gamma\frac{1}{r}\frac{\partial r}{\partial n}F_1^\alpha(x_3^p+r_{,3}r)\mathrm{d}\Gamma\right]\right\}
\end{aligned}\tag{6.62}
$$

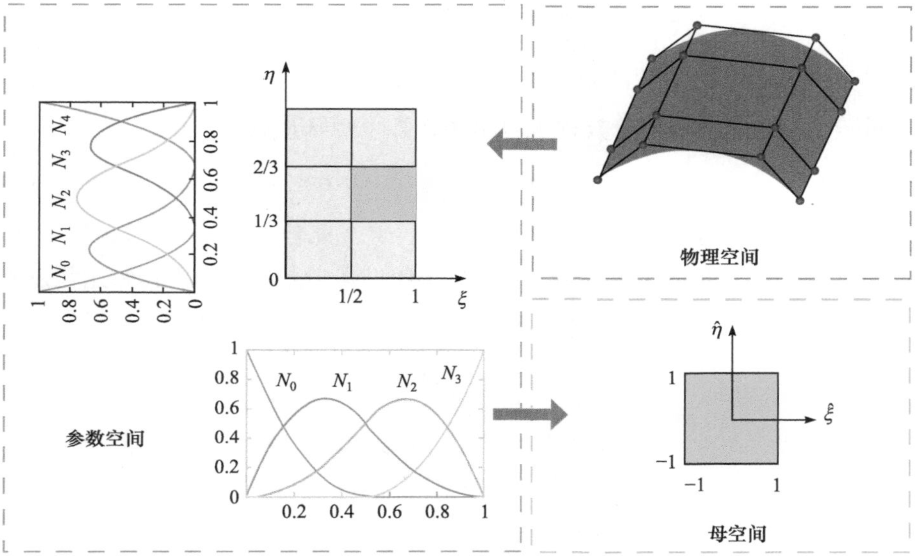

图 6.5　IGA 中的空间变换

式(6.62)中，Φ_{ab}^{-1} 是 $\boldsymbol{\Phi}^{-1}$ 中位于第 a 行和第 b 列处的元素，而 $\boldsymbol{\Phi}$ 是由式(6.51)和式(6.52)形成的类似于式(5.19)或式(5.20)中的系数矩阵 $\boldsymbol{\Phi}$ 。

三维离散的应变积分方程可以写成如下形式：

$$
\begin{aligned}
\varepsilon(\boldsymbol{p})=&-\sum_{e=1}^{N_e}\sum_{l=1}^{(\hat p+1)(\hat q+1)}\left\{\int_{-1}^1\int_{-1}^1T^{\varepsilon*}\left[\boldsymbol{Q}(\hat\xi,\hat\eta),\boldsymbol{p}\right]R_l^e(\hat\xi,\hat\eta)J^e(\hat\xi,\hat\eta)\mathrm{d}\hat\xi\mathrm{d}\hat\eta\right\}\boldsymbol{u}_l^e\\
&+\sum_{e=1}^{N_e}\sum_{l=1}^{(\hat p+1)(\hat q+1)}\left\{\int_{-1}^1\int_{-1}^1U^{\varepsilon*}\left[\boldsymbol{Q}(\hat\xi,\hat\eta),\boldsymbol{p}\right]R_l^e(\hat\xi,\hat\eta)J^e(\hat\xi,\hat\eta)\mathrm{d}\hat\xi\mathrm{d}\hat\eta\right\}\boldsymbol{f}_l^e\\
&+\hat{\boldsymbol{f}}^\varepsilon(\boldsymbol{p})+\boldsymbol{W}(\boldsymbol{p})\boldsymbol{\sigma}^{\mathrm{m}}\\
&+\frac{1}{2}\sum_{e=1}^{N_e}\sum_{l=1}^{(\hat p+1)(\hat q+1)}\left\{\int_{-1}^1\int_{-1}^1\varepsilon_{ilj}^*\left[\boldsymbol{Q}(\hat\xi,\hat\eta),\boldsymbol{p}\right]n_kR_l^e(\hat\xi,\hat\eta)J^e(\hat\xi,\hat\eta)\mathrm{d}\hat\xi\mathrm{d}\hat\eta\right.\\
&\left.+\varepsilon_{ilk}^*\left[\boldsymbol{Q}(\hat\xi,\hat\eta),\boldsymbol{p}\right]n_jR_l^e(\hat\xi,\hat\eta)J^e(\hat\xi,\hat\eta)\mathrm{d}\hat\xi\mathrm{d}\hat\eta\right\}\sigma_{jk}^{\mathrm{m}}(\boldsymbol{Q})
\end{aligned}\tag{6.63}
$$

6.2.7 方程组的求解和迭代过程

三维线性黏弹性问题的位移积分方程的矩阵形式为

$$\boldsymbol{H}_{\mathrm{b}}\boldsymbol{u}_{\mathrm{b}}(t_k)=\boldsymbol{G}_{\mathrm{b}}\boldsymbol{f}_{\mathrm{b}}(t_k)+\boldsymbol{B}_{\mathrm{b}}\boldsymbol{\sigma}^{\mathrm{m}}(t_k)+\hat{\boldsymbol{f}}_{\mathrm{b}} \tag{6.64}$$

$$\boldsymbol{u}_{\mathrm{I}}(t_k)=-\boldsymbol{H}_{\mathrm{I}}\boldsymbol{u}_{\mathrm{b}}(t_k)+\boldsymbol{G}_{\mathrm{I}}\boldsymbol{f}_{\mathrm{b}}(t_k)+\hat{\boldsymbol{f}}_{\mathrm{I}}+\boldsymbol{B}_{\mathrm{I}}\boldsymbol{\sigma}^{\mathrm{m}}(t_k) \tag{6.65}$$

其中，矩阵 $\boldsymbol{H}$、$\boldsymbol{G}$、$\boldsymbol{B}$ 及向量 $\hat{\boldsymbol{f}}$ 是由边界积分和域积分获得的；下标 I 和 b 分别表示内点和边界点。代入边界条件，上面两个方程可以转换成如下求解形式：

$$\boldsymbol{AX}(t_k)=\boldsymbol{F}+\hat{\boldsymbol{f}}+\boldsymbol{B\sigma}^{\mathrm{m}}(t_k) \tag{6.66}$$

同样，内点的应变积分方程也可以写成如下矩阵形式：

$$\boldsymbol{\varepsilon}_{\mathrm{I}}(t_k)=-\boldsymbol{H}^{\varepsilon}\boldsymbol{u}_{\mathrm{b}}(t_k)+\boldsymbol{G}^{\varepsilon}\boldsymbol{f}_{\mathrm{b}}(t_k)+\hat{\boldsymbol{f}}_{\mathrm{I}}^{\varepsilon}+\boldsymbol{W\sigma}^{\mathrm{m}}(t_k)+\boldsymbol{Z\sigma}_{\mathrm{b}}^{\mathrm{m}}(t_k) \tag{6.67}$$

当 $t=0$ 时，没有记忆应力，即 $\boldsymbol{\sigma}^{\mathrm{m}}(t=0)=\boldsymbol{0}$。此外，矩阵 $\boldsymbol{B}$、$\boldsymbol{A}$、$\boldsymbol{H}$、$\boldsymbol{H}^{\varepsilon}$、$\boldsymbol{G}$、$\boldsymbol{G}^{\varepsilon}$、$\boldsymbol{W}$、$\boldsymbol{Z}$ 以及向量 $\boldsymbol{F}$、$\hat{\boldsymbol{f}}$、$\hat{\boldsymbol{f}}_{\mathrm{I}}^{\varepsilon}$ 在迭代过程中是不变的。具体迭代过程如图 6.6 所示。每一步迭代的收敛性条件如下：

$$\left\|\frac{u_{i+1}(t_k)-u_i(t_k)}{u_{i+1}(t_k)}\right\|<\mathrm{eps} \tag{6.68}$$

其中，eps 是给定的计算精度阈值；下标 i 表示迭代步数。

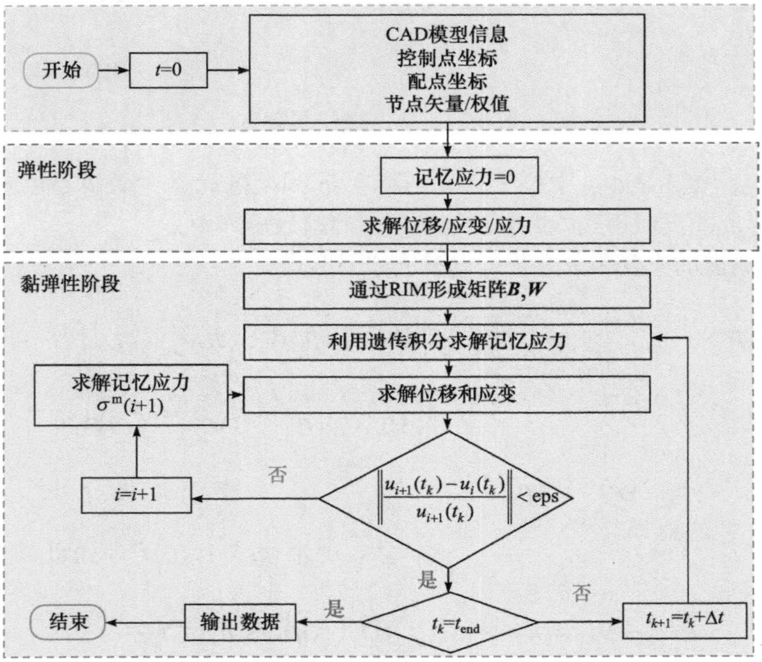

图 6.6　求解黏弹性问题的等几何边界元法流程图

6.3 数 值 算 例

通过四个三维数值算例，讨论了等几何边界元法在求解含体力的黏弹性问题时的有效性。从宏观力学角度看，固体推进剂是线性黏弹性材料。第一个算例没有考虑体力的影响，对于其他的算例，假设模型的密度为 1kg/m^3，重力加速度为 9.8m/s^2。

本节中所有算例的黏弹性材料参数均采用 Prony 级数近似，其形式为

$$G(t)=G_0+G_1\exp\left(-\frac{1}{\gamma}t\right)\ \ (\text{Pa}) \tag{6.69}$$

$$K(t)=K(0)=20000\ \text{Pa} \tag{6.70}$$

其中，$G_0=100\text{Pa}$，$G_1=9900\text{Pa}$，$\gamma=0.4170\text{s}$。

6.3.1 三维黏弹性立方体模型

如图 6.7 所示，我们考虑一个长、宽、高都为 2m 的黏弹性立方体模型。模型中心点的坐标为(0, 0, 0)，边界条件如下：

$$\begin{cases} u_1=0\text{m}, & \Gamma\in(x_1=-1\text{m}) \\ u_2=0\text{m}, & \Gamma\in(x_2=-1\text{m}) \\ u_3=0\text{m}, & \Gamma\in(x_3=-1\text{m}) \\ f_3=1\text{Pa}, & \Gamma\in(x_3=1\text{m}) \end{cases} \tag{6.71}$$

其中，Γ 表示立方体的表面。

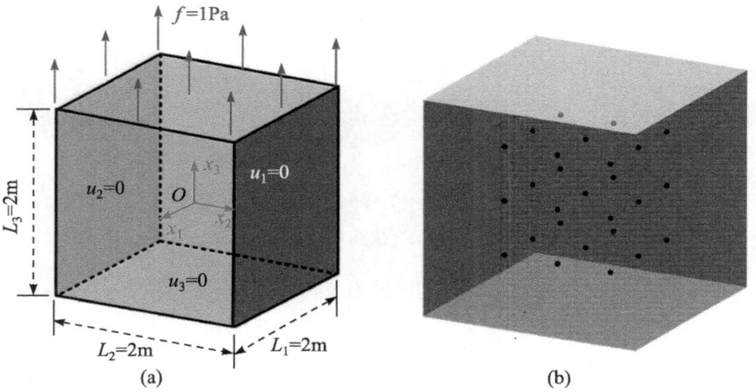

图 6.7 立方体模型
(a) 模型尺寸和边界条件；(b) 27 个内点分布

本算例的时间步长选为 $\Delta t = 0.05\text{ s}$。图 6.7 (b)给出了 27 个应用内点的分布。为了验证算法的有效性，我们将其与商业软件 ANSYS 的计算结果进行对比。图 6.8 给出了 FEM 网格，由 8000 个 8 节点单元组成。

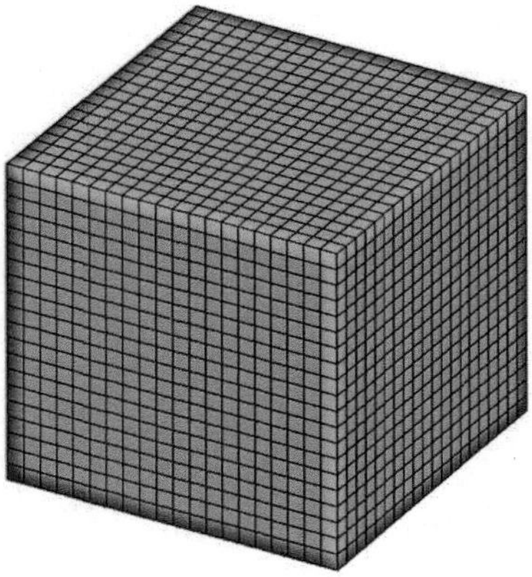

图 6.8 立方体的 FEM 网格

1. h 细化对计算结果的影响

这里利用黏弹性立方体模型讨论 h 细化对计算结果的影响。我们采用了 6 张 NURBS 片来描述这个模型。对这个立方体模型进行了三种细化，分别为：1 次细化；2 次细化；3 次细化。图 6.9 给出了不同 h 细化次数下的控制点分布及网格。三种 h 细化后模型的位移自由度分别为 78、162 和 294。每一张 NURBS 片两个方向上基函数的阶次都是 2。每一张 NURBS 片的初始节点矢量为：$\boldsymbol{U} = \boldsymbol{V} = \{0,0,0,1,1,1\}$ 。

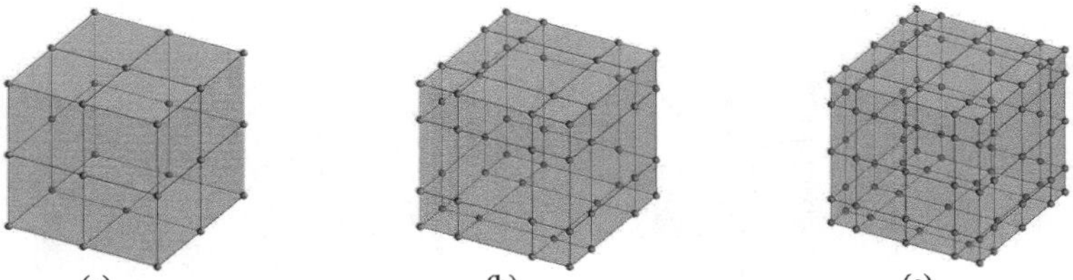

(a) (b) (c)

图 6.9 不同细化下模型的控制点和网格

(a) 1 次细化；(b) 2 次细化；(c) 3 次细化

图 6.10 中可以清楚地看出，在利用等几何边界元法分析三维黏弹性力学问题时，h 细化对计算结果具有一定的影响，也就是说增加一定的自由度是必要的。当 h 细化数为 1 时，计算结果的误差随着时间的增加而增大，这是因为该方法采用了等参元的概念，在利用等几何边界元法分析三维黏弹性力学问题时，少数控制点可以准确描述几何形状，无离散误差，但不能准确描述物理量。当细化次数

大于 1 时，等几何边界元法计算的数值结果与有限元法计算的数值结果吻合较好。值得注意的是，得到相同精度的数值结果，等几何边界元的自由度比有限元要少得多。因此，对于简单的模型，等几何边界元法的计算效率比有限元法高。

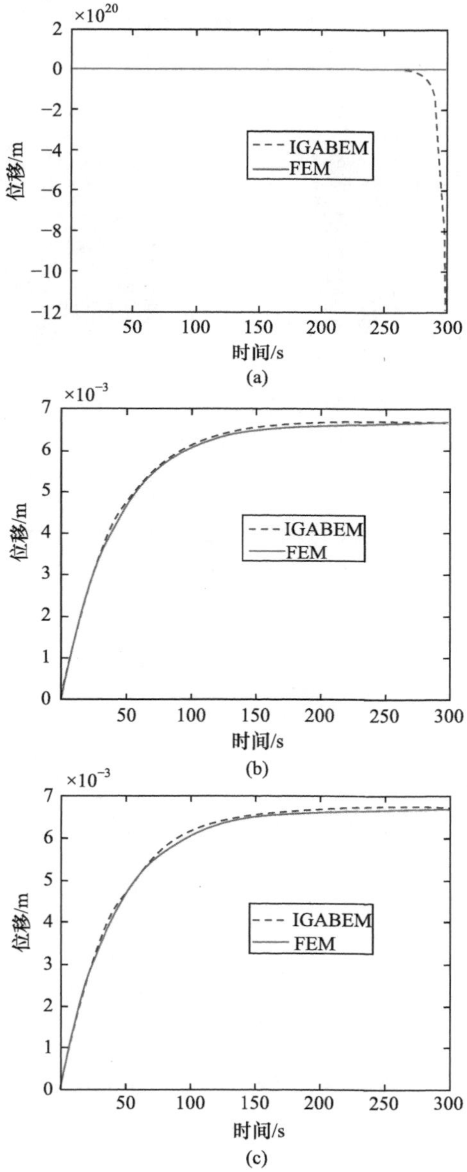

图 6.10　不同细化条件下 A 点(0,0,1m)在 x_3 方向的位移

(a) 1 次细化；(b)2 次细化；(c)3 次细化

2. p 细化对计算结果的影响

在这一部分，我们将利用三维黏弹性立方体模型讨论 NURBS 阶次对计算结果的影响。如图 6.9(c)和图 6.11 所示，对 $\hat{p}=\hat{q}=2$、$\hat{p}=\hat{q}=3$ 和 $\hat{p}=\hat{q}=4$ 的 NURBS 片进行 3 次细化。3 阶和 4 阶 NURBS 片两个方向的初始节点矢量分别为：$\boldsymbol{U}=\boldsymbol{V}=\{0,0,0,0,1,1,1,1\}$ 和 $\boldsymbol{U}=\boldsymbol{V}=\{0,0,0,0,0,1,1,1,1,1\}$。

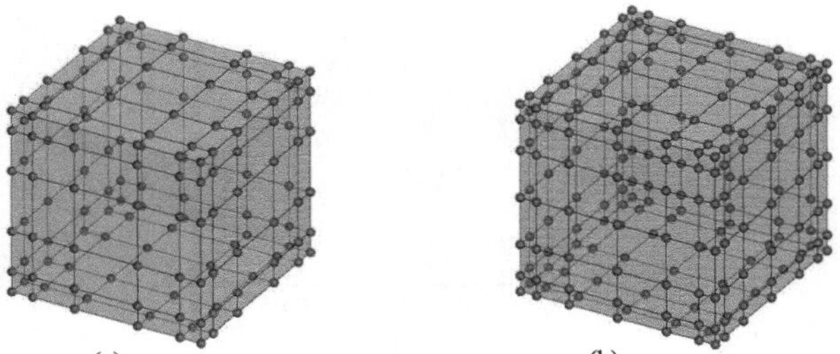

(a) (b)

图 6.11 对不同阶次的 NURBS 片进行 3 次细化

(a) $\hat{p}=\hat{q}=3$；(b) $\hat{p}=\hat{q}=4$

由图 6.12 可以清楚地看出，当使用等几何边界元法分析这类模型简单的线性黏弹性问题时，NURBS 基函数的阶次对该方法的影响很小。因此，考虑到该方法的计算效率，在下面的例子中，我们使用阶次为 2 的 NURBS 基函数。

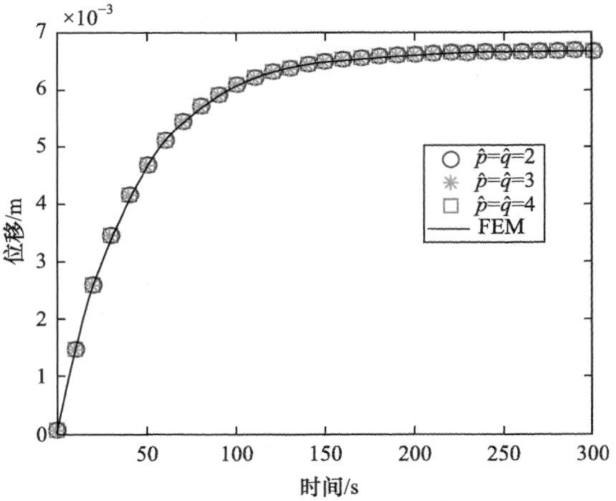

图 6.12 不同基函数阶次条件下 A 点(0,0,1m)在 x_3 方向的位移

3. 内点数对计算结果的影响

在这里，我们利用三维黏弹性立方体模型讨论内点数对计算结果的影响。相同的模型，相同的 NURBS 基函数阶次和相同的细化次数，分布了不同数量的内

点，如图 6.7(b)和图 6.13 所示。

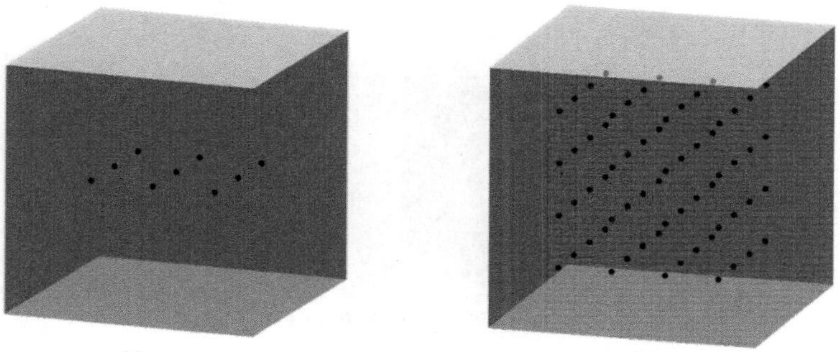

(a)　　(b)

图 6.13　两种不同数量的内点分布

(a) $N_{\mathrm{I}}=9$；(b) $N_{\mathrm{I}}=64$

如图 6.14 所示，不难发现，利用等几何边界元法分析这类模型简单的线性黏弹性问题时，内点的数量对计算结果的影响很小，即使只使用 9 个内点，仍然可以得到相对准确的计算结果。

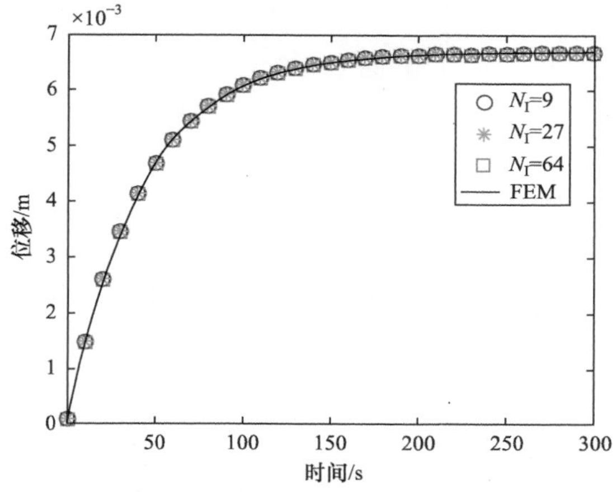

图 6.14　不同内点数量条件下 A 点(0,0,1m)在 x_3 方向的位移

6.3.2　三维黏弹性哑铃模型

在本例中，考虑一个单轴拉伸的三维黏弹性推进剂哑铃试件。由于模型的几何形状和边界条件都是对称的，因此可以将模型简化为等效的 1/8 计算模型，如图 6.15 所示。模型的具体尺寸如图 6.16 所示。

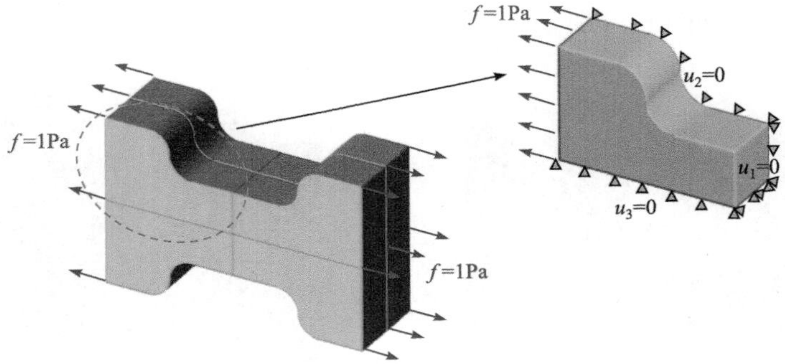

图 6.15　1/8 哑铃试件等效模型及边界条件

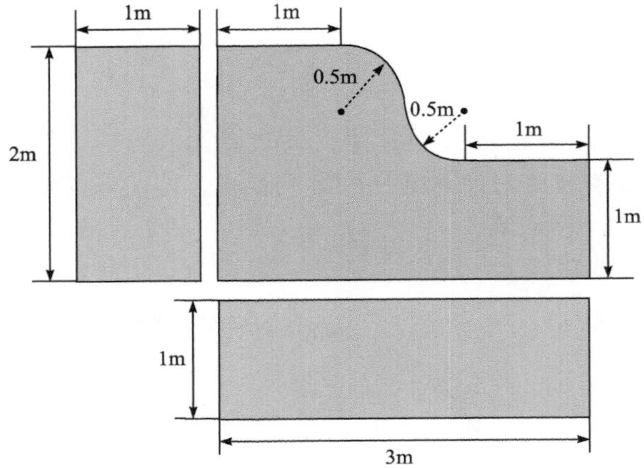

图 6.16　模型的几何尺寸

采用 6 张 NURBS 片描述 1/8 哑铃试件模型，图 6.17 展示了每张 NURBS 片的编号、对应的控制点网格和 117 个内点。表 6.1 给出了每张 NURBS 片两个方向的初始节点矢量。各个 NURBS 片的控制点和权值在文献[43]的附录中给出。

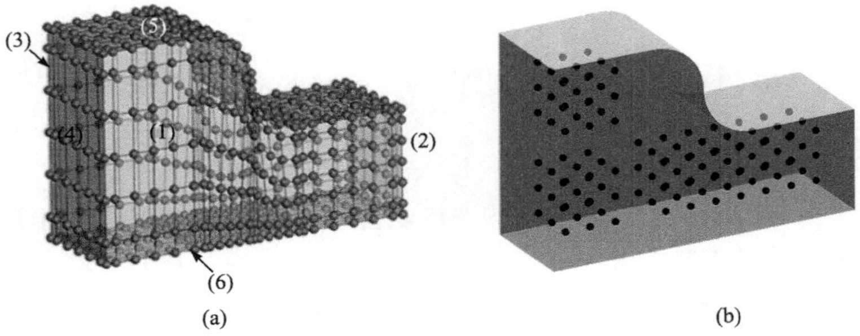

图 6.17　1/8 哑铃模型
(a) 控制点分布及 NURBS 网格；(b) 117 个内点分布

表 6.1　1/8 哑铃标本模型每个 NURBS 片的初始节点矢量

编号	参数方向	节点矢量
1,3,5,6	ξ	$\boldsymbol{U}=\{0,0,0,1/4,1/4,1/2,1/2,3/4,3/4,1,1,1\}$
	η	$\boldsymbol{V}=\{0,0,0,1/4,1/4,1/2,1/2,3/4,3/4,1,1,1\}$
2,4	ξ	$\boldsymbol{U}=\{0,0,0,1,1,1\}$
	η	$\boldsymbol{V}=\{0,0,0,1,1,1\}$

我们采用这个算例模拟黏弹性材料的加载和卸载过程。时间步长选取为 $\Delta t=0.05\text{s}$。为验证等几何边界元法的有效性，分别给出了三种不同的边界条件如下：

$$\begin{cases} f=1\text{Pa}, & 0\leqslant t<10\text{s} \\ f=0\text{Pa}, & 10\leqslant t\leqslant 300\text{s} \end{cases} \tag{6.72}$$

$$\begin{cases} f=1\text{Pa}, & 0\leqslant t<50\text{s} \\ f=0\text{Pa}, & 50\leqslant t\leqslant 300\text{s} \end{cases} \tag{6.73}$$

$$\begin{cases} f=1\text{Pa}, & 0\leqslant t<100\text{s} \\ f=0\text{Pa}, & 100\leqslant t\leqslant 300\text{s} \end{cases} \tag{6.74}$$

从图 6.18 可以看出，当 t = 10s、50s 和 100s 时，监测点 A 位移的快速下降反映了式(6.3)中的线弹性特性。结果表明，等几何边界元法(IGABEM)可以用于分析线性黏弹性材料的蠕变和恢复行为。

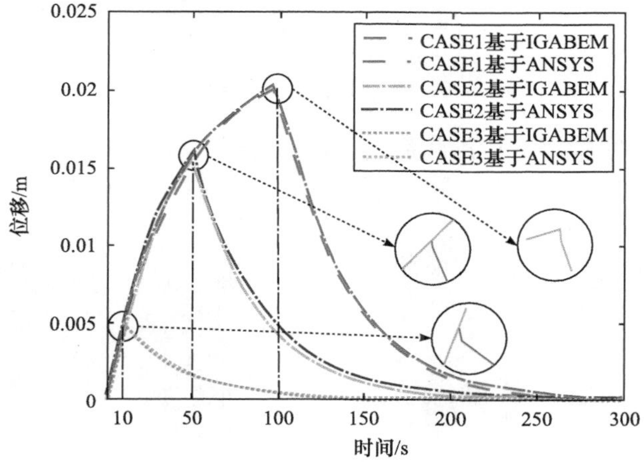

图 6.18　A 点(0,0,1m)在 x_1 方向的位移变化

6.3.3 三维黏弹性厚壁圆筒模型

本算例考虑一个均匀内压作用下的三维黏弹性厚壁圆筒推进剂药柱，模型下边界在 x_3 方向上施加位移约束。由于模型几何形状和边界条件是对称的，因此该模型可简化为等效的 1/4 模型，如图 6.19 所示。该模型的几何尺寸如图 6.20 所示。均匀分布的内部压力为 1Pa。

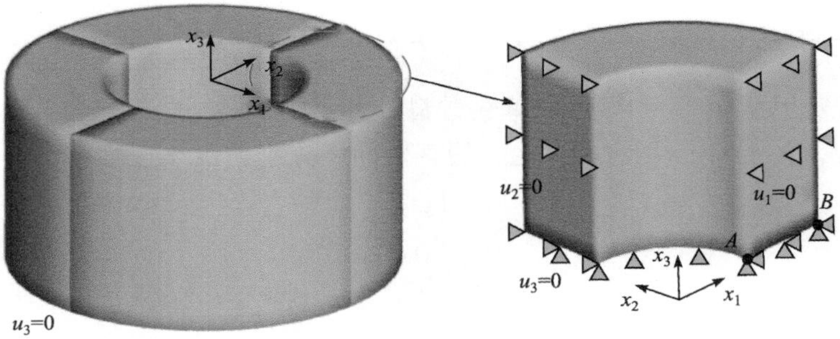

图 6.19 1/4 厚壁圆筒的等效模型及边界条件

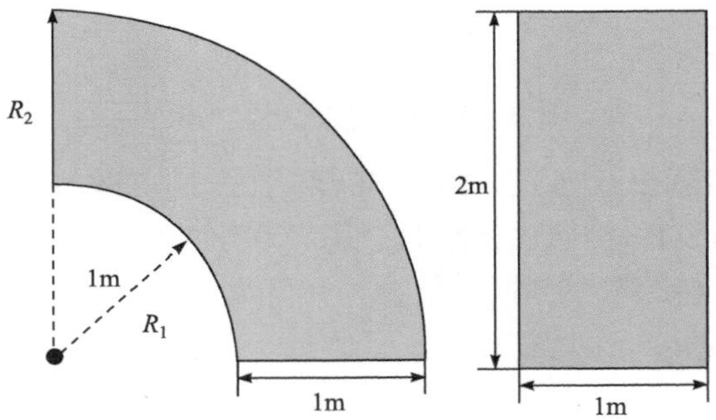

图 6.20 模型上表面和侧面的尺寸

图 6.21 给出了模型 5 次细化后的控制点分布、NURBS 网格和 486 个内点分布。每张 NURBS 片的两个方向的初始节点矢量为 $\boldsymbol{U}=\boldsymbol{V}=\{0,0,0,1,1,1\}$ 。

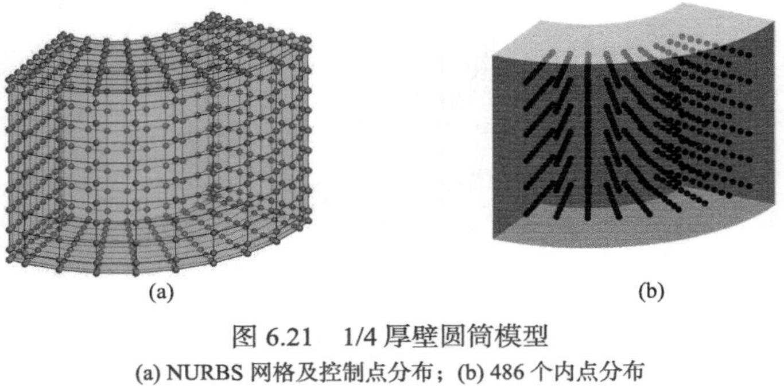

(a) (b)

图 6.21 1/4 厚壁圆筒模型

(a) NURBS 网格及控制点分布；(b) 486 个内点分布

这个问题的位移解析解可以写为[43]

$$u(r)=\frac{R_2{}^2 f(9KR_1{}^2+3Kr^2+2q_0^{\mathrm{G}}r^2)}{9Kq_0^{\mathrm{G}}r(R_{\mathrm{a}}+R_2)(R_1-R_2)} - \frac{R_2{}^2 f\exp\left(-\dfrac{q_0^{\mathrm{G}}t}{q_1^{\mathrm{G}}}\right)(3R_1{}^2q_1^{\mathrm{G}}+q_1^{\mathrm{G}}r^2-3R_1{}^2q_0^{\mathrm{G}}q_1^{\mathrm{G}}-p_0^{\mathrm{G}}q_0^{\mathrm{G}}r^2)}{3q_0^{\mathrm{G}}q_0^{\mathrm{G}}r(R_1+R_2)(R_1-R_2)} \tag{6.75}$$

其中，r 表示模型中任意点到中心的距离；R_1 和 R_2 分别表示厚壁圆筒的内径和外径。此外，与 Prony 级数相关的其他参数可以表示为如下形式：

$$\begin{cases} p_0^{\mathrm{G}}=\gamma_1 \\ q_0^{\mathrm{G}}=2G_0 \\ q_1^{\mathrm{G}}=2\gamma_1(G_0+G_1) \end{cases} \tag{6.76}$$

我们选取的时间步长为 $\Delta t=0.005\mathrm{s}$ 。图 6.22 为使用等几何边界元法获得的 t=0s 到 t=3s 时刻点 A(1m,0,0)和 B(2m,0,0)的计算结果。不难看出，等几何边界元法得到的数值结果与解析解吻合较好，A 点和 B 点各时刻的径向位移相对误差如图 6.23 所示。图 6.24 为均匀内压作用下模型不同时刻的变形图及径向位移云图。图 6.25 显示了三个不同时刻的 von Mises 应力云图。von Mises 应力的计算公式如下：

$$\sigma_{\mathrm{M}}=\sqrt{(\sigma_{11}-\sigma_{22})^2+(\sigma_{22}-\sigma_{33})^2+(\sigma_{33}-\sigma_{11})^2/2} \tag{6.77}$$

其中，σ_{11}、σ_{22}、σ_{33} 为三个方向的主应力。可以看出，由于记忆应力的影响，模型的应力随时间的增加而增加。

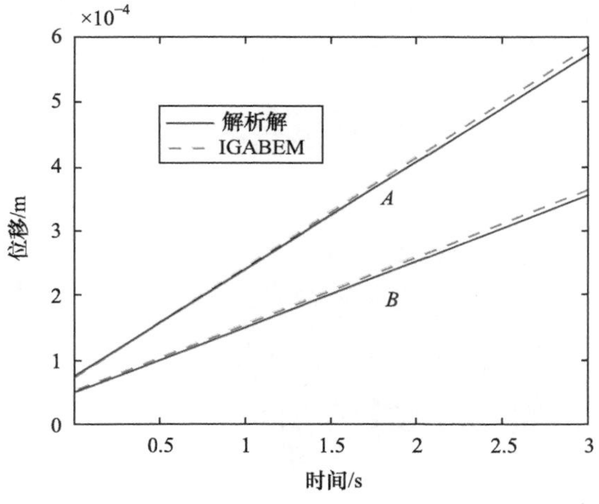

图 6.22 A 点(1m,0,0)和 B 点(2m,0,0)的径向位移

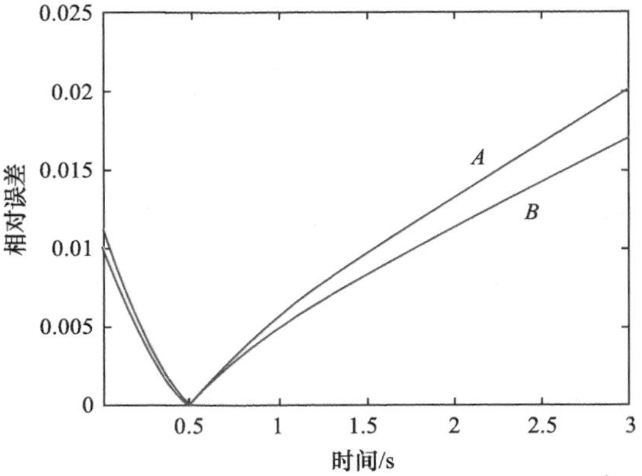

图 6.23 *A* 点和 *B* 点每一时刻径向位移的相对误差

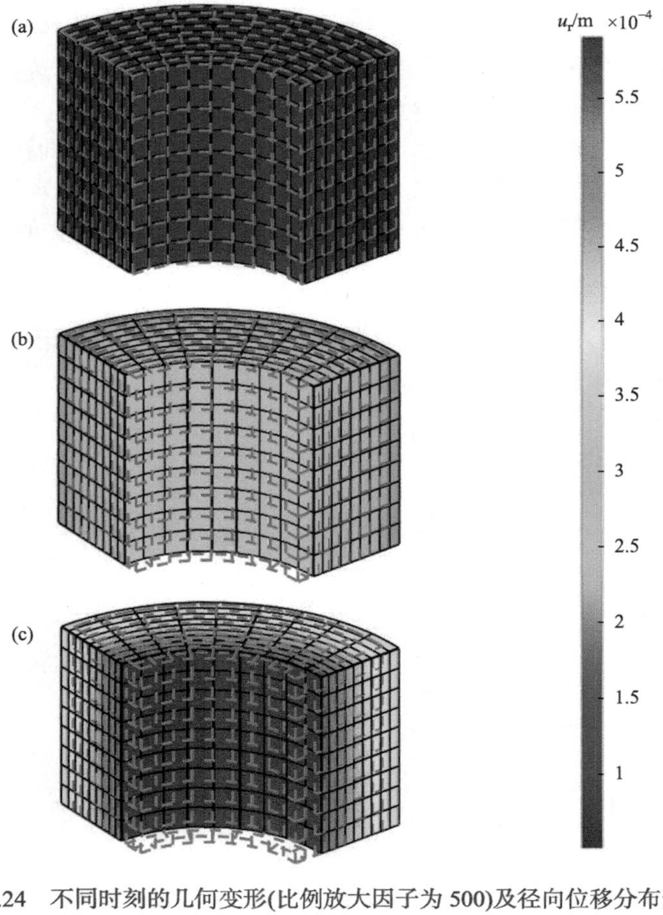

图 6.24 不同时刻的几何变形(比例放大因子为 500)及径向位移分布云图

(a) t =0.01s；(b) t =1.5s；(c) t =3s

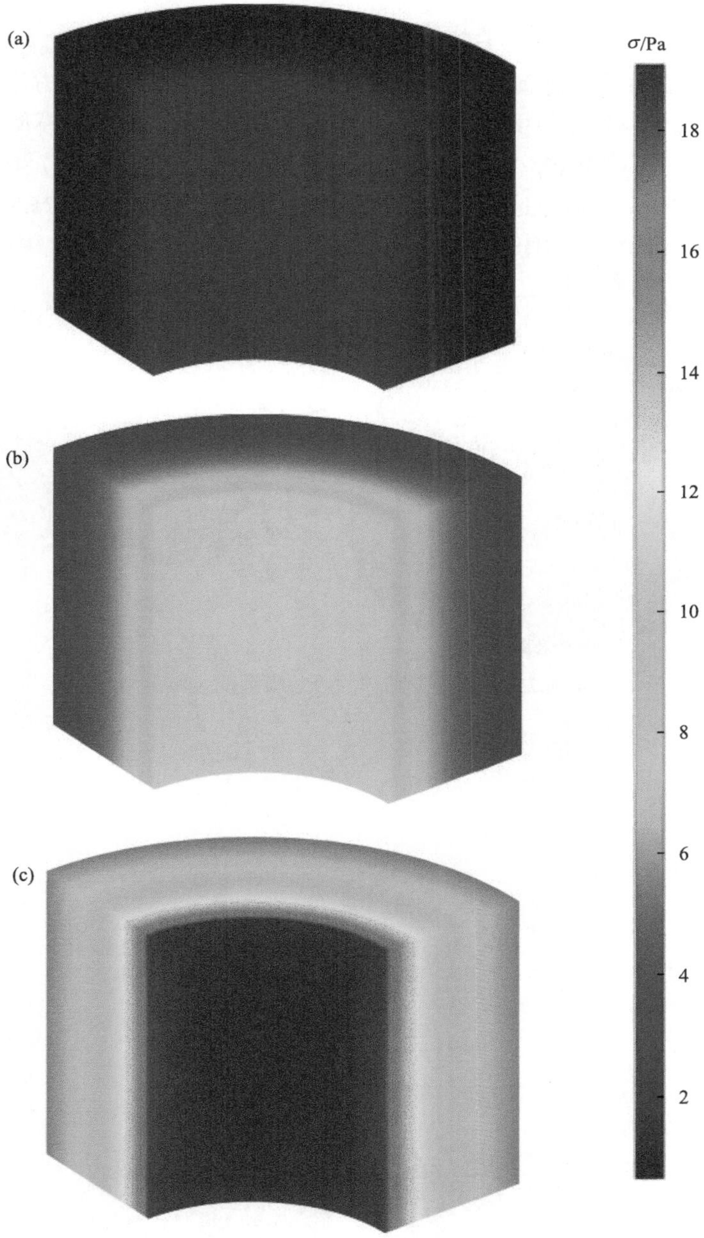

图 6.25　模型在不同时刻的 von Mises 应力分布

(a) $t = 0.01$s；(b) $t = 1.5$s；(c) $t = 3$s

6.3.4 三维黏弹性星形药柱模型

本算例考虑一个三维黏弹性星形推进剂药柱，该模型考虑重力作用。下边界在 x_3 方向施加位移约束。由于模型和边界条件为对称的，因此该模型可以简化为 1/4 星形药柱的等效模型，如图 6.26 所示。模型具体尺寸在图 6.27 中给出。我们采用 6 张 NURBS 片描述这个 1/4 等效模型。图 6.28 给出了 NURBS 编号和 224 个内点分布。每个 NURBS 片两个方向的初始节点矢量在表 6.2 中给出。每个 NURBS 片的控制点和权值可在文献[43]中找到。

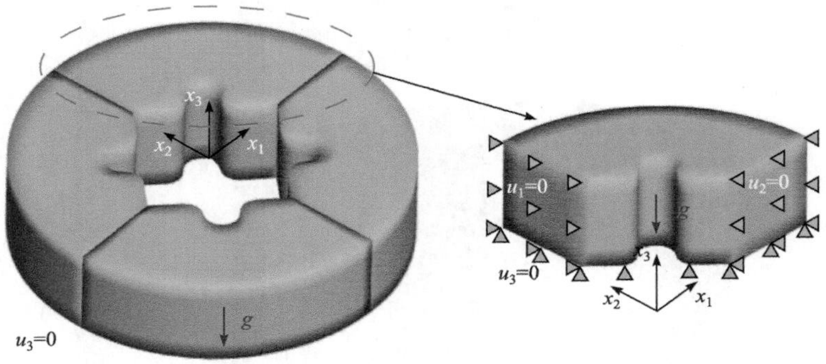

图 6.26 1/4 星形药柱的等效模型及边界条件

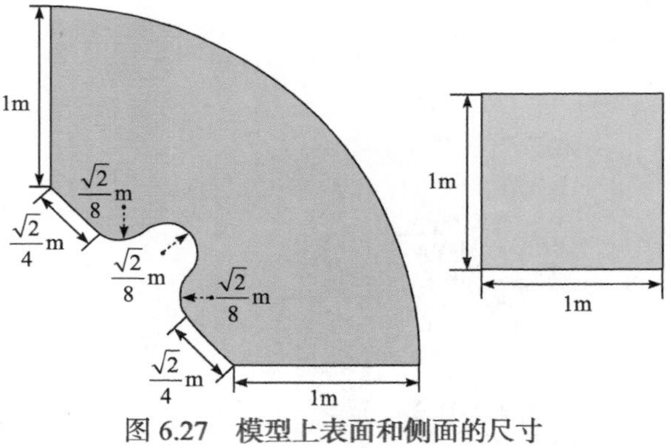

图 6.27 模型上表面和侧面的尺寸

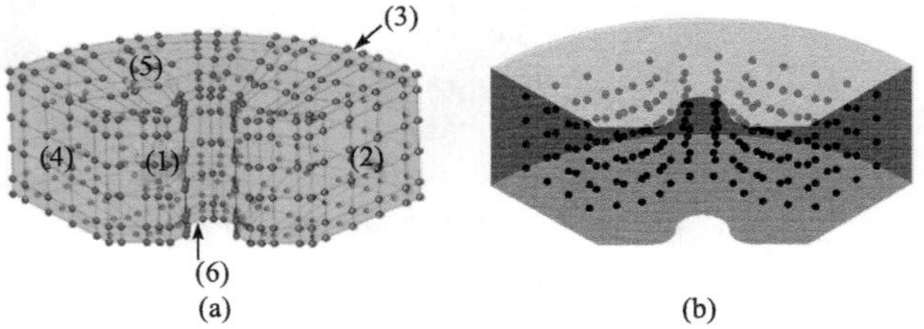

图 6.28 1/4 等效星形药柱模型

(a) NURBS 网格及控制点分布；(b) 224 个内点分布

表 6.2　1/4 星形药柱模型每个 NURBS 片的初始节点矢量

编号	参数方向	节点矢量
1,3,5,6	ξ	$\boldsymbol{U}=\{0,0,0,1/6,1/6,1/3,1/3,1/2,1/2,2/3,2/3,5/6,5/6,1,1,1\}$
	η	$\boldsymbol{V}=\{0,0,0,1/2,1,1,1\}$
2,4	ξ	$\boldsymbol{U}=\{0,0,0,1/2,1,1,1\}$
	η	$\boldsymbol{V}=\{0,0,0,1/2,1,1,1\}$

为了验证算法的有效性，我们将 IGABEM 与 ANSYS 软件的计算结果进行对比。图 6.29 给出了 FEM 网格，由 37500 个 8 节点单元组成。计算时选取的时间步长为 $\Delta t = 0.01\text{s}$ 。

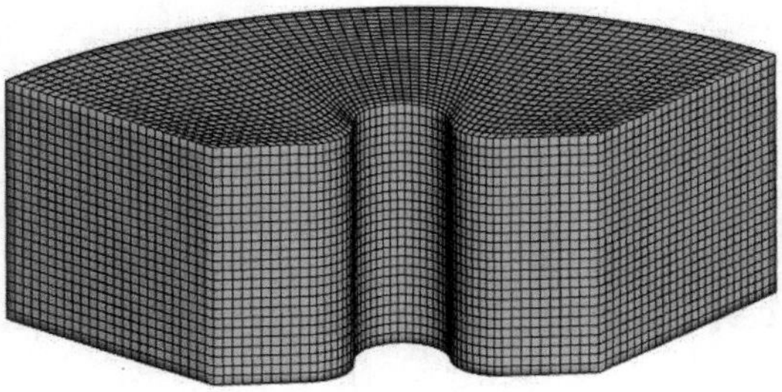

图 6.29　FEM 网格

图 6.30 给出了 A 点(1m, 0, 1m)在 x_3 方向的位移随时间的变化。不难发现，IGABEM 计算的数值结果与 ANSYS 计算的结果吻合较好。但在获得相同精度的数值结果的前提下，IGABEM 的自由度远少于 FEM。此外，我们给出了 t=0.01s,

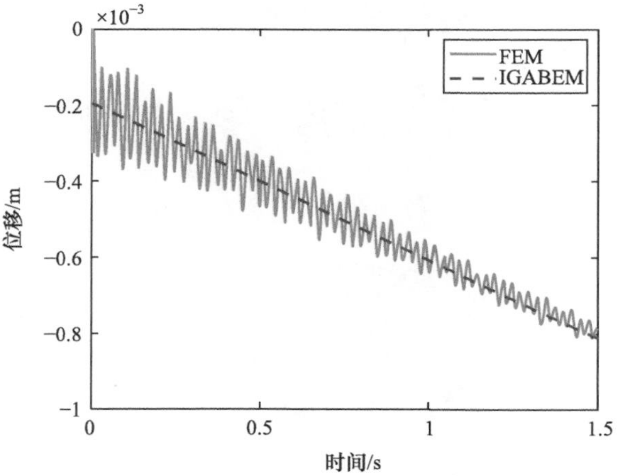

图 6.30　A 点在 x_3 方向的位移

0.75s 和 1.5s 三个时刻的径向位移变形云图(图 6.31)、主应变 ε_{33} 云图(图 6.32)和 von Mises 应力云图(图 6.33)。

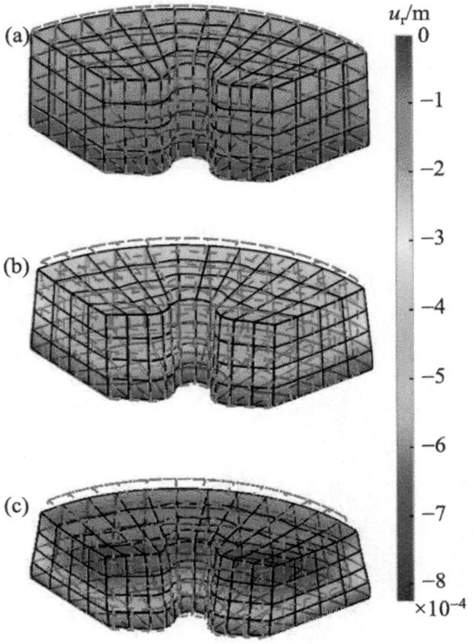

图 6.31 不同时刻的几何变形(比例放大因子为 200)及径向位移分布云图

(a) t =0.01s；(b) t =0.75s；(c) t =1.5s

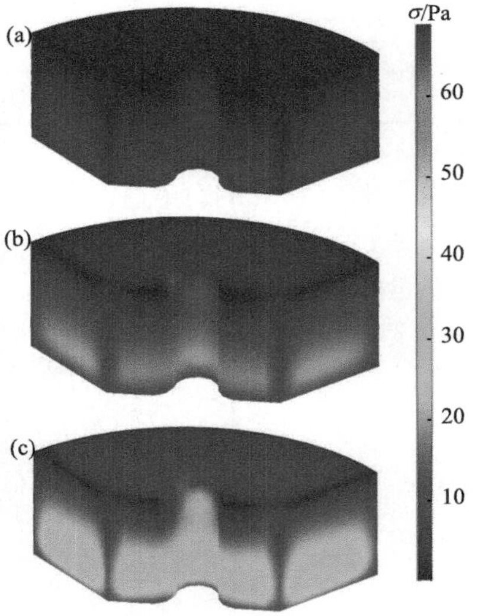

图 6.32 模型在不同时刻的主应变 ε_{33} 的分布云图

(a) t =0.01s；(b) t =0.75s；(c) t =1.5s

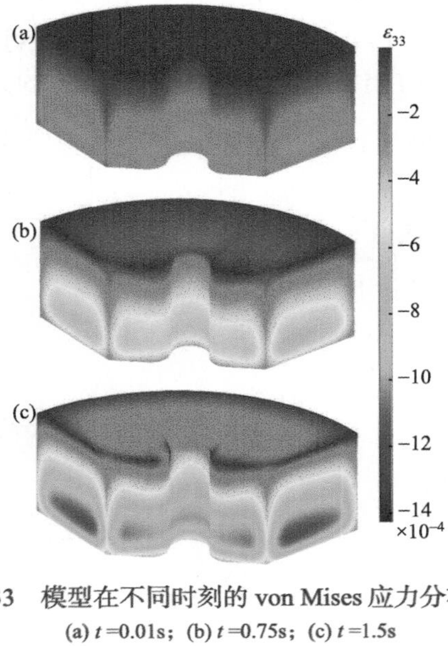

图 6.33 模型在不同时刻的 von Mises 应力分布云图

(a) t =0.01s；(b) t =0.75s；(c) t =1.5s

6.4 小 结

本章提出了求解含有体力项的三维黏弹性问题的等几何边界元法。该算法可用于分析黏弹性材料(如固体火箭推进剂和橡胶材料)的力学性能。

本章使用 Kelvin 基本解推导了黏弹性力学问题的位移边界域积分方程,以及正则化应变和应力边界域积分方程。利用径向积分法将位移、应变和应力积分方程中由记忆应力和体力引起的域积分转化为相应的等效边界积分。这样的变换不仅保留了等几何边界元法仅对边界离散的优点,而且降低了积分的奇异性。通过控制点与配点之间的简单变换,可以直接用刚体位移法求解等几何边界元法中的强奇异积分。为了避免用边界积分方程求解边界应力/应变时出现超奇异积分,本章提出了改进的面力恢复法,用于求解模型表面节点的应力/应变。

参 考 文 献

[1] Srinatha H R, Lewis R W. A finite element method for thermoviscoelastic analysis of plane problems[J]. Computer Methods in Applied Mechanics and Engineering, 1981, 25: 21-33.

[2] Pavan R C, Oliveira B F, Maghous S, et al. A model for anisotropic viscoelastic damage in composites[J]. Composite Structures, 2010, 92: 1223-1228.

[3] Mesquita A D, Coda H B. A boundary element methodology for viscoelastic analysis: part Ⅰ with cells[J]. Applied Mathematical Modelling, 2007, 31: 1149-1170.

[4] Mesquita A D, Coda H B. A boundary element methodology for viscoelastic analysis: part Ⅱ without cells[J]. Applied Mathematical Modelling, 2007, 31: 1171-1185.

[5] Marques S P C, Creus G J. Computational Viscoelasticity[M]. London: Springer, 2012.

[6] Sun D Y, Dai R, Liu X Y, et al. RI-IGABEM for 2D viscoelastic problems and its application to solid propellant grains[J]. Computer Methods in Applied Mechanics and Engineering, 2021, 378: 113737.

[7] Neto C G R, Santiago J A F, Telles J C F, et al. An accurate Galerkin-BEM approach for the modeling of quasi-static viscoelastic problems[J]. Engineering Analysis with Boundary Elements, 2021, 130: 94-108.

[8] Gual L, Schanz M. A comparative study of three boundary element approaches to calculate the transient response of viscoelastic solids with unbounded domains[J]. Computer Methods in Applied Mechanics and Engineering, 1999, 179 : 111-123.

[9] Neto A R, Leonel E D. Nonlinear IGABEM formulations for the mechanical modelling of 3D reinforced structures[J]. Applied Mathematical Modelling, 2022, 102 : 62-100.

[10] Hughes T J R, Cottrell J A, Bazilevs Y. Isogeometric analysis: CAD, finite elements, NURBS, exact geometry and mesh refinement[J]. Computer Methods in Applied Mechanics and Engineering, 2005, 194: 4135-4195.

[11] Falco C D, Reali A, Vazquez R. GeoPDEs: a research tool for isogeometric analysis of PDEs

[J]. Advances in Engineering Software, 2011, 42: 1020-1034.

[12] Nguyen V P, Anitescu C, Bordas S P A, et al. Isogeometric analysis: an overview and computer implementation aspects[J]. Mathematics and Computers in Simulation, 2015, 117: 89-116.

[13] Politis C, Ginnis A I, Kaklis P D, et al. An isogeometric BEM for exterior potential-flow problems in the plane[C]//Proceedings of the 2009 ACM Symposium on Solid and Physical Modeling, San Francisco, California, USA, 2009.

[14] Simpson R N, Bordas S P A, Trevelyan J, et al. A two-dimensional isogeometric boundary element method for elastostatic analysis[J]. Computer Methods in Applied Mechanics and Engineering, 2012, 209: 87-100.

[15] Simpson R N, Bordas S P A, Lian H J, et al. An isogeometric boundary element method for elastostatic analysis: 2D implementation aspects[J]. Composite Structures, 2013, 118: 2-12.

[16] Xu C, Dai R, Dong C Y, et al. RI-IGABEM based on generalized-α method in 2D and 3D elastodynamic problems[J]. Computer Methods in Applied Mechanics and Engineering, 2021, 383: 113890.

[17] Chen L L, Lian H J, Liu Z, et al. Structural shape optimization of three dimensional acoustic problems with isogeometric boundary element methods[J]. Computer Methods in Applied Mechanics and Engineering, 2019, 355: 926-951.

[18] Lian H J, Kerfriden P, Bordas S P A. Implementation of regularized isogeometric boundary element methods for gradient-based shape optimization in two-dimensional linear elasticity[J]. International Journal for Numerical Methods in Engineering, 2016, 106 : 972-1017.

[19] Beer G, Marussig B, Duenser C. The Isogeometric Boundary Element Method[M]. Austria: Springer, 2020.

[20] Kusama T, Mitsui Y. Boundary element method applied to linear viscoelastic analysis[J]. Applied Mathematical Modelling, 1982, 6: 285-290.

[21] Lian H J. Shape optimisation directly from CAD: an isogeometric boundary element approach[J]. Applied Thermal Engineering, 2015, 75: 93-102.

[22] Gao X W, Davies T G. Boundary Element Programming in Mechanics[M]. Cambridge: Cambridge University Press, 2002.

[23] Cui M, Xu B B, Feng W Z, et al. A radial integration boundary element method for solving transient heat conduction problems with heat sources and variable thermal conductivity[J]. Numerical Heat Transfer Part B—Fundamentals, 2018, 73: 1-18.

[24] Feng W Z, Yang K, Cui M, et al. Analytically-integrated radial integration BEM for solving three-dimensional transient heat conduction problems[J]. International Communications in Heat and Mass Transfer, 2016, 79: 21-30.

[25] Cruse T A. Boundary integral equation method for three-dimensional elastic fracture mechanics[M]. AFOSR-TR-75-0813, ADA 011660, Pratt and Whitney Aircraft, Connecticut, 1975.

[26] Rizzo F J, Shippy D J. An advanced boundary integral equation method for three dimensionalthermoelasticity[J]. International Journal for Numerical Methods in Engineering, 1977, 11: 1753-1768.

[27] Danson D J. A boundary element formulation for problems in linear isotropic elasticity with body forces[M]// Brebbia C A. Boundary Element Methods. Berlin: Springer, 1981.
[28] Nardini D, Brebbia C A. A new approach for free vibration analysis using boundary elements[J]. Applied Mathematical Modelling, 1983, 7: 157-162.
[29] Neves A C, Brebbia C A. The multiple reciprocity boundary element method in elasticity: a new approach for transforming domain integrals to the boundary[J]. International Journal for Numerical Methods in Engineering, 1991, 31: 709-727.
[30] Partridge P W, Brebbia C A, Wrobel L C. The Dual Reciprocity Boundary Element Method[M]. Southampton: Computational Mechanics Publications, 1992.
[31] Yu B, Cao G Y, Huo W D, et al. Isogeometric dual reciprocity boundary element method for solving transient heat conduction problems with heat sources[J]. Journal of Computational and Applied Mathematics, 2021, 385 : 113197.
[32] Feng W Z, Gao L F, Du J M, et al. A meshless interface integral BEM for solving heat conduction in multi-non-homogeneous media with multiple heat sources[J]. International Communications in Heat and Mass Transfer, 2019, 104 : 70-82.
[33] Feng W Z, Li H Y, Gao L F, et al. Hypersingular flux interface integral equation for multi-medium heat transfer analysis[J]. International Journal of Heat and Mass Transfer, 2019, 138: 852-865.
[34] Dong C Y, Lo S H, Cheung Y K. Numerical solution for elastic inclusion problems by domain integral equation with integration by means of radial basis functions[J]. Engineering Analysis with Boundary Elements, 2004, 28: 623-632.
[35] Xu C, Dong C Y. RI-IGABEM in inhomogeneous heat conduction problems[J]. Engineering Analysis with Boundary Elements, 2021, 124: 21-236.
[36] Xu C, Dong C Y, Dai R. RI-IGABEM based on PIM in transient heat conduction problems of FGMs[J]. Computer Methods in Applied Mechanics and Engineering, 2020, 374: 113601.
[37] Gao X W. An effective method for numerical evaluation of general 2D and 3D high order singular boundary integrals[J]. Computer Methods in Applied Mechanics and Engineering, 2010, 199 (45) : 2856-2864.
[38] Brebbia C A, Telles J C F, Wrobel L C. Boundary Element Techniques: Theory and Applications in Engineering[M]. New York: Springer, 1984.
[39] Matsumoto T, Tanaka M, Hirata H. Boundary stress calculation using regularized boundary integral equation for displacement gradients[J]. International Journal for Numerical Methods in Engineering, 1993, 36: 783-797.
[40] Banerjee P K, Raveendra S T. Advanced boundary element analysis of two-and three-dimensional problems of elasto-plasticity[J]. International Journal for Numerical Methods in Engineering, 1986, 23: 985-1002.
[41] Liu J, Peng H F, Gao X W, et al. A traction-recovery method for evaluating boundary stresses on thermal elasticity problems of FGMs[J]. Engineering Analysis with Boundary Elements, 2015, 61: 226-231.
[42] Muki R, Sternberg E. On transient thermal stresses in viscoelastic materials with

temperature-dependent properties[J]. Journal of Applied Mechanics-Transactions of the ASME, 1961, 28: 193-207.

[43] Xu C, Zhan Y S, Dai R, et al. RI-IGABEM for 3D viscoelastic problems with body force[J]. Computer Methods in Applied Mechanics and Engineering, 2022, 394: 114911.

第7章

多层复合材料结构的非相适应界面的等几何边界元分析

7.1 引言

多层复合结构[1-3]在实际工程中广泛存在，每层都有不同的材料(如黏弹性材料和弹性材料等)，每层的厚度也不同。SRM 燃烧室是典型的多材料、多层结构，其一般由推进剂药柱、绝热层、衬层和发动机壳体组成[4-6]。燃烧室是 SRM 的主要动力源，也是一个相对薄弱的部分。其推进剂药柱为黏弹性材料，在储存过程中主要受内压、温度载荷和重力的影响。推进剂药柱形状的改变会改变燃烧室的表面几何形状或阻塞气体通道，从而降低运动性能或引起自爆[7]。因此，对 SRM 燃烧室进行结构完整性分析具有重要的工程意义。近年来，一些数值方法，如 FEM[8,9]、FDM[10,11]、BEM[12,13]和 FE-BE 耦合算法[14,15]被广泛应用于黏性-弹性结构的结构完整性分析。其中，FEM 由于其理论简单，易于实现，是分析复合结构最有效的方法[16]。然而，在衬层和保温层的有限元分析中，由于这些结构太薄，需要将模型划分为许多单元来保证计算精度，这将消耗大量的存储空间和计算成本。与 FEM 相比，BEM 的理论比较复杂，但它具有边界离散化和精度高的优点。特别是在求解薄壁结构问题时，它更有优势[17,18]。因此，边界元法也适用于 SRM 燃烧室等多材料、多尺度结构的分析。

现有的研究大多采用传统的拉格朗日函数进行几何形状逼近和物理场插值，但存在以下不足：①几何误差是不可避免的；②CAD 建模和 CAE 分析之间的联系不是无缝的；③网格细化过程是复杂的；④单元之间只能满足 C^0 连续，不能满足高阶连续。等几何分析(IGA)理论完美地解决了这些问题[19]。IGA 的核心思想是使用 CAD 程序中常用的 NURBS 函数来表示几何变量和场变量，其优点是，精确描述几何形状、形函数具有非负高阶次和单元具有高连续性[20]。但是，CAD 软件只提供模型的边界信息。如果采用 IGAFEM 进行分析，则根据边界信息进行进一步运算以获得内部网格是一个复杂的过程。而 BEM 只需要边界信息来分析模型。因此，与 IGAFEM 相比，IGABEM[21,22]与 CAD 软件的结合更加自然，真正实现了 CAD 与 CAE 之间的无缝连接。IGABEM 已被应用到了很多领域[23-32]。本章采用非相适应界面的 IGABEM 对温度载荷和体力作用下的多层弹性-黏弹性复合结构进行了结构完整性分析。

本章采用径向积分法将积分方程中的域积分转换为等效边界积分，利用幂级数展开法[23]求解弱奇异积分；通过一个控制点到配点的转换方法，将刚体位移法用于强奇异积分的求解。此外，为了避免在求解边界节点应变时求解超奇异积分，我们提出了二维及三维问题的热黏弹性力学问题的 TRM。

分析含有薄壁结构的多层复合结构时，如何有效求解薄壁结构是非常重要的[24-27]。由于薄壁结构的厚度和长宽尺寸差距较大，如果采用 FEM 进行分析，则一般需要将其划分为大量的单元，以保证计算精度，这需要大量的计算成本。虽然我们可以用壳单元来分析这个问题，但不能得到单元厚度方向的物理量[28]。IGABEM 在薄壁结构分析中有其独特的优势。它不要求单元的长度和宽度接近单元的厚度，并且可以通过后处理过程精确地获得厚度方向的物理量。IGABEM 在求解薄壁结构问题时需要精确地处理拟奇异积分。拟奇异积分是指那些源点接近但不在积分单元上的积分。求解拟奇异积分的常用方法包括全局正则化法[29,30]、自适应单元子分法[31,32]、解析或半解析法[33,34]、坐标转换法[35-37]等。本章提出了一种基于层次四叉树分解算法的自适应积分方法来处理拟奇异积分。自适应积分过程由误差公式和四叉树分解方法驱动，在保证计算精度的前提下给出最优高斯点分布，从而达到计算精度和效率之间的平衡。

对于多层复合结构来说，不同层的几何形状和尺寸是不同的。因此，CAD 软件建立的复合结构模型，其不同层耦合界面的网格一般是不相适应的(即耦合接口两侧的网格数量或顺序不一致)，甚至耦合界面可能包含大量的修剪曲面[38]。解决耦合界面不协调问题的关键是在两个耦合界面上建立位移连续和面力平衡关系。许多学者针对传统 BEM 或 FEM 的非一致性界面耦合问题提出了一系列的解决方法[39-43]。这些方法都是建立在传统拉格朗日函数的基础上，有的引入了额外的自由度，有的推导了相关的变分公式。这些方法不能有效地应用于以 NURBS 函数为形函数的等几何耦合算法。Yang 等[44,45]利用虚拟节点插入技术实现了 IGAFEM 和 IGABEM 之间非相适应界面耦合问题。该方法是基于 NURBS 曲面细分的特点和虚功原理实现的。Zhan 等[46]利用该方法对 SRM 燃烧室进行了结构完整性分析。本章利用基于非相适应界面的 IGABEM 求解了多层复合结构的热黏弹性问题。

本章 7.2 节介绍线性热黏弹性问题的等几何边界元法。7.3 节通过 5 个例子验证了算法的正确性。

7.2　三维热黏弹性力学问题的等几何边界元法

7.2.1　热黏弹性力学问题的本构方程及记忆应力

各向同性、线性黏弹性材料热黏弹性力学问题的本构方程可以写为[47]

$$\begin{aligned}\sigma_{il}(\boldsymbol{x},t)=&\delta_{il}\int_{-\infty}^{t}\left[K(\gamma-\gamma')-\frac{2}{3}G(\gamma-\gamma')\right]\frac{\partial\varepsilon_{nn}(\boldsymbol{x},t')}{\partial t'}\mathrm{d}t'\\&+2\int_{-\infty}^{t}G(\gamma-\gamma')\frac{\partial\varepsilon_{il}(\boldsymbol{x},t')}{\partial t'}\mathrm{d}t'\\&-3\alpha\delta_{il}\int_{-\infty}^{t}K(\gamma-\gamma')\frac{\partial\Delta T(\boldsymbol{x},t')}{\partial t'}\mathrm{d}t'\end{aligned}\tag{7.1}$$

其中，γ 表示温度变化的等效时间，其形式为 $\gamma=\int_0^t 1/\{a_T[T(\boldsymbol{x},\tau)]\}\mathrm{d}\tau$，而 γ' 的形式为 $\gamma'=\int_0^{t'}1/\{a_T[T(\boldsymbol{x},\tau)]\}\mathrm{d}\tau$，这里，$a_T$ 是移位因子，α 是热膨胀系数；K 和 G 分别表示体积模量和剪切模量；σ_{il} 和 ε_{il} (对于二维问题，i, l=1, 2；对于三维问题，i, l =1, 2, 3)分别表示应力和应变分量；对于二维问题，$\varepsilon_{nn}=\varepsilon_{11}+\varepsilon_{22}$，对于三维问题，$\varepsilon_{nn}=\varepsilon_{11}+\varepsilon_{22}+\varepsilon_{33}$；$\delta_{il}$ 表示 Kronecker-delta 函数；t' 表示时间变量；t 表示加载时间；ΔT 表示温度变化。Williams 等[48]提出了一个经验公式，即 William-Landel-Ferry 方程，具体形式如下：

$$\lg a_T=-\frac{C_1(T-T_{\mathrm{ref}})}{C_2+T-T_{\mathrm{ref}}}\tag{7.2}$$

其中，C_1、C_2 表示材料常数；T_{ref} 表示参考温度。

式(7.1)中的应力可以分成三部分，即弹性应力、记忆应力和热应力：

$$\sigma=\sigma^{\mathrm{e}}-\sigma^{\mathrm{m}}-\sigma^{\mathrm{T}}\tag{7.3}$$

其中，

$$\sigma_{il}^{\mathrm{e}}(\boldsymbol{x},t)=\delta_{il}\left[K(0)-\frac{2}{3}G(0)\right]\varepsilon_{nn}(\boldsymbol{x},t)+2G(0)\varepsilon_{il}(\boldsymbol{x},t)\tag{7.4}$$

$$\begin{aligned}\sigma_{il}^{\mathrm{m}}=&2\int_0^t\frac{\partial G(\gamma-\gamma')}{\partial t'}\varepsilon_{il}(\boldsymbol{x},t')\mathrm{d}t'+\delta_{il}\int_0^t\frac{\partial K(\gamma-\gamma')}{\partial t'}\varepsilon_{nn}(\boldsymbol{x},t)\mathrm{d}t'\\&-\delta_{il}\frac{2}{3}\int_0^t\frac{\partial G(\gamma-\gamma')}{\partial t'}\varepsilon_{nn}(\boldsymbol{x},t')\mathrm{d}t'\end{aligned}\tag{7.5}$$

$$\sigma_{il}^{\mathrm{T}}=\delta_{il}3\alpha K(0)\Delta T(\boldsymbol{x},t)-\delta_{il}3\alpha_0\int_0^t\frac{\partial K(\gamma-\gamma')}{\partial t'}\Delta T(\boldsymbol{x},t')\mathrm{d}t' \tag{7.6}$$

其中，$K(0)$ 和 $G(0)$ 表示弹性状态下的体积模量和剪切模量。

当得到温度场和弹性应变后，弹性应力和热应力可分别由式(7.4)和式(7.6)来计算。然而，在求解记忆应力时，需要计算遗传积分。假设体积模量 K 为常数，因此 $\int_0^t\frac{\partial K(t-t')}{\partial t'}\varepsilon_{nn}(\boldsymbol{x},t')\mathrm{d}t'=0$。此时，式(7.5)可以化简为如下形式：

$$\sigma_{il}^{\mathrm{m}}=\int_0^t\frac{\partial G(t-t')}{\partial t'}\hat{\varepsilon}_{il}(\boldsymbol{x},t')\mathrm{d}t' \tag{7.7}$$

其中，对于二维问题，

$$\begin{Bmatrix}\hat{\varepsilon}_{11}\\ \hat{\varepsilon}_{22}\\ \hat{\varepsilon}_{12}\\ \hat{\varepsilon}_{21}\end{Bmatrix}=\begin{bmatrix}\frac{4}{3} & -\frac{2}{3} & 0 & 0\\ -\frac{2}{3} & \frac{4}{3} & 0 & 0\\ 0 & 0 & 2 & 0\\ 0 & 0 & 0 & 2\end{bmatrix}\begin{Bmatrix}\varepsilon_{11}\\ \varepsilon_{22}\\ \varepsilon_{12}\\ \varepsilon_{21}\end{Bmatrix} \tag{7.8}$$

对于三维问题，

$$\begin{Bmatrix}\hat{\varepsilon}_{11}\\ \hat{\varepsilon}_{22}\\ \hat{\varepsilon}_{33}\\ \hat{\varepsilon}_{12}\\ \hat{\varepsilon}_{13}\\ \hat{\varepsilon}_{21}\\ \hat{\varepsilon}_{23}\\ \hat{\varepsilon}_{31}\\ \hat{\varepsilon}_{32}\end{Bmatrix}=\begin{bmatrix}\frac{4}{3} & -\frac{2}{3} & -\frac{2}{3} & 0 & 0 & 0 & 0 & 0 & 0\\ -\frac{2}{3} & \frac{4}{3} & -\frac{2}{3} & 0 & 0 & 0 & 0 & 0 & 0\\ -\frac{2}{3} & -\frac{2}{3} & \frac{4}{3} & 0 & 0 & 0 & 0 & 0 & 0\\ 0 & 0 & 0 & 2 & 0 & 0 & 0 & 0 & 0\\ 0 & 0 & 0 & 0 & 2 & 0 & 0 & 0 & 0\\ 0 & 0 & 0 & 0 & 0 & 2 & 0 & 0 & 0\\ 0 & 0 & 0 & 0 & 0 & 0 & 2 & 0 & 0\\ 0 & 0 & 0 & 0 & 0 & 0 & 0 & 2 & 0\\ 0 & 0 & 0 & 0 & 0 & 0 & 0 & 0 & 2\end{bmatrix}\begin{Bmatrix}\varepsilon_{11}\\ \varepsilon_{22}\\ \varepsilon_{33}\\ \varepsilon_{12}\\ \varepsilon_{13}\\ \varepsilon_{21}\\ \varepsilon_{23}\\ \varepsilon_{31}\\ \varepsilon_{32}\end{Bmatrix} \tag{7.9}$$

为了简化计算和节省计算机内存，$G(t)$ 可以展开成 Prony 级数形式：

$$G(t)=G_0+\sum_{w=1}^{\bar{m}}G_w\exp\left(-t/\tau_w^G\right) \tag{7.10}$$

其中，τ_w^G 表示松弛时间；G_0 和 G_w 表示材料参数；$\bar{m}$ 表示 Prony 级数展开的项数。

$$
\begin{aligned}
\boldsymbol{\sigma}^{\mathrm{m}} &= \int_0^{t_S} \frac{\partial G(t_S - t')}{\partial t'} \hat{\boldsymbol{\varepsilon}}(\boldsymbol{x}, t') \mathrm{d}t' \\
&= \sum_{w=1}^{\bar{m}} G_w \boldsymbol{q}_{w,S} + \frac{1}{2}[G(0) - G(t_S - t'_{S-1})] \hat{\boldsymbol{\varepsilon}}(\boldsymbol{x}, t'_{S-1}) \\
&\quad + \frac{1}{2}[G(0) - G(t_S - t'_{S-1})] \hat{\boldsymbol{\varepsilon}}(\boldsymbol{x}, t'_S)
\end{aligned} \tag{7.11}
$$

其中，

$$
\boldsymbol{q}_{w,S} = \exp\left[-(t_S - t'_{S-1}) / \tau_w^G\right] \left(\left\{1 - \exp\left[-(t'_{S-1} - t'_{S-2}) / \tau_w^G\right]\right\} \hat{\boldsymbol{\varepsilon}}^*(\boldsymbol{x}, t'_{S-1}) + \boldsymbol{q}_{w,S-1}\right) \tag{7.12}
$$

其中，$\hat{\boldsymbol{\varepsilon}}^*(\boldsymbol{x}, t'_{j+1}) = \{[\hat{\boldsymbol{\varepsilon}}(\boldsymbol{x}, t'_j) + \hat{\boldsymbol{\varepsilon}}(\boldsymbol{x}, t'_{j+1})]\}/2$。此外，$\boldsymbol{q}_{w,1}$ 和 $\boldsymbol{q}_{w,2}$ 为 **0** 向量。

7.2.2　边界域积分方程

对于热黏弹性力学问题，控制方程为

$$
\sigma_{jk,k} + b_j = 0, \quad x_j \in \Omega \tag{7.13}
$$

边界条件为

$$
\begin{cases}
u_j = \bar{u}_j, & x_j \in \Gamma_{\mathrm{b1}} \\
f_j = \bar{f}_j, & x_j \in \Gamma_{\mathrm{b2}} \\
T = \bar{T}, & x_j \in \Gamma_1 \\
q = \bar{q}, & x_j \in \Gamma_2
\end{cases} \tag{7.14}
$$

其中，对于二维问题 $j = 1,2$；对于三维问题 $j = 1,2,3$。Ω 是求解的区域，$\Gamma_{\mathrm{b1}} \cup \Gamma_{\mathrm{b2}} = \Gamma$ 是 Ω 域的边界。Γ_{b1} 是位移边界，Γ_{b2} 是面力边界。Γ_1 是边界上给定温度 $\bar{T}$ 的部分，Γ_2 是边界上给定热流 $\bar{q}$ 的部分。b_j、u_j、f_j 和 x_j 分别表示体力分量、位移分量、面力分量和坐标分量。$\bar{u}_j$ 和 $\bar{f}_j$ 分别是边界 Γ_{b1} 和 Γ_{b2} 上位移和面力的给定值。

引入权函数 U^* 和 T^*，则式（7.13）和式（7.14）的加权余量式为

$$
\int_\Omega \sigma_{kj,k} U_j^* \mathrm{d}\Omega + \int_\Omega b_j U_j^* \mathrm{d}\Omega = -\int_{\Gamma_{\mathrm{b1}}} (u_j - \bar{u}_j) T_j^* \mathrm{d}\Gamma + \int_{\Gamma_{\mathrm{b2}}} (f_j - \bar{f}_j) U_j^* \mathrm{d}\Gamma \tag{7.15}
$$

其中，

$$
\begin{cases}
U_j^*(\boldsymbol{q}) = U_{ij}^*(\boldsymbol{q}, \boldsymbol{p}) e_i^*(\boldsymbol{p}) \\
T_j^*(\boldsymbol{q}) = T_{ij}^*(\boldsymbol{q}, \boldsymbol{p}) e_i^*(\boldsymbol{p})
\end{cases} \tag{7.16}
$$

式中，U_j^* 和 T_j^* 表示加权场的位移和面力；$e_i^*(\boldsymbol{p})$ 表示在 $\boldsymbol{p}$ 点沿 i 方向施加的单位

集中力。式(7.15)中的第一个积分可以写成如下形式：

$$\int_{\Omega}\sigma_{kj,k}U_j^*\mathrm{d}\Omega=\int_{\Omega}\left(\sigma_{kj}U_j^*\right)_{,k}\mathrm{d}\Omega-\int_{\Omega}\sigma_{kj}U_{j,k}^*\mathrm{d}\Omega \tag{7.17}$$

由高斯公式，式(7.17)中的 $\int_{\Omega}(\sigma_{kj}U_j^*)_{,k}\mathrm{d}\Omega$ 可以写成

$$\int_{\Omega}\left(\sigma_{kj}U_j^*\right)_{,k}\mathrm{d}\Omega=\int_{\Gamma}\sigma_{kj}U_j^*n_k\mathrm{d}\Gamma=\int_{\Gamma}\left(\sigma_{kj}n_k\right)U_j^*\mathrm{d}\Gamma=\int_{\Gamma}f_jU_j^*\mathrm{d}\Gamma \tag{7.18}$$

因此，式(7.17)可以写成

$$\int_{\Omega}\sigma_{kj,k}U_j^*\mathrm{d}\Omega=\int_{\Gamma}f_jU_j^*\mathrm{d}\Gamma-\int_{\Omega}\sigma_{kj}U_{j,k}^*\mathrm{d}\Omega \tag{7.19}$$

将式(7.19)和式(7.3)代入式(7.15)，可得

$$\begin{aligned}&-\int_{\Omega}\sigma_{kj}^{\mathrm{e}}U_{j,k}^*\mathrm{d}\Omega+\int_{\Omega}\sigma_{kj}^{\mathrm{m}}U_{j,k}^*\mathrm{d}\Omega+\int_{\Omega}\sigma_{kj}^{\mathrm{T}}U_{j,k}^*\mathrm{d}\Omega+\int_{\Omega}b_jU_j^*\mathrm{d}\Omega\\&=-\int_{\Gamma_{\mathrm{b1}}}\left(u_j-\bar{u}_j\right)T_j^*\mathrm{d}\Gamma-\int_{\Gamma_{\mathrm{b2}}}\bar{f}_jU_j^*\mathrm{d}\Gamma-\int_{\Gamma_{\mathrm{b1}}}f_jU_j^*\mathrm{d}\Gamma\end{aligned} \tag{7.20}$$

式(7.20)中的第一项积分 $\int_{\Omega}\sigma_{kj}^{\mathrm{e}}U_{j,k}^*\mathrm{d}\Omega$ 可以通过 Betti 功互易定理和高斯公式转换为如下形式：

$$\begin{aligned}\int_{\Omega}\sigma_{kj}^{\mathrm{e}}U_{j,k}^*\mathrm{d}\Omega&=\int_{\Omega}\sigma_{kj}^{\mathrm{e}*}u_{j,k}\mathrm{d}\Omega=\int_{\Omega}\left(\sigma_{kj}^{\mathrm{e}*}u_j\right)_{,k}\mathrm{d}\Omega-\int_{\Omega}\sigma_{kj,k}^{\mathrm{e}*}u_j\mathrm{d}\Omega\\&=\int_{\Gamma}\sigma_{kj}^{\mathrm{e}*}u_jn_k\mathrm{d}\Gamma-\int_{\Omega}\sigma_{kj,k}^{\mathrm{e}*}u_j\mathrm{d}\Omega\\&=\int_{\Gamma}T_j^*u_j\mathrm{d}\Gamma-\int_{\Omega}\sigma_{kj,k}^{\mathrm{e}*}u_j\mathrm{d}\Omega\end{aligned} \tag{7.21}$$

式(7.20)中的第二项积分 $\int_{\Omega}\sigma_{kj}^{\mathrm{m}}U_{j,k}^*\mathrm{d}\Omega$ 可以写成如下等效形式：

$$\int_{\Omega}\sigma_{kj}^{\mathrm{m}}U_{j,k}^*\mathrm{d}\Omega=\int_{\Omega}\sigma_{kj}^{\mathrm{m}}\frac{1}{2}\left(U_{j,k}^*+U_{k,j}^*\right)\mathrm{d}\Omega=\int_{\Omega}\sigma_{jk}^{\mathrm{m}}\varepsilon_{jk}^*\mathrm{d}\Omega \tag{7.22}$$

将式(7.21)和式(7.22)代入式(7.20)，可得

$$\int_{\Omega}\sigma_{kj,k}^{\mathrm{e}*}u_j\mathrm{d}\Omega+\int_{\Omega}\sigma_{jk}^{\mathrm{m}}\varepsilon_{jk}^*\mathrm{d}\Omega+\int_{\Omega}\sigma_{jk}^{\mathrm{T}}U_{j,k}^*\mathrm{d}\Omega+\int_{\Omega}b_jU_j^*\mathrm{d}\Omega=\int_{\Gamma}u_jT_j^*\mathrm{d}\Gamma-\int_{\Gamma}f_jU_j^*\mathrm{d}\Gamma \tag{7.23}$$

对于式(7.23)中的第一项积分 $\int_{\Omega}\sigma_{kj,k}^{\mathrm{e}*}u_j\mathrm{d}\Omega$，由基本解的性质可得

$$\int_{\Omega}\sigma_{kj,k}^{\mathrm{e}*}u_j\mathrm{d}\Omega=-\int_{\Omega}\delta(\boldsymbol{q},\boldsymbol{p})\delta_{ij}e_i^*(\boldsymbol{p})u_j(\boldsymbol{q})\mathrm{d}\Omega \tag{7.24}$$

将式(7.16)和式(7.24)代入式(7.23)，可得

$$-\int_{\Omega}\delta(\boldsymbol{q},\boldsymbol{p})\delta_{ij}e_i^*(\boldsymbol{p})u_j(\boldsymbol{q})\mathrm{d}\Omega+\int_{\Omega}\sigma_{jk}^{\mathrm{m}}(\boldsymbol{q})\varepsilon_{ijk}^*(\boldsymbol{q},\boldsymbol{p})e_i^*(\boldsymbol{p})\mathrm{d}\Omega$$

$$+\int_{\Omega}3\alpha K(0)\Delta T(\boldsymbol{q})\delta_{jk}U^*_{ij,k}(\boldsymbol{q},\boldsymbol{p})e^*_i(\boldsymbol{p})\mathrm{d}\Omega+\int_{\Omega}b_j(\boldsymbol{q})U^*_{ij}(\boldsymbol{q},\boldsymbol{p})e^*_i(\boldsymbol{p})\mathrm{d}\Omega$$
$$=\int_{\Gamma}u_j(\boldsymbol{Q})T^*_{ij}(\boldsymbol{Q},\boldsymbol{p})e^*_i(\boldsymbol{p})\mathrm{d}\Gamma-\int_{\Gamma}f_j(\boldsymbol{Q})U^*_{ij}(\boldsymbol{Q},\boldsymbol{p})e^*_i(\boldsymbol{p})\mathrm{d}\Gamma \tag{7.25}$$

其中，$\delta_{jk}U^*_{ij,k}=U^*_{ik,k}$。为了统一文中的角标，用$U^*_{ij,j}$替代$U^*_{ik,k}$，它的具体形式如下：

$$U^*_{ij,j}=\frac{-(1-2\nu)}{4(\beta-1)\pi(1-\nu)G(0)}\frac{r_{,i}}{r^{(\beta-1)}} \tag{7.26}$$

式中，$\beta=2$（二维问题）或 3（三维问题）。

式(7.25)中的单位向量对于所有积分都是相同的，而且每个分量都是相互独立的。此外$-\int_{\Omega}\delta(\boldsymbol{q},\boldsymbol{p})\delta_{ij}u_j(\boldsymbol{q})\mathrm{d}\Omega=-u_i(\boldsymbol{p})$，因此，式(7.25)可以写成如下形式：

$$\begin{aligned}u_i(\boldsymbol{p})=&\int_{\Gamma}f_j(\boldsymbol{Q})U^*_{ij}(\boldsymbol{Q},\boldsymbol{p})\mathrm{d}\Gamma(\boldsymbol{Q})-\int_{\Gamma}u_j(\boldsymbol{Q})T^*_{ij}(\boldsymbol{Q},\boldsymbol{p})\mathrm{d}\Gamma(\boldsymbol{Q})\\&+\int_{\Omega}b_j(\boldsymbol{q})U^*_{ij}(\boldsymbol{q},\boldsymbol{p})\mathrm{d}\Omega(\boldsymbol{q})+\int_{\Omega}\sigma^m_{jk}(\boldsymbol{q})\varepsilon^*_{ijk}(\boldsymbol{q},\boldsymbol{p})\mathrm{d}\Omega(\boldsymbol{q})\\&+\int_{\Omega}\Delta T(\boldsymbol{q})\varPsi_i(\boldsymbol{q},\boldsymbol{p})\mathrm{d}\Omega(\boldsymbol{q})\end{aligned} \tag{7.27}$$

其中，$\boldsymbol{p}$ 表示域内的源点；$\boldsymbol{q}$ 和 $\boldsymbol{Q}$ 分别表示域内和边界上的场点；基本解U^*_{ij}、T^*_{ij}、ε^*_{ijk}和核函数$\varPsi_i$可以写成如下形式：

$$\begin{cases}U^*_{ij}=\dfrac{1}{8(\beta-1)\pi G(0)(1-\nu)r^{(\beta-2)}}\left\{(3-4\nu)\delta_{ij}\left[(\beta-3)\ln\left(\dfrac{1}{r}\right)+(\beta-2)\right]+r_{,i}r_{,j}\right\}\\T^*_{ij}=-\dfrac{1}{4\pi(\beta-1)(1-\nu)r^{(\beta-1)}}\left\{\dfrac{\partial r}{\partial n}[(1-2\nu)\delta_{ij}+\beta r_{,i}r_{,j}]+(1-2\nu)\left(-r_{,j}n_i+r_{,i}n_j\right)\right\}\\\varepsilon^*_{ijk}=-\dfrac{1}{[8\pi(1-\nu)G(0)](\beta-1)r^{(\beta-1)}}[(1-2\nu)\left(\delta_{ik}r_{,j}+\delta_{ij}r_{,k}\right)-\delta_{jk}r_{,i}+\beta r_{,i}r_{,j}r_{,k}]\\\varPsi_i=-\dfrac{(1+\nu)\alpha}{2(\beta-1)\pi(1-\nu)r^{(\beta-1)}}r_{,i}\end{cases} \tag{7.28}$$

其中，对于二维问题，i, j, k=1,2；对于三维问题，i, j, k=1,2,3；$r_{,i}=\partial r/\partial x_i=(x^Q_i-x^p_i)/r(\boldsymbol{Q},\boldsymbol{p})$ 表示 r 对坐标 x_i 的偏导数；ν 表示材料的泊松比；n 表示边界的外法向，而且$\dfrac{\partial r}{\partial n}=r_{,i}n_i$。对于边界源点的边界域积分方程可以写成如下形式：

$$\begin{aligned}c_{ij}(\boldsymbol{P})u_j(\boldsymbol{P})=&\int_{\Gamma}f_j(\boldsymbol{Q})U^*_{ij}(\boldsymbol{Q},\boldsymbol{P})\mathrm{d}\Gamma(\boldsymbol{Q})-\int_{\Gamma}u_j(\boldsymbol{Q})T^*_{ij}(\boldsymbol{Q},\boldsymbol{P})\mathrm{d}\Gamma(\boldsymbol{Q})\\&+\int_{\Omega}b_j(\boldsymbol{q})U^*_{ij}(\boldsymbol{q},\boldsymbol{P})\mathrm{d}\Omega(\boldsymbol{q})+\int_{\Omega}\sigma^{\mathrm{m}}_{jk}(\boldsymbol{q})\varepsilon^*_{ijk}(\boldsymbol{q},\boldsymbol{P})\mathrm{d}\Omega(\boldsymbol{q})\end{aligned}$$

$$+\int_{\Omega}\Delta T(\boldsymbol{q})\Psi_i(\boldsymbol{q},\boldsymbol{P})\mathrm{d}\Omega(\boldsymbol{q}) \tag{7.29}$$

其中，$c_{ij}(\boldsymbol{P})$ 与边界上的源点 $\boldsymbol{P}$ 的位置有关。

7.2.3 内点和边界点应变

由于记忆应力与每个时间点的应变有关，因此需要计算内部应变和边界应变。为避免直接求解超奇异积分，采用热黏弹性问题正则化应变积分方程求解内点应变，采用热黏弹性问题的面力恢复法求解边界点应变。

1. 热黏弹性力学问题内点正则化应变积分方程

热黏弹性力学问题的内点积分方程可以由式(7.27)对源点 $\boldsymbol{p}$ 求导获得

$$\begin{aligned}\varepsilon_{il}(\boldsymbol{p})=&\int_{\Gamma}f_j(\boldsymbol{Q})U_{ilj}^{\varepsilon*}(\boldsymbol{Q},\boldsymbol{p})\mathrm{d}\Gamma(\boldsymbol{Q})-\int_{\Gamma}u_j(\boldsymbol{Q})T_{ilj}^{\varepsilon*}(\boldsymbol{Q},\boldsymbol{p})\mathrm{d}\Gamma(\boldsymbol{Q})\\&+\int_{\Omega}b_j(\boldsymbol{q})U_{ilj}^{\varepsilon*}(\boldsymbol{q},\boldsymbol{p})\mathrm{d}\Omega(\boldsymbol{q})\\&+\left[\int_{\Omega-\Omega_\zeta}\sigma_{jk}^{\mathrm{m}}(\boldsymbol{q})\varepsilon_{iljk}^{*}(\boldsymbol{q},\boldsymbol{p})\mathrm{d}\Omega(\boldsymbol{q})+\int_{\Omega_\zeta}\sigma_{jk}^{\mathrm{m}}(\boldsymbol{q})\varepsilon_{iljk}^{*}(\boldsymbol{q},\boldsymbol{p})\mathrm{d}\Omega(\boldsymbol{q})\right]\\&+\left[\int_{\Omega-\Omega_\zeta}\Delta T(\boldsymbol{q})\Psi_{il}(\boldsymbol{q},\boldsymbol{p})\mathrm{d}\Omega(\boldsymbol{q})+\int_{\Omega_\zeta}\Delta T(\boldsymbol{q})\Psi_{il}(\boldsymbol{q},\boldsymbol{p})\mathrm{d}\Omega(\boldsymbol{q})\right]\end{aligned} \tag{7.30}$$

其中，Ω_ζ 表示半径为 ζ 的圆或球；$U_{ilj}^{\varepsilon*}$、$T_{ilj}^{\varepsilon*}$、ε_{iljk}^{*} 和 Ψ_{il} 可以写成如下形式：

$$\begin{cases}U_{ilj}^{\varepsilon*}=\dfrac{1}{[8\pi(1-\nu)G(0)](\beta-1)r^{(\beta-1)}}\left[(1-2\nu)(\delta_{jl}r_{,i}+\delta_{ij}r_{,l})-\delta_{il}r_{,j}+3r_{,i}r_{,j}r_{,l}\right]\\T_{ilj}^{\varepsilon*}=\dfrac{-1}{4\pi(1-\nu)(\beta-1)r^{\beta}}\Big\{(1-2\nu)n_j(\delta_{il}-3r_{,i}r_{,l})-n_i[(1-2\nu)\delta_{jl}+\beta\nu r_{,j}r_{,l}]\\\qquad-n_l[(1-2\nu)\delta_{ij}+\beta\nu r_{,i}r_{,j}]-\beta\dfrac{\partial r}{\partial n}[\nu\delta_{ij}r_{,l}+\nu\delta_{jl}r_{,i}+\delta_{il}r_{,j}-(\beta+2)r_{,i}r_{,j}r_{,l}]\Big\}\\\varepsilon_{iljk}^{*}=\dfrac{1}{8\pi G(1-\nu)(\beta-1)r^{\beta}}\Big\{(1-2\nu)(\delta_{ij}\delta_{kl}+\delta_{ik}\delta_{jl})-\delta_{il}\delta_{jk}\\\qquad+\beta\nu(\delta_{ij}r_{,k}r_{,l}+\delta_{ik}r_{,j}r_{,l}+\delta_{jl}r_{,i}r_{,k}+\delta_{kl}r_{,i}r_{,j})\\\qquad+\beta(\delta_{jk}r_{,i}r_{,l}+\delta_{il}r_{,j}r_{,k})-\beta(\beta+2)r_{,i}r_{,j}r_{,k}r_{,l}\Big\}\\\Psi_{il}=\dfrac{-(1-\nu)\alpha}{2(\beta-1)\pi(1-\nu)r^{\beta}}(\beta r_{,i}r_{,l}-\delta_{il})\end{cases} \tag{7.31}$$

为了降低域积分的奇异性，可在域 Ω 内以源点 $\boldsymbol{p}$ 为中心挖掉一个半径为 ζ 的

圆或球 Ω_ζ (图 7.1)，由此对域积分进行正则化，其形式为

$$\int_{\Omega-\Omega_\zeta}\sigma_{jk}^{\mathrm{m}}(\boldsymbol{q})\varepsilon_{iljk}^{*}(\boldsymbol{q},\boldsymbol{p})\mathrm{d}\Omega(\boldsymbol{q})=\int_{\Omega-\Omega_\zeta}[\sigma_{jk}^{\mathrm{m}}(\boldsymbol{q})-\sigma_{jk}^{\mathrm{m}}(\boldsymbol{p})]\varepsilon_{iljk}^{*}(\boldsymbol{q},\boldsymbol{p})\mathrm{d}\Omega(\boldsymbol{q})+\int_{\Omega-\Omega_\zeta}\sigma_{jk}^{\mathrm{m}}(\boldsymbol{p})\varepsilon_{iljk}^{*}(\boldsymbol{q},\boldsymbol{p})\mathrm{d}\Omega(\boldsymbol{q}) \tag{7.32}$$

$$\int_{\Omega-\Omega_\zeta}\Delta T(\boldsymbol{q})\Psi_{il}(\boldsymbol{q},\boldsymbol{p})\mathrm{d}\Omega(\boldsymbol{q})=\int_{\Omega-\Omega_\zeta}[\Delta T(\boldsymbol{q})-\Delta T(\boldsymbol{p})]\Psi_{il}(\boldsymbol{q},\boldsymbol{p})\mathrm{d}\Omega(\boldsymbol{q})+\Delta T(\boldsymbol{p})\int_{\Omega-\Omega_\zeta}\Psi_{il}(\boldsymbol{q},\boldsymbol{p})\mathrm{d}\Omega(\boldsymbol{q}) \tag{7.33}$$

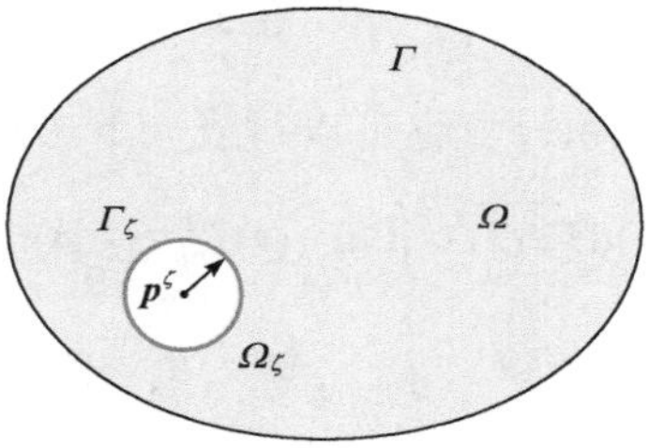

图 7.1 从域 Ω 中挖出一个半径为 ζ 的域 Ω_ζ

在域 Ω_ζ 内，假设 $\sigma^{\mathrm{m}}(\boldsymbol{q})=\sigma^{\mathrm{m}}(\boldsymbol{p})$ 和 $\Delta T(\boldsymbol{q})=\Delta T(\boldsymbol{p})$。因此，式(7.30)中的最后四项域积分可以转化为如下形式

$$\begin{aligned}&\int_{\Omega-\Omega_\zeta}\sigma_{jk}^{\mathrm{m}}(\boldsymbol{q})\varepsilon_{iljk}^{*}(\boldsymbol{q},\boldsymbol{p})\mathrm{d}\Omega(\boldsymbol{q})+\int_{\Omega_\zeta}\sigma_{jk}^{\mathrm{m}}(\boldsymbol{q})\varepsilon_{iljk}^{*}(\boldsymbol{q},\boldsymbol{p})\mathrm{d}\Omega(\boldsymbol{q})\\&=\int_{\Omega-\Omega_\zeta}[\sigma_{jk}^{\mathrm{m}}(\boldsymbol{q})-\sigma_{jk}^{\mathrm{m}}(\boldsymbol{p})]\varepsilon_{iljk}^{*}(\boldsymbol{q},\boldsymbol{p})\mathrm{d}\Omega(\boldsymbol{q})\\&\quad+\int_{\Omega-\Omega_\zeta}\sigma_{jk}^{\mathrm{m}}(\boldsymbol{p})\varepsilon_{iljk}^{*}(\boldsymbol{q},\boldsymbol{p})\mathrm{d}\Omega(\boldsymbol{q})+\int_{\Omega_\zeta}\sigma_{jk}^{\mathrm{m}}(\boldsymbol{q})\varepsilon_{iljk}^{*}(\boldsymbol{q},\boldsymbol{p})\mathrm{d}\Omega(\boldsymbol{q})\\&=\int_{\Omega-\Omega_\zeta}[\sigma_{jk}^{\mathrm{m}}(\boldsymbol{q})-\sigma_{jk}^{\mathrm{m}}(\boldsymbol{p})]\varepsilon_{iljk}^{*}(\boldsymbol{q},\boldsymbol{p})\mathrm{d}\Omega(\boldsymbol{q})+\sigma_{jk}^{\mathrm{m}}(\boldsymbol{p})\int_{\Omega}\varepsilon_{iljk}^{*}(\boldsymbol{q},\boldsymbol{p})\mathrm{d}\Omega(\boldsymbol{q})\end{aligned} \tag{7.34}$$

和

$$\begin{aligned}&\int_{\Omega-\Omega_\zeta}\Delta T(\boldsymbol{q})\Psi_{il}(\boldsymbol{q},\boldsymbol{p})\mathrm{d}\Omega(\boldsymbol{q})+\int_{\Omega_\zeta}\Delta T(\boldsymbol{q})\Psi_{il}(\boldsymbol{q},\boldsymbol{p})\mathrm{d}\Omega(\boldsymbol{q})\\&=\int_{\Omega-\Omega_\zeta}[\Delta T(\boldsymbol{q})-\Delta T(\boldsymbol{p})]\Psi_{il}(\boldsymbol{q},\boldsymbol{p})\mathrm{d}\Omega(\boldsymbol{q})\\&\quad+\int_{\Omega-\Omega_\zeta}\Delta T(\boldsymbol{p})\Psi_{il}(\boldsymbol{q},\boldsymbol{p})\mathrm{d}\Omega(\boldsymbol{q})+\int_{\Omega_\zeta}\Delta T(\boldsymbol{q})\Psi_{il}(\boldsymbol{q},\boldsymbol{p})\mathrm{d}\Omega(\boldsymbol{q})\\&=\int_{\Omega-\Omega_\zeta}[\Delta T(\boldsymbol{q})-\Delta T(\boldsymbol{p})]\Psi_{il}(\boldsymbol{q},\boldsymbol{p})\mathrm{d}\Omega(\boldsymbol{q})+\Delta T(\boldsymbol{p})\int_{\Omega}\Psi_{il}(\boldsymbol{q},\boldsymbol{p})\mathrm{d}\Omega(\boldsymbol{q})\end{aligned} \tag{7.35}$$

其中，

$$\int_{\Omega}\varepsilon_{iljk}^{*}(\boldsymbol{q},\boldsymbol{p})\mathrm{d}\Omega(\boldsymbol{q})=\int_{\Omega}\frac{1}{2}\left[\frac{\partial\varepsilon_{ilj}^{*}(\boldsymbol{q},\boldsymbol{p})}{\partial x_{k}^{q}}+\frac{\partial\varepsilon_{ilk}^{*}(\boldsymbol{q},\boldsymbol{p})}{\partial x_{j}^{q}}\right]\mathrm{d}\Omega(\boldsymbol{q})$$
$$=\int_{\Gamma}\frac{1}{2}\left[\varepsilon_{ilj}^{*}(\boldsymbol{Q},\boldsymbol{p})n_{k}+\varepsilon_{ilk}^{*}(\boldsymbol{Q},\boldsymbol{p})n_{j}\right]\mathrm{d}\Gamma(\boldsymbol{Q}) \tag{7.36}$$

和

$$\int_{\Omega}\Psi_{il}(\boldsymbol{q},\boldsymbol{p})\mathrm{d}\Omega(\boldsymbol{q})=\int_{\Omega}\frac{1}{2}\left(\frac{\partial\Psi_{i}(\boldsymbol{q},\boldsymbol{p})}{\partial x_{l}^{p}}+\frac{\partial\Psi_{l}(\boldsymbol{q},\boldsymbol{p})}{\partial x_{i}^{p}}\right)\mathrm{d}\Omega(\boldsymbol{q})$$
$$=-\int_{\Gamma}\frac{1}{2}[\Psi_{i}(\boldsymbol{Q},\boldsymbol{p})n_{l}+\Psi_{l}(\boldsymbol{Q},\boldsymbol{p})n_{l}]\mathrm{d}\Gamma(\boldsymbol{Q}) \tag{7.37}$$

因此，正则化内点应变积分方程可以写成如下形式：

$$\begin{aligned}\varepsilon_{il}(\boldsymbol{p})=&\int_{\Gamma}f_{j}(\boldsymbol{Q})U_{ilj}^{\varepsilon*}(\boldsymbol{Q},\boldsymbol{p})\mathrm{d}\Gamma(\boldsymbol{Q})-\int_{\Gamma}u_{j}(\boldsymbol{Q})T_{ilj}^{\varepsilon*}(\boldsymbol{Q},\boldsymbol{p})\mathrm{d}\Gamma(\boldsymbol{Q})\\&+\int_{\Omega}b_{j}(\boldsymbol{q})U_{ilj}^{\varepsilon*}(\boldsymbol{q},\boldsymbol{p})\mathrm{d}\Omega(\boldsymbol{q})+\int_{\Omega-\Omega_{\zeta}}[\sigma_{jk}^{m}(\boldsymbol{q})-\sigma_{jk}^{m}(\boldsymbol{p})]\varepsilon_{iljk}^{*}(\boldsymbol{q},\boldsymbol{p})\mathrm{d}\Omega(\boldsymbol{q})\\&+\frac{1}{2}\sigma_{jk}^{m}(\boldsymbol{Q})\int_{\Gamma}[\varepsilon_{ilj}^{*}(\boldsymbol{Q},\boldsymbol{p})n_{k}+\varepsilon_{ilk}^{*}(\boldsymbol{Q},\boldsymbol{p})n_{j}]\mathrm{d}\Gamma(\boldsymbol{Q})\\&+\int_{\Omega-\Omega_{\zeta}}[\Delta T(\boldsymbol{q})-\Delta T(\boldsymbol{p})]\Psi_{il}(\boldsymbol{q},\boldsymbol{p})\mathrm{d}\Omega(\boldsymbol{q})\\&+\frac{1}{2}\Delta T(\boldsymbol{Q})\int_{\Gamma}[\Psi_{i}(\boldsymbol{Q},\boldsymbol{p})n_{l}+\Psi_{l}(\boldsymbol{Q},\boldsymbol{p})n_{i}]\mathrm{d}\Gamma(\boldsymbol{Q})\end{aligned} \tag{7.38}$$

2. 二维及三维热黏弹性面力恢复法

对于边界点的应变，可以采用如式(7.38)所示的应变边界域积分方程，但需要求解超奇异积分。因此，为了避免计算超奇异积分，我们使用一种改进的面力恢复法来求解热黏弹性问题的边界点应变。

边界源点的位移和面力可以写成如下形式：

$$\begin{cases}\boldsymbol{u}(\boldsymbol{P})=\boldsymbol{\Phi}\boldsymbol{u}(\boldsymbol{Q})\\\boldsymbol{f}(\boldsymbol{P})=\boldsymbol{\Phi}\boldsymbol{f}(\boldsymbol{Q})\end{cases} \tag{7.39}$$

其中，矩阵 $\boldsymbol{\Phi}$ 表示从控制点 $\boldsymbol{Q}$ 到配点 $\boldsymbol{P}$ 的转化矩阵。对于二维问题，局部单位向量(图 7.2(a))可以由如下公式获得：

$$\begin{cases}\hat{e}_{1}=\dfrac{\boldsymbol{n}}{|\boldsymbol{n}|}\\\hat{e}_{2}=\dfrac{\boldsymbol{m}}{|\boldsymbol{m}|}\end{cases} \tag{7.40}$$

其中，$\boldsymbol{n}$ 表示法向量；$\boldsymbol{m}$ 表示切向量。因此，可以得到整体坐标系到局部坐标系的转换矩阵：

$$\boldsymbol{A}=\begin{Bmatrix}\hat{\boldsymbol{e}}_1\\ \hat{\boldsymbol{e}}_2\end{Bmatrix}=\begin{bmatrix}n_1 & n_2\\ -n_2 & n_1\end{bmatrix} \tag{7.41}$$

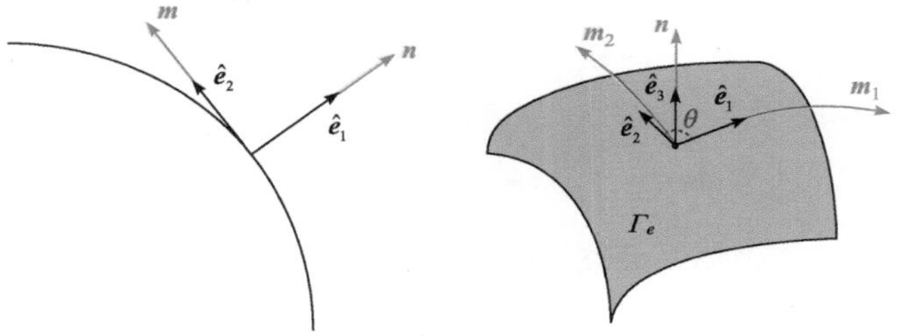

图 7.2　二维和三维问题的局部坐标系

(a) 二维问题；(b) 三维问题

局部坐标系中的应力张量可以写成

$$\begin{cases}\hat{\sigma}_{11}^{\mathrm{e}}=\hat{\sigma}_{11}+\hat{\sigma}_{11}^{\mathrm{m}}+\hat{\sigma}_{11}^{\mathrm{T}}=\hat{t}_1+\hat{\sigma}_{11}^{\mathrm{m}}+\hat{\sigma}_{11}^{\mathrm{T}}\\ \hat{\sigma}_{12}^{\mathrm{e}}=\hat{\sigma}_{21}^{\mathrm{e}}=\hat{\sigma}_{21}+\hat{\sigma}_{21}^{\mathrm{m}}+\hat{\sigma}_{21}^{\mathrm{T}}=\hat{t}_2+\hat{\sigma}_{21}^{\mathrm{m}}+\hat{\sigma}_{21}^{\mathrm{T}}\\ \hat{\sigma}_{22}^{\mathrm{e}}=[E/(1-\nu^2)]\hat{\varepsilon}_{22}+[\nu/(1-\nu)]\hat{\sigma}_{11}^{\mathrm{e}}\end{cases} \tag{7.42}$$

其中，

$$\hat{\varepsilon}_{22}=\frac{\partial\hat{u}_2}{\partial\xi}\frac{\partial\xi}{\partial\hat{x}_2}=A_{2i}\frac{\partial u_i}{\partial\xi}\frac{\partial\xi}{\partial\hat{x}_2}=A_{2i}\frac{\partial u_i}{\partial\xi}\frac{1}{|\boldsymbol{m}|} \tag{7.43}$$

$$\begin{bmatrix}\hat{\boldsymbol{\sigma}}_{11}^{\mathrm{m}} & \hat{\boldsymbol{\sigma}}_{12}^{\mathrm{m}}\\ \hat{\boldsymbol{\sigma}}_{21}^{\mathrm{m}} & \hat{\boldsymbol{\sigma}}_{22}^{\mathrm{m}}\end{bmatrix}=\boldsymbol{A}\begin{bmatrix}\boldsymbol{\sigma}_{11}^{\mathrm{m}} & \boldsymbol{\sigma}_{12}^{\mathrm{m}}\\ \boldsymbol{\sigma}_{21}^{\mathrm{m}} & \boldsymbol{\sigma}_{22}^{\mathrm{m}}\end{bmatrix} \tag{7.44}$$

$$\begin{bmatrix}\hat{\boldsymbol{\sigma}}_{11}^{\mathrm{T}} & \hat{\boldsymbol{\sigma}}_{12}^{\mathrm{T}}\\ \hat{\boldsymbol{\sigma}}_{21}^{\mathrm{T}} & \hat{\boldsymbol{\sigma}}_{22}^{\mathrm{T}}\end{bmatrix}=\boldsymbol{A}\begin{bmatrix}\boldsymbol{\sigma}_{11}^{\mathrm{T}} & \boldsymbol{\sigma}_{12}^{\mathrm{T}}\\ \boldsymbol{\sigma}_{21}^{\mathrm{T}} & \boldsymbol{\sigma}_{22}^{\mathrm{T}}\end{bmatrix} \tag{7.45}$$

求解局部坐标系下的弹性应力后，由公式 $\sigma_{il}^{\mathrm{e}}=A_{ji}A_{kl}\hat{\sigma}_{jk}^{\mathrm{e}}$ 可得到整体坐标系下的弹性应力。此外，热黏弹性力学问题的应变可由下式求解：

$$\varepsilon_{il}=\frac{1}{2G(0)}\sigma_{il}^{\mathrm{e}}-\delta_{il}\frac{\nu}{E}\sigma_{nn}^{\mathrm{e}} \tag{7.46}$$

对于三维问题，局部单位切向量 $\hat{\boldsymbol{e}}_i\,(i=1,2)$ 和局部单位法向量 $\hat{\boldsymbol{e}}_3$ (图 7.2(b))可以由如下公式获得：

$$\begin{cases} \hat{\boldsymbol{e}}_1 = \boldsymbol{m}_1 / |\boldsymbol{m}_1| \\ \hat{\boldsymbol{e}}_2 = \hat{\boldsymbol{e}}_1 \times \hat{\boldsymbol{e}}_3 \\ \hat{\boldsymbol{e}}_3 = \boldsymbol{n} / |\boldsymbol{n}| \end{cases} \tag{7.47}$$

其中，

$$\begin{cases} \boldsymbol{m}_1(\hat{\xi}_1, \hat{\xi}_2) = \partial \boldsymbol{r} / \partial \xi_1 \\ \boldsymbol{m}_2(\hat{\xi}_1, \hat{\xi}_2) = \partial \boldsymbol{r} / \partial \xi_2 \\ \boldsymbol{n}(\hat{\xi}_1, \hat{\xi}_2) = \boldsymbol{m}_1(\hat{\xi}_1, \hat{\xi}_2) \times \boldsymbol{m}_2(\hat{\xi}_1, \hat{\xi}_2) / |\boldsymbol{m}_1(\hat{\xi}_1, \hat{\xi}_2) \times \boldsymbol{m}_2(\hat{\xi}_1, \hat{\xi}_2)| \end{cases} \tag{7.48}$$

三维问题从局部坐标系到整体坐标系的变换矩阵为

$$\boldsymbol{A} = \begin{Bmatrix} \hat{\boldsymbol{e}}_1 \\ \hat{\boldsymbol{e}}_2 \\ \hat{\boldsymbol{e}}_3 \end{Bmatrix} \tag{7.49}$$

此外，局部弹性应力可由下式获得：

$$\begin{cases} \hat{\sigma}_{13}^{\mathrm{e}} = \hat{\sigma}_{13} + \hat{\sigma}_{13}^{\mathrm{m}} = \hat{f}_1 + \hat{\sigma}_{13}^{\mathrm{m}} + \hat{\sigma}_{13}^{\mathrm{T}} \\ \hat{\sigma}_{23}^{\mathrm{e}} = \hat{\sigma}_{23} + \hat{\sigma}_{23}^{\mathrm{m}} = \hat{f}_2 + \hat{\sigma}_{23}^{\mathrm{m}} + \hat{\sigma}_{23}^{\mathrm{T}} \\ \hat{\sigma}_{33}^{\mathrm{e}} = \hat{\sigma}_{22} + \hat{\sigma}_{33}^{\mathrm{m}} = \hat{f}_3 + \hat{\sigma}_{33}^{\mathrm{m}} + \hat{\sigma}_{33}^{\mathrm{T}} \\ \hat{\sigma}_{11}^{\mathrm{e}} = \dfrac{E}{1-\nu^2}(\hat{\varepsilon}_{11} + \nu \hat{\varepsilon}_{22}) + \dfrac{\nu}{1-\nu}\hat{\sigma}_{33}^{\mathrm{e}} \\ \hat{\sigma}_{12}^{\mathrm{e}} = 2G(0)\hat{\varepsilon}_{12} \\ \hat{\sigma}_{22}^{\mathrm{e}} = \dfrac{E}{1-\nu^2}(\hat{\varepsilon}_{22} + \nu \hat{\varepsilon}_{11}) + \dfrac{\nu}{1-\nu}\hat{\sigma}_{33}^{\mathrm{e}} \end{cases} \tag{7.50}$$

其中，

$$\hat{\varepsilon}_{il} = \frac{\partial \hat{u}_i}{\partial \hat{x}_l} = \frac{\partial \hat{u}_i}{\partial \xi_j} \frac{\partial \xi_j}{\partial \hat{x}_l} \tag{7.51}$$

$$\begin{bmatrix} \hat{\boldsymbol{\sigma}}_{11}^{\mathrm{m}} & \hat{\boldsymbol{\sigma}}_{12}^{\mathrm{m}} & \hat{\boldsymbol{\sigma}}_{13}^{\mathrm{m}} \\ \hat{\boldsymbol{\sigma}}_{21}^{\mathrm{m}} & \hat{\boldsymbol{\sigma}}_{22}^{\mathrm{m}} & \hat{\boldsymbol{\sigma}}_{23}^{\mathrm{m}} \\ \hat{\boldsymbol{\sigma}}_{31}^{\mathrm{m}} & \hat{\boldsymbol{\sigma}}_{32}^{\mathrm{m}} & \hat{\boldsymbol{\sigma}}_{33}^{\mathrm{m}} \end{bmatrix} = \boldsymbol{A} \begin{bmatrix} \boldsymbol{\sigma}_{11}^{\mathrm{m}} & \boldsymbol{\sigma}_{12}^{\mathrm{m}} & \boldsymbol{\sigma}_{13}^{\mathrm{m}} \\ \boldsymbol{\sigma}_{21}^{\mathrm{m}} & \boldsymbol{\sigma}_{22}^{\mathrm{m}} & \boldsymbol{\sigma}_{23}^{\mathrm{m}} \\ \boldsymbol{\sigma}_{31}^{\mathrm{m}} & \boldsymbol{\sigma}_{32}^{\mathrm{m}} & \boldsymbol{\sigma}_{33}^{\mathrm{m}} \end{bmatrix} \tag{7.52}$$

$$\begin{bmatrix} \hat{\boldsymbol{\sigma}}_{11}^{\mathrm{T}} & \hat{\boldsymbol{\sigma}}_{12}^{\mathrm{T}} & \hat{\boldsymbol{\sigma}}_{13}^{\mathrm{T}} \\ \hat{\boldsymbol{\sigma}}_{21}^{\mathrm{T}} & \hat{\boldsymbol{\sigma}}_{22}^{\mathrm{T}} & \hat{\boldsymbol{\sigma}}_{23}^{\mathrm{T}} \\ \hat{\boldsymbol{\sigma}}_{31}^{\mathrm{T}} & \hat{\boldsymbol{\sigma}}_{32}^{\mathrm{T}} & \hat{\boldsymbol{\sigma}}_{33}^{\mathrm{T}} \end{bmatrix} = \boldsymbol{A} \begin{bmatrix} \boldsymbol{\sigma}_{11}^{\mathrm{T}} & \boldsymbol{\sigma}_{12}^{\mathrm{T}} & \boldsymbol{\sigma}_{13}^{\mathrm{T}} \\ \boldsymbol{\sigma}_{21}^{\mathrm{T}} & \boldsymbol{\sigma}_{22}^{\mathrm{T}} & \boldsymbol{\sigma}_{23}^{\mathrm{T}} \\ \boldsymbol{\sigma}_{31}^{\mathrm{T}} & \boldsymbol{\sigma}_{32}^{\mathrm{T}} & \boldsymbol{\sigma}_{33}^{\mathrm{T}} \end{bmatrix} \tag{7.53}$$

与二维问题类似，整体坐标系下三维问题的弹性应力可由公式 $\sigma_{il}^{\mathrm{e}}=A_{ji}A_{kl}\hat{\sigma}_{jk}^{\mathrm{e}}$ 获得。此外，热黏弹性力学问题的应变可由下式求解：

$$\varepsilon_{il}=\frac{1}{2G(0)}\sigma_{il}^{\mathrm{e}}-\delta_{il}\left\{\frac{1}{2G_0}\left[K(0)-\frac{2}{3}G(0)\right]\right\}\varepsilon_{nn} \tag{7.54}$$

7.2.4 利用径向积分法将域积分转换为等效边界积分

由于使用 Kelvin 基本解来推导热黏弹性力学问题的边界域积分方程，因此积分方程中必然存在域积分。为了保证等几何边界元法仅对边界离散化的优势，可利用径向积分法将域积分转换为等效边界积分。

1. 已知核函数

由体力项引起的域积分 $\int_{\Omega}b_j(\boldsymbol{q})U_{ij}^{*}(\boldsymbol{p},\boldsymbol{q})\mathrm{d}\Omega(\boldsymbol{q})$ 和 $\int_{\Omega}b_j(\boldsymbol{q})U_{ilj}^{\varepsilon*}(\boldsymbol{q},\boldsymbol{p})\mathrm{d}\Omega(\boldsymbol{q})$ 中的核函数都是已知的，因此，可以直接利用径向积分法将这两个域积分转换为如下形式的等效边界积分：

$$\begin{cases}\displaystyle\int_{\Omega}b_j(\boldsymbol{q})U_{ij}^{*}(\boldsymbol{p},\boldsymbol{q})\mathrm{d}\Omega(\boldsymbol{q})=\int_{\Gamma}\frac{1}{r^{(\beta-1)}(\boldsymbol{p},\boldsymbol{Q})}\frac{\partial r}{\partial n}F_i^{\mathrm{b}1}(\boldsymbol{p},\boldsymbol{Q})\mathrm{d}\Gamma(\boldsymbol{Q})\\ \displaystyle\int_{\Omega}b_j(\boldsymbol{q})U_{ilj}^{\varepsilon*}(\boldsymbol{q},\boldsymbol{p})\mathrm{d}\Omega(\boldsymbol{q})=\int_{\Gamma}\frac{1}{r^{(\beta-1)}(\boldsymbol{p},\boldsymbol{Q})}\frac{\partial r}{\partial n}F_{il}^{\mathrm{b}2}(\boldsymbol{p},\boldsymbol{Q})\mathrm{d}\Gamma(\boldsymbol{Q})\end{cases} \tag{7.55}$$

其中，

$$\begin{cases}F_i^{\mathrm{b}1}=\displaystyle\int_0^r U_{ij}^{*}b_j r^{(\beta-1)}\mathrm{d}r\\ F_{il}^{\mathrm{b}2}=\displaystyle\int_0^r U_{ilj}^{\varepsilon*}b_j r^{(\beta-1)}\mathrm{d}r\end{cases} \tag{7.56}$$

2. 未知核函数

域积分 $\int_{\Omega}\sigma_{jk}^{\mathrm{m}}(\boldsymbol{q})\varepsilon_{ijk}^{*}(\boldsymbol{q},\boldsymbol{p})\mathrm{d}\Omega(\boldsymbol{q})$ 、$\int_{\Omega-\Omega_{\xi}}[\sigma_{jk}^{\mathrm{m}}(\boldsymbol{q})-\sigma_{jk}^{\mathrm{m}}(\boldsymbol{p})]\varepsilon_{iljk}^{*}(\boldsymbol{q},\boldsymbol{p})\mathrm{d}\Omega(\boldsymbol{q})$ 、$\int_{\Omega}\Delta T(\boldsymbol{q})\Psi_i(\boldsymbol{q},\boldsymbol{p})\mathrm{d}\Omega(\boldsymbol{q})$ 和 $\int_{\Omega-\Omega_{\xi}}[\Delta T(\boldsymbol{q})-\Delta T(\boldsymbol{p})]\varepsilon_{iljk}^{*}(\boldsymbol{q},\boldsymbol{p})\mathrm{d}\Omega(\boldsymbol{q})$ 中含有未知核函数 $\sigma_{jk}^{\mathrm{m}}(\boldsymbol{q})$ 或 $\Delta T(\boldsymbol{q})$，因此，不能直接利用径向积分法将这些域积分转换为等效边界积分。这些未知的核函数需要用紧致径向基函数展开成如下形式：

$$\left.\begin{aligned}\sigma_{jk}^{\mathrm{m}}(\boldsymbol{q})&=\sum_{A=1}^{N_A}\alpha_{jk}^{A}\phi^{A}(R)+a_{jk}^{0}+\sum_{\overline{m}=1}^{\overline{k}}a_{jk}^{\overline{m}}x_{\overline{m}}^{q}\\ \Delta T(\boldsymbol{q})&=\sum_{A=1}^{N_A}\chi^{A}\phi^{A}(R)+b^{0}+\sum_{\overline{m}=1}^{\overline{k}}b^{\overline{m}}x_{\overline{m}}^{q}\end{aligned}\right\}\overline{k}=2(\text{二维});\ \overline{k}=3(\text{三维}) \tag{7.57}$$

为了使方程可解，系数α和χ还需要满足如下等式：

$$\left.\begin{aligned}\sum_{A=1}^{N_A}\alpha_{jk}^{A}&=\sum_{A=1}^{N_A}\alpha_{jk}^{A}x_{\overline{m}}^{A}=0\\ \sum_{A=1}^{N_A}\chi^{A}&=\sum_{A=1}^{N_A}\chi^{A}x_{\overline{m}}^{A}=0\end{aligned}\right\}\overline{m}=1,2(\text{二维});\ \overline{m}=1,2,3(\text{三维}) \tag{7.58}$$

式中，A 表示应用点；$N_A=N_{\mathrm{b}}+N_{\mathrm{I}}$ 表示应用点的个数，应用点由内点和边界配点组成，N_{b} 表示边界配点的个数，N_{I} 表示内点的个数。如图 7.3 所示，$R=\|\boldsymbol{q}-\boldsymbol{A}\|$ 表示应用点到场点的距离。此外，α_{jk}^{A}、$a_{jk}^{\overline{m}}$、χ^{A} 和 $b^{\overline{m}}$ 为待定系数。本章采用四阶样条紧支径向基函数来近似未知量，具体形式如下：

$$\phi(R/d_A)=\begin{cases}1-6\left(\dfrac{R}{d_A}\right)^2+8\left(\dfrac{R}{d_A}\right)^3-3\left(\dfrac{R}{d_A}\right)^4, & 0\leqslant R\leqslant d_A\\ 0, & R>d_A\end{cases} \tag{7.59}$$

其中，d_A 表示应用点 A 处的紧支域的大小。R、$\overline{R}$ 和 r 的关系为

$$\begin{cases}R=\sqrt{r^2+2sr+\overline{R}^2}\\ \overline{R}=\sqrt{\overline{R}_i\overline{R}_i}\\ \overline{R}_i=x_i^p-x_i^A\\ s=r_{,i}\overline{R}_i\end{cases} \tag{7.60}$$

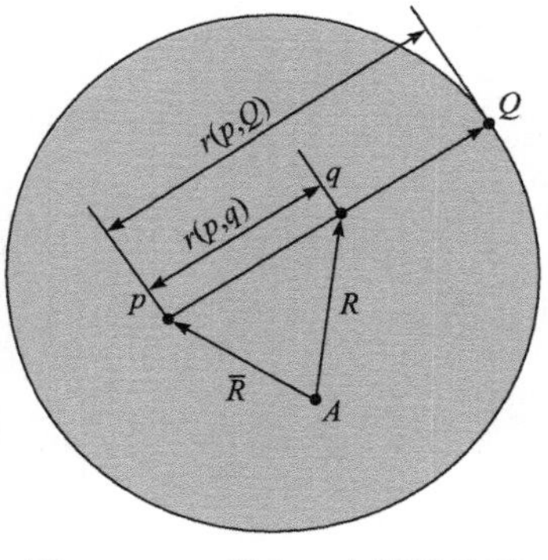

图 7.3　R、$\overline{R}$ 和 r 之间的关系

使用四阶样条紧支径向基函数展开未知核函数后，可以利用径向积分法将域积分 $\int_{\Omega}\Delta T(\boldsymbol{q})\Psi_i(\boldsymbol{q},\boldsymbol{p})\mathrm{d}\Omega(\boldsymbol{q})$ 和 $\int_{\Omega}\sigma_{jk}^{\mathrm{m}}(\boldsymbol{q})\varepsilon_{ijk}^{*}(\boldsymbol{q},\boldsymbol{p})\mathrm{d}\Omega(\boldsymbol{q})$ 转换为如下形式的等效边界积分：

$$\begin{aligned}\int_{\Omega}\sigma_{jk}^{\mathrm{m}}(\boldsymbol{q})\varepsilon_{ijk}^{*}(\boldsymbol{q},\boldsymbol{p})\mathrm{d}\Omega(\boldsymbol{q})=&\sum_{A=1}^{N_A}\alpha_{jk}^{A}\int_{\Gamma}\frac{1}{r^2(\boldsymbol{Q},\boldsymbol{p})}\frac{\partial r}{\partial n}F_{ijk}^{A}(\boldsymbol{q},\boldsymbol{p})\mathrm{d}\Gamma(\boldsymbol{Q})\\&+a_{jk}^{0}\int_{\Gamma}\frac{1}{r^2(\boldsymbol{Q},\boldsymbol{p})}\frac{\partial r}{\partial n}F_{ijk}^{0}(\boldsymbol{q},\boldsymbol{p})\mathrm{d}\Gamma(\boldsymbol{Q})\\&+\sum_{\bar{m}=1}^{\bar{k}}a_{jk}^{\bar{m}}\int_{\Gamma}\frac{1}{r^2(\boldsymbol{Q},\boldsymbol{p})}\frac{\partial r}{\partial n}F_{ijk}^{\bar{m}}(\boldsymbol{q},\boldsymbol{p})\mathrm{d}\Gamma(\boldsymbol{Q})\end{aligned}\tag{7.61}$$

$$\begin{aligned}\int_{\Omega}\Delta T(\boldsymbol{q})\Psi_i(\boldsymbol{q},\boldsymbol{p})\mathrm{d}\Omega(\boldsymbol{q})=&\sum_{A=1}^{N_A}\chi^{A}\int_{\Gamma}\frac{1}{r^{(\beta-1)}(\boldsymbol{Q},\boldsymbol{p})}\frac{\partial r}{\partial n}K_{i}^{A}(\boldsymbol{q},\boldsymbol{p})\mathrm{d}\Gamma(\boldsymbol{Q})\\&+b^{0}\int_{\Gamma}\frac{1}{r^{(\beta-1)}(\boldsymbol{Q},\boldsymbol{p})}\frac{\partial r}{\partial n}K_{i}^{0}(\boldsymbol{q},\boldsymbol{p})\mathrm{d}\Gamma(\boldsymbol{Q})\\&+\sum_{\bar{m}=1}^{\bar{k}}b^{\bar{m}}\int_{\Gamma}\frac{1}{r^{(\beta-1)}(\boldsymbol{Q},\boldsymbol{p})}\frac{\partial r}{\partial n}K_{i}^{\bar{m}}(\boldsymbol{q},\boldsymbol{p})\mathrm{d}\Gamma(\boldsymbol{Q})\end{aligned}\tag{7.62}$$

其中，

$$\begin{cases}F_{ijk}^{A}=\int_{0}^{r(\boldsymbol{p},\boldsymbol{Q})}\varepsilon_{ijk}^{*}(\boldsymbol{q},\boldsymbol{p})\phi^{A}(R)r^{(\beta-1)}(\boldsymbol{q},\boldsymbol{p})\mathrm{d}r(\boldsymbol{q})\\F_{ijk}^{0}=\int_{0}^{r(\boldsymbol{p},\boldsymbol{Q})}\varepsilon_{ijk}^{*}(\boldsymbol{q},\boldsymbol{p})r^{(\beta-1)}(\boldsymbol{q},\boldsymbol{p})\mathrm{d}r(\boldsymbol{q})\\F_{ijk}^{\bar{m}}=\int_{0}^{r(\boldsymbol{p},\boldsymbol{Q})}\varepsilon_{ijk}^{*}(\boldsymbol{q},\boldsymbol{p})(\chi_{\bar{m}}^{\boldsymbol{p}}+r_{,\bar{m}}r)r^{(\beta-1)}(\boldsymbol{q},\boldsymbol{p})\mathrm{d}r(\boldsymbol{q})\end{cases}\tag{7.63}$$

$$\begin{cases}K_{i}^{A}=\int_{0}^{r(\boldsymbol{p},\boldsymbol{Q})}\Psi_i(\boldsymbol{q},\boldsymbol{p})\phi^{A}(R)r^{(\beta-1)}(\boldsymbol{q},\boldsymbol{p})\mathrm{d}r(\boldsymbol{q})\\K_{i}^{0}=\int_{0}^{r(\boldsymbol{p},\boldsymbol{Q})}\Psi_i(\boldsymbol{q},\boldsymbol{p})r^{(\beta-1)}(\boldsymbol{q},\boldsymbol{p})\mathrm{d}r(\boldsymbol{q})\\K_{i}^{\bar{m}}=\int_{0}^{r(\boldsymbol{p},\boldsymbol{Q})}\Psi_i(\boldsymbol{q},\boldsymbol{p})(x_{\bar{m}}^{\boldsymbol{p}}+r_{,\bar{m}}r)r^{(\beta-1)}(\boldsymbol{q},\boldsymbol{p})\mathrm{d}r(\boldsymbol{q})\end{cases}\tag{7.64}$$

类似地，可以使用相同的方法将域积分 $\int_{\Omega-\Omega_\zeta}[\sigma_{jk}^{\mathrm{m}}(\boldsymbol{q})-\sigma_{jk}^{\mathrm{m}}(\boldsymbol{p})]\varepsilon_{iljk}^{*}(\boldsymbol{q},\boldsymbol{p})\mathrm{d}\Omega(\boldsymbol{q})$ 和 $\int_{\Omega-\Omega_\zeta}[\Delta T(\boldsymbol{q})-\Delta T(\boldsymbol{p})]\varepsilon_{iljk}^{*}(\boldsymbol{q},\boldsymbol{p})\mathrm{d}\Omega(\boldsymbol{q})$ 转换为其相应的等效边界积分。

7.2.5　边界积分方程的等几何分析

采用 NURBS 作为等几何边界元法的形函数。与常规边界元法不同，等几何边界元法可以精确地描述模型的几何形状。利用 NURBS 作为形函数描述二维和三维模型的节点坐标、位移和面力，其形式如下：

$$
\begin{cases}
\boldsymbol{x}(\xi)=\sum_{i=1}^{\hat{p}+1}\Phi_i^{\hat{p}}(\xi)\boldsymbol{Q}_i \\
\boldsymbol{u}(\xi,t)=\sum_{i=1}^{\hat{p}+1}\Phi_i^{\hat{p}}(\xi)\boldsymbol{u}_i(t), \quad \text{二维} \\
\boldsymbol{f}(\xi,t)=\sum_{i=1}^{\hat{p}+1}\Phi_i^{\hat{p}}(\xi)\boldsymbol{f}_i(t)
\end{cases}
\tag{7.65}
$$

和

$$
\begin{cases}
\boldsymbol{x}(\xi,\eta)=\sum_{i=1}^{(\hat{p}+1)}\sum_{j=1}^{(\hat{q}+1)}\Phi_{i,j}^{\hat{p},\hat{q}}(\xi,\eta)\boldsymbol{Q}_{i,j} \\
\boldsymbol{u}(\xi,\eta,t)=\sum_{i=1}^{(\hat{p}+1)}\sum_{j=1}^{(\hat{q}+1)}\Phi_{i,j}^{\hat{p},\hat{q}}(\xi,\eta)\boldsymbol{u}_{i,j}(t), \quad \text{三维} \\
\boldsymbol{f}(\xi,\eta,t)=\sum_{i=1}^{(\hat{p}+1)}\sum_{j=1}^{(\hat{q}+1)}\Phi_{i,j}^{\hat{p},\hat{q}}(\xi,\eta)\boldsymbol{f}_{i,j}(t)
\end{cases}
\tag{7.66}
$$

其中，$\Phi_i^{\hat{p}}$ 和 $\Phi_{i,j}^{\hat{p},\hat{q}}$ 表示一维和二维 NURBS 形函数；$\hat{p}$ 和 $\hat{q}$ 表示两个方向形函数的阶次；$\boldsymbol{Q}$ 表示控制点的坐标。在此基础上，二维热黏弹性力学问题的位移积分方程可以离散成如下形式：

$$
\begin{aligned}
\boldsymbol{u}(\boldsymbol{p},t)=&\sum_{e=1}^{N_e}\sum_{\bar{l}=1}^{\hat{p}+1}\left\{\int_{-1}^{1}\boldsymbol{U}^*\left[\boldsymbol{Q}(\hat{\xi}),\boldsymbol{p}\right]\Phi_{\bar{l}}^e(\hat{\xi})J^e(\hat{\xi})\mathrm{d}\hat{\xi}\right\}\boldsymbol{f}_{\bar{l}}^e(\boldsymbol{Q},t) \\
&-\sum_{e=1}^{N_e}\sum_{\bar{l}=1}^{\hat{p}+1}\left\{\int_{-1}^{1}\boldsymbol{T}^*\left[\boldsymbol{Q}(\hat{\xi}),\boldsymbol{p}\right]\Phi_{\bar{l}}^e(\hat{\xi})J^e(\hat{\xi})\mathrm{d}\hat{\xi}\right\}\boldsymbol{u}_{\bar{l}}^e(\boldsymbol{Q},t) \\
&+\boldsymbol{b}(\boldsymbol{p})+\boldsymbol{B}(\boldsymbol{p})\boldsymbol{\sigma}^{\mathrm{m}}(\boldsymbol{A},t)+\boldsymbol{D}(\boldsymbol{p})\Delta\boldsymbol{T}(\boldsymbol{A},t)
\end{aligned}
\tag{7.67}
$$

其中，$\boldsymbol{u}(\boldsymbol{p},t)$ 是一个 2×1 阶的位移向量；$\boldsymbol{U}^*$ 和 $\boldsymbol{T}^*$ 是一个 2×2 阶的基本解矩阵；局部坐标 $\hat{\xi}$ 的范围是−1～1；N_e 表示单元的个数；雅可比转换行列式 $J^e(\hat{\xi})$ 可以写成如下形式：

$$
J^e(\hat{\xi})=\frac{\mathrm{d}\Gamma}{\mathrm{d}\xi}\frac{\mathrm{d}\xi}{\mathrm{d}\hat{\xi}}=\sqrt{\left(\frac{\mathrm{d}x_1}{\mathrm{d}\xi}\right)^2+\left(\frac{\mathrm{d}x_2}{\mathrm{d}\xi}\right)^2}\cdot\frac{\xi_2-\xi_1}{2}
\tag{7.68}
$$

式(7.67)中，$\boldsymbol{b}(\boldsymbol{p})$ 表示 2×1 阶的体力向量；$\boldsymbol{\sigma}^{\mathrm{m}}(\boldsymbol{A},t)$ 和 $\Delta\boldsymbol{T}(\boldsymbol{A},t)$ 分别表示应用点记忆应力和温度变化向量；矩阵 $\boldsymbol{B}(\boldsymbol{p})$ 的具体形式在第 6 章中已给出；$\boldsymbol{D}(\boldsymbol{p})$ 是一

个 $2\times N_A$ 阶的矩阵，第 $i(=1,2)$行的第 j 个量 D_j 可以写成如下形式：

$$D_j=\left\{\Phi_{1j}^{-1}\int_\Gamma\frac{1}{r}\frac{\partial r}{\partial n}K_i^1\mathrm{d}\Gamma+\Phi_{2j}^{-1}\int_\Gamma\frac{1}{r}\frac{\partial r}{\partial n}K_i^2\mathrm{d}\Gamma+\cdots+\Phi_{Nj}^{-1}\int_\Gamma\frac{1}{r}\frac{\partial r}{\partial n}K_i^{N_A}\mathrm{d}\Gamma\right.$$
$$+\Phi_{(N+1)j}^{-1}\int_\Gamma\frac{1}{r}\frac{\partial r}{\partial n}K_i^0\mathrm{d}\Gamma+\Phi_{(N+2)j}^{-1}\left[\int_\Gamma\frac{1}{r}\frac{\partial r}{\partial n}K_i^1(x_1^p+r_{,1}r)\mathrm{d}\Gamma\right]$$
$$\left.+\Phi_{(N+3)j}^{-1}\left[\int_\Gamma\frac{1}{r}\frac{\partial r}{\partial n}K_i^2(x_2^p+r_{,2}r)\mathrm{d}\Gamma\right]\right\}\tag{7.69}$$

式中，Φ_{ab}^{-1} 是 $\boldsymbol{\Phi}^{-1}$ 中位于第 a 行和第 b 列处的元素；而 $\boldsymbol{\Phi}$ 是由式(7.57)和式(7.58)形成的类似于式(5.19)或式(5.20)中的系数矩阵 $\boldsymbol{\Phi}$ 。

二维问题的应变积分方程的离散形式为

$$\boldsymbol{\varepsilon}(\boldsymbol{p},t)=-\sum_{e=1}^{N_e}\sum_{\bar{l}=1}^{(\hat{p}+1)}\left\{\int_{-1}^{1}\boldsymbol{T}^{\varepsilon*}\left[\boldsymbol{Q}(\hat{\xi}),\boldsymbol{p}\right]\Phi_{\bar{l}}^{e}(\hat{\xi})J^{e}(\hat{\xi})\mathrm{d}\hat{\xi}\right\}\boldsymbol{u}_{\bar{l}}^{e}(\boldsymbol{Q},t)$$
$$+\sum_{e=1}^{N_e}\sum_{\bar{l}=1}^{(\hat{p}+1)}\left\{\int_{-1}^{1}\boldsymbol{U}^{\varepsilon*}\left[\boldsymbol{Q}(\hat{\xi}),\boldsymbol{p}\right]\Phi_{\bar{l}}^{e}(\hat{\xi})J^{e}(\hat{\xi})\mathrm{d}\hat{\xi}\right\}\boldsymbol{f}_{\bar{l}}^{e}(\boldsymbol{Q},t)$$
$$+\boldsymbol{b}^{\varepsilon}(\boldsymbol{p})+\boldsymbol{W}(\boldsymbol{p})\boldsymbol{\sigma}^{\mathrm{m}}(\boldsymbol{A},t)+\boldsymbol{Y}(\boldsymbol{p})\Delta\boldsymbol{T}(\boldsymbol{A},t)+\boldsymbol{Z}(\boldsymbol{p},t)\tag{7.70}$$

其中，$\boldsymbol{\varepsilon}(\boldsymbol{p},t)$ 为 4×1 阶的应变向量；$\boldsymbol{U}^{\varepsilon*}$ 和 $\boldsymbol{T}^{\varepsilon*}$ 为 4×4 阶的基本解矩阵；$\boldsymbol{b}^{\varepsilon}(\boldsymbol{p})$ 为 4×1 阶的体力矩阵；矩阵 $\boldsymbol{W}(\boldsymbol{p})$ 的具体形式在第 6 章已给出；矩阵 $\boldsymbol{Y}(\boldsymbol{p})$ 也可以由式(7.69)获得；向量 $\boldsymbol{Z}(\boldsymbol{p},t)$ 可以写成 $\boldsymbol{Z}(\boldsymbol{p},t)=\{Z_{11},Z_{22},Z_{12},Z_{21}\}^{\mathrm{T}}$ ，其形式为

$$Z_{il}=\frac{1}{2}\sum_{e=1}^{N_e}\sum_{\bar{l}=1}^{(\hat{p}+1)}\left\{\int_{-1}^{1}\varepsilon_{ilj}^{*}\left[\boldsymbol{Q}(\hat{\xi}),\boldsymbol{p}\right]n_k\Phi_{\bar{l}}^{e}(\hat{\xi})J^{e}(\hat{\xi})\mathrm{d}\hat{\xi}+\varepsilon_{ilk}^{*}\left[\boldsymbol{Q}(\hat{\xi}),\boldsymbol{p}\right]n_j\Phi_{\bar{l}}^{e}(\hat{\xi})J^{e}(\hat{\xi})\mathrm{d}\hat{\xi}\right\}\sigma_{jk}^{m}(\boldsymbol{Q},t)$$
$$+\frac{1}{2}\sum_{e=1}^{N_e}\sum_{\bar{l}=1}^{(\hat{p}+1)}\left\{\int_{-1}^{1}\Psi_i\left[\boldsymbol{Q}(\hat{\xi}),\boldsymbol{p}\right]n_l\Phi_{\bar{l}}^{e}(\hat{\xi})J^{e}(\hat{\xi})\mathrm{d}\hat{\xi}+\Psi_l\left[\boldsymbol{Q}(\hat{\xi}),\boldsymbol{p}\right]n_i\Phi_{\bar{l}}^{e}(\hat{\xi})J^{e}(\hat{\xi})\mathrm{d}\hat{\xi}\right\}\Delta T(\boldsymbol{Q},t)\tag{7.71}$$

类似于二维问题，三维热黏弹性力学问题的位移积分方程的离散形式为

$$\boldsymbol{u}(\boldsymbol{p},t)=-\sum_{e=1}^{N_e}\sum_{\bar{l}=1}^{(\hat{p}+1)(\hat{q}+1)}\left\{\int_{-1}^{1}\int_{-1}^{1}\boldsymbol{T}^{*}\left[\boldsymbol{Q}(\hat{\xi},\hat{\eta}),\boldsymbol{p}\right]\Phi_{\bar{l}}^{e}(\hat{\xi},\hat{\eta})J^{e}(\hat{\xi},\hat{\eta})\mathrm{d}\hat{\xi}\mathrm{d}\hat{\eta}\right\}\boldsymbol{u}_{\bar{l}}^{e}(\boldsymbol{Q},t)$$
$$+\sum_{e=1}^{N_e}\sum_{\bar{l}=1}^{(\hat{p}+1)(\hat{q}+1)}\left\{\int_{-1}^{1}\int_{-1}^{1}\boldsymbol{U}^{*}\left[\boldsymbol{Q}(\hat{\xi},\hat{\eta}),\boldsymbol{p}\right]\Phi_{\bar{l}}^{e}(\hat{\xi},\hat{\eta})J^{e}(\hat{\xi},\hat{\eta})\mathrm{d}\hat{\xi}\mathrm{d}\hat{\eta}\right\}\boldsymbol{f}_{\bar{l}}^{e}(\boldsymbol{Q},t)$$
$$+\boldsymbol{b}(\boldsymbol{p})+\boldsymbol{B}(\boldsymbol{p})\boldsymbol{\sigma}^{\mathrm{m}}(\boldsymbol{A},t)+\boldsymbol{D}(\boldsymbol{p})\Delta\boldsymbol{T}(\boldsymbol{A},t)\tag{7.72}$$

其中，

$$J^{e}(\hat{\xi},\hat{\eta})=\frac{\mathrm{d}^2\Gamma}{\mathrm{d}\hat{\xi}\mathrm{d}\hat{\eta}}=\frac{\mathrm{d}^2\Gamma}{\mathrm{d}\xi\mathrm{d}\eta}\frac{\mathrm{d}\xi}{\mathrm{d}\hat{\xi}}\frac{\mathrm{d}\eta}{\mathrm{d}\hat{\eta}}\tag{7.73}$$

三维问题的应变积分方程的离散形式为

$$
\begin{aligned}
\boldsymbol{\varepsilon}(\boldsymbol{p},t) = &-\sum_{e=1}^{N_e}\sum_{\bar{l}=1}^{(\hat{p}+1)(\hat{q}+1)}\left\{\int_{-1}^{1}\int_{-1}^{1}\boldsymbol{T}^{\varepsilon*}\left[\boldsymbol{Q}(\hat{\xi},\hat{\eta}),\boldsymbol{p}\right]\Phi_{\bar{l}}^{e}(\hat{\xi},\hat{\eta})J^{e}(\hat{\xi},\hat{\eta})\mathrm{d}\hat{\xi}\mathrm{d}\hat{\eta}\right\}\boldsymbol{u}_{\bar{l}}^{e}(\boldsymbol{Q},t)\\
&+\sum_{e=1}^{N_e}\sum_{\bar{l}=1}^{(\hat{p}+1)(\hat{q}+1)}\left\{\int_{-1}^{1}\int_{-1}^{1}\boldsymbol{U}^{\varepsilon*}\left[\boldsymbol{Q}(\hat{\xi},\hat{\eta}),\boldsymbol{p}\right]\Phi_{\bar{l}}^{e}(\hat{\xi},\hat{\eta})J^{e}(\hat{\xi},\hat{\eta})\mathrm{d}\hat{\xi}\mathrm{d}\hat{\eta}\right\}\boldsymbol{f}_{\bar{l}}^{e}(\boldsymbol{Q},t)\\
&+\boldsymbol{b}^{\varepsilon}(\boldsymbol{p})+\boldsymbol{W}(\boldsymbol{p})\boldsymbol{\sigma}^{\mathrm{m}}(\boldsymbol{A},t)+\boldsymbol{Y}(\boldsymbol{p})\Delta\boldsymbol{T}(\boldsymbol{A},t)+\boldsymbol{Z}(\boldsymbol{p},t)
\end{aligned}
\tag{7.74}
$$

7.2.6 非相适应界面的处理方法

如图 7.4 所示，在构建多层结构时，通过 CAD 得到的各层 NURBS 网格一般是非相适应的。我们使用虚拟结点插入技术，在两个子域之间的非相适应界面上建立位移(或面力)之间的关系。将控制点分别插入两个 NURBS 片中，以建立 NURBS 片之间的连接。将新的节点矢量 $\hat{\boldsymbol{U}}=\left\{\hat{\xi}_1,\hat{\xi}_2,\cdots,\hat{\xi}_{s_1}\right\}$ 、$\hat{\boldsymbol{V}}=\left\{\hat{\eta}_1,\hat{\eta}_2,\cdots,\hat{\eta}_{s_2}\right\}$ 插入 $\boldsymbol{U}=\left\{\xi_1,\xi_2,\cdots,\xi_{n+\hat{p}+1}\right\}$ 和 $\boldsymbol{V}=\left\{\eta_1,\eta_2,\cdots,\eta_{m+\hat{q}+1}\right\}$ 中，可以得到如下形式的扩展节点矢量：

$$\tilde{\boldsymbol{U}}=\left\{\xi_1,\xi_2,\cdots,\xi_{n+s_1+\hat{p}+1}\right\} \tag{7.75}$$

$$\tilde{\boldsymbol{V}}=\left\{\eta_1,\eta_2,\cdots,\eta_{m+s_2+\hat{q}+1}\right\} \tag{7.76}$$

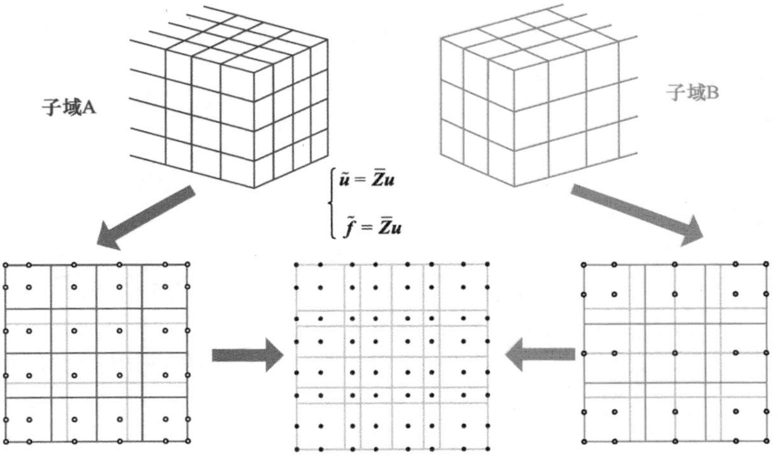

图 7.4 非相适应界面耦合过程

另外，两个方向上新增的控制点 $\tilde{\boldsymbol{Q}}_\xi,\tilde{\boldsymbol{Q}}_\eta$ 如下：

$$\begin{cases}\tilde{\boldsymbol{Q}}_\xi=\boldsymbol{M}_\xi\boldsymbol{Q}_\xi\\ \tilde{\boldsymbol{Q}}_\eta=\boldsymbol{M}_\eta\boldsymbol{Q}_\eta\end{cases} \tag{7.77}$$

其中，$\boldsymbol{Q}_{\xi}$ 和 $\boldsymbol{Q}_{\eta}$ 分别是控制点在两个方向 ξ 和 η 上的参数坐标；矩阵 $\boldsymbol{M}_{\xi}$ 和 $\boldsymbol{M}_{\eta}$ 分别是两个方向 ξ 和 η 上的局部细化算子，可以通过求解 Bezier 提取算子的方法[49]得到。NURBS 片扩展后的控制点可以通过下式获得：

$$\tilde{\boldsymbol{Q}}_{[(n+s_1)\times(m+s_2)]}=\boldsymbol{Z}\boldsymbol{Q}_{(n\times m)} \tag{7.78}$$

其中，

$$\boldsymbol{Z}=(\boldsymbol{L}_{\tilde{Q}})^{-1}\boldsymbol{M}_{\xi\eta}\boldsymbol{L}_{Q} \tag{7.79}$$

式中，

$$\boldsymbol{L}_{\tilde{Q}}=\begin{bmatrix}\omega_{\tilde{Q}_1} & & & \\ & \omega_{\tilde{Q}_2} & & \\ & & \ddots & \\ & & & \omega_{\tilde{Q}_{(n+s_1)(m+s_2)}}\end{bmatrix},\quad \boldsymbol{L}_{Q}=\begin{bmatrix}\omega_{Q_1} & & & \\ & \omega_{Q_2} & & \\ & & \ddots & \\ & & & \omega_{Q_{mn}}\end{bmatrix} \tag{7.80}$$

和

$$\boldsymbol{M}_{\xi\eta}=\boldsymbol{M}_{\xi}\otimes\boldsymbol{M}_{\eta} \tag{7.81}$$

这里，$\omega_{\tilde{Q}_i}$ 和 ω_{Q_j} 分别是控制点 $\tilde{\boldsymbol{Q}}_i$ 和 $\boldsymbol{Q}_j$ 的权值，其中 $i=1,2,\cdots,(n+s_1)(m+s_2)$，$j=1,2,\cdots,mn$。算子 $\otimes$ 在两个矩阵上的实现方法可以在参考文献[50]中找到。类似于式(7.78)，扩展的位移和面力可以由下式获得：

$$\begin{cases}\tilde{\boldsymbol{u}}=\bar{\boldsymbol{Z}}\boldsymbol{u}\\ \tilde{\boldsymbol{f}}=\bar{\boldsymbol{Z}}\boldsymbol{u}\end{cases} \tag{7.82}$$

其中，矩阵 $\bar{\boldsymbol{Z}}$ 是矩阵 $\boldsymbol{Z}$ 的扩充。

7.2.7 基于四叉树的自适应积分算法求解拟奇异积分

当源点与单元非常接近时，需要在单元上放置更多的高斯点。高斯点的个数取决于两个方向 ξ 和 η 的积分阶次，它可由下式获得[50]：

$$\begin{cases}m_{\xi}=\left|\sqrt{\dfrac{2}{3}\lambda+\dfrac{2}{5}}\left[-\dfrac{1}{10}\ln\left(\dfrac{e_{\text{int}}}{2}\right)\right]\left[\left(\dfrac{8L_{\xi}}{3d}\right)^{\frac{3}{4}}+1\right]\right|\\ m_{\eta}=\left|\sqrt{\dfrac{2}{3}\lambda+\dfrac{2}{5}}\left[-\dfrac{1}{10}\ln\left(\dfrac{e_{\text{int}}}{2}\right)\right]\left[\left(\dfrac{8L_{\eta}}{3d}\right)^{\frac{3}{4}}+1\right]\right|\end{cases} \tag{7.83}$$

其中，λ 表示奇异积分的阶次；e_{int} 表示目标积分的误差水平；d 表示源点到积分单元的最短距离；L 表示积分单元的边长，其可以由下式获得[50]：

$$\begin{cases} L_\xi = \int_{\xi_i}^{\xi_{i+1}} \sqrt{\sum_{l=1}^{3}\left[\sum_{z=1}^{(\hat{p}+1)(\hat{q}+1)} \frac{\partial \Phi_z(\xi,\eta_c)}{\partial \xi} Q_l^z\right]^2} \mathrm{d}\xi \\ L_\eta = \int_{\eta_j}^{\eta_{j+1}} \sqrt{\sum_{l=1}^{3}\left[\sum_{z=1}^{(\hat{p}+1)(\hat{q}+1)} \frac{\partial \Phi_z(\xi_c,\eta)}{\partial \eta} Q_l^z\right]^2} \mathrm{d}\eta \end{cases} \tag{7.84}$$

其中，Φ 表示形函数；Q 表示控制点；(ξ_c,η_c)为单元$[\xi_i,\xi_{i+1}]\times[\eta_j,\eta_{j+1}]$的参数中心，它可以由下式求得：

$$\begin{cases} \xi_c = \dfrac{\xi_i + \xi_{i+1}}{2} \\ \eta_c = \dfrac{\eta_j + \eta_{j+1}}{2} \end{cases} \tag{7.85}$$

如图 7.5 所示，采用 Newton-Raphson 法得到从源点到积分单元的最短距离。通过第 k 次迭代，距离可以写成如下形式：

$$\boldsymbol{d}^k = \sum_{z=1}^{(\hat{p}+1)(\hat{q}+1)} \Phi_z^e(\xi^k,\eta^k)\boldsymbol{Q}_z^e - \boldsymbol{x}_P \tag{7.86}$$

其中，$\boldsymbol{x}_P$ 表示源点。

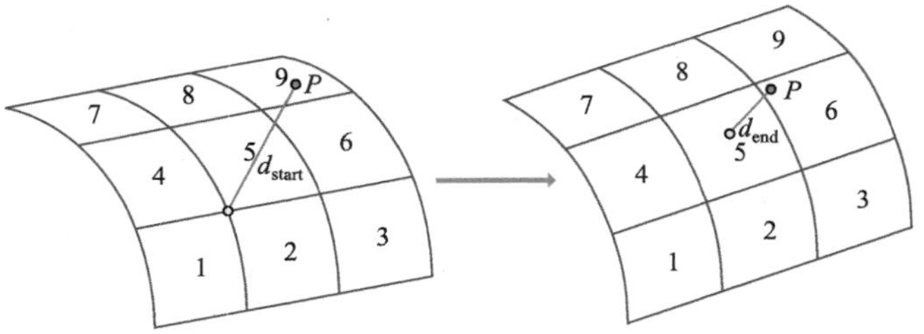

图 7.5 利用迭代法寻找源点 P 到单元 5 的最短距离

第$(k+1)$次迭代的距离向量可以用泰勒级数展开表示为

$$\boldsymbol{d}^{k+1} = \boldsymbol{d}^k + \frac{\partial \boldsymbol{d}^k}{\partial \xi}\Delta\xi + \frac{\partial \boldsymbol{d}^k}{\partial \eta}\Delta\eta \tag{7.87}$$

令 $\boldsymbol{d}^{k+1}=0$，式(7.87)可以写成如下形式：

$$\mathcal{K}_{\text{iter}}^k \begin{Bmatrix} \Delta\xi \\ \Delta\eta \end{Bmatrix} = -\boldsymbol{d}^k \tag{7.88}$$

其中，

$$\mathcal{K}_{\text{iter}}^{k}=\left[\sum_{z=1}^{(\hat{p}+1)(\hat{q}+1)}\frac{\partial \Phi_{z}^{e}(\xi^{k},\eta^{k})}{\partial \xi}\boldsymbol{Q}_{z}^{e},\quad \sum_{z=1}^{(\hat{p}+1)(\hat{q}+1)}\frac{\partial \Phi_{z}^{(e)}\left(\xi^{k},\eta^{k}\right)}{\partial \eta}\boldsymbol{Q}_{z}^{e}\right] \tag{7.89}$$

由于三维 NURBS 曲面只涉及两个独立的参数坐标，因此可以采用最小二乘近似来获得参数增量。迭代表达式如下：

$$(\mathcal{K}_{\text{iter}}^{k})^{\mathrm{T}}\mathcal{K}_{\text{iter}}^{k}\begin{Bmatrix}\Delta\xi\\ \Delta\eta\end{Bmatrix}=-\left(\mathcal{K}_{\text{iter}}^{k}\right)^{\mathrm{T}}\boldsymbol{d}^{k} \tag{7.90}$$

第$(k+1)$次迭代的 ξ^{k+1} 和 η^{k+1} 形式如下：

$$\xi^{k+1}=\xi^{k}+\Delta\xi,\quad \eta^{k+1}=\eta^{k}+\Delta\eta \tag{7.91}$$

经过迭代，可以通过 $d=\left\|\boldsymbol{d}^{k}\right\|$ 得到从源点到单元的最短距离。为了进一步减少积分单元中的高斯点数量，我们采用四叉树层次分解方法驱动自适应积分过程。两个方向 ξ 和 η 的积分阶次可由式(7.83)得到。如果积分阶次大于指定值，则将单元细分为四个相等的正交子单元。重复此过程，直到每个方向上所需的所有积分阶次都小于或等于预定值。当源点与积分单元非常接近时，积分表现出拟奇异性质，自适应积分方法也能有效地处理这类积分。

7.2.8　方程的求解和迭代过程

如图 7.6 所示，层 A 为黏弹性材料，层 B、C 和 D 为弹性材料。不同材料的二维和三维位移边界积分方程可以写成如下矩阵形式：

$$\boldsymbol{H}_{\mathrm{A}}^{\mathrm{b}}\boldsymbol{u}_{\mathrm{A}}^{\mathrm{b}}(t_k)=\boldsymbol{G}_{\mathrm{A}}^{\mathrm{b}}\boldsymbol{f}_{\mathrm{A}}^{\mathrm{b}}(t_k)+\boldsymbol{B}_{\mathrm{A}}^{\mathrm{b}}\boldsymbol{\sigma}_{\mathrm{A}}^{\mathrm{m}}(t_k)+\overline{\boldsymbol{b}}_{\mathrm{A}}^{\mathrm{b}}+\boldsymbol{D}_{\mathrm{A}}^{\mathrm{b}}\Delta\boldsymbol{T}_{\mathrm{A}} \tag{7.92}$$

$$\boldsymbol{H}_{\mathrm{B}}^{\mathrm{b}}\boldsymbol{u}_{\mathrm{B}}^{\mathrm{b}}=\boldsymbol{G}_{\mathrm{B}}^{\mathrm{b}}\boldsymbol{f}_{\mathrm{B}}^{\mathrm{b}}+\overline{\boldsymbol{b}}_{\mathrm{B}}^{\mathrm{b}}+\boldsymbol{D}_{\mathrm{B}}^{\mathrm{b}}\Delta\boldsymbol{T}_{\mathrm{B}} \tag{7.93}$$

$$\boldsymbol{H}_{\mathrm{C}}^{\mathrm{b}}\boldsymbol{u}_{\mathrm{C}}^{\mathrm{b}}=\boldsymbol{G}_{\mathrm{C}}^{\mathrm{b}}\boldsymbol{f}_{\mathrm{C}}^{\mathrm{b}}+\overline{\boldsymbol{b}}_{\mathrm{C}}^{\mathrm{b}}+\boldsymbol{D}_{\mathrm{C}}^{\mathrm{b}}\Delta\boldsymbol{T}_{\mathrm{C}} \tag{7.94}$$

$$\boldsymbol{H}_{\mathrm{D}}^{\mathrm{b}}\boldsymbol{u}_{\mathrm{D}}^{\mathrm{b}}=\boldsymbol{G}_{\mathrm{D}}^{\mathrm{b}}\boldsymbol{f}_{\mathrm{D}}^{\mathrm{b}}+\overline{\boldsymbol{b}}_{\mathrm{D}}^{\mathrm{b}}+\boldsymbol{D}_{\mathrm{D}}^{\mathrm{b}}\Delta\boldsymbol{T}_{\mathrm{D}} \tag{7.95}$$

其中，上标 b 表示“边界”；矩阵 $\boldsymbol{H}$ 、$\boldsymbol{G}$ 、$\boldsymbol{B}$ 、$\boldsymbol{D}$ 和向量 $\overline{\boldsymbol{b}}$ 由边界积分和域积分获得。为了求解式(7.92)～式(7.95)，我们需要对其进行如下形式的分块：

$$[\boldsymbol{H}_{\mathrm{A1}}^{\mathrm{b}}\quad \boldsymbol{H}_{\mathrm{A2}}^{\mathrm{b}}\quad \boldsymbol{H}_{\mathrm{A3}}^{\mathrm{b}}]\begin{Bmatrix}\boldsymbol{u}_{\mathrm{A1}}^{\mathrm{b}}(\text{known})\\ \boldsymbol{u}_{\mathrm{A2}}^{\mathrm{b}}(\text{unknown})\\ \boldsymbol{u}_{\mathrm{A3}}^{\mathrm{b}}(\text{couplingAB})\end{Bmatrix}=[\boldsymbol{G}_{\mathrm{A1}}^{\mathrm{b}}\quad \boldsymbol{G}_{\mathrm{A2}}^{\mathrm{b}}\quad \boldsymbol{G}_{\mathrm{A3}}^{\mathrm{b}}]\begin{Bmatrix}\boldsymbol{f}_{\mathrm{A1}}^{\mathrm{b}}(\text{unknown})\\ \boldsymbol{f}_{\mathrm{A2}}^{\mathrm{b}}(\text{known})\\ \boldsymbol{f}_{\mathrm{A3}}^{\mathrm{b}}(\text{couplingAB})\end{Bmatrix}+\boldsymbol{B}_{\mathrm{A}}^{\mathrm{b}}\boldsymbol{\sigma}_{\mathrm{A}}^{\mathrm{m}}(t_k)+\overline{\boldsymbol{b}}_{\mathrm{A}}^{\mathrm{b}}+\boldsymbol{D}_{\mathrm{A}}^{\mathrm{b}}\Delta\boldsymbol{T}_{\mathrm{A}} \tag{7.96}$$

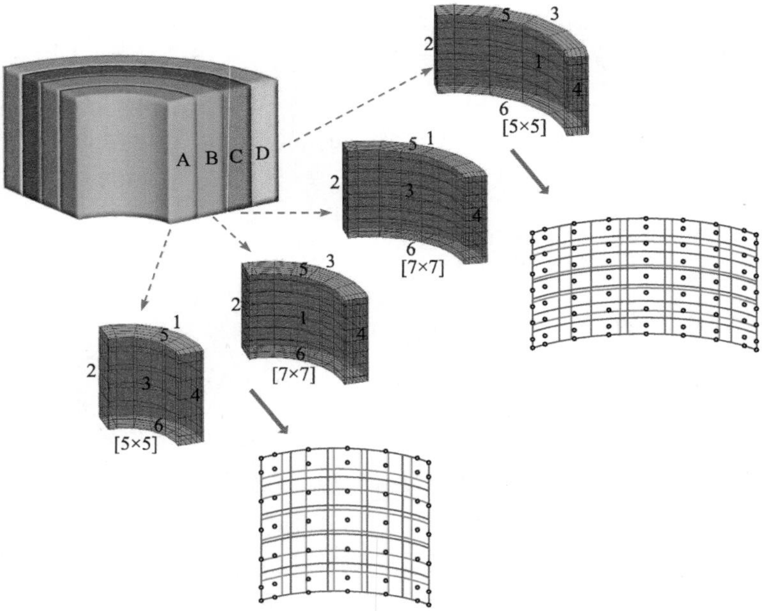

图 7.6　不同层材料模型示意图

$$[\boldsymbol{H}_{\mathrm{B1}}^{\mathrm{b}} \quad \boldsymbol{H}_{\mathrm{B2}}^{\mathrm{b}} \quad \boldsymbol{H}_{\mathrm{B3}}^{\mathrm{b}} \quad \boldsymbol{H}_{\mathrm{B4}}^{\mathrm{b}}]\begin{Bmatrix} \boldsymbol{u}_{\mathrm{B1}}^{\mathrm{b}}(\text{couplingAB}) \\ \boldsymbol{u}_{\mathrm{B2}}^{\mathrm{b}}(\text{known}) \\ \boldsymbol{u}_{\mathrm{B3}}^{\mathrm{b}}(\text{unknown}) \\ \boldsymbol{u}_{\mathrm{B4}}^{\mathrm{b}}(\text{couplingBC}) \end{Bmatrix} = [\boldsymbol{G}_{\mathrm{B1}}^{\mathrm{b}} \quad \boldsymbol{G}_{\mathrm{B2}}^{\mathrm{b}} \quad \boldsymbol{G}_{\mathrm{B3}}^{\mathrm{b}} \quad \boldsymbol{G}_{\mathrm{B4}}^{\mathrm{b}}]\begin{Bmatrix} \boldsymbol{f}_{\mathrm{B1}}^{\mathrm{b}}(\text{couplingAB}) \\ \boldsymbol{f}_{\mathrm{B2}}^{\mathrm{b}}(\text{unknown}) \\ \boldsymbol{f}_{\mathrm{B3}}^{\mathrm{b}}(\text{known}) \\ \boldsymbol{f}_{\mathrm{B4}}^{\mathrm{b}}(\text{couplingBC}) \end{Bmatrix} + \overline{\boldsymbol{b}}_{\mathrm{B}}^{\mathrm{b}} + \boldsymbol{D}_{\mathrm{B}}^{\mathrm{b}}\Delta \boldsymbol{T}_{\mathrm{B}} \tag{7.97}$$

$$[\boldsymbol{H}_{\mathrm{C1}}^{\mathrm{b}} \quad \boldsymbol{H}_{\mathrm{C2}}^{\mathrm{b}} \quad \boldsymbol{H}_{\mathrm{C3}}^{\mathrm{b}} \quad \boldsymbol{H}_{\mathrm{C4}}^{\mathrm{b}}]\begin{Bmatrix} \boldsymbol{u}_{\mathrm{C1}}^{\mathrm{b}}(\text{couplingBC}) \\ \boldsymbol{u}_{\mathrm{C2}}^{\mathrm{b}}(\text{known}) \\ \boldsymbol{u}_{\mathrm{C3}}^{\mathrm{b}}(\text{unknown}) \\ \boldsymbol{u}_{\mathrm{C4}}^{\mathrm{b}}(\text{couplingCD}) \end{Bmatrix} = [\boldsymbol{G}_{\mathrm{C1}}^{\mathrm{b}} \quad \boldsymbol{G}_{\mathrm{C2}}^{\mathrm{b}} \quad \boldsymbol{G}_{\mathrm{C3}}^{\mathrm{b}} \quad \boldsymbol{G}_{\mathrm{C4}}^{\mathrm{b}}]\begin{Bmatrix} \boldsymbol{f}_{\mathrm{C1}}^{\mathrm{b}}(\text{couplingBC}) \\ \boldsymbol{f}_{\mathrm{C2}}^{\mathrm{b}}(\text{unknown}) \\ \boldsymbol{f}_{\mathrm{C3}}^{\mathrm{b}}(\text{known}) \\ \boldsymbol{f}_{\mathrm{C4}}^{\mathrm{b}}(\text{couplingCD}) \end{Bmatrix} + \overline{\boldsymbol{b}}_{\mathrm{C}}^{\mathrm{b}} + \boldsymbol{D}_{\mathrm{C}}^{\mathrm{b}}\Delta \boldsymbol{T}_{\mathrm{C}} \tag{7.98}$$

$$[\boldsymbol{H}_{\mathrm{D1}}^{\mathrm{b}} \quad \boldsymbol{H}_{\mathrm{D2}}^{\mathrm{b}} \quad \boldsymbol{H}_{\mathrm{D3}}^{\mathrm{b}}]\begin{Bmatrix} \boldsymbol{u}_{\mathrm{D1}}^{\mathrm{b}}(\text{couplingCD}) \\ \boldsymbol{u}_{\mathrm{D2}}^{\mathrm{b}}(\text{known}) \\ \boldsymbol{u}_{\mathrm{D3}}^{\mathrm{b}}(\text{unknown}) \end{Bmatrix} = [\boldsymbol{G}_{\mathrm{D1}}^{\mathrm{b}} \quad \boldsymbol{G}_{\mathrm{D2}}^{\mathrm{b}} \quad \boldsymbol{G}_{\mathrm{D3}}^{\mathrm{b}}]\begin{Bmatrix} \boldsymbol{f}_{\mathrm{D1}}^{\mathrm{b}}(\text{couplingCD}) \\ \boldsymbol{f}_{\mathrm{D2}}^{\mathrm{b}}(\text{unknown}) \\ \boldsymbol{f}_{\mathrm{D3}}^{\mathrm{b}}(\text{known}) \end{Bmatrix} + \overline{\boldsymbol{b}}_{\mathrm{D}}^{\mathrm{b}} + \boldsymbol{D}_{\mathrm{D}}^{\mathrm{b}}\Delta \boldsymbol{T}_{\mathrm{D}} \tag{7.99}$$

代入已知边界条件后，式(7.96)～式(7.99)可以联立成如下公式：

$$
\begin{bmatrix}
-\boldsymbol{G}_{\mathrm{A1}}^{\mathrm{b}} & \boldsymbol{H}_{\mathrm{A2}}^{\mathrm{b}} & \boldsymbol{H}_{\mathrm{A3}}^{\mathrm{b}} & \mathbf{0} & \mathbf{0} & \mathbf{0} & \mathbf{0} & \mathbf{0} & \mathbf{0} & \mathbf{0} & \mathbf{0} & -\boldsymbol{G}_{\mathrm{A3}}^{\mathrm{b}} & \mathbf{0} & \mathbf{0} \\
\mathbf{0} & \mathbf{0} & \mathbb{H}_{\mathrm{B1}}^{\mathrm{b}} & -\boldsymbol{G}_{\mathrm{B2}}^{\mathrm{b}} & \boldsymbol{H}_{\mathrm{B3}}^{\mathrm{b}} & \boldsymbol{H}_{\mathrm{B4}}^{\mathrm{b}} & \mathbf{0} & \mathbf{0} & \mathbf{0} & \mathbf{0} & \mathbf{0} & \mathbb{G}_{\mathrm{B1}}^{\mathrm{b}} & -\boldsymbol{G}_{\mathrm{B4}}^{\mathrm{b}} & \mathbf{0} \\
\mathbf{0} & \mathbf{0} & \mathbf{0} & \mathbf{0} & \mathbf{0} & \mathbf{H}_{\mathrm{C1}}^{\mathrm{b}} & -\boldsymbol{G}_{\mathrm{C2}}^{\mathrm{b}} & \boldsymbol{H}_{\mathrm{C3}}^{\mathrm{b}} & \boldsymbol{H}_{\mathrm{C4}}^{\mathrm{b}} & \mathbf{0} & \mathbf{0} & \mathbf{0} & \boldsymbol{G}_{\mathrm{C1}}^{\mathrm{b}} & -\boldsymbol{G}_{\mathrm{C4}}^{\mathrm{b}} \\
\mathbf{0} & \mathbf{0} & \mathbf{0} & \mathbf{0} & \mathbf{0} & \mathbf{0} & \mathbf{0} & \mathbf{0} & \mathbb{H}_{\mathrm{D1}}^{\mathrm{b}} & -\boldsymbol{G}_{\mathrm{D2}}^{\mathrm{b}} & \boldsymbol{H}_{\mathrm{D3}}^{\mathrm{b}} & \mathbf{0} & \mathbf{0} & \mathbb{G}_{\mathrm{D1}}^{\mathrm{b}}
\end{bmatrix}
\begin{Bmatrix}
\boldsymbol{f}_{\mathrm{A1}}^{\mathrm{b}} \\ \boldsymbol{u}_{\mathrm{A2}}^{\mathrm{b}} \\ \boldsymbol{u}_{\mathrm{A3}}^{\mathrm{b}} \\ \boldsymbol{f}_{\mathrm{B2}}^{\mathrm{b}} \\ \boldsymbol{u}_{\mathrm{B3}}^{\mathrm{b}} \\ \boldsymbol{u}_{\mathrm{B4}}^{\mathrm{b}} \\ \boldsymbol{f}_{\mathrm{C2}}^{\mathrm{b}} \\ \boldsymbol{u}_{\mathrm{C3}}^{\mathrm{b}} \\ \boldsymbol{u}_{\mathrm{C4}}^{\mathrm{b}} \\ \boldsymbol{f}_{\mathrm{D2}}^{\mathrm{b}} \\ \boldsymbol{u}_{\mathrm{D3}}^{\mathrm{b}} \\ \boldsymbol{f}_{\mathrm{A3}}^{\mathrm{b}} \\ \boldsymbol{f}_{\mathrm{B4}}^{\mathrm{b}} \\ \boldsymbol{f}_{\mathrm{C4}}^{\mathrm{b}}
\end{Bmatrix}
$$

$$
=\begin{bmatrix}
-\boldsymbol{H}_{\mathrm{A1}}^{\mathrm{b}} & \boldsymbol{G}_{\mathrm{A2}}^{\mathrm{b}} & \mathbf{0} & \mathbf{0} & \mathbf{0} & \mathbf{0} & \mathbf{0} & \mathbf{0} \\
\mathbf{0} & \mathbf{0} & -\boldsymbol{H}_{\mathrm{B2}}^{\mathrm{b}} & \boldsymbol{G}_{\mathrm{B3}}^{\mathrm{b}} & \mathbf{0} & \mathbf{0} & \mathbf{0} & \mathbf{0} \\
\mathbf{0} & \mathbf{0} & \mathbf{0} & \mathbf{0} & -\boldsymbol{H}_{\mathrm{C2}}^{\mathrm{b}} & \boldsymbol{G}_{\mathrm{C3}}^{\mathrm{b}} & \mathbf{0} & \mathbf{0} \\
\mathbf{0} & \mathbf{0} & \mathbf{0} & \mathbf{0} & \mathbf{0} & \mathbf{0} & -\boldsymbol{H}_{\mathrm{D2}}^{\mathrm{b}} & \boldsymbol{G}_{\mathrm{D3}}^{\mathrm{b}}
\end{bmatrix}
\begin{Bmatrix}
\boldsymbol{u}_{\mathrm{A1}}^{\mathrm{b}} \\ \boldsymbol{f}_{\mathrm{A2}}^{\mathrm{b}} \\ \boldsymbol{u}_{\mathrm{B2}}^{\mathrm{b}} \\ \boldsymbol{f}_{\mathrm{B3}}^{\mathrm{b}} \\ \boldsymbol{u}_{\mathrm{C2}}^{\mathrm{b}} \\ \boldsymbol{f}_{\mathrm{C3}}^{\mathrm{b}} \\ \boldsymbol{u}_{\mathrm{D2}}^{\mathrm{b}} \\ \boldsymbol{f}_{\mathrm{D3}}^{\mathrm{b}}
\end{Bmatrix}
+\begin{Bmatrix}
\boldsymbol{B}_{\mathrm{A}}^{\mathrm{b}}\boldsymbol{\sigma}_{\mathrm{A}}^{\mathrm{m}} \\ \mathbf{0} \\ \mathbf{0} \\ \mathbf{0}
\end{Bmatrix}
+\begin{Bmatrix}
\bar{\boldsymbol{b}}_{\mathrm{A}}^{\mathrm{b}} \\ \bar{\boldsymbol{b}}_{\mathrm{B}}^{\mathrm{b}} \\ \bar{\boldsymbol{b}}_{\mathrm{C}}^{\mathrm{b}} \\ \bar{\boldsymbol{b}}_{\mathrm{D}}^{\mathrm{b}}
\end{Bmatrix}
+\begin{Bmatrix}
\boldsymbol{D}_{\mathrm{A}}^{\mathrm{b}}\Delta\boldsymbol{T}_{\mathrm{A}} \\ \boldsymbol{D}_{\mathrm{B}}^{\mathrm{b}}\Delta\boldsymbol{T}_{\mathrm{B}} \\ \boldsymbol{D}_{\mathrm{C}}^{\mathrm{b}}\Delta\boldsymbol{T}_{\mathrm{C}} \\ \boldsymbol{D}_{\mathrm{D}}^{\mathrm{b}}\Delta\boldsymbol{T}_{\mathrm{D}}
\end{Bmatrix}
\tag{7.100}
$$

其中，

$$
\begin{cases}
\mathbb{H}_{\mathrm{B1}}^{\mathrm{b}} = \bar{\mathbb{Z}}_{\mathrm{AB}}\boldsymbol{H}_{\mathrm{B1}}^{\mathrm{b}} \\
\mathbb{G}_{\mathrm{B1}}^{\mathrm{b}} = \bar{\mathbb{Z}}_{\mathrm{AB}}\boldsymbol{G}_{\mathrm{B1}}^{\mathrm{b}} \\
\mathbb{H}_{\mathrm{D1}}^{\mathrm{b}} = \bar{\mathbb{Z}}_{\mathrm{CD}}\boldsymbol{H}_{\mathrm{D1}}^{\mathrm{b}} \\
\mathbb{G}_{\mathrm{D1}}^{\mathrm{b}} = \bar{\mathbb{Z}}_{\mathrm{CD}}\boldsymbol{G}_{\mathrm{D1}}^{\mathrm{b}}
\end{cases}
\tag{7.101}
$$

式中，

$$
\begin{cases}
\bar{\mathbb{Z}}_{\mathrm{AB}} = (\bar{\boldsymbol{Z}}_{\mathrm{AB}})^{+}\,\bar{\boldsymbol{Z}}_{\mathrm{AB}} \\
\bar{\mathbb{Z}}_{\mathrm{CD}} = (\bar{\boldsymbol{Z}}_{\mathrm{CD}})^{+}\,\bar{\boldsymbol{Z}}_{\mathrm{CD}}
\end{cases}
\tag{7.102}
$$

这里上标“+”表示 Moore-Penrose 伪逆。此外，应变积分方程可以写成如下的矩阵形式：

$$\boldsymbol{\varepsilon}_{\mathrm{A}}^{\mathrm{I}}(t_k) = -\boldsymbol{H}_{\mathrm{A}}^{\varepsilon}\boldsymbol{u}_{\mathrm{b}}(t_k) + \boldsymbol{G}_{\mathrm{A}}^{\varepsilon}\boldsymbol{f}_{\mathrm{b}}(t_k) + \hat{\boldsymbol{b}}_{\mathrm{A}}^{\varepsilon} + \boldsymbol{W}\boldsymbol{\sigma}^{\mathrm{m}}(t_k) + \boldsymbol{Y}\Delta\boldsymbol{T}_{\mathrm{A}}(t_k) + \boldsymbol{Z}(t_k) \tag{7.103}$$

当$t = 0$时，没有记忆应力。具体的迭代求解过程可参考图 7.7。每一步迭代需要满足如下收敛条件：

$$\left\|\frac{\boldsymbol{u}_{i+1}(t_k) - \boldsymbol{u}_i(t_k)}{\boldsymbol{u}_{i+1}(t_k)}\right\| < \mathrm{eps} \tag{7.104}$$

其中，eps 为给定的精度阈值。

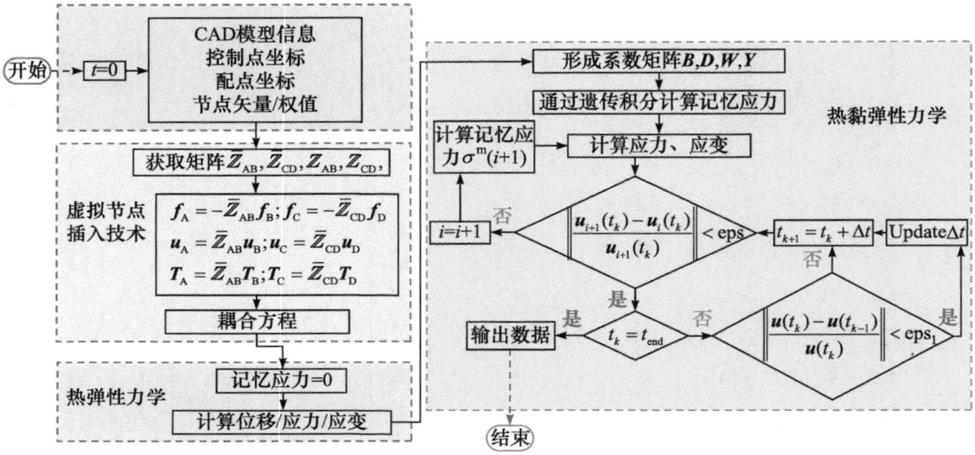

图 7.7 用等几何边界元法求解热黏弹性力学问题的流程图

内点位移可由下式求得：

$$\begin{cases} \boldsymbol{u}_{\mathrm{A}}^{\mathrm{I}}(t_k) = -\boldsymbol{H}_{\mathrm{A}}^{\mathrm{I}}\boldsymbol{u}_{\mathrm{A}}^{\mathrm{I}}(t_k) + \boldsymbol{G}_{\mathrm{A}}^{\mathrm{I}}\boldsymbol{f}_{\mathrm{A}}^{\mathrm{I}}(t_k) + \bar{\boldsymbol{b}}_{\mathrm{A}}^{\mathrm{I}} + \boldsymbol{B}_{\mathrm{A}}^{\mathrm{I}}\boldsymbol{\sigma}_{\mathrm{A}}^{\mathrm{m}}(t_k) + \boldsymbol{D}_{\mathrm{A}}^{\mathrm{I}}\Delta\boldsymbol{T}_{\mathrm{A}} \\ \boldsymbol{u}_{\mathrm{B}}^{\mathrm{I}} = -\boldsymbol{H}_{\mathrm{B}}^{\mathrm{I}}\boldsymbol{u}_{\mathrm{B}}^{\mathrm{I}} + \boldsymbol{G}_{\mathrm{B}}^{\mathrm{I}}\boldsymbol{f}_{\mathrm{B}}^{\mathrm{I}} + \bar{\boldsymbol{b}}_{\mathrm{B}}^{\mathrm{I}} + \boldsymbol{D}_{\mathrm{B}}^{\mathrm{I}}\Delta\boldsymbol{T}_{\mathrm{B}} \\ \boldsymbol{u}_{\mathrm{C}}^{\mathrm{I}} = -\boldsymbol{H}_{\mathrm{C}}^{\mathrm{I}}\boldsymbol{u}_{\mathrm{C}}^{\mathrm{I}} + \boldsymbol{G}_{\mathrm{C}}^{\mathrm{I}}\boldsymbol{f}_{\mathrm{C}}^{\mathrm{I}} + \bar{\boldsymbol{b}}_{\mathrm{C}}^{\mathrm{I}} + \boldsymbol{D}_{\mathrm{C}}^{\mathrm{I}}\Delta\boldsymbol{T}_{\mathrm{C}} \\ \boldsymbol{u}_{\mathrm{D}}^{\mathrm{I}} = -\boldsymbol{H}_{\mathrm{D}}^{\mathrm{I}}\boldsymbol{u}_{\mathrm{D}}^{\mathrm{I}} + \boldsymbol{G}_{\mathrm{D}}^{\mathrm{I}}\boldsymbol{f}_{\mathrm{D}}^{\mathrm{I}} + \bar{\boldsymbol{b}}_{\mathrm{D}}^{\mathrm{I}} + \boldsymbol{D}_{\mathrm{D}}^{\mathrm{I}}\Delta\boldsymbol{T}_{\mathrm{D}} \end{cases} \tag{7.105}$$

考虑到等几何边界元法的计算效率，我们使用一种自适应时间步长选择方案。当位移变化较大时，采用较小的时间步长。当位移变化较慢时，采用较大的时间步长也可以得到满意的结果。首先给出较短的初始时间步长，然后根据各时刻位移的变化来改变时间步长，这样既保证了计算精度，又提高了计算效率。

7.3 数值算例

在本节中，我们使用等几何边界元法进行数值研究。通过 5 个二维和三维数值算例，以表明非相适应界面等几何边界元法解决热弹性-黏弹性问题的正确性。数值算例的参考温度为 0℃。

所有算例的黏弹性材料参数相同，其中剪切模量用 Prony 级数近似，体积模量为常数，其形式为

$$\begin{cases} G(t) = G_0 + G_1 \exp\left(-\dfrac{1}{\gamma}t\right) \quad (\mathrm{Pa}) \\ K(t) = K(0) = 20000\mathrm{Pa} \end{cases} \tag{7.106}$$

其中，$G_0 = 100\mathrm{Pa}$、$G_1 = 9900\mathrm{Pa}$ 和 $\gamma = 0.4170\mathrm{s}$。

7.3.1 二维矩形板的热黏弹性力学问题

如图 7.8(a)所示，考虑长 10m、宽 6m 的黏弹性矩形板。右边界施加 $f = 10\,\mathrm{N/m^2}$ 的均匀面力。本算例选取 24 个内点，其分布见图 7.8 (b)。图 7.9 给出了 800 个 4 节点 FEM 网格。

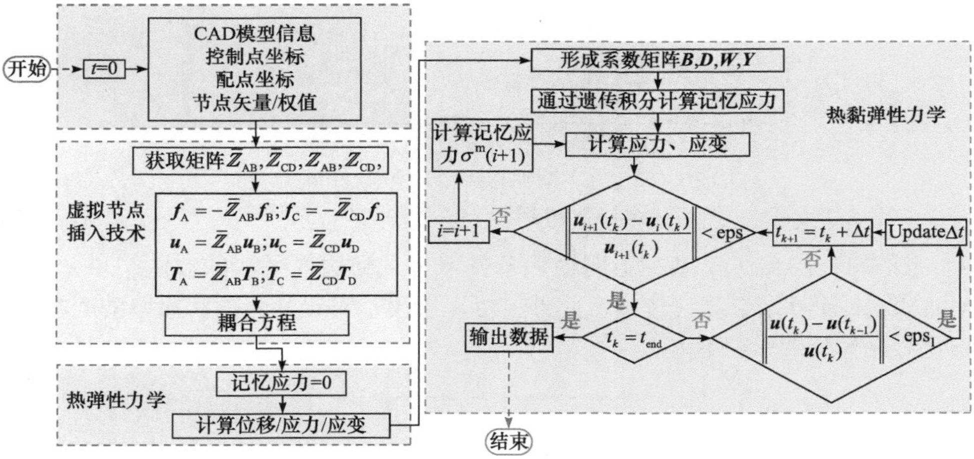

图 7.8 矩形板

(a) 模型的几何尺寸和边界条件；(b) 内点的分布

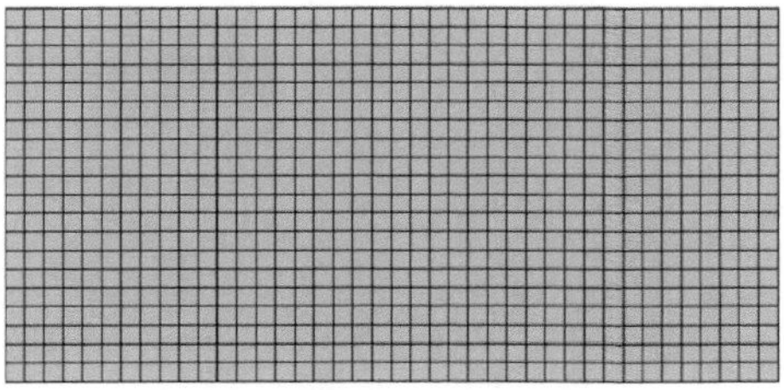

图 7.9 FEM 网格

模型的温度为 100℃，即ΔT=100℃。为了验证等几何边界元法在求解热黏弹性问题上的有效性，本章通过该算例讨论 p 细化、h 细化和内点数对计算结果的影响。

1. h 细化对计算结果的影响

使用本算例讨论 h 细化对计算结果的影响。如图 7.10 所示，我们分别对模型进行了 1 次细化、3 次细化、5 次细化和 11 次细化。NURBS 基函数的阶次为 2，几何的初始节点矢量为

$$\Xi=\left\{0,0,0,\frac{1}{6},\frac{1}{6},\frac{1}{3},\frac{1}{3},\frac{1}{2},\frac{1}{2},\frac{2}{3},\frac{2}{3},\frac{5}{6},\frac{5}{6},1,1,1\right\} \tag{7.107}$$

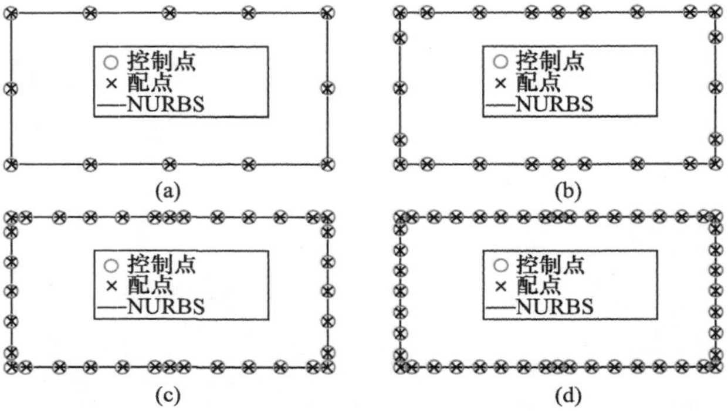

图 7.10 模型在不同细化条件下控制点和配点分布

(a) 1 次细化；(b) 3 次细化；(c) 5 次细化；(d) 7 次细化

图 7.11 给出了模型在不同细化条件下等几何边界元法的数值结果，其与 ANSYS 的计算结果吻合较好。此外，可以看出，对于几何形状简单的问题，h 细化对计算结果影响不大。

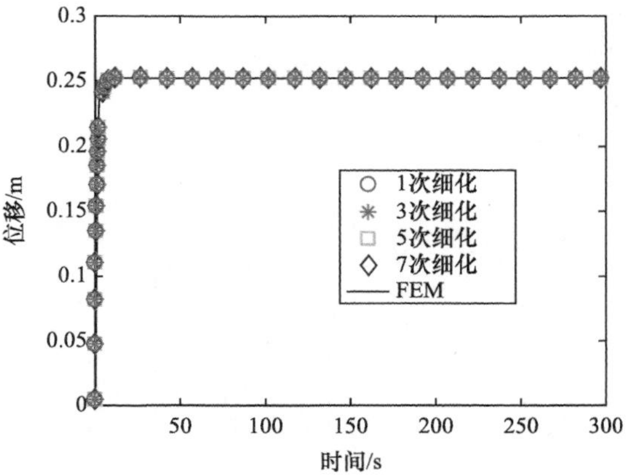

图 7.11 不同细化下 A 点 x_1 方向位移随时间的变化

2. p 细化对计算结果的影响

使用矩形板模型讨论基函数升阶对计算结果的影响。图 7.10(a)和图 7.12 分别给出了不同阶次 NURBS 基函数(分别为 $\hat{p}=2$ 、 $\hat{p}=3$ 和 $\hat{p}=4$)对模型的描述，即控制点和配点的分布。

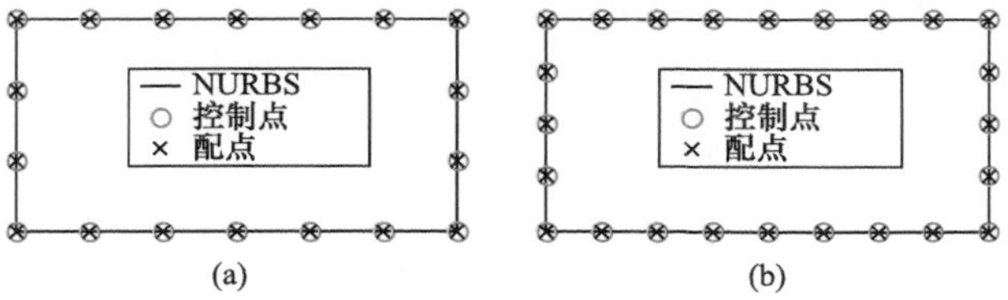

图 7.12　NURBS 不同阶次条件下模型控制点及配点分布

(a) $\hat{p}=3$ ；(b) $\hat{p}=4$

图 7.13 给出了不同阶次 NURBS 基函数的等几何边界元法计算的 A 点 x_1 方向位移随时间的变化，与 ANSYS 软件的计算结果吻合较好。可以看出，基函数的阶次对计算结果影响不大。为了保证计算效率，下面的例子中我们将使用二阶 NURBS 基函数的等几何边界元法进行分析。

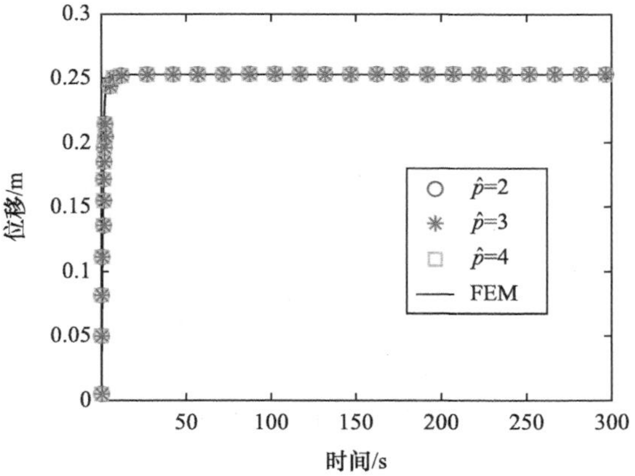

图 7.13　不同 NURBS 阶次下 A 点 x_1 方向位移随时间的变化

3. 内点数的影响

使用二维矩形板讨论内点数对计算结果的影响。如图 7.14 所示，在相同的控制点下，使用三种不同数目（8、16 和 24）的内点进行分析。

从图 7.15 中可以看出，对于模型几何形状简单的问题，内点数对计算结果影响不大。即使只使用 8 个内点，仍然可以得到相对准确的结果。

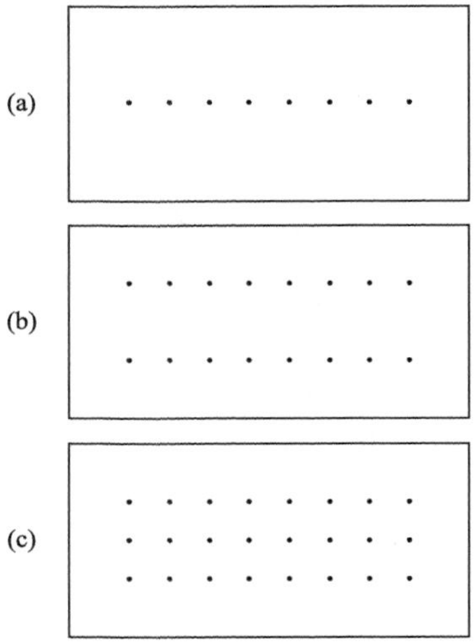

图 7.14　三种不同数量的内点分布

(a) 8；(b) 16；(c) 24

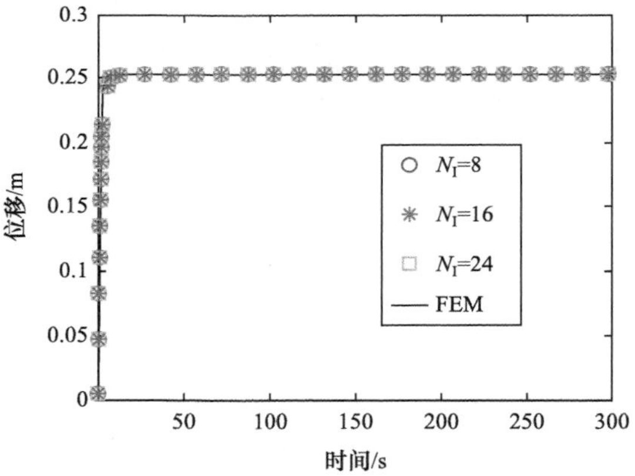

图 7.15　不同内点数下 A 点 x_1 方向位移随时间的变化

7.3.2　二维哑铃板的热黏弹性力学问题

在本例中，我们研究一个二维单轴拉伸的哑铃板模型。该模型为对称模型，其可简化为等效 1/4 计算模型，如图 7.16 所示。1/4 哑铃试件尺寸如图 7.17 所示。几何图形的节点矢量如下所示：

$$\Xi=\left\{0,0,0,\frac{1}{10},\frac{1}{10},\frac{1}{5},\frac{1}{5},\frac{3}{10},\frac{3}{10},\frac{2}{5},\frac{2}{5},\frac{1}{2},\frac{1}{2},\frac{3}{5},\frac{3}{5},\frac{7}{10},\frac{7}{10},\frac{4}{5},\frac{4}{5},\frac{9}{10},\frac{9}{10},1,1,1\right\} \tag{7.108}$$

通过本算例讨论温度载荷的影响。为了验证等几何边界元法的正确性，将计算结果与 ANSYS 软件进行比较。图 7.18 给出了有限元(FEM)网格和等几何边界元(IGABEM)网格以及内点分布。为了描述该模型，IGABEM 采用 9 个 NURBS 单元，FEM 采用 1111 个 4 节点四边形单元。

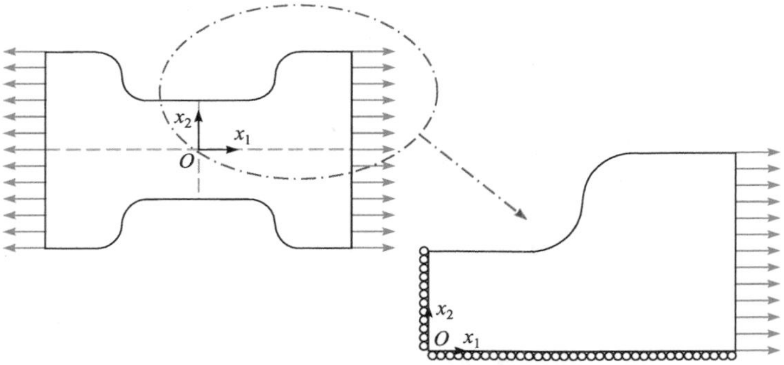

图 7.16　1/4 哑铃试件等效模型及边界条件

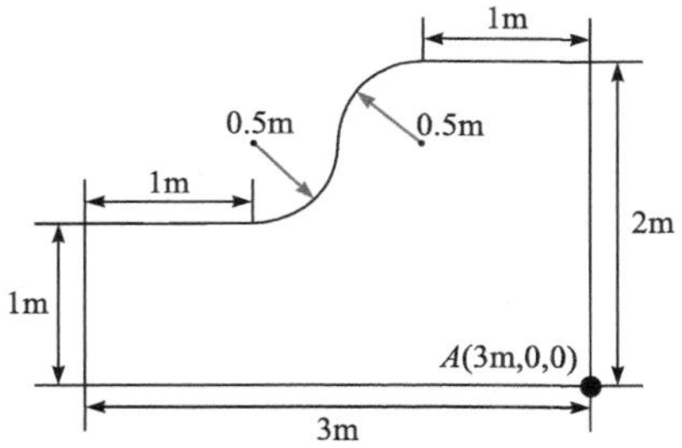

图 7.17　1/4 哑铃试件尺寸

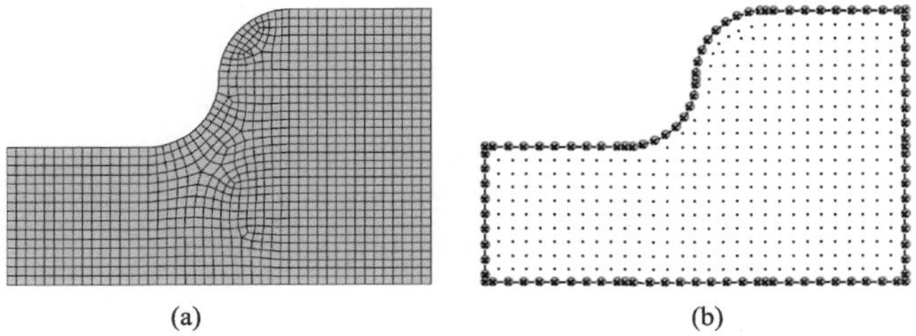

图 7.18　FEM 网格及 IGABEM 控制点、配单和内点

(a) FEM 网格；(b) IGABEM 网格及内点

图 7.19 给出了不同温度载荷下(即ΔT=0℃，ΔT=20℃ 和 ΔT=40℃) IGABEM 和 ANSYS 软件计算的 A 点 x_1 方向位移随时间的变化，其中，Δu_1 和Δu_2 分别表示 IGABEM 在 20℃和 40℃下计算的 A 点在 x_1 方向上相对于 0℃的位移差。结果表明，温度对计算结果有较大影响。此外，温度变化并没有改变黏弹性材料的短期和长期行为，而是缩短或延长了材料的松弛时间。在不同温度下，最终时刻的位移是不同的。这是由于温度变化引起材料的膨胀，而且温差越大，膨胀现象越明

显。图 7.20 给出了 $t=10\mathrm{s}$ 时，不同温度载荷下，IGABEM 得到的未变形和变形轮廓。

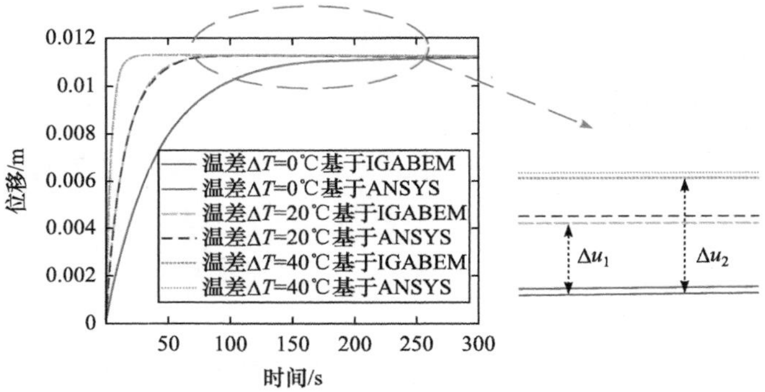

图 7.19 不同温差下 A 点 x_1 方向位移随时间的变化

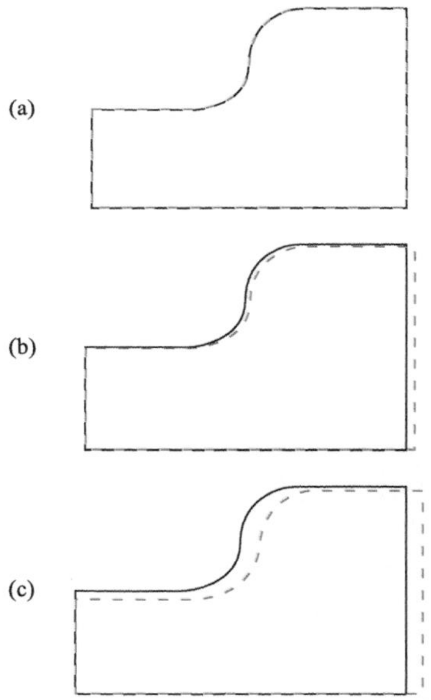

图 7.20 $t=10\mathrm{s}$ 时，不同温度载荷下，IGABEM 得到的未变形和变形轮廓

(a) ΔT=0℃；(b) ΔT=20℃；(c) ΔT=40℃

7.3.3 二维多层圆筒的热黏弹性力学问题

在本例中，我们利用双层厚壁圆筒的二维热黏弹性模型来讨论时间步长和外弹性圆柱的影响。由于模型的几何形状和边界条件是对称的，因此我们可以将其简化为等效的 1/4 双层厚壁圆筒，如图 7.21 所示。

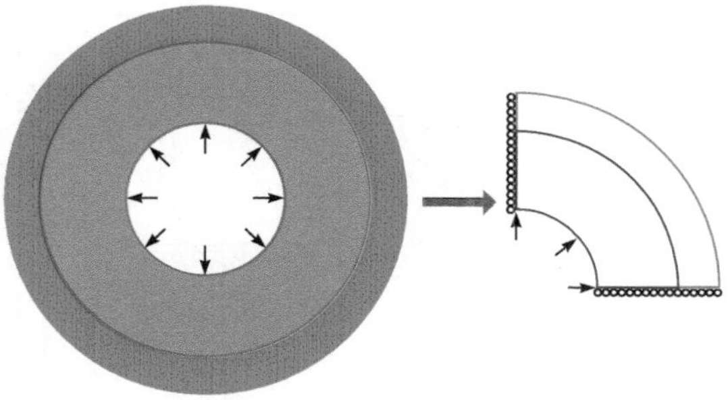

图 7.21　1/4 双层厚壁圆筒的等效模型

1. 时间步长对计算结果的影响

我们利用单层黏弹性厚壁圆筒讨论时间步长对计算结果的影响。如图 7.22 所示，选取 81 个内点。内圆半径为 5m，外圆半径为 10m，内压为 10Pa。模型的初始节点矢量为

$$\boldsymbol{\Xi}=\left\{0,0,0,\frac{1}{4},\frac{1}{4},\frac{1}{2},\frac{1}{2},\frac{3}{4},\frac{3}{4},1,1,1\right\} \tag{7.109}$$

为了清楚地说明问题，我们选取了 5 种不同的时间步长，分别是 0.03、0.3、1、3 和自适应时间步长。本算例在极坐标形式下的位移解析解可以在参考文献[51]中找到。

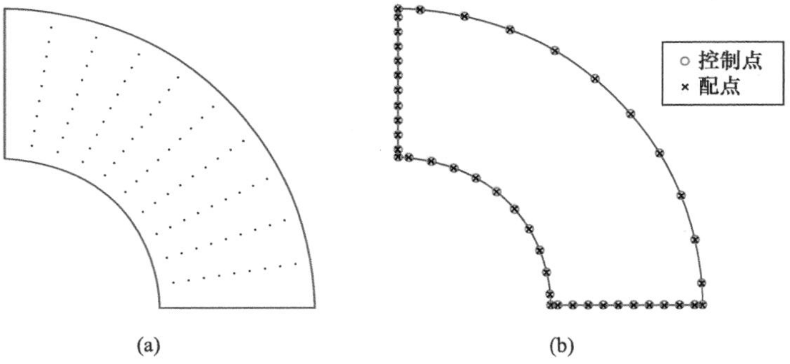

图 7.22　1/4 厚壁圆筒模型
(a) 81 个内点；(b) 控制点与配点分布

从图 7.23 可以看出，时间步长对计算结果影响较大，最大绝对误差(ABSERR)随着时间步长增大而增大(表 7.1)，其中 $\text{ABSERR}=\left|u_{\text{numerical},i}(\boldsymbol{x},t)-u_{\text{exact},i}(\boldsymbol{x},t)\right|$。值得注意的是，如果时间步长过小，则需要更多的迭代步数，这会降低计算效率。我们选择如图 7.23 所示的自适应时间步长方案来求解黏弹性问题的蠕变行为，该方案是平衡计算效率和精度的最佳方案。

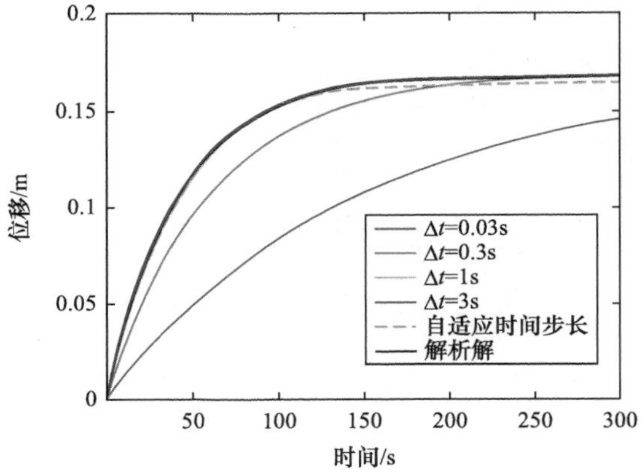

图 7.23　不同时间步长下 A 点 x_1 方向位移随时间的变化

表 7.1　不同时间步长下所用的时间步数及结果的最大绝对误差

	$\Delta t=0.03$s	$\Delta t=0.3$s	$\Delta t=1$s	$\Delta t=3$s	自适应
步数	10000	1000	300	100	398
最大绝对误差	0.0018	0.0021	0.028	0.078	0.0022

2. 外壁厚度对计算结果的影响

在这里，我们讨论外层弹性材料壳体对内层黏弹性材料的作用。外部弹性壳体的杨氏模量为 28000Pa，泊松比为 0.4。图 7.24 给出了三个不同厚度的外部弹性壳体(2.5m、5m 和 7.5m)。为了验证等几何边界元法的正确性，将计算结果与 ANSYS 软件进行对比。模型的 FEM 网格如图 7.25 所示。

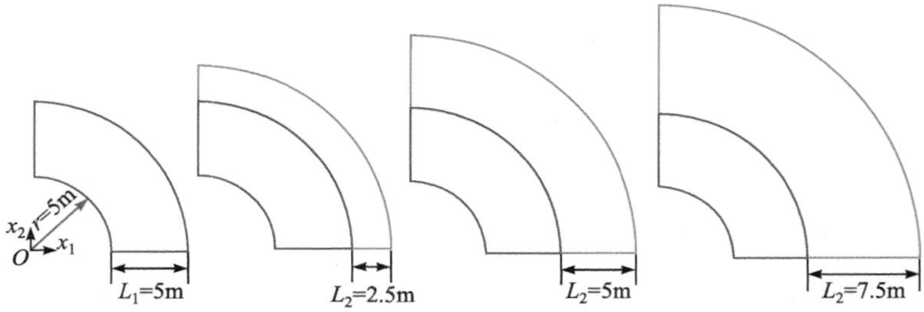

图 7.24　双层厚壁圆筒的尺寸

从图 7.26 可以看出，在分析多材料双层厚壁圆筒问题时，IGABEM 的数值结果与 ANSYS 软件的计算结果吻合较好。此外，外部弹性壳体对整个结构的变形有很大的影响。如图 7.27 所示，当整个结构中存在外部弹性壳体时，黏弹性材料部分的变形比单层黏弹性材料变形减小了一个数量级。

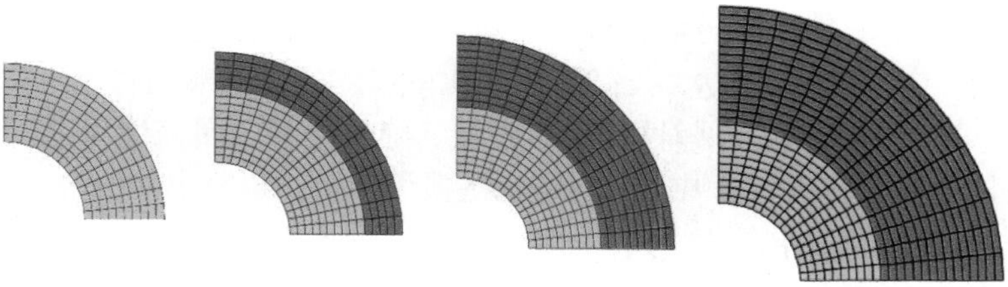

图 7.25　模型的 FEM 网格

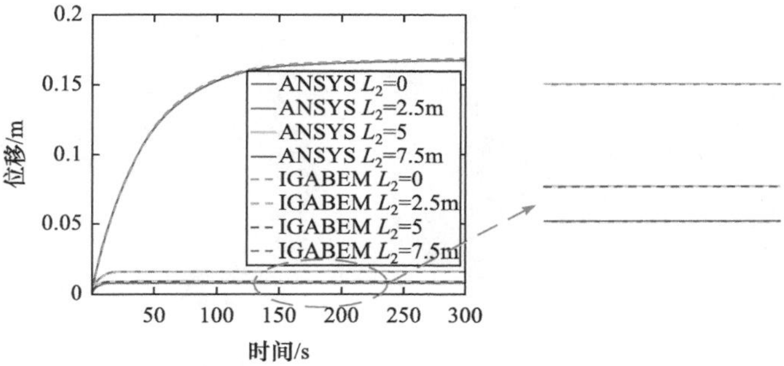

图 7.26　不同外壁厚度条件下 A 点 x_1 方向位移随时间的变化

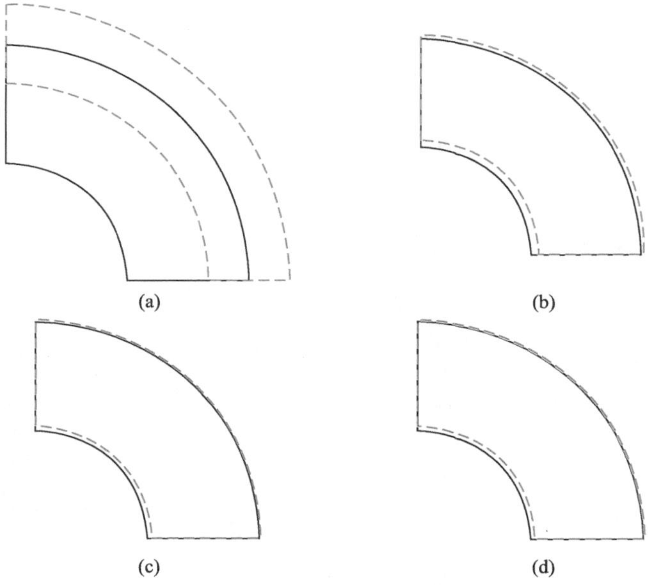

图 7.27　在 $t = 300\text{s}$ 时刻，IGABEM 得到不同外壁厚度下的未变形和变形轮廓

(a) $L_2 = 0$；(b) $L_2 = 2.5\text{m}$；(c) $L_2 = 5\text{m}$；(d) $L_2 = 7.5\text{m}$

7.3.4 三维含圆柱形药柱的 SRM 燃烧室的热黏弹性力学问题

SRM 燃烧室由黏弹性药柱、衬层、绝热层和金属壳体组成，是典型的多层、多材料结构。如图 7.28 所示，我们采用简化模型对复杂加载条件下 SRM 燃烧室进行研究。推进剂药柱为黏弹性材料，衬层、绝热层和金属壳体为弹性材料。弹性材料的材料参数在表 7.2 中给出。含圆柱形药柱的 SRM 燃烧室的 1/4 等效模型尺寸在图 7.29 中给出。模型的上下边界在 x_3 方向固定，内压为 $f = 1\text{N}/\text{m}^2$，温度载荷为ΔT=20℃。

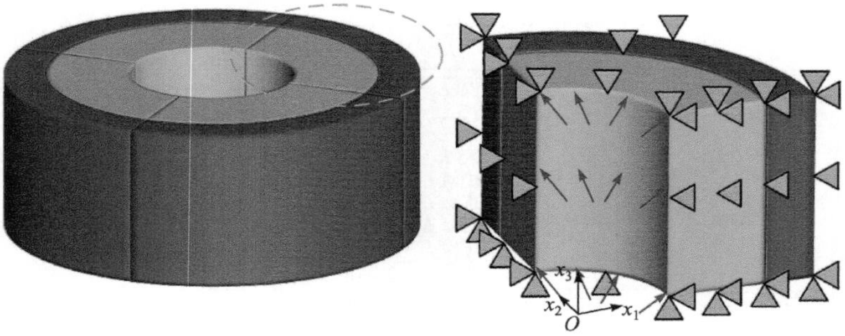

图 7.28 含圆柱形药柱的 SRM 燃烧室的 1/4 等效模型及边界条件

表 7.2 外层结构的材料参数

材料参数	衬层	隔热层	壳体
热膨胀系数/(1/K)	3×10^{-5}	3×10^{-5}	1×10^{-5}
杨氏模量/ Pa	2.57×10^{4}	2.59×10^{4}	1×10^{5}
泊松比	2/7	2/7	0.3

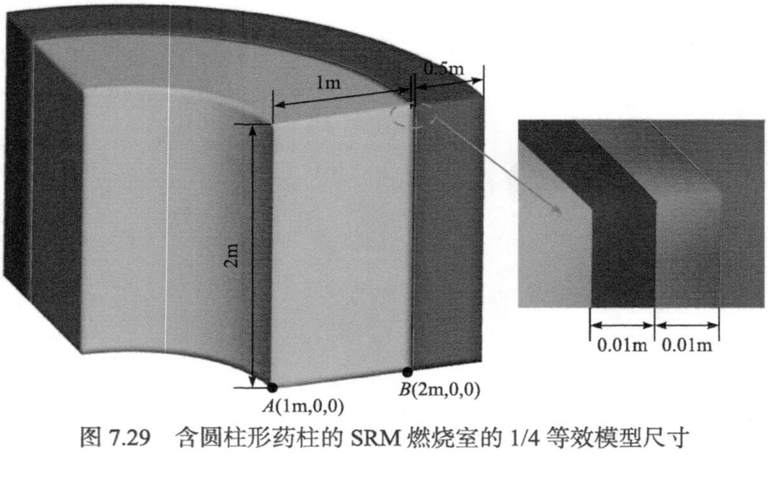

图 7.29 含圆柱形药柱的 SRM 燃烧室的 1/4 等效模型尺寸

图 7.30 给出了 SRM 燃烧室的不同细化网格，其中蓝色、紫色、绿色、红色网格分别代表药柱(A 子域)、衬层(B 子域)、绝热层(C 子域)和金属外壳(D 子域)。A～D 子域的网格分别为 $8\times8\times8$、$10\times1\times10$、$10\times1\times10$ 和 $9\times5\times9$。每个子域由 6 个 NURBS 片构成，每个子域中的 NURBS 编号如图 7.30 所示。值得注意的是，在该多层模型中存在两个非相适应界面，即界面 AB 和界面 CD，这两个界面的虚拟网格如图 7.30 所示。此外，子域 A 内点分布在如图 7.31 所示。界面 BC 为相适应界面。

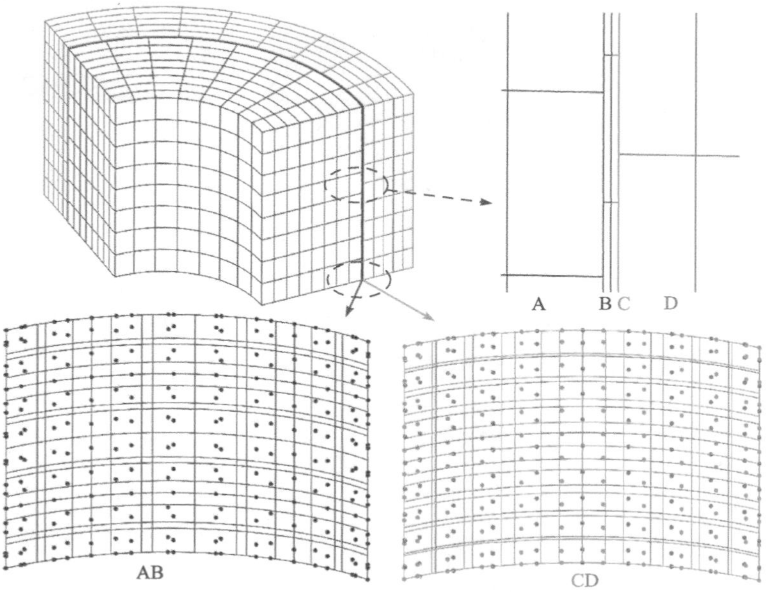

图 7.30　含圆柱形药柱的 SRM 燃烧室非相适应节点 NURBS 网格划分
其中蓝色、紫色、绿色和红色网格分别代表药柱、衬层、绝热层和金属壳体网格

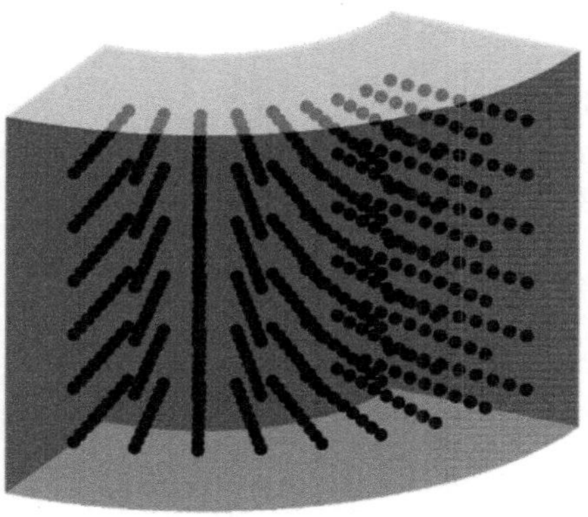

图 7.31　药柱内 486 个内点的分布

我们将等几何边界元法的计算结果与 ANSYS 软件进行对比。如图 7.32 所示，采用 8 节点线性六面体 FEM 单元对模型进行离散，子域 A～D 的单元数分别为 32000、40000、40000 和 8000。

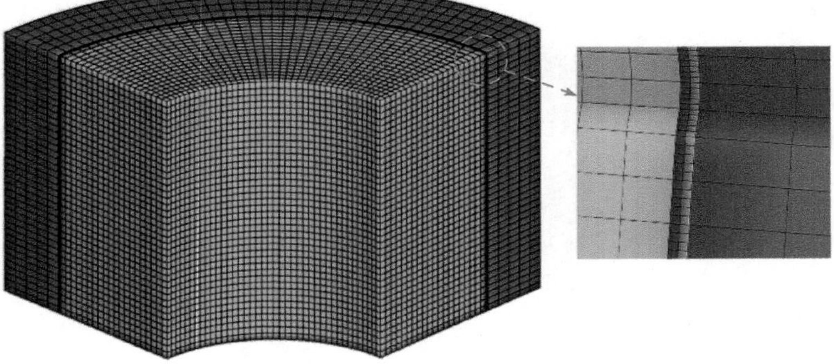

图 7.32　1/4 SRM 燃烧室 FEM 网格

图 7.33 给出了无温度载荷和有温度载荷下点 A 和点 B 随时间变化的径向位移。与 ANSYS 软件的计算结果进行比较，发现等几何边界元法可以获得理想的计算结果，且所需自由度远低于传统有限元法。

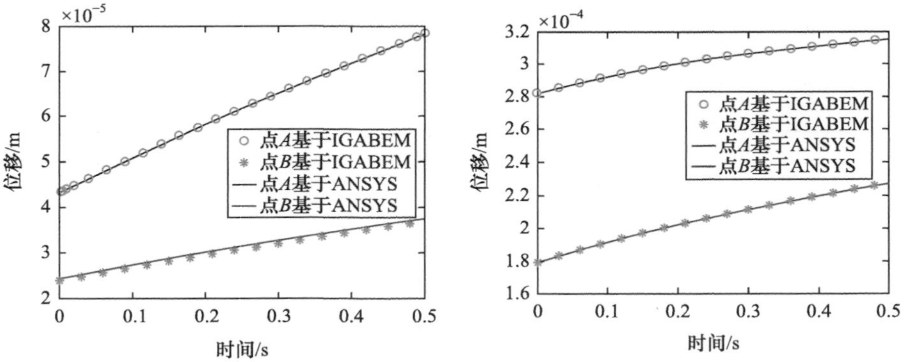

(a)　　(b)

图 7.33　点 A(1m,0,0)和点 B(2m,0,0)随时间变化的位移

(a) 无温度载荷；(b) 有温度载荷

图 7.34 为均匀内压作用下模型在不同时刻径向的变形云图，虚线为原始轮廓线，实线为比例放大因子为 1000 时的变形轮廓线。由此不难看出，温度载荷对 SRM 燃烧室的变形有较大的影响。当药柱结构变形时，药柱的燃烧表面积会发生变化，可能导致燃烧行为异常。

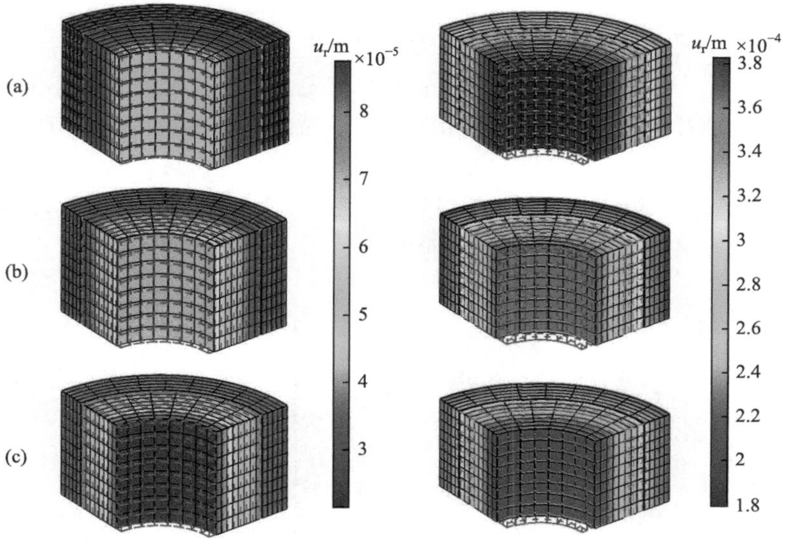

图 7.34　模型在 3 个不同时刻的几何变形图和位移云图

(a) t=0.01s；(b) t=0.25s；(c) t=0.5s；左为无温度载荷的几何变形；右为有温度载荷的几何变形

图 7.35 为无温度载荷和有温度载荷时，均匀内压作用下模型不同时刻的 von Mises 应力云图。von Mises 应力计算公式如下：

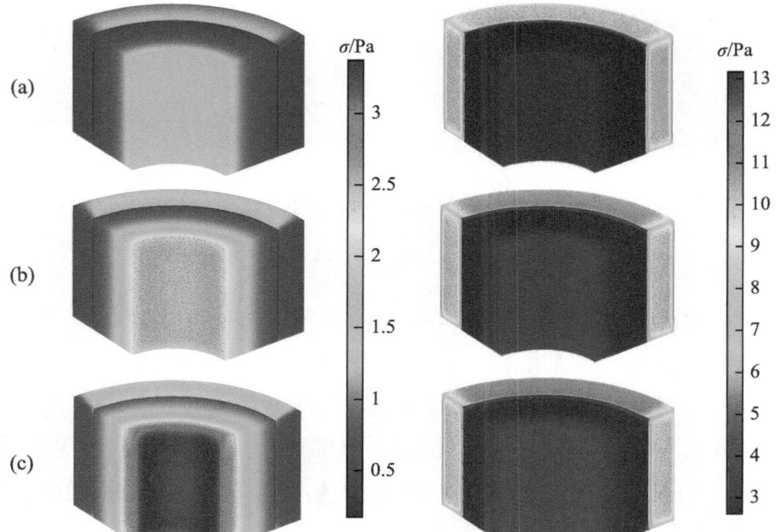

图 7.35　模型在 3 个不同时刻的 von Mises 应力分布云图

(a) t=0.01s；(b) t=1.5s；(c) t=3s；左图为无温度载荷时的 von Mises 应力分布；右图为有温度载荷时的 von Mises 应力分布

$$\sigma_M = \sqrt{[(\sigma_{11}-\sigma_{22})^2+(\sigma_{22}-\sigma_{33})^2+(\sigma_{33}-\sigma_{11})^2]/2} \tag{7.110}$$

其中，σ_{11}、σ_{22} 和 σ_{33} 为三个主应力分量。首先，由于记忆应力的影响，模型的应力随时间的增加而增加。其次，由于每个子域的材料参数不同，应力是不连续的，我们也可以找出不同载荷下最大 von Mises 应力的区域。为了更清楚地显示径向应力的变化，我们给出了在无温度载荷(图 7.36)和有温度载荷(图 7.37)时直线($x_3 = 2\text{m}$, $\theta = 45°$)上的 von Mises 应力。

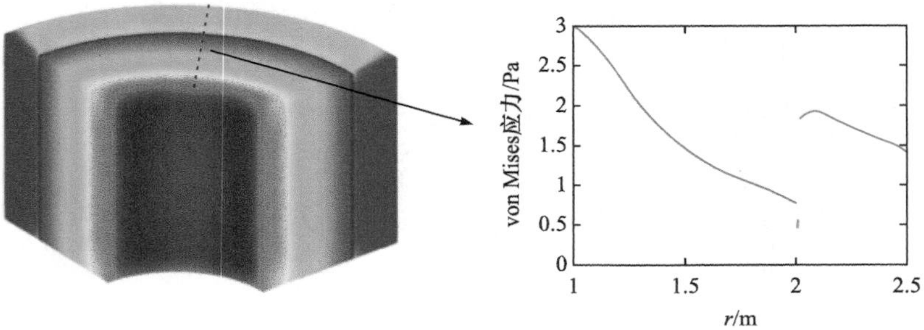

图 7.36　无温度载荷时的 von Mises 应力曲线($x_3 = 2\text{m}$, $\theta = 45°$)

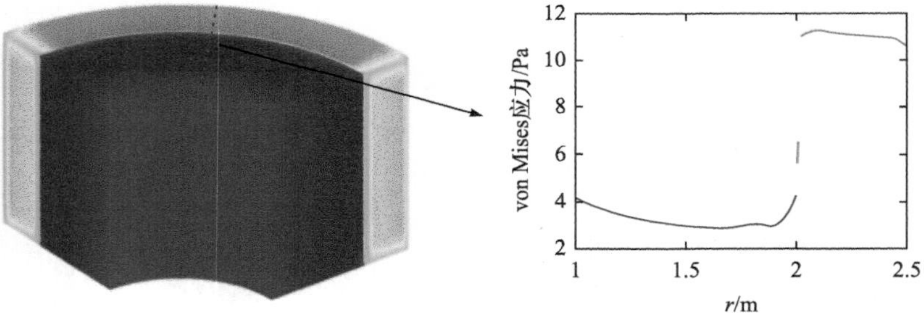

图 7.37　有温度载荷时的 von Mises 应力曲线($x_3 = 2\text{m}$, $\theta = 45°$)

7.3.5　三维含有星形药柱的 SRM 燃烧室的热黏弹性力学问题

与圆柱形药柱相比，星形药柱的燃烧面积更大。因此，星形药柱是 SRM 燃烧室中最常用的药柱。在本例中，讨论了含星形药柱的 SRM 燃烧室在内压、温度载荷和体力作用下的情况。图 7.38 给出了 SRM 燃烧室的 1/4 简化模型。每一层材料参数与 7.3.4 节算例相同。模型上下表面上的 x_3 方向的位移被约束。此外，内边界施加 $f = 1\text{N}/\text{m}^2$ 的均匀压力。重力加速度为 9.8m/s^2。施加的温度载荷为 ΔT=20℃。简化模型的尺寸在图 7.39 中给出。

与 7.3.4 节算例类似，图 7.40 给出了 SRM 燃烧室细化网格，其中蓝色、紫色、绿色和红色网格分别代表药柱(A 子域)、衬层(B 子域)、绝热层(C 子域)和金属壳体(D 子域)。每个子域由 6 个 NURBS 片组成，每个子域的 NURBS 片编号与

7.3.4 节算例相同。该模型中有两个非相适应界面 AB 和 CD，这两个界面的虚拟网格在图 7.40 中给出。

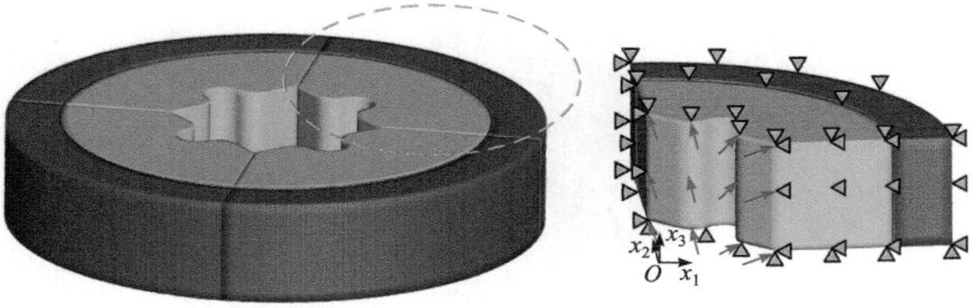

图 7.38　含星形药柱的 SRM 燃烧室的 1/4 等效模型及边界条件

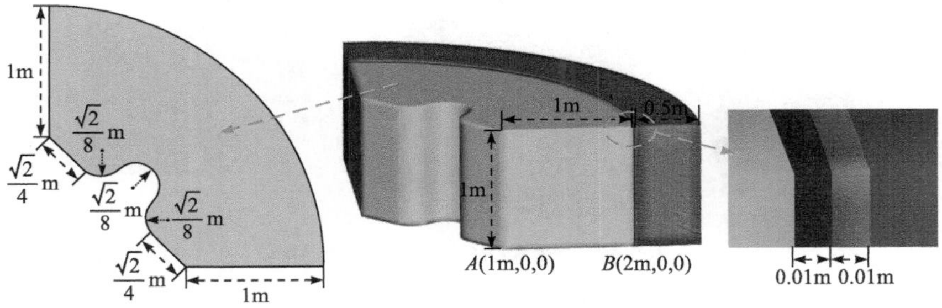

图 7.39　含星形药柱的 SRM 燃烧室的 1/4 等效模型几何尺寸

星形药柱的 224 个内点分布如图 7.41 所示。为了验证等几何边界元法的正确性，将其计算结果与 ANSYS 软件进行了比较。如图 7.42 所示，星形药柱使用 103365 个四节点四面体拉格朗日单元进行离散，其余部分采用 8 节点线性六面体拉格朗日单元，子域 B～D 的单元个数分别为 20000、20000 和 4000。

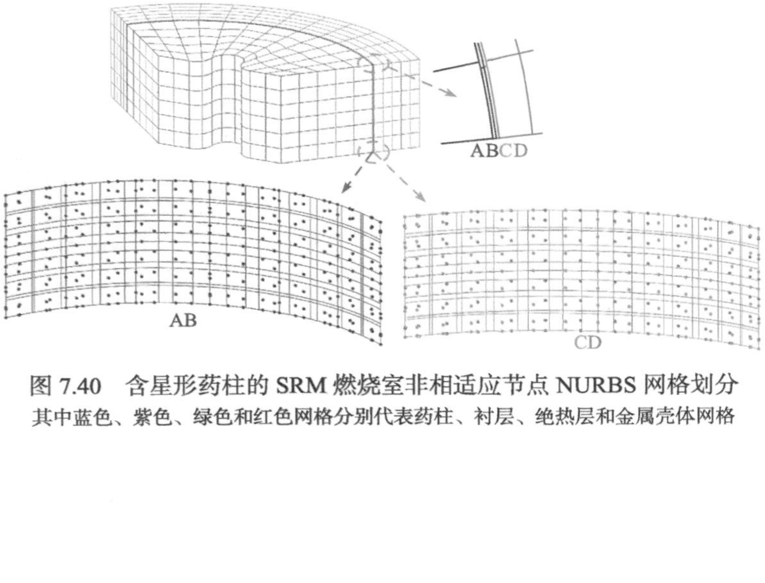

图 7.40　含星形药柱的 SRM 燃烧室非相适应节点 NURBS 网格划分
其中蓝色、紫色、绿色和红色网格分别代表药柱、衬层、绝热层和金属壳体网格

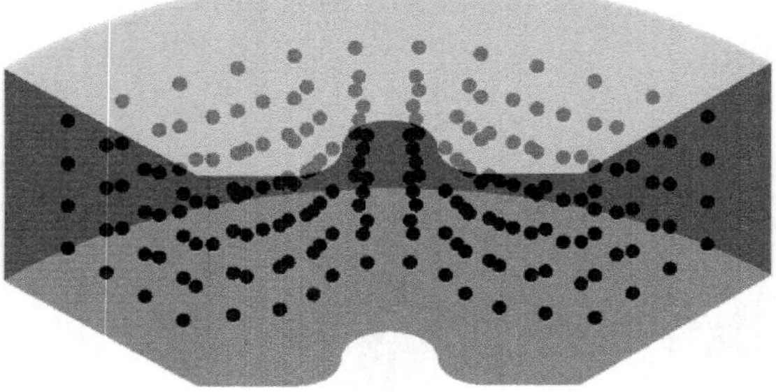

图 7.41 药柱内 224 个内点的分布

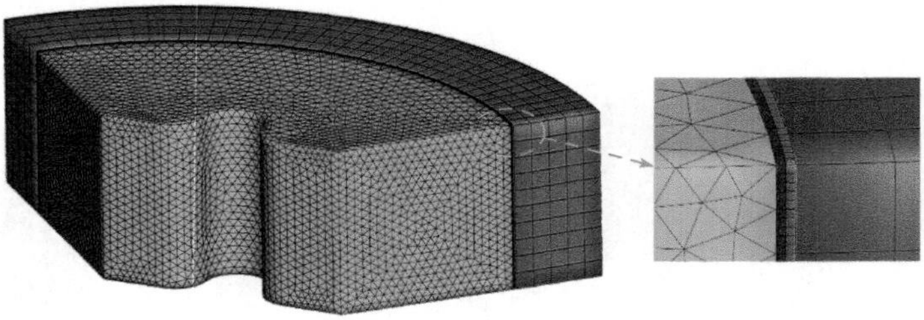

图 7.42 1/4 SRM 燃烧室 FEM 网格

图 7.43 给出了 1/4 SRM 燃烧室在无体力作用和有体力作用下点 A(1m,0,0)和点 B(2m,0,0)随时间变化的位移。可以看出，IGABEM 的数值解和 ANSYS 软件的结果吻合较好。图 7.44 为模型在均匀内部压力和温度载荷作用下不同时刻 x_3 方向的变形云图。可见，体力对 SRM 燃烧室变形的影响很大。此外，图 7.45 给出了 SRM 燃烧室不同时刻无体力和有体力时的 von Mises 应力分布。为了更好地说明体力对应力分布的影响，我们给出了模型同一表面中三条曲线上的 von Mises 应力分布，如图 7.46 和图 7.47 所示。

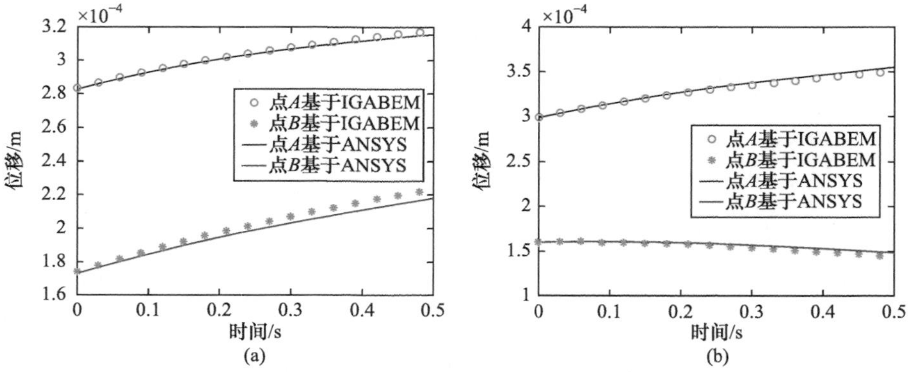

图 7.43 点 A(1m,0,0)和点 B(2m,0,0)随时间变化的位移

(a) 无体力项；(b) 有体力项

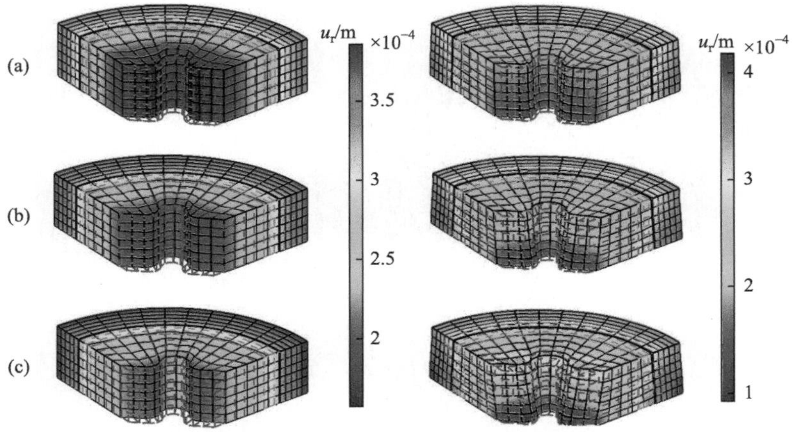

图 7.44　模型在 3 个不同时刻的几何变形图和位移云图

(a) t=0.01s；(b) t=0.25s；(c) t=0.5s；左为无体力项的几何变形；右为有体力项的几何变形

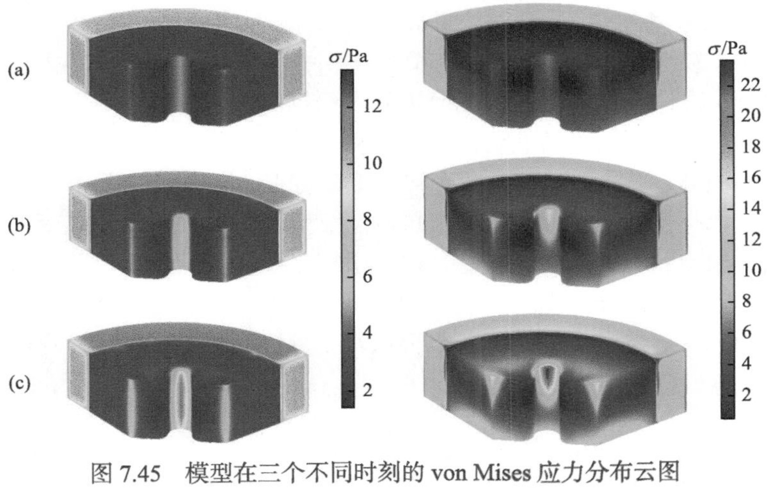

图 7.45　模型在三个不同时刻的 von Mises 应力分布云图

(a) t=0.01s；(b) t=1.5s；(c) t=3s；左图为无体力项时的 von Mises 应力分布；右图为有体力项时的 von Mises 应力分布

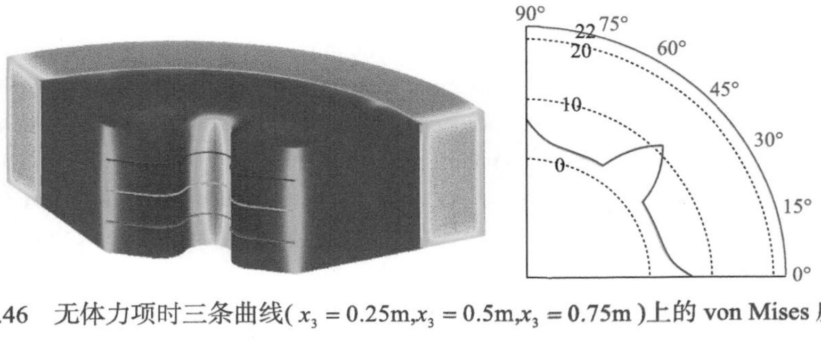

图 7.46　无体力项时三条曲线($x_3 = 0.25\text{m}, x_3 = 0.5\text{m}, x_3 = 0.75\text{m}$)上的 von Mises 应力

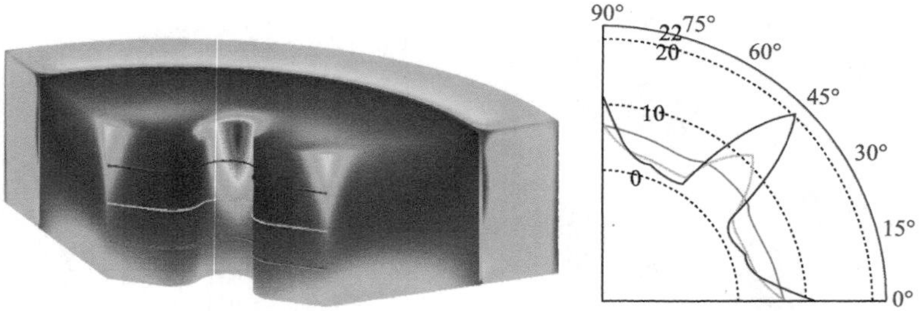

图 7.47　有体力项时三条曲线($x_3 = 0.25\text{m}, x_3 = 0.5\text{m}, x_3 = 0.75\text{m}$)上的 von Mises 应力

7.4 小　结

为了解决多层复合结构热黏弹性-弹性力学问题，本章给出了非相适应界面的等几何边界元法。该算法可用于 SRM 燃烧室等实际工程结构的分析。

基于 Kelvin 基本解，我们详细推导了热黏弹性力学问题的位移边界积分方程和正则化应变积分方程。利用径向积分法将积分方程中的域积分转化为相应的等效边界积分。这样不仅保留了等几何边界元法仅离散边界的优点，而且减小了积分的奇异性。考虑到编程的准确性、效率和便捷性，我们采用不同的方法求解不同阶的奇异积分。其中，强奇异积分可以通过控制点与配点之间的简单变换，用刚体位移技术直接求解。本章针对热黏弹性模型表面节点应力和应变求解问题给出了修正的二维和三维面力恢复法，避免了拟奇异积分的计算。采用基于四叉树细分技术的自适应积分方案求解近奇异积分。采用虚拟节点插入技术实现了非相适应界面耦合，该技术与问题无关，仅依赖于界面两侧的 NURBS 网格，具有编程简单、鲁棒性强的特点。

参 考 文 献

[1] Fish J, Shek K. Multiscale analysis of composite materials and structures[J]. Composites Science and Technology, 2000, 60 : 2547-2556.

[2] Bordas S P A. Isogeometric analysis of laminated composite plates using the higher-order shear deformation theory[J]. Mechanics of Advanced Materials and Structures, 2015, 22: 451-469.

[3] Patton A, Antolín P, Dufour J E, et al. Accurate equilibrium-based interlaminar stress recovery for isogeometric laminated composite Kirchhoff plates[J]. Composite Structures, 2021, 256: 112976.

[4] Montesano J, Behdinan K, Greatrix D R, et al. Internal chamber modeling of a solid rocket motor: effects of coupled structural and acoustic oscillations on combustion[J]. Journal of Sound and Vibration, 2008, 311: 20-38.

[5]Biagi M, Mauries A, Perugini P, et al. Structural qualification of P80 solid rocket motor composite case[J]. AIAA Journal, 2012, 17: 2011-5958.

[6] Bing M. Development of a multiple-use solid rocket motor[J]. AIAA Journal, 1989: 2424.
[7] Yildinm H C, Oezupek S. Structural assessment of a solid propellant rocket motor: effects of aging and damage[J]. Aerospace Science and Technology, 2011, 15: 635-641.
[8] Srinatha H R, Lewis R W. A finite element method for thermoviscoelastic analysis of plane problems[J]. Computer Methods in Applied Mechanics and Engineering, 1981, 25: 21-33.
[9] Wang Z, Smith D E. Numerical analysis on viscoelastic creep responses of aligned short fiber reinforced composites[J]. Composite Structures, 2019, 229: 111394.
[10] Zak A R. Structural analysis of realistic solid propellant materials[J]. Journal of Spacecraft Rockets, 1968, 5: 270-275.
[11] Lee E H, Rogers T G. Solution of viscoelastic stress analysis problems using measured creep or relaxation functions[J]. Aerospace Science and Technology, 1963, 30: 127-133.
[12] Mesquita A D, Coda H B. A boundary element methodology for viscoelastic analysis: part Ⅰ with cells[J]. Applied Mathematical Modelling, 2007, 31: 1149-1170.
[13] Mesquita A D, Coda H B. A boundary element methodology for viscoelastic analysis: part Ⅱ without cells[J]. Applied Mathematical Modelling, 2007, 31: 1171-1185.
[14] Mesquita A D, Coda H B. New methodology for the treatment of two dimensional viscoelasticcoupling problems[J]. Computer Methods in Applied Mechanics and Engineering, 2003, 192: 1911-1927.
[15] Mesquita A D, Coda H B. A two-dimensional BEM/FEM coupling applied to viscoelastic analysis of composite domains[J]. International Journal for Numerical Methods in Engineering, 2003, 57: 251-270.
[16] Marques S P C, Creus G J. Computational Viscoelasticity[M]. London: Springer, 2012.
[17] Gong Y P, Trevelyan J, Hattori G, et al. Hybrid nearly singular integration for three-dimensional isogeometric boundary element analysis of coatings and other thin structures[J]. Computer Methods in Applied Mechanics and Engineering, 2019, 367: 642-673.
[18] Gong Y P, Dong C Y, Qin F, et al. Hybrid nearly singular integration for three-dimensional isogeometric boundary element analysis of coatings and other thin structures[J]. Computer Methods in Applied Mechanics and Engineering, 2020, 367: 113099.
[19] Nguyen V P, Anitescu C, Bordas S P A, et al. Isogeometric analysis: an overview and computer implementation aspects[J]. Mathematics and Computers in Simulation, 2015, 117: 89-116.
[20] Li K, Qian X. Isogeometric analysis and shape optimization via boundary integral[J]. Computer-Aided Design, 2011, 43: 1427-1437.
[21] Simpson R N, Bordas S P A, Trevelyan J, et al. A two-dimensional isogeometric boundary element method for elastostatic analysis[J]. Computer Methods in Applied Mechanics and Engineering, 2012, 209: 87-100.
[22] Simpson R N, Bordas S P A, Lian H J, et al. An isogeometric boundary element method for elastostatic analysis: 2D implementation aspects[J]. Composite Structures, 2013, 118: 2-12.
[23] Gong Y P, Dong C Y, Qin X C. An isogeometric boundary element method for three dimensional potential problems[J]. Journal of Computational and Applied Mathematics, 2017, 313:454-468.

[24] Zhu W, Cai M, Yang L, et al. The effect of morphology of thermally grown oxide on the stress field in a turbine blade with thermal barrier coatings[J]. Surface & Coatings Technology, 2015, 276: 160-167.

[25] Wang L, Li D C, Yang J S, et al. Modeling of thermal properties and failure of thermal barrier coatings with the use of finite element methods: a review[J]. Journal of the European Ceramic Society, 2016, 36: 1313-1331.

[26] Zhu W, Wang J W, Yang L, et al. Modeling and simulation of the temperature and stress fields in a 3D turbine blade coated with thermal barrier coatings[J]. Surface & Coatings Technology, 2017, 315: 443-453.

[27] Shi D Q, Song J N, Li S L, et al. Cracking behaviors of EB-PVD thermal barrier coating under temperature gradient[J]. Ceramics International, 2019, 45: 18518-18528.

[28] Benson D J, Bazilevs Y, Hsu M C, et al. Isogeometric shell analysis: the Reissner-Mindlin shell[J]. Computer Methods in Applied Mechanics and Engineering, 2010, 199: 276-289.

[29] Sladek V, Sladek J, Tanaka M. Regularization of hypersingular and nearly singular integrals in the potential theory and elasticity[J]. International Journal for Numerical Methods in Engineering, 1993, 36: 1609-1628.

[30] Schulz H, Schwab C, Wendl W L. The computation of potentials near and on the boundary by an extraction technique for boundary element methods[J]. Computer Methods in Applied Mechanics and Engineering, 1998, 157: 225-238.

[31] Telles J C F. A self-adaptive coordinate transformation for efficient numerical evaluation of general boundary element integral[J]. International Journal for Numerical Methods in Engineering, 1987, 24: 959-973.

[32] Gao X W, Yang K, Wang J. An adaptive element subdivision technique for evaluation of various 2D singular boundary integrals[J]. Engineering Analysis with Boundary Elements, 2008, 32: 692-696.

[33] Niu Z R, Wendland W L, Wang X X, et al. A semi-analytical algorithm for the evaluation of the nearly singular integrals in three dimensional boundary element methods[J]. Computer Methods in Applied Mechanics and Engineering, 2005, 194: 1057-1074.

[34] Zhang Y M, Gu Y, Chen J T. Stress analysis for multilayered coating systems using semi-analytical BEM with geometric non-linearities[J]. Computational Mechanics, 2011, 47: 493-504.

[35] Sladek V, Sladek J, Tanaka M. Optimal transformations of the integration variables in computation of singular integrals in BEM[J]. International Journal for Numerical Methods in Engineering, 2000, 47: 1263-1283.

[36] Ye W J. A new transformation technique for evaluating nearly singular integrals[J]. Computational Mechanics, 2008, 42: 457-466.

[37] Zhang Y M, Qu W Z, Chen J T. BEM analysis of thin structures for thermoelastic problems[J]. Engineering Analysis with Boundary Elements, 2013, 37: 441-452.

[38] Marussig B, Hughes T J R. A Review of trimming in isogeometric analysis: challenges, data exchange and simulation aspects[J]. Archives of Computational Methods in Engineering, 2018,

25: 1059-1127.

[39] Hsiao G C, Schnack E, Wendland W L. A hybrid coupled finite-boundary element method in elasticity[J]. Computer Methods in Applied Mechanics and Engineering, 1999, 173: 287-316.

[40] Hsiao G C, Schnack E, Wendland W L. Hybrid coupled finite-boundary element methods for elliptic systems of second order[J]. Computer Methods in Applied Mechanics and Engineering, 2000, 190: 431-485.

[41] Schnack E, Türke K. Domain decomposition with BEM and FEM[J]. International Journal for Numerical Methods in Engineering, 1997, 40: 2593-2610.

[42] González J A, Park K C, Felippa C A. FEM and BEM coupling in elastostatics using localized Lagrange multipliers[J]. International Journal for Numerical Methods in Engineering, 2007, 69: 2058-2074.

[43] González J A, Rodríguez T L, Park K C, et al. The nsBETI method: an extension of the FETI method to non-symmetrical BEM-FEM coupled problems[J]. International Journal for Numerical Methods in Engineering, 2013, 93: 1015-1039.

[44] Yang H S, Dong C Y, Wu Y H. Non-conforming interface coupling and symmetric iterative solution in isogeometric FE-BE analysis [J]. Computer Methods in Applied Mechanics and Engineering, 2021, 373: 113561.

[45] 杨华实. 含缺陷薄壁结构及混合维度耦合问题的等几何边界元法研究[D]. 北京：北京理工大学, 2022.

[46] Zhan Y S, Xu C, Yang H S, et al. Isogeometric FE-BE method with non-conforming coupling interface for solving elasto-thermoviscoelastic problems[J]. Engineering Analysis with Boundary Elements, 2022, 141: 199-221.

[47] Muki R, Sternberg E. On transient thermal stresses in viscoelastic materials with temperature-dependent properties[J]. International Journal of Applied Mechanics, 1961, 28: 193-207.

[48] Williams M L, Landel R F, Ferry J D. The temperature dependence of relaxation mechanisms in amorphous polymers and other glass-forming liquids[J]. Journal of the American Chemical Society, 1955, 77: 3701-3707.

[49] Scott M A, Borden M J, Verhoosel C V, et al. Isogeometric finite element data structures based on Bézierextraction of T-splines[J]. International Journal for Numerical Methods in Engineering, 2011, 87: 126-156.

[50] 公颜鹏. 等几何边界元法的基础性研究及应用[D]. 北京：北京理工大学, 2019.

[51] Sun D Y, Dai R, Liu X Y, et al. RI-IGABEM for 2D viscoelastic problems and its application to solid propellant grains[J]. Computer Methods in Applied Mechanics and Engineering, 2021, 378: 113737.

第 8 章

非相适应界面力学问题的等几何有限元-边界元耦合分析

8.1 引　言

等几何有限元法已被广泛地应用于结构静动力学、非线性大变形及弹塑性分析等研究领域，能够弥补等几何边界元法在研究非线性材料行为方面的不足，而等几何边界元法在研究应力集中、断裂力学和声场等问题时具有天然的优势，同时又能克服等几何有限元法在数值计算时需要将 CAD 分析模型的边界表征向适合于分析的实体表征进行转换的困难。因此，研究等几何有限元法与等几何边界元法的耦合算法，可以发挥每种方法各自的优点，达到相辅相成的目的。传统有限元法与常规边界元法耦合的历史最早可以追溯到 Zienkiewicz 等[1]的研究，并很快应用于各种工程问题，如动力分析[2]、弹塑性分析[3]、断裂力学[4]、损伤增长[5]和声-结构相互作用[6,7]等。等几何有限元-边界元耦合方法的研究相对有限，主要集中在声-结构或流体-结构耦合领域，包括声-壳耦合分析[8-10]、磁力耦合问题[11]和剪切流中动力分析[12]等。有限元法通过建立结点外力与位移关系的系统方程，得到稀疏对称的正定矩阵。以配点建立的边界元法则以结点位移和边界上结点面力为基本变量，得到稠密的非对称满秩系数矩阵。因此，耦合分析实施的关键是在耦合界面上建立控制变量的协调方程和面力平衡方程，即建立这两类方程中未知变量之间的关系。

针对求解规模不大且有限元和边界元子域尺度相当的问题，我们可以划分相同的耦合界面网格，轻易实现点对点的强连接。但一般情况下，从 CAD 系统中导出的耦合界面两侧的等几何有限元和等几何边界元网格是不相匹配的(非相适应或非完全重合、界面两侧网格的数量和阶次都不一致)，或是基于不同类型的样条函数建立的(如 NURBS 和非结构性 PHT 样条、LR-B 样条等)，甚至耦合界面还可能包含大量的剪裁曲线和剪裁曲面。因此，本章将研究等几何有限元-边界元耦合分析中非相适应界面耦合约束建立的稳定算法。针对耦合方法形成的非对称耦合系数矩阵问题，将对称迭代方法引入耦合矩阵的求解中，并探究对称迭代方法的有效性和收敛性。

8.2　三维弹性力学问题的等几何有限元求解公式

给定三维参数空间中的三个非递减节点矢量 $\boldsymbol{\Xi}=\{\xi_1,\xi_2,\cdots,\xi_{n+p+1}\}$、$\boldsymbol{\Psi}=\{\eta_1,\eta_2,\cdots,\eta_{m+q+1}\}$ 和 $\boldsymbol{\mathcal{M}}=\{\zeta_1,\zeta_2,\cdots,\zeta_{l+r+1}\}$，以及相关的控制点坐标 $\boldsymbol{P}_{i,j,k}$，NURBS 三维实体可以看作是三变量 NURBS 基函数的线性组合，组合系数为相关的控制点，其表达式为

$$\boldsymbol{V}(\xi,\eta,\zeta)=\sum_{i=1}^{n}\sum_{j=1}^{m}\sum_{k=1}^{l}R_{i,j,k}(\xi,\eta,\zeta)\boldsymbol{P}_{i,j,k} \tag{8.1}$$

式中，n、m 和 l 分别是三个参数方向 ξ、η 和 ζ 上基函数的个数；三变量 NURBS 基函数 $R_{i,j,k}(\xi,\eta,\zeta)$ 由三个参数方向上的单变量 B 样条基函数 $N_{i,p}$、$M_{j,q}$ 和 $L_{k,r}$ 加权而成，其表达式为[13]

$$R_{i,j,k}(\xi,\eta,\zeta)=\frac{N_{i,p}(\xi)M_{j,q}(\eta)L_{k,r}(\zeta)\omega_{i,j,k}}{\sum_{i'=1}^{n}\sum_{j'=1}^{m}\sum_{k'=1}^{l}N_{i',p}(\xi)M_{j',q}(\eta)L_{k',r}(\zeta)\omega_{i',j',k'}} \tag{8.2}$$

其中，p、q 和 r 分别是三个参数方向 ξ、η 和 ζ 上基函数的阶次；$\omega_{i,j,k}$ 是与控制点 $\boldsymbol{P}_{i,j,k}$ 对应的正权值。

根据 Cox-de Boor 递推公式[13,14]，单变量 B 样条基函数定义为

$$\begin{aligned}
&N_{i,0}(\xi)=\begin{cases}1, & \xi_i\leqslant\xi<\xi_{i+1}\\ 0, & \text{其他}\end{cases}\\
&N_{i,p}(\xi)=\frac{\xi-\xi_i}{\xi_{i+p}-\xi_i}N_{i,p-1}(\xi)+\frac{\xi_{i+p+1}-\xi}{\xi_{i+p+1}-\xi_{i+1}}N_{i+1,p-1}(\xi)\quad(p\geqslant 1)
\end{aligned} \tag{8.3}$$

以及其对参数坐标的一阶导数表达为

$$N'_{i,p}(\xi)=\frac{p}{\xi_{i+p}-\xi_i}N_{i,p-1}(\xi)+\frac{p}{\xi_{i+p+1}-\xi_{i+1}}N_{i+1,p-1}(\xi) \tag{8.4}$$

因此，可以推导三变量 NURBS 基函数对参数坐标的一阶导数为

$$\frac{\partial R_{i,j,k}}{\partial\xi}=\frac{N'_{i,p}(\xi)M_{j,q}(\eta)L_{k,r}(\zeta)\omega_{i,j,k}-R_{i,j,k}\left[\sum_{i'=1}^{n}\sum_{j'=1}^{m}\sum_{k'=1}^{l}N'_{i',p}(\xi)M_{j',q}(\eta)L_{k',r}(\zeta)\omega_{i',j',k'}\right]}{\sum_{i'=1}^{n}\sum_{j'=1}^{m}\sum_{k'=1}^{l}N_{i',p}(\xi)M_{j',q}(\eta)L_{k',r}(\zeta)\omega_{i',j',k'}}$$

$$\frac{\partial R_{i,j,k}}{\partial \eta}=\frac{N_{i,p}(\xi)M'_{j,q}(\eta)L_{k,r}(\zeta)\omega_{i,j,k}-R_{i,j,k}\left[\sum_{i'=1}^{n}\sum_{j'=1}^{m}\sum_{k'=1}^{l}N_{i',p}(\xi)M'_{j',q}(\eta)L_{k',r}(\zeta)\omega_{i',j',k'}\right]}{\sum_{i'=1}^{n}\sum_{j'=1}^{m}\sum_{k'=1}^{l}N_{i',p}(\xi)M_{j',q}(\eta)L_{k',r}(\zeta)\omega_{i',j',k'}}$$

$$\frac{\partial R_{i,j,k}}{\partial \zeta}=\frac{N_{i,p}(\xi)M_{j,q}(\eta)L'_{k,r}(\zeta)\omega_{i,j,k}-R_{i,j,k}\left[\sum_{i'=1}^{n}\sum_{j'=1}^{m}\sum_{k'=1}^{l}N_{i',p}(\xi)M_{j',q}(\eta)L'_{k',r}(\zeta)\omega_{i',j',k'}\right]}{\sum_{i'=1}^{n}\sum_{j'=1}^{m}\sum_{k'=1}^{l}N_{i',p}(\xi)M_{j',q}(\eta)L_{k',r}(\zeta)\omega_{i',j',k'}}$$

(8.5)

式中，$N'_{i,p}(\xi)$、$M'_{j,q}(\eta)$ 和 $L'_{k,r}(\zeta)$ 通过式(8.4)计算得到。

定义在二维参数空间中的 NURBS 曲面表示为

$$\boldsymbol{S}(\xi,\eta)=\sum_{i=1}^{n}\sum_{j=1}^{m}R_{i,j}(\xi,\eta)\boldsymbol{P}_{i,j} \tag{8.6}$$

式中，双变量 NURBS 基函数 $R_{i,j}(\xi,\eta)$ 表示为[13]

$$R_{i,j}(\xi,\eta)=\frac{N_{i,p}(\xi)M_{j,q}(\eta)\omega_{i,j}}{\sum_{i'=1}^{n}\sum_{j'=1}^{m}N_{i',p}(\xi)M_{j',q}(\eta)\omega_{i',j'}} \tag{8.7}$$

其中，$\omega_{i,j}$ 是与控制点 $\boldsymbol{P}_{i,j}$ 对应的正权值。

考虑三维区域 $\varOmega$ 上的等几何弹性力学问题，其区域的边界表示为 $\varGamma$，在本质边界 $\varGamma_{\mathrm{u}}$ 上指定位移边界条件为 $\boldsymbol{u}=\bar{\boldsymbol{u}}$，在自然边界 $\varGamma_{\mathrm{t}}$ 上指定面力边界条件为 $\boldsymbol{t}=\bar{\boldsymbol{t}}$，该问题以 Voigt 记法表达的弱形式为[15]

$$\int_{\varOmega}(\delta\boldsymbol{\varepsilon})^{\mathrm{T}}\boldsymbol{\mathcal{D}}\boldsymbol{\varepsilon}\mathrm{d}\varOmega-\int_{\varOmega}(\delta\boldsymbol{u})^{\mathrm{T}}\boldsymbol{b}\mathrm{d}\varOmega-\int_{\varGamma_{\mathrm{t}}}(\delta\boldsymbol{u})^{\mathrm{T}}\bar{\boldsymbol{t}}\,\mathrm{d}\varGamma_{\mathrm{t}}=0 \tag{8.8}$$

式中，$\boldsymbol{\varepsilon}$ 和 $\boldsymbol{b}$ 分别代表应变张量和体力矢量；$\boldsymbol{\mathcal{D}}$ 是材料本构矩阵。

IGA 继承了等参的思想，即 NURBS 形函数在精确描述几何体形状的同时也用来插值未知的位移场。因此，近似位移场表达为

$$\boldsymbol{u}(\xi,\eta,\zeta)=\sum_{i=1}^{n}\sum_{j=1}^{m}\sum_{k=1}^{l}R_{i,j,k}(\xi,\eta,\zeta)\boldsymbol{u}_{\boldsymbol{P}_{i,j,k}} \tag{8.9}$$

然后，可以得到离散的求解代数方程组

$$\boldsymbol{K}\boldsymbol{u}=\boldsymbol{f} \tag{8.10}$$

其中，$\boldsymbol{u}$ 代表与控制点相关的位移矢量；$\boldsymbol{K}$ 和 $\boldsymbol{f}$ 分别是对称稀疏的全局刚度矩阵

和全局外力矢量，它们由各个单元的刚度矩阵 $\boldsymbol{K}^e$ 和节点外力矢量 $\boldsymbol{f}^e$ 组装而成，其表达式为

$$\begin{aligned}\boldsymbol{K}^e &= \int_{\Omega^e} \boldsymbol{B}_e^{\mathrm{T}} \boldsymbol{\mathcal{D}} \boldsymbol{B}_e |\boldsymbol{J}_1||\boldsymbol{J}_2| \mathrm{d}\Omega^e \\ \boldsymbol{f}^e &= \int_{\Omega^e} \boldsymbol{R}_e^{\mathrm{T}} \boldsymbol{b} |\boldsymbol{J}_1||\boldsymbol{J}_2| \mathrm{d}\Omega^e + \int_{\Gamma_t^e} \boldsymbol{R}_e^{\mathrm{T}} \bar{\boldsymbol{t}} \left|\boldsymbol{J}_1^{\Gamma_t^e}\right| \left|\boldsymbol{J}_2^{\Gamma_t^e}\right| \mathrm{d}\Gamma_t^e\end{aligned} \tag{8.11}$$

式中，$\boldsymbol{B}_e$ 是通过几何方程得到的应变-位移矩阵，其表达式为

$$\boldsymbol{B}_e = \begin{bmatrix} R_{1,x}^e & 0 & 0 & & R_{(p+1)(q+1)(r+1),x}^e & 0 & 0 \\ 0 & R_{1,y}^e & 0 & \cdots & 0 & R_{(p+1)(q+1)(r+1),y}^e & 0 \\ 0 & 0 & R_{1,z}^e & & 0 & 0 & R_{(p+1)(q+1)(r+1),z}^e \\ R_{1,y}^e & R_{1,x}^e & 0 & \cdots & R_{(p+1)(q+1)(r+1),y}^e & R_{(p+1)(q+1)(r+1),x}^e & 0 \\ 0 & R_{1,z}^e & R_{1,y}^e & & 0 & R_{(p+1)(q+1)(r+1),z}^e & R_{(p+1)(q+1)(r+1),y}^e \\ R_{1,z}^e & 0 & R_{1,x}^e & \cdots & R_{(p+1)(q+1)(r+1),z}^e & 0 & R_{(p+1)(q+1)(r+1),x}^e \end{bmatrix} \tag{8.12}$$

$\boldsymbol{J}_1\left(\boldsymbol{J}_1^{\Gamma_t^e}\right)$ 和 $\boldsymbol{J}_2\left(\boldsymbol{J}_2^{\Gamma_t^e}\right)$ 是相关的雅可比转换矩阵，分别表示从物理空间到参数空间、从参数空间到积分空间的转换。

8.3　三维弹性力学问题的等几何边界元求解公式

8.3.1　边界积分方程

忽略体积力的情况下，三维弹性体的位移边界积分方程表示为[16]

$$c_{ij}(\boldsymbol{P}_{\mathrm{s}}) u_j(\boldsymbol{P}_{\mathrm{s}}) + \int_\Gamma T_{ij}(\boldsymbol{P}_{\mathrm{s}}, \boldsymbol{Q}) u_j(\boldsymbol{Q}) \mathrm{d}\Gamma = \int_\Gamma U_{ij}(\boldsymbol{P}_{\mathrm{s}}, \boldsymbol{Q}) t_j(\boldsymbol{Q}) \mathrm{d}\Gamma \tag{8.13}$$

其中，$\boldsymbol{P}_{\mathrm{s}} \in \Gamma$ 和 $\boldsymbol{Q} \in \Gamma$ 分别称作源点(配点)和场点(积分点)；方程左端的积分为柯西主值意义下的奇异积分；$c_{ij}(\boldsymbol{P}_{\mathrm{s}})$ 是与源点处边界几何形状相关的自由项，对于光滑边界处的源点，其值为 $c_{ij} = \frac{1}{2}\delta_{ij}$，这里 δ 为 Kronecker 符号；U_{ij} 和 T_{ij} 分别是线弹性材料的位移和面力基本解(也称核函数)，其表达式为[16]

$$\begin{aligned} U_{ij}(\boldsymbol{P}_{\mathrm{s}}, \boldsymbol{Q}) &= \frac{1}{16\pi\mu(1-\nu)r}\left[(3-4\nu)\delta_{ij} + r_{,i}r_{,j}\right] \\ T_{ij}(\boldsymbol{P}_{\mathrm{s}}, \boldsymbol{Q}) &= -\frac{1}{8\pi(1-\nu)r^2}\left\{\frac{\partial r}{\partial n}\left[(1-2\nu)\delta_{ij} + 3r_{,i}r_{,j}\right] - (1-2\nu)\left(r_{,i}n_j - r_{,j}n_i\right)\right\} \end{aligned} \tag{8.14}$$

式中，$\mu=\dfrac{E}{2(1+\nu)}$为剪切模量，这里 E 和 ν 分别为弹性模量和泊松比；$r=\|\boldsymbol{Q}-\boldsymbol{P}_{\mathrm{s}}\|$ 代表源点和场点之间的物理距离，其沿坐标方向和场点法线方向的偏导数为

$$r_{,i}=\frac{r_i}{r},\quad r_i=\boldsymbol{Q}_i-\boldsymbol{P}_{si},\quad \frac{\partial r}{\partial n}=\nabla r\cdot\boldsymbol{n}=r_{,i}n_i \tag{8.15}$$

其中，$\boldsymbol{n}$ 是场点 $\boldsymbol{Q}$ 处的法向量；n_i 是其第 i 个分量。

8.3.2 等几何多片表达

由于 NURBS 曲面的张量积结构特点，很难使用单个 NURBS 片来描述三维对象的整个边界。因此，复杂几何体的表示需要多个 NURBS 片组合而成。本章的主要目的是实现有限元和边界元子域之间非相适应界面的耦合。我们使用相同阶次和边界网格相兼容的多个 NURBS 片来构造等几何边界元模型的边界，即控制点沿片的边界是重合的。该思路的程序实现只需将每个片上控制点的局部编号与多片构成的整体区域上的全局编号联系起来，即构建一个全局的几何连接矩阵。图 8.1 给出了由六个 NURBS 片构成的圆柱体表面网格的实施过程。

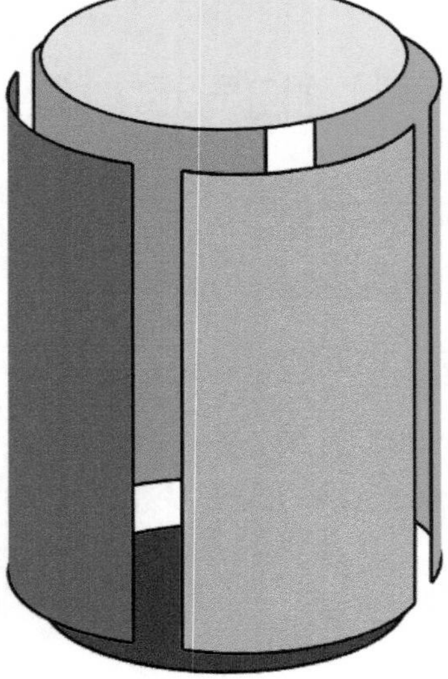

(a)

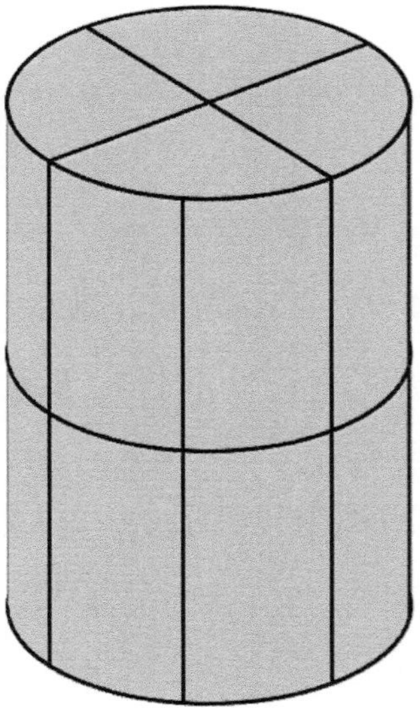

(b)

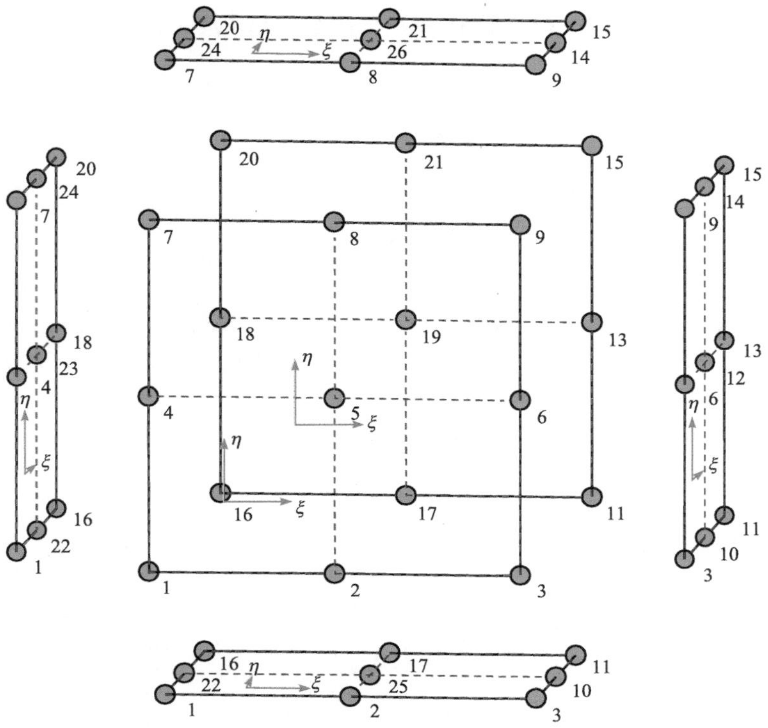

(c)

图 8.1　圆柱体边界的六个边兼容 NURBS 片

(a) 物理域；(b) 一次细化后的网格；(c) 每个片上的参数方向以及控制点的全局编号

8.3.3　等几何离散公式

为了利用 NURBS 基函数将式(8.13)中的边界积分方程离散化，需要利用参数空间中的 Greville 横坐标(Greville abscissa, GA)来获得物理空间中配点的位置。对于 NURBS 曲面，两个参数方向上的 GA 坐标表示为[17]

$$
\begin{aligned}
\xi_i' &= \frac{\xi_{i+1}+\xi_{i+2}+\cdots+\xi_{i+p}}{p}, \quad i=1,2,\cdots,n \\
\eta_j' &= \frac{\eta_{j+1}+\eta_{j+2}+\cdots+\eta_{j+q}}{q}, \quad j=1,2,\cdots,m
\end{aligned}
\tag{8.16}
$$

其中，$\left(\xi_i',\eta_j'\right)$是一个配点对应的 GA 参数坐标，其他字母的含义与式(8.6)中的意义相同，此处不再重复解释。根据 GA 参数坐标，应用等几何映射可以得到配点的物理坐标，且其得到的配点数量与控制点数量相同。

采用多个边兼容 NURBS 片的表达方式，在第 k 个片上点 $\boldsymbol{x}^{(k)}(\xi,\eta)$ 处的位移

场和面力场可通过控制点处的值由 NURBS 函数插值近似为

$$
\begin{aligned}
\boldsymbol{u}^{(k)}(\xi,\eta)&=\sum_{e=1}^{n_e^{(k)}}\sum_{A=1}^{n_A^{(k)}}R_A^{(k)(e)}(\xi,\eta)\boldsymbol{u}_A^{(k)(e)}\\
\boldsymbol{t}^{(k)}(\xi,\eta)&=\sum_{e=1}^{n_e^{(k)}}\sum_{A=1}^{n_A^{(k)}}R_A^{(k)(e)}(\xi,\eta)\boldsymbol{t}_A^{(k)(e)}
\end{aligned}
\tag{8.17}
$$

式中，$n_e^{(k)}$ 是第 k 个片上的 NURBS 单元个数；$n_A^{(k)}$ 是第 k 个片上第 e 个单元相关基函数的个数；$\boldsymbol{u}_A^{(k)(e)}$ 和 $\boldsymbol{t}_A^{(k)(e)}$ 分别是与控制点 $\boldsymbol{P}_A$ 相关的位移和面力系数。

将离散式(8.17)代入式(8.13)中，可得到离散形式的边界积分方程为

$$
\begin{aligned}
&c_{ij}(\boldsymbol{P}_{\mathrm{s}})\sum_{A=1}^{n_A^{(k_0)}}R_A^{(k_0)(e_0)}(\xi',\eta')u_{Aj}^{(k_0)(e_0)}+\sum_{k=1}^{n_{\mathrm{patch}}}\sum_{e=1}^{n_e^{(k)}}\sum_{A=1}^{n_A^{(k)}}\mathbb{H}_{ij}^{(k)(e)(A)}(\boldsymbol{P}_{\mathrm{s}},\boldsymbol{Q})u_{Aj}^{(k)(e)}\\
&=\sum_{k=1}^{n_{\mathrm{patch}}}\sum_{e=1}^{n_e^{(k)}}\sum_{A=1}^{n_A^{(k)}}\mathbb{G}_{ij}^{(k)(e)(A)}(\boldsymbol{P}_{\mathrm{s}},\boldsymbol{Q})t_{Aj}^{(k)(e)}
\end{aligned}
\tag{8.18}
$$

其中，$u_{Aj}^{(k)(e)}$ 和 $t_{Aj}^{(k)(e)}$ 分别代表第 k 个片上第 e 个单元与第 A 个控制点相关的位移和面力矢量的第 j 个分量；$\mathbb{H}_{ij}^{(k)(e)(A)}(\boldsymbol{P}_{\mathrm{s}},\boldsymbol{Q})$ 和 $\mathbb{G}_{ij}^{(k)(e)(A)}(\boldsymbol{P}_{\mathrm{s}},\boldsymbol{Q})$ 表示为

$$
\begin{aligned}
\mathbb{H}_{ij}^{(k)(e)(A)}(\boldsymbol{P}_{\mathrm{s}},\boldsymbol{Q})&=\int_{\Gamma_e^{(k)}}T_{ij}(\boldsymbol{P}_{\mathrm{s}},\boldsymbol{Q})R_A^{(k)(e)}(\overline{\xi},\overline{\eta})\mathrm{d}\Gamma_e^{(k)}\\
\mathbb{G}_{ij}^{(k)(e)(A)}(\boldsymbol{P}_{\mathrm{s}},\boldsymbol{Q})&=\int_{\Gamma_e^{(k)}}U_{ij}(\boldsymbol{P}_{\mathrm{s}},\boldsymbol{Q})R_A^{(k)(e)}(\overline{\xi},\overline{\eta})\mathrm{d}\Gamma_e^{(k)}
\end{aligned}
\tag{8.19}
$$

式中，n_{patch} 为 NURBS 片的个数；(ξ',η') 和 $(\overline{\xi},\overline{\eta})$ 分别为配点和场点对应的参数坐标；k_0 和 e_0 是配点所在的片和单元的编号。

在循环所有的配点之后，得到以矩阵形式表达的线性代数方程组

$$
\boldsymbol{H}\boldsymbol{u}=\boldsymbol{G}\boldsymbol{t} \tag{8.20}
$$

其中，$\boldsymbol{H}$ 和 $\boldsymbol{G}$ 是由式(8.18)构造的影响系数矩阵；$\boldsymbol{u}$ 和 $\boldsymbol{t}$ 分别是包含整个边界的全局位移和面力矢量。在替换好已知的边界条件后，式(8.20)可以写成

$$
\mathcal{A}\mathcal{X}=\mathcal{B}\mathcal{Y}=f_{\mathrm{b}} \tag{8.21}
$$

式中，$\mathcal{X}$ 包含所有的未知位移和面力；$\mathcal{Y}$ 包含所有已知的位移和面力；系数矩阵 $\mathcal{A}$ 由矩阵 $\boldsymbol{H}$ 和 $\boldsymbol{G}$ 移项而得，且其为非对称和稠密的满秩矩阵。一旦未知的位移和面力求得后，就可直接通过域内位移和应力边界积分方程获得任意内部点的位移和应力值，而边界上的应力值则通过面力恢复法[16]获得。

域内任意一点 $\boldsymbol{y}$ 处的位移边界积分方程表示为

$$u_i(\boldsymbol{y})=\int_\Gamma U_{ij}(\boldsymbol{y},\boldsymbol{Q})t_j(\boldsymbol{Q})\mathrm{d}\Gamma-\int_\Gamma T_{ij}(\boldsymbol{y},\boldsymbol{Q})u_j(\boldsymbol{Q})\mathrm{d}\Gamma \tag{8.22}$$

式中，$U_{ij}(\boldsymbol{y},\boldsymbol{Q})$ 和 $T_{ij}(\boldsymbol{y},\boldsymbol{Q})$ 分别通过式(8.14)计算。

引入广义胡克定律，域内任意一点 $\boldsymbol{y}$ 处的应力边界积分方程表示为[16]

$$\sigma_{ij}(\boldsymbol{y})=\int_\Gamma D_{ijk}(\boldsymbol{y},\boldsymbol{Q})t_k(\boldsymbol{Q})\mathrm{d}\Gamma-\int_\Gamma S_{ijk}(\boldsymbol{y},\boldsymbol{Q})u_k(\boldsymbol{Q})\mathrm{d}\Gamma \tag{8.23}$$

其中，基本解 D_{ijk} 和 S_{ijk} 定义为[16]

$$\begin{aligned}
D_{ijk}(\boldsymbol{y},\boldsymbol{Q})&=\frac{1}{8\pi(1-\nu)r^2}\left[(1-2\nu)\left(r_{,j}\delta_{ki}+r_{,i}\delta_{kj}-r_{,k}\delta_{ij}\right)+3r_{,i}r_{,j}r_{,k}\right]\\
S_{ijk}(\boldsymbol{y},\boldsymbol{Q})&=\frac{\mu}{4\pi(1-\nu)r^3}\left\{3\frac{\partial r}{\partial n}\left[(1-2\nu)r_{,k}\delta_{ij}+\nu\left(r_{,j}\delta_{ki}+r_{,i}\delta_{kj}\right)-5r_{,i}r_{,j}r_{,k}\right]\right.\\
&\quad\left.+3\nu\left(n_i r_{,j}r_{,k}+n_j r_{,i}r_{,k}\right)+(1-2\nu)\left(3n_k r_{,i}r_{,j}+n_j\delta_{ki}+n_i\delta_{kj}\right)-(1-4\nu)n_k\delta_{ij}\right\}
\end{aligned} \tag{8.24}$$

式中，δ 为 Kronecker 符号；$\mu=\dfrac{E}{2(1+\nu)}$ 为剪切模量，以及

$$r_i=\boldsymbol{Q}_i-\boldsymbol{y}_i,\quad r=\sqrt{r_i r_i},\quad r_{,i}=\frac{r_i}{r},\quad \frac{\partial r}{\partial n}=\nabla r\cdot\boldsymbol{n}=r_{,i}n_i \tag{8.25}$$

8.3.4　改进的幂级数展开法

IGABEM 中数值积分的精确计算对解的精度有重要影响。根据配点的位置，曲面积分可以分为奇异积分和离散后的非奇异积分。排除配点的单元称为非奇异单元。这种积分可以用高斯求积法精确计算。包含配点的单元称为奇异单元，对该单元进行数值积分时会发生奇异积分。本章在文献[18]和[19]的基础上，对幂级数展开法进行了一些改进(M-PSEM)，可以有效地计算各种类型的样条函数奇异积分。

M-PSEM 消除奇异性的基本实现策略如下：①通过径向积分法将物理空间中的 NURBS 曲面奇异积分转化为沿母单元四条边的线积分之和[20]；②将核函数的非奇异部分和物理距离展开为母空间中局部距离的幂级数展开形式；③利用显式解析法提取奇异积分的有限部分。

在三维 IGABEM 中遇到的所有奇异积分都可以写为如下的统一形式[19]：

$$I^e(\boldsymbol{P}_\mathrm{s}) = \int_{\Omega_e} \frac{\bar{f}(\boldsymbol{Q},\boldsymbol{P}_\mathrm{s})}{r^\lambda(\boldsymbol{Q},\boldsymbol{P}_\mathrm{s})}\mathrm{d}\Omega_e(\boldsymbol{Q}) = \int_{-1}^{1}\int_{-1}^{1} \frac{\bar{f}(\boldsymbol{Q},\boldsymbol{P}_\mathrm{s})}{r^\lambda(\boldsymbol{Q},\boldsymbol{P}_\mathrm{s})}|\boldsymbol{J}_e|\mathrm{d}\tilde{\xi}\mathrm{d}\tilde{\eta} \tag{8.26}$$

其中，$\bar{f}(\boldsymbol{Q},\boldsymbol{P}_\mathrm{s})$代表非奇异部分；$\boldsymbol{J}_e$代表从物理空间到积分空间的雅可比矩阵。当上角标$\lambda$取值为 1、2 和 3 时，上述奇异积分分别称为广义条件下的弱奇异积分、柯西主值意义下的强奇异和 Hadamard 有限部分的超奇异积分。

利用径向积分法[20]，可以将上式中的奇异积分描述为一系列线积分的和，表示为

$$I^e(\boldsymbol{P}_\mathrm{s}) = \int_L \frac{1}{\rho(\boldsymbol{Q},\boldsymbol{P}_\mathrm{s})}\frac{\partial\rho(\boldsymbol{Q},\boldsymbol{P}_\mathrm{s})}{\partial\boldsymbol{n}'(\boldsymbol{Q})}F(\boldsymbol{Q},\boldsymbol{P}_\mathrm{s})\mathrm{d}L(\boldsymbol{Q}) \tag{8.27}$$

式中，

$$F(\boldsymbol{Q},\boldsymbol{P}_\mathrm{s}) = \lim_{\rho_e(\varepsilon)\to 0}\int_{\rho_e(\varepsilon)}^{\rho(\boldsymbol{Q},\boldsymbol{P}_\mathrm{s})} \frac{\bar{f}(\boldsymbol{q},\boldsymbol{P}_\mathrm{s})}{r^\lambda(\boldsymbol{q},\boldsymbol{P}_\mathrm{s})}|\boldsymbol{J}_e|\rho\mathrm{d}\rho \tag{8.28}$$

$\dfrac{\partial\rho}{\partial\boldsymbol{n}'(\boldsymbol{Q})} = \nabla\rho\cdot\boldsymbol{n}'$，这里$\boldsymbol{n}'(\boldsymbol{Q})$是母空间中单元边界线上的单位法向量；$\rho(\boldsymbol{Q},\boldsymbol{P}_\mathrm{s})$是母空间中的局部距离，它们的关系如图 8.2 所示。

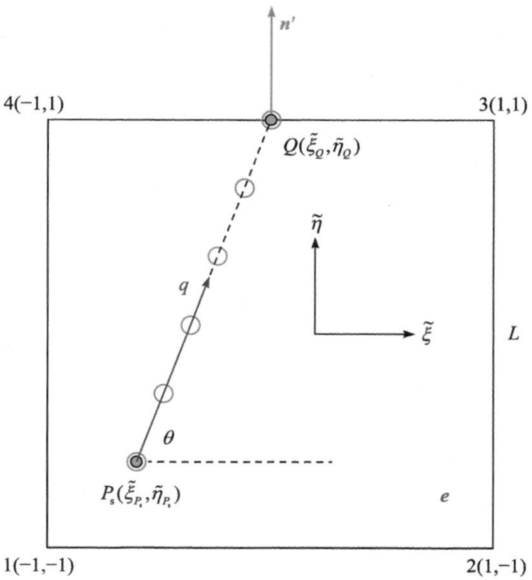

图 8.2　母空间单元中涉及的变量

从图 8.2 中，我们可以得到如下的关系：

$$\rho(\boldsymbol{Q},\boldsymbol{P}_{\mathrm{s}})=\sqrt{\left(\tilde{\xi}_{\boldsymbol{Q}}-\tilde{\xi}_{\boldsymbol{P}_{\mathrm{s}}}\right)^2+\left(\tilde{\eta}_{\boldsymbol{Q}}-\tilde{\eta}_{\boldsymbol{P}_{\mathrm{s}}}\right)^2}$$
$$\frac{\partial\rho(\boldsymbol{Q},\boldsymbol{P}_{\mathrm{s}})}{n'(\boldsymbol{Q})}=\frac{\left(\tilde{\xi}_{\boldsymbol{Q}}-\tilde{\xi}_{\boldsymbol{P}_{\mathrm{s}}}\right)\cdot n'_{\tilde{\xi}}+\left(\tilde{\eta}_{\boldsymbol{Q}}-\tilde{\eta}_{\boldsymbol{P}_{\mathrm{s}}}\right)\cdot n'_{\tilde{\eta}}}{\rho(\boldsymbol{Q},\boldsymbol{P}_{\mathrm{s}})} \tag{8.29}$$

接下来，我们将总体坐标系下物理距离 r 表示为局部距离 ρ 的幂级数展开形式，并写为

$$\frac{r}{\rho}=\bar{\rho}=\sum_{i=0}^{N}C_i\rho^i=\sqrt{\sum_{i=0}^{M}G_i\rho^i} \tag{8.30}$$

随后，需要确定 M 的值。在给定源点和场点后，母空间中的坐标 $\tilde{\xi}$ 和 $\tilde{\eta}$ 是局部距离 ρ 的线性函数。其次，IGA 中的参数空间与母空间之间的投影转换也是线性变换。此外，NURBS 基函数是参数坐标的有理多项式。因此，给定沿两个参数方向的 NURBS 阶次 p 和 q ，M 的计算式为

$$M=2(p+q)-2 \tag{8.31}$$

然后，我们要求出系数 G_i 和 C_i 。传统方法中，会在如图 8.2 所示的线段 $P_{\mathrm{s}}Q$ 上均匀选取 $M+1$ 个点，然后计算出这些点处的 $\boldsymbol{r}$ 和 ρ 的值。此处值得注意的是，传统方法中 r 的计算是通过拉格朗日函数插值得到的，存在明显的几何误差，而在 IGA 中，r 的值是通过 NURBS 基函数精确求出的。最后通过待定系数法获取这些系数，其中会涉及如下所示的由局部距离 ρ 构成的范德蒙德(Vandermonde)矩阵

$$\begin{bmatrix} 1 & \rho_1^1 & \rho_1^2 & \cdots & \rho_1^{M-1} \\ 1 & \rho_2^1 & \rho_2^2 & \cdots & \rho_2^{M-1} \\ 1 & \rho_3^1 & \rho_3^2 & \cdots & \rho_3^{M-1} \\ \vdots & \vdots & \vdots & \ddots & \vdots \\ 1 & \rho_M^1 & \rho_M^2 & \cdots & \rho_M^{M-1} \end{bmatrix} \tag{8.32}$$

当高阶 NURBS 形函数($\geqslant 3$)应用于数值分析时，或者在母空间中源点到单元边界的距离非常小时(注意，该距离并未达到可以忽略的极小值)，则 Vandermonde 矩阵中的 ρ_1^{M-1} 与其周围的项将变得非常小，甚至小于机器精度(2.2204×10^{-16})，从而导致 Vandermonde 矩阵几乎是奇异的，以至于无法求解。另一方面，在线弹性 IGABEM 中通常仅需要处理到超奇异积分($\lambda=3$)。因此，只需使用到 G_i 和 C_i 中的前两个系数。接下来，我们使用改进的方法来获得这些系数。

当 $\boldsymbol{q}$ 点与源点 $\boldsymbol{P}_{\mathrm{s}}$ 十分接近时，我们可以将物理距离 r 表示为局部距离 ρ 的幂级数展开形式，忽略三阶及以上的高阶小量后，可以得到

$$
\begin{aligned}
x_i^{\boldsymbol{q}} - x_i^{\boldsymbol{P}_s} = & \frac{\partial x_i^{\boldsymbol{q}}}{\partial \tilde{\xi}}\left(\tilde{\xi}_{\boldsymbol{q}} - \tilde{\xi}_{\boldsymbol{P}_s}\right) + \frac{\partial x_i^{\boldsymbol{q}}}{\partial \tilde{\eta}}\left(\tilde{\eta}_{\boldsymbol{q}} - \tilde{\eta}_{\boldsymbol{P}_s}\right) \\
& + \frac{1}{2}\frac{\partial^2 x_i^{\boldsymbol{q}}}{\partial \tilde{\xi}^2}\left(\tilde{\xi}_{\boldsymbol{q}} - \tilde{\xi}_{\boldsymbol{P}_s}\right)^2 + \frac{1}{2}\frac{\partial^2 x_i^{\boldsymbol{q}}}{\partial \tilde{\eta}^2}\left(\tilde{\eta}_{\boldsymbol{q}} - \tilde{\eta}_{\boldsymbol{P}_s}\right)^2 \\
& + \frac{\partial^2 x_i^{\boldsymbol{q}}}{\partial \tilde{\xi}\partial \tilde{\eta}}\left(\tilde{\xi}_{\boldsymbol{q}} - \tilde{\xi}_{\boldsymbol{P}_s}\right)\left(\tilde{\eta}_{\boldsymbol{q}} - \tilde{\eta}_{\boldsymbol{P}_s}\right)
\end{aligned} \tag{8.33}
$$

式(8.33)中的一阶导数和二阶导数表示物理坐标对母空间坐标的导数。将下列关系代入式(8.33)

$$
\begin{aligned}
\tilde{\xi}_{\boldsymbol{q}} - \tilde{\xi}_{\boldsymbol{P}_s} &= \rho \cdot \rho_{,\tilde{\xi}} \\
\tilde{\eta}_{\boldsymbol{q}} - \tilde{\eta}_{\boldsymbol{P}_s} &= \rho \cdot \rho_{,\tilde{\eta}}
\end{aligned} \tag{8.34}
$$

可以得到

$$
x_i^{\boldsymbol{q}} - x_i^{\boldsymbol{P}_s} = \mathcal{D}_1 \rho + \mathcal{D}_2 \rho^2 \tag{8.35}
$$

其中，

$$
\begin{aligned}
\mathcal{D}_1 &= \frac{\partial x_i^{\boldsymbol{q}}}{\partial \tilde{\xi}} \rho_{,\tilde{\xi}} + \frac{\partial x_i^{\boldsymbol{q}}}{\partial \tilde{\eta}} \rho_{,\tilde{\eta}} \\
\mathcal{D}_2 &= \frac{1}{2}\frac{\partial^2 x_i^{\boldsymbol{q}}}{\partial \tilde{\xi}^2} \rho_{,\tilde{\xi}}{}^2 + \frac{1}{2}\frac{\partial^2 x_i^{\boldsymbol{q}}}{\partial \tilde{\eta}^2} \rho_{,\tilde{\eta}}{}^2 + \frac{\partial^2 x_i^{\boldsymbol{q}}}{\partial \tilde{\xi}\partial \tilde{\eta}} \rho_{,\tilde{\xi}} \rho_{,\tilde{\eta}}
\end{aligned} \tag{8.36}
$$

再由关系 $r = \sqrt{\left(x_i^{\boldsymbol{q}} - x_i^{\boldsymbol{P}_s}\right) \cdot \left(x_i^{\boldsymbol{q}} - x_i^{\boldsymbol{P}_s}\right)}$，可得

$$
\frac{r^2}{\rho^2} = \mathcal{D}_1 \cdot \mathcal{D}_1 + 2\mathcal{D}_1 \cdot \mathcal{D}_2 \rho + \mathcal{D}_2 \cdot \mathcal{D}_2 \rho^2 \tag{8.37}
$$

因此可求出需要的系数，并表示为

$$
\begin{aligned}
C_0 &= \sqrt{G_0} = \left\| \mathcal{D}_1 \right\| \\
C_1 &= \frac{G_1}{2C_0} = \frac{\mathcal{D}_1 \cdot \mathcal{D}_2}{C_0}
\end{aligned} \tag{8.38}
$$

将式(8.30)代入式(8.28)中可得

$$
F\left(\boldsymbol{Q}, \boldsymbol{P}_{\mathrm{s}}\right) = \lim_{\rho_e(\varepsilon) \to 0} \int_{\rho_e(\varepsilon)}^{\rho(\boldsymbol{Q}, \boldsymbol{P}_{\mathrm{s}})} \frac{\overline{f}\left(\boldsymbol{q}, \boldsymbol{P}_{\mathrm{s}}\right)\left|\boldsymbol{J}_e\right|}{\rho^{\lambda-1}\left(\boldsymbol{q}, \boldsymbol{P}_{\mathrm{s}}\right)\overline{\rho}\left(\boldsymbol{q}, \boldsymbol{P}_{\mathrm{s}}\right)^{\lambda}} \mathrm{d}\rho \tag{8.39}
$$

式(8.39)中的非奇异部分可以表示为局部距离 ρ 的幂级数展开形式，并记为

$$
\overline{F}(\rho) = \frac{\overline{f}\left(\boldsymbol{q}, \boldsymbol{P}_{\mathrm{s}}\right)\left|\boldsymbol{J}_e\right|}{\overline{\rho}\left(\boldsymbol{q}, \boldsymbol{P}_{\mathrm{s}}\right)^{\lambda}} = \sum_{k=0}^{K} B^k \rho^k \tag{8.40}
$$

将式(8.40)代入式(8.39)中可得

$$F\left(\boldsymbol{Q},\boldsymbol{P}_{\mathrm{s}}\right)=\sum_{k=0}^{K}B^{k}E_{k} \tag{8.41}$$

式(8.41)中，$E_k=\lim\limits_{\rho_e(\varepsilon)\to 0}\int_{\rho_e(\varepsilon)}^{\rho(\boldsymbol{Q},\boldsymbol{P}_{\mathrm{s}})}\rho^{k-\lambda+1}\left(\boldsymbol{q},\boldsymbol{P}_{\mathrm{s}}\right)\mathrm{d}\rho$，其有限项部分为[18]

$$E_k=\begin{cases}\dfrac{1}{k-\lambda+2}\left[\dfrac{1}{\rho^{\lambda-k-2}\left(\boldsymbol{Q},\boldsymbol{P}_{\mathrm{s}}\right)}-H_{\lambda-k-2}\right], & 0\leqslant k\leqslant\lambda-3\\ \ln\rho\left(\boldsymbol{Q},\boldsymbol{P}_{\mathrm{s}}\right)-\ln H_0, & k=\lambda-2\\ \dfrac{\rho^{k-\lambda+2}\left(\boldsymbol{Q},\boldsymbol{P}_{\mathrm{s}}\right)}{k-\lambda+2}, & k>\lambda-2\end{cases} \tag{8.42}$$

其中，$H_0=\dfrac{1}{C_0},H_1=\dfrac{C_1}{C_0}$。

8.3.5　自适应积分

非奇异积分通过高斯求积法进行计算。曲面高斯求积公式可以表示为

$$\begin{aligned}I&=\int_{-1}^{1}\int_{-1}^{1}f\left(\tilde{\xi},\tilde{\eta}\right)\mathrm{d}\tilde{\xi}\mathrm{d}\tilde{\eta}\\&=\sum_{i=1}^{m_1}\sum_{j=1}^{m_2}f\left(\tilde{\xi}_i,\tilde{\eta}_j\right)\mathcal{W}_i\mathcal{W}_j+E_1+E_2\end{aligned} \tag{8.43}$$

式中，$\tilde{\xi}_i$ 和 $\tilde{\eta}_j$ 是高斯积分点坐标；m_1 和 m_2 是两个母空间坐标方向上的积分阶次；$\mathcal{W}_i$ 和 $\mathcal{W}_j$ 是积分点处的权值；E_1 和 E_2 是两个坐标方向上的积分误差，它们相对误差的上界表示为[21]

$$\frac{E_i}{I}\leqslant 2\left(\frac{L_i}{4d}\right)^{2m_i}\frac{\left(2m_i+\lambda+1\right)!}{\left(2m_i\right)!\left(\lambda-1\right)!} \tag{8.44}$$

其中，λ 代表奇异积分的阶次；L_i 是被积单元的长度；d 是源点到该积分单元的最短距离，它们的计算过程详见文献[19]和[21]。

在给定目标积分误差水平 e_{int} 下，所需的积分阶次表示为[19,21,22]

$$m_i=\sqrt{\frac{2}{3}\lambda+\frac{2}{5}\left[-\frac{1}{10}\ln\left(\frac{e_{\mathrm{int}}}{2}\right)\right]\left[\left(\frac{8L_i}{3d}\right)^{\frac{3}{4}}+1\right]} \tag{8.45}$$

本章结合上述积分误差方程与四叉树逐层分解方法来驱动自适应积分过程。首先，给定目标积分误差水平和每个积分单元中指定的最大高斯阶次，计

算每个参数方向上积分单元的长度以及从配点到该单元的最短距离，然后根据式(8.45)获得所需的高斯阶次。如果其值大于指定值，则该单元通过四叉树层次分解被细分为四个相等的积分子单元。重复上述过程，直到每个方向上所需的高斯阶次均小于或等于预定值。图 8.3 展示了当源点逐渐靠近积分单元，和 $\lambda=3$ 时的单元细分过程。我们可以看到，除非采用自适应积分方案，否则就需要更多的高斯点。

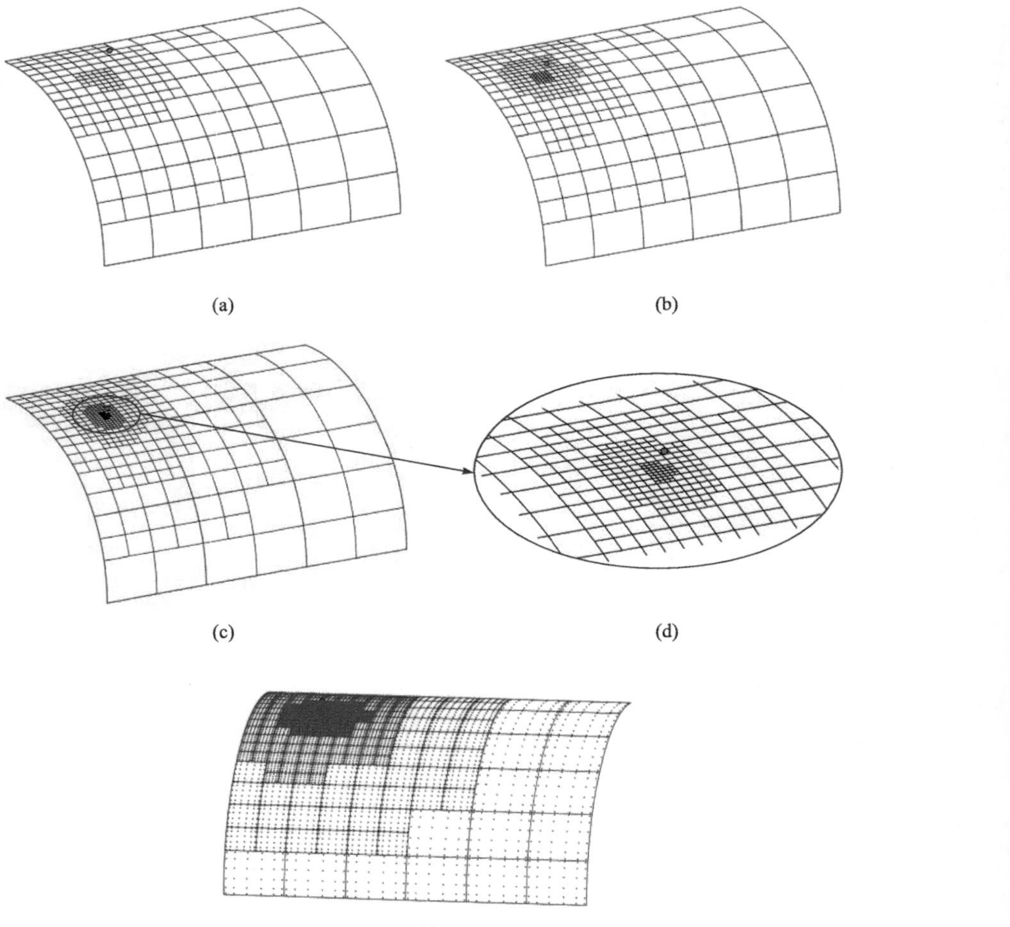

图 8.3　当源点逐渐靠近积分单元，和 $\lambda=3$ 时的单元细分过程

蓝色线表示 NURBS 网格(圆柱体的半径为 1)：(a) 源点到单元的距离为 0.2；(b) 源点到单元的距离为 0.1；(c) 源点到单元的距离为 0.05；(d) 图(c)的局部详细视图；(e) 与距离为 0.1 的源点相关的积分点分布，其中蓝星代表高斯点

8.4　非相适应界面耦合

等几何有限元法建立了结点外力和位移方程组，而等几何边界元法涉及的基本变量是结点位移和结点面力。因此，耦合这两种方法的核心是建立这两类方程中的未知场变量之间的联系。不失一般性，考虑如图 8.4 所示求解区域 Ω，其包含有限元子域 Ω^{fem} 和边界元子域 Ω^{bem}，以及耦合界面 Γ_{I}。

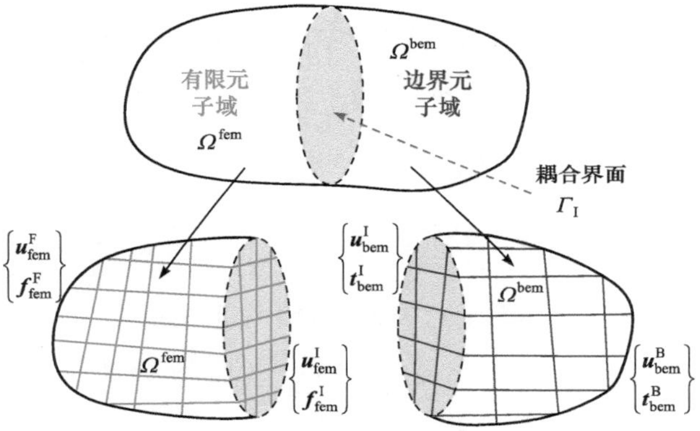

图 8.4　等几何有限元-边界元耦合区域示意图

所需的平衡和连续耦合条件为

$$
\begin{aligned}
\boldsymbol{f}_{\text{I}}^{\text{F}} &= -\boldsymbol{t}_{\text{I}}^{\text{B}} \\
\boldsymbol{u}_{\text{I}}^{\text{F}} &= \boldsymbol{u}_{\text{I}}^{\text{B}}
\end{aligned}
\tag{8.46}
$$

式中，$\boldsymbol{t}_{\text{I}}^{\text{B}}$ 为耦合界面上边界元子域面力 $\boldsymbol{t}_{\text{bem}}^{\text{I}}$ 所在控制点处所形成的等效合力。

8.4.1　面力平衡耦合约束

这里利用虚功原理建立了界面上的控制力矢量 $\boldsymbol{f}_{\text{fem}}^{\text{I}}$ 和控制面力矢量 $\boldsymbol{t}_{\text{bem}}^{\text{I}}$ 之间的关系，即 $\boldsymbol{f}_{\text{fem}}^{\text{I}}$ 和 $\boldsymbol{t}_{\text{bem}}^{\text{I}}$ 在任意虚位移 $\delta\boldsymbol{u}^{\text{I}}$ 上产生的虚功相等。界面上任意一点处的位移和面力表示为

$$
\begin{aligned}
\boldsymbol{u}^{\text{I}} &= \boldsymbol{R}_{\text{fem}}^{\text{I}}\boldsymbol{u}_{\text{fem}}^{\text{I}} \\
\boldsymbol{t}^{\text{I}} &= \boldsymbol{R}_{\text{bem}}^{\text{I}}\boldsymbol{t}_{\text{bem}}^{\text{I}}
\end{aligned}
\tag{8.47}
$$

其中，$\boldsymbol{u}_{\text{fem}}^{\text{I}}$ 是耦合界面上属于有限元子域的控制点处的位移矢量；$\boldsymbol{R}_{\text{fem}}^{\text{I}}$ 和 $\boldsymbol{R}_{\text{bem}}^{\text{I}}$ 分别是对应于有限元和边界元子域的 NURBS 基函数矩阵，表示为

$$\boldsymbol{R}_{\text{fem}}^{\text{I}}=\begin{bmatrix} R_1^{\text{fem}} & 0 & 0 & \cdots & R_{n_{\text{f}}}^{\text{fem}} & 0 & 0 \\ 0 & R_1^{\text{fem}} & 0 & \cdots & 0 & R_{n_{\text{f}}}^{\text{fem}} & 0 \\ 0 & 0 & R_1^{\text{fem}} & \cdots & 0 & 0 & R_{n_{\text{f}}}^{\text{fem}} \end{bmatrix}$$
$$\boldsymbol{R}_{\text{bem}}^{\text{I}}=\begin{bmatrix} R_1^{\text{bem}} & 0 & 0 & \cdots & R_{n_{\text{b}}}^{\text{bem}} & 0 & 0 \\ 0 & R_1^{\text{bem}} & 0 & \cdots & 0 & R_{n_{\text{b}}}^{\text{bem}} & 0 \\ 0 & 0 & R_1^{\text{bem}} & \cdots & 0 & 0 & R_{n_{\text{b}}}^{\text{bem}} \end{bmatrix} \tag{8.48}$$

式中，n_{f} 和 n_{b} 分别为有限元和边界元子域中与耦合界面相关的控制点数量。

由 $\boldsymbol{f}_{\text{fem}}^{\text{I}}$ 和 $\boldsymbol{t}_{\text{bem}}^{\text{I}}$ 产生的虚功表示为

$$\delta\mathcal{W}^{\text{fem}}=\delta\left(\boldsymbol{u}_{\text{fem}}^{\text{I}}\right)^{\text{T}}\boldsymbol{f}_{\text{fem}}^{\text{I}}$$
$$\delta\mathcal{W}^{\text{bem}}=\int_{\Gamma_{\text{I}}}\delta\left(\boldsymbol{u}^{\text{I}}\right)^{\text{T}}\boldsymbol{t}^{\text{I}}\text{d}\Gamma_{\text{I}}=\int_{\Gamma_{\text{I}}}\delta\left(\boldsymbol{u}_{\text{fem}}^{\text{I}}\right)^{\text{T}}\left(\boldsymbol{R}_{\text{fem}}^{\text{I}}\right)^{\text{T}}\boldsymbol{R}_{\text{bem}}^{\text{I}}\boldsymbol{t}_{\text{bem}}^{\text{I}}\text{d}\Gamma_{\text{I}} \tag{8.49}$$

因此，可获得耦合界面上的控制力矢量与面力之间的关系为

$$\boldsymbol{f}_{\text{fem}}^{\text{I}}=-\boldsymbol{M}\boldsymbol{t}_{\text{bem}}^{\text{I}} \tag{8.50}$$

其中，$\boldsymbol{M}$ 是转换矩阵，其表达式为

$$\boldsymbol{M}=\int_{\Gamma_{\text{I}}}\left(\boldsymbol{R}_{\text{fem}}^{\text{I}}\right)^{\text{T}}\boldsymbol{R}_{\text{bem}}^{\text{I}}\text{d}\Gamma_{\text{I}} \tag{8.51}$$

应注意的是，对于非相适应的耦合界面，$\boldsymbol{M}$ 不是一个方阵。在程序实施的过程中，耦合界面两侧的积分点必须位于相同位置。因此，需要点的反求算法将更密集耦合网格一侧的积分点映射到另一侧粗糙的耦合网格上去。

8.4.2 位移连续耦合约束

基于 Coox 和其合作者[23]提出的方法，使用虚拟节点插入技术，构造两个子域之间非相适应界面上的位移约束。该方法的实施与求解问题的类型无关，只涉及界面两侧节点向量和 NURBS 权重的几何信息。本章将上述方法推广到非相适应 NURBS 曲面的情况，并将其应用于三维等几何有限元-边界元耦合分析问题。

对于定义在节点矢量 $\boldsymbol{\varXi}=\left\{\xi_1,\xi_2,\cdots,\xi_{n+p+1}\right\}$ 上的 B 样条曲线，我们通过插入 r 个非递减的节点集 $\hat{\boldsymbol{\varXi}}=\left\{\hat{\xi}_1,\hat{\xi}_2,\cdots,\hat{\xi}_r\right\}$ (允许插入重复节点)，可以得到新的扩展的节点矢量 $\tilde{\boldsymbol{\varXi}}=\left\{\tilde{\xi}_1=\xi_1,\cdots,\tilde{\xi}_{n+r+p+1}=\xi_{n+p+1}\right\}$。新的 $n+r$ 个控制点 $\tilde{\boldsymbol{P}}_{\xi}=\left\{\tilde{\boldsymbol{P}}_1;\tilde{\boldsymbol{P}}_2;\cdots;\tilde{\boldsymbol{P}}_{n+r}\right\}$

可以表示为初始控制点 $\boldsymbol{P}_\xi = \{\boldsymbol{P}_1; \boldsymbol{P}_2; \cdots; \boldsymbol{P}_n\}$ 的线性组合，记为

$$\tilde{\boldsymbol{P}}_\xi = \boldsymbol{C}_\xi \boldsymbol{P}_\xi \tag{8.52}$$

式中，$\boldsymbol{C}_\xi$ 是参数 ξ 方向上的线性局部细分算子，其可以通过类似于求解 Bezier 提取算子的方法[24]得出。

由初始节点矢量和插入节点集合，线性局部细分算子 $\boldsymbol{C}_\xi$ 的计算式为

$$\boldsymbol{C}_\xi = \boldsymbol{C}_\xi^{(r)} \boldsymbol{C}_\xi^{(r-1)} \boldsymbol{C}_\xi^{(r-2)} \cdots \boldsymbol{C}_\xi^{(2)} \boldsymbol{C}_\xi^{(1)} \tag{8.53}$$

其中，

$$\boldsymbol{C}_\xi^{(j)} = \begin{bmatrix} \alpha_1^{(j)} & 0 & 0 & \cdots & & & 0 \\ 1-\alpha_2^{(j)} & \alpha_2^{(j)} & 0 & & \cdots & & 0 \\ 0 & 1-\alpha_3^{(j)} & \alpha_3^{(j)} & 0 & & \cdots & 0 \\ \vdots & & & \ddots & & & \vdots \\ 0 & \cdots & & & 0 & 1-\alpha_{n+j-1}^{(j)} & \alpha_{n+j-1}^{(j)} \\ 0 & & \cdots & & & 0 & 1-\alpha_{n+j}^{(j)} \end{bmatrix} \quad (j = 1, 2, \cdots, r) \tag{8.54}$$

式中，

$$\alpha_i^{(j)} = \begin{cases} 1, & 1 \leqslant i \leqslant k_j - p \\ \dfrac{\hat{\xi}_j - \xi_i}{\xi_{i+p} - \xi_i}, & k_j - p + 1 \leqslant i \leqslant k_j \quad (i = 1, 2, \cdots, n+j) \\ 0, & i \geqslant k_j + 1 \end{cases} \tag{8.55}$$

其中，k_j 是节点 $\hat{\xi}_j$ 在初始节点矢量 $\boldsymbol{\varXi}$ 中的节点区间跨度编号。

按照同样的道理，给定初始节点矢量 $\boldsymbol{\varPsi} = \{\eta_1, \eta_2, \cdots, \eta_{m+q+1}\}$ 和插入 s 个非递减的节点集 $\hat{\boldsymbol{\varPsi}} = \{\hat{\eta}_1, \hat{\eta}_2, \cdots, \hat{\eta}_s\}$ (仍然允许插入重复节点)，可以得到新的扩展的节点矢量 $\tilde{\boldsymbol{\varPsi}} = \{\tilde{\eta}_1 = \eta_1, \cdots, \tilde{\eta}_{m+s+q+1} = \eta_{m+q+1}\}$，从而得到参数 η 方向上的线性局部细分算子 $\boldsymbol{C}_\eta$。由于 B 样条曲面是由 ξ 和 η 两个参数方向上 B 样条曲线张量积而成，因此其线性局部细分算子 $\boldsymbol{C}_{\mathrm{BS}}$ 表示为

$$\boldsymbol{C}_{\mathrm{BS}} = \boldsymbol{C}_\eta \otimes \boldsymbol{C}_\xi \tag{8.56}$$

式中，算子 $\otimes$ 对于两个矩阵的运算规则为

$$M \otimes N = \begin{bmatrix} M_{11}N & M_{12}N & \cdots \\ M_{21}N & M_{22}N & \\ \vdots & & \ddots \end{bmatrix} \tag{8.57}$$

NURBS 曲面是 B 样条曲面的有理形式，通过对 B 样条曲面的控制点加权而成。给定初始的控制点网格 $\boldsymbol{P}=\left\{\boldsymbol{P}_{i,j}\right\}(i=1,2,\cdots,n;\ j=1,2,\cdots,m)$ 和曲面细分之后的控制点网格 $\tilde{\boldsymbol{P}}=\left\{\tilde{\boldsymbol{P}}_{\tilde{i},\tilde{j}}\right\}\left(\tilde{i}=1,2,\cdots,n+r;\ \tilde{j}=1,2,\cdots,m+s\right)$，利用细分算子 $\mathcal{D}$ 可以得到如下的关系：

$$\tilde{\boldsymbol{P}}=\mathcal{D}\boldsymbol{P}=\left(\boldsymbol{W}_{\tilde{\boldsymbol{P}}}\right)^{-1}\boldsymbol{C}_{\mathrm{BS}}\boldsymbol{W}_{\boldsymbol{P}}\boldsymbol{P} \tag{8.58}$$

其中，$\boldsymbol{W}_{\boldsymbol{P}}$ 和 $\boldsymbol{W}_{\tilde{\boldsymbol{P}}}$ 是对角矩阵，分别包含控制点 $\boldsymbol{P}$ 和 $\tilde{\boldsymbol{P}}$ 的权值，它们写为

$$\boldsymbol{W}_{\boldsymbol{P}}=\begin{bmatrix} \omega_{P_1} & & & \\ & \omega_{P_2} & & \\ & & \ddots & \\ & & & \omega_{P_{nm}} \end{bmatrix},\quad \boldsymbol{W}_{\tilde{\boldsymbol{P}}}=\begin{bmatrix} \omega_{\tilde{P}_1} & & & \\ & \omega_{\tilde{P}_2} & & \\ & & \ddots & \\ & & & \omega_{\tilde{P}_{(n+r)(m+s)}} \end{bmatrix} \tag{8.59}$$

接下来，考虑如图 8.5 所示的有限元和边界元子域之间的非相适应耦合界面。耦合过程实施的先决条件是耦合界面两侧的 NURBS 基函数的阶次相同，我们总是可以通过升阶(p 细分)来达到要求。

绿色和黄色填充圆圈分别代表属于有限元和边界元子域耦合界面上的控制点。灰线代表物理空间中新的虚拟插入的节点线在物理空间中的投影。通过虚拟插入位于另一侧曲面节点向量中而不包含在其自身节点向量中的新节点，直到两侧的耦合曲面被细化到一致为止，得到细化后更新的虚拟控制点与原始控制点之间的关系为

$$\begin{aligned} \tilde{\boldsymbol{P}} &= \mathcal{D}_{\mathrm{fem}}\boldsymbol{P}_{\mathrm{fem}} \\ \tilde{\boldsymbol{P}} &= \mathcal{D}_{\mathrm{bem}}\boldsymbol{P}_{\mathrm{bem}} \end{aligned} \tag{8.60}$$

式中，$\mathcal{D}_{\mathrm{fem}}$ 和 $\mathcal{D}_{\mathrm{bem}}$ 分别对应于有限元和边界元子域的局部虚拟细分算子。

上述约束也适用于这些控制点处控制变量(自由度)之间的关系。因此，耦合界面上的位移约束描述为

$$\mathcal{T}_{\mathrm{fem}}^{\mathrm{I}}\boldsymbol{u}_{\mathrm{fem}}^{\mathrm{I}}-\mathcal{T}_{\mathrm{bem}}^{\mathrm{I}}\boldsymbol{u}_{\mathrm{bem}}^{\mathrm{I}}=\boldsymbol{0} \tag{8.61}$$

其中，上角标 I 代表界面；$\mathcal{T}_{\mathrm{fem}}^{\mathrm{I}}$ 和 $\mathcal{T}_{\mathrm{bem}}^{\mathrm{I}}$ 分别是 $\mathcal{D}_{\mathrm{fem}}$ 和 $\mathcal{D}_{\mathrm{bem}}$ 的扩展形式(每个控制点处位移自由度的数目)。

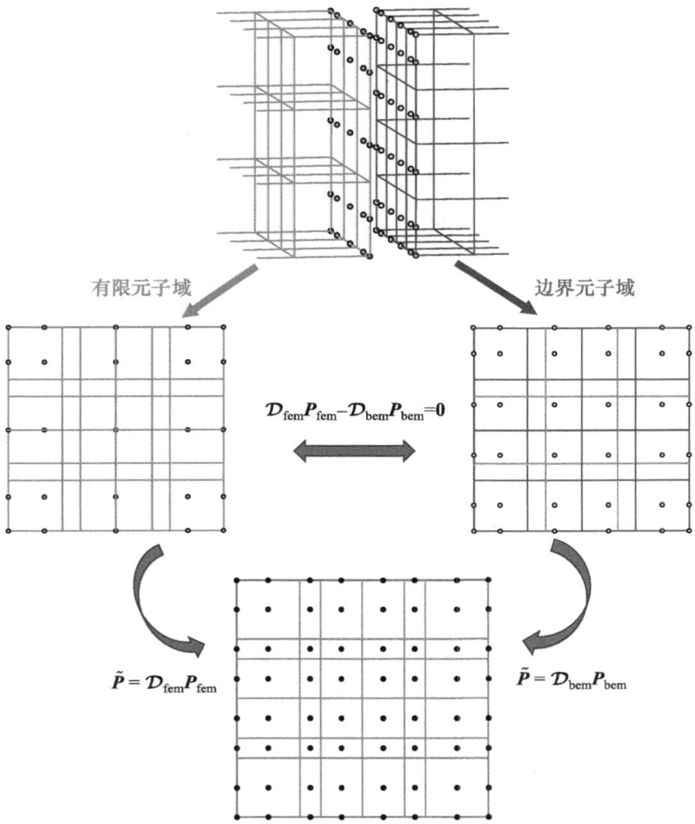

图 8.5　非相适应界面上控制点耦合约束的实施过程

8.4.3　耦合系统方程

在这里，我们约定上标 F、B 和 I 分别代表有限元子域、边界元子域和耦合界面。将自由度划分为与耦合界面相关的自由度和其余自由度两个部分，则有限元部分的系统方程可以描述为

$$\begin{bmatrix} \boldsymbol{K}_{\text{fem}}^{\text{FF}} & \boldsymbol{K}_{\text{fem}}^{\text{FI}} \\ \boldsymbol{K}_{\text{fem}}^{\text{IF}} & \boldsymbol{K}_{\text{fem}}^{\text{II}} \end{bmatrix} \begin{Bmatrix} \boldsymbol{u}_{\text{fem}}^{\text{F}} \\ \boldsymbol{u}_{\text{fem}}^{\text{I}} \end{Bmatrix} = \begin{Bmatrix} \boldsymbol{f}_{\text{fem}}^{\text{F}} \\ \boldsymbol{f}_{\text{fem}}^{\text{I}} \end{Bmatrix} \tag{8.62}$$

同样地，边界元部分的系统方程可以描述为

$$\begin{bmatrix} \boldsymbol{H}_{\text{bem}}^{\text{II}} & \boldsymbol{H}_{\text{bem}}^{\text{IB}} \\ \boldsymbol{H}_{\text{bem}}^{\text{BI}} & \boldsymbol{H}_{\text{bem}}^{\text{BB}} \end{bmatrix} \begin{Bmatrix} \boldsymbol{u}_{\text{bem}}^{\text{I}} \\ \boldsymbol{u}_{\text{bem}}^{\text{B}} \end{Bmatrix} = \begin{bmatrix} \boldsymbol{G}_{\text{bem}}^{\text{II}} & \boldsymbol{G}_{\text{bem}}^{\text{IB}} \\ \boldsymbol{G}_{\text{bem}}^{\text{BI}} & \boldsymbol{G}_{\text{bem}}^{\text{BB}} \end{bmatrix} \begin{Bmatrix} \boldsymbol{t}_{\text{bem}}^{\text{I}} \\ \boldsymbol{t}_{\text{bem}}^{\text{B}} \end{Bmatrix} \tag{8.63}$$

当耦合界面两侧的网格相一致时，我们可以得到一对一协调的位移连续条件 $\boldsymbol{u}_{\text{fem}}^{\text{I}} = \boldsymbol{u}_{\text{bem}}^{\text{I}} = \boldsymbol{u}^{\text{I}}$，引入面力平衡条件，并与有限元和边界元部分的系统方程耦合

后，得到全局的耦合系统方程为

$$\begin{bmatrix} \boldsymbol{K}_{\text{fem}}^{\text{FF}} & \boldsymbol{K}_{\text{fem}}^{\text{FI}} & \boldsymbol{0} & \boldsymbol{0} \\ \boldsymbol{K}_{\text{fem}}^{\text{IF}} & \boldsymbol{K}_{\text{fem}}^{\text{II}} & \boldsymbol{M} & \boldsymbol{0} \\ \boldsymbol{0} & \boldsymbol{H}_{\text{bem}}^{\text{II}} & -\boldsymbol{G}_{\text{bem}}^{\text{II}} & \boldsymbol{H}_{\text{bem}}^{\text{IB}} \\ \boldsymbol{0} & \boldsymbol{H}_{\text{bem}}^{\text{BI}} & -\boldsymbol{G}_{\text{bem}}^{\text{BI}} & \boldsymbol{H}_{\text{bem}}^{\text{BB}} \end{bmatrix} \begin{Bmatrix} \boldsymbol{u}_{\text{fem}}^{\text{F}} \\ \boldsymbol{u}^{\text{I}} \\ \boldsymbol{t}_{\text{bem}}^{\text{I}} \\ \boldsymbol{u}_{\text{bem}}^{\text{B}} \end{Bmatrix} = \begin{Bmatrix} \boldsymbol{f}_{\text{fem}}^{\text{F}} \\ \boldsymbol{0} \\ \boldsymbol{G}_{\text{bem}}^{\text{IB}} \boldsymbol{t}_{\text{bem}}^{\text{B}} \\ \boldsymbol{G}_{\text{bem}}^{\text{BB}} \boldsymbol{t}_{\text{bem}}^{\text{B}} \end{Bmatrix} \tag{8.64}$$

对于非相适应界面网格，采用静态矩阵凝聚法求解耦合方程。根据文献[23]，建议选择与具有更密集界面网格的子域相关的未知变量作为非独立自由度进行压缩，这种选择不仅可以最小化未知自变量的数量，而且还可以满足最小二乘近似意义下的位移耦合约束。当 FE 子域具有更细的界面网格时，位移耦合约束关系表示为

$$\boldsymbol{u}_{\text{bem}}^{\text{I}} = \boldsymbol{T}_{\text{bf}} \boldsymbol{u}_{\text{fem}}^{\text{I}} = \left(\boldsymbol{\mathcal{T}}_{\text{bem}}^{\text{I}}\right)^{+} \boldsymbol{\mathcal{T}}_{\text{fem}}^{\text{I}} \boldsymbol{u}_{\text{fem}}^{\text{I}} \tag{8.65}$$

式中，上角标 “+”代表 Moore-Penrose 伪逆。施加边界元子域对应的边界条件后，得到全局的耦合系统方程为

$$\begin{bmatrix} \boldsymbol{K}_{\text{fem}}^{\text{FF}} & \boldsymbol{K}_{\text{fem}}^{\text{FI}} & \boldsymbol{0} & \boldsymbol{0} \\ \boldsymbol{K}_{\text{fem}}^{\text{IF}} & \boldsymbol{K}_{\text{fem}}^{\text{II}} & \boldsymbol{M} & \boldsymbol{0} \\ \boldsymbol{0} & \boldsymbol{H}_{\text{bem}}^{\text{II}} \boldsymbol{T}_{\text{bf}} & -\boldsymbol{G}_{\text{bem}}^{\text{II}} & \boldsymbol{\mathcal{A}}_{\text{bem}}^{\text{IB}} \\ \boldsymbol{0} & \boldsymbol{H}_{\text{bem}}^{\text{BI}} \boldsymbol{T}_{\text{bf}} & -\boldsymbol{G}_{\text{bem}}^{\text{BI}} & \boldsymbol{\mathcal{A}}_{\text{bem}}^{\text{BB}} \end{bmatrix} \begin{Bmatrix} \boldsymbol{u}_{\text{fem}}^{\text{F}} \\ \boldsymbol{u}_{\text{fem}}^{\text{I}} \\ \boldsymbol{t}_{\text{bem}}^{\text{I}} \\ \boldsymbol{\mathcal{X}}_{\text{bem}}^{\text{B}} \end{Bmatrix} = \begin{Bmatrix} \boldsymbol{f}_{\text{fem}}^{\text{F}} \\ \boldsymbol{0} \\ \boldsymbol{\mathcal{B}}_{\text{bem}}^{\text{IB}} \boldsymbol{\mathcal{Y}}_{\text{bem}}^{\text{B}} \\ \boldsymbol{\mathcal{B}}_{\text{bem}}^{\text{IB}} \boldsymbol{\mathcal{Y}}_{\text{bem}}^{\text{B}} \end{Bmatrix} \tag{8.66}$$

同样地，得到另一种情况下的位移耦合约束关系为

$$\boldsymbol{u}_{\text{fem}}^{\text{I}} = \boldsymbol{T}_{\text{fb}} \boldsymbol{u}_{\text{bem}}^{\text{I}} = \left(\boldsymbol{\mathcal{T}}_{\text{fem}}^{\text{I}}\right)^{+} \boldsymbol{\mathcal{T}}_{\text{bem}}^{\text{I}} \boldsymbol{u}_{\text{bem}}^{\text{I}} \tag{8.67}$$

在施加边界元子域对应的边界条件后，同样得到全局的耦合系统方程为

$$\begin{bmatrix} \boldsymbol{K}_{\text{fem}}^{\text{FF}} & \boldsymbol{K}_{\text{fem}}^{\text{FI}} \boldsymbol{T}_{\text{fb}} & \boldsymbol{0} & \boldsymbol{0} \\ \boldsymbol{T}_{\text{fb}}^{\text{T}} \boldsymbol{K}_{\text{fem}}^{\text{IF}} & \boldsymbol{T}_{\text{fb}}^{\text{T}} \boldsymbol{K}_{\text{fem}}^{\text{II}} \mathbf{T}_{\text{fb}} & \boldsymbol{T}_{\text{fb}}^{\text{T}} \boldsymbol{M} & \boldsymbol{0} \\ \boldsymbol{0} & \boldsymbol{H}_{\text{bem}}^{\text{II}} & -\boldsymbol{G}_{\text{bem}}^{\text{II}} & \boldsymbol{\mathcal{A}}_{\text{bem}}^{\text{IB}} \\ \boldsymbol{0} & \boldsymbol{H}_{\text{bem}}^{\text{BI}} & -\boldsymbol{G}_{\text{bem}}^{\text{BI}} & \boldsymbol{\mathcal{A}}_{\text{bem}}^{\text{BB}} \end{bmatrix} \begin{Bmatrix} \boldsymbol{u}_{\text{fem}}^{\text{F}} \\ \boldsymbol{u}_{\text{bem}}^{\text{I}} \\ \boldsymbol{t}_{\text{bem}}^{\text{I}} \\ \boldsymbol{\mathcal{X}}_{\text{bem}}^{\text{B}} \end{Bmatrix} = \begin{Bmatrix} \boldsymbol{f}_{\text{fem}}^{\text{F}} \\ \boldsymbol{0} \\ \boldsymbol{\mathcal{B}}_{\text{bem}}^{\text{IB}} \boldsymbol{\mathcal{Y}}_{\text{bem}}^{\text{B}} \\ \boldsymbol{\mathcal{B}}_{\text{bem}}^{\text{IB}} \boldsymbol{\mathcal{Y}}_{\text{bem}}^{\text{B}} \end{Bmatrix} \tag{8.68}$$

8.4.4 对称迭代求解

本章将对称迭代有限元-边界元耦合方法[25,26]推广到等几何耦合分析领域。在该方法中，除与耦合界面对应的自由度外，与边界元子域相关的所有自由度都被压缩。

在施加边界元子域对应的边界条件后，边界元子域的系统方程可以写作

$$\begin{bmatrix} \boldsymbol{H}_{\text{bem}}^{\text{II}} & \boldsymbol{\mathcal{A}}_{\text{bem}}^{\text{IB}} \\ \boldsymbol{H}_{\text{bem}}^{\text{BI}} & \boldsymbol{\mathcal{A}}_{\text{bem}}^{\text{BB}} \end{bmatrix} \begin{Bmatrix} \boldsymbol{u}_{\text{bem}}^{\text{I}} \\ \boldsymbol{\mathcal{X}}_{\text{bem}}^{\text{B}} \end{Bmatrix} = \begin{Bmatrix} \boldsymbol{f}_{\text{b}}^{\text{I}} \\ \boldsymbol{f}_{\text{b}} \end{Bmatrix} + \begin{bmatrix} \boldsymbol{G}_{\text{bem}}^{\text{II}} \\ \boldsymbol{G}_{\text{bem}}^{\text{BI}} \end{bmatrix} \left\{ \boldsymbol{t}_{\text{bem}}^{\text{I}} \right\} \tag{8.69}$$

其中，$\boldsymbol{f}_{\text{b}}^{\text{I}} = \boldsymbol{\mathcal{B}}_{\text{bem}}^{\text{IB}} \boldsymbol{\mathcal{Y}}_{\text{bem}}^{\text{B}}$，$\boldsymbol{f}_{\text{b}} = \boldsymbol{\mathcal{B}}_{\text{bem}}^{\text{BB}} \boldsymbol{\mathcal{Y}}_{\text{bem}}^{\text{B}}$。

通过部分消去子矩阵 $\boldsymbol{\mathcal{A}}_{\text{bem}}^{\text{IB}}$，式(8.69)可以转化为如下形式：

$$\begin{bmatrix} \bar{\boldsymbol{H}}_{\text{bem}}^{\text{II}} & \boldsymbol{0} \\ \bar{\boldsymbol{H}}_{\text{bem}}^{\text{BI}} & \bar{\boldsymbol{\mathcal{A}}}_{\text{bem}}^{\text{BB}} \end{bmatrix} \begin{Bmatrix} \boldsymbol{u}_{\text{bem}}^{\text{I}} \\ \boldsymbol{\mathcal{X}}_{\text{bem}}^{\text{B}} \end{Bmatrix} = \begin{Bmatrix} \bar{\boldsymbol{f}}_{\text{b}}^{\text{I}} \\ \bar{\boldsymbol{f}}_{\text{b}} \end{Bmatrix} + \begin{bmatrix} \bar{\boldsymbol{G}}_{\text{bem}}^{\text{II}} \\ \bar{\boldsymbol{G}}_{\text{bem}}^{\text{BI}} \end{bmatrix} \left\{ \boldsymbol{t}_{\text{bem}}^{\text{I}} \right\} \tag{8.70}$$

因此，得到仅与耦合界面上自由度相关的边界元子域方程为

$$\bar{\boldsymbol{H}}_{\text{bem}}^{\text{II}} \boldsymbol{u}_{\text{bem}}^{\text{I}} = \bar{\boldsymbol{f}}_{\text{b}}^{\text{I}} + \bar{\boldsymbol{G}}_{\text{bem}}^{\text{II}} \boldsymbol{t}_{\text{bem}}^{\text{I}} \tag{8.71}$$

从而可以得到边界元子域耦合界面上面力与位移的关系

$$\boldsymbol{t}_{\text{bem}}^{\text{I}} = -\left(\bar{\boldsymbol{G}}_{\text{bem}}^{\text{II}} \right)^{-1} \left(\bar{\boldsymbol{f}}_{\text{b}}^{\text{I}} - \bar{\boldsymbol{H}}_{\text{bem}}^{\text{II}} \boldsymbol{u}_{\text{bem}}^{\text{I}} \right) \tag{8.72}$$

将式(8.65)和式(8.72)代入式(8.50)中可得

$$\boldsymbol{f}_{\text{fem}}^{\text{I}} = \boldsymbol{M} \left(\bar{\boldsymbol{G}}_{\text{bem}}^{\text{II}} \right)^{-1} \bar{\boldsymbol{f}}_{\text{b}}^{\text{I}} - \boldsymbol{M} \left(\bar{\boldsymbol{G}}_{\text{bem}}^{\text{II}} \right)^{-1} \bar{\boldsymbol{H}}_{\text{bem}}^{\text{II}} \boldsymbol{T}_{\text{bf}} \boldsymbol{u}_{\text{fem}}^{\text{I}} \tag{8.73}$$

采用如下的记号：

$$\begin{aligned} \boldsymbol{K}_{\text{b}}^{\text{I}} &= \boldsymbol{M} \left(\bar{\boldsymbol{G}}_{\text{bem}}^{\text{II}} \right)^{-1} \bar{\boldsymbol{H}}_{\text{bem}}^{\text{II}} \boldsymbol{T}_{\text{bf}} \\ \boldsymbol{F}_{\text{b}}^{\text{I}} &= \boldsymbol{M} \left(\bar{\boldsymbol{G}}_{\text{bem}}^{\text{II}} \right)^{-1} \bar{\boldsymbol{f}}_{\text{b}}^{\text{I}} \end{aligned} \tag{8.74}$$

然后将式(8.73)代入式(8.62)中，可以得到最终的耦合系统方程为

$$\begin{bmatrix} \boldsymbol{K}_{\text{fem}}^{\text{FF}} & \boldsymbol{K}_{\text{fem}}^{\text{FI}} \\ \boldsymbol{K}_{\text{fem}}^{\text{IF}} & \boldsymbol{K}_{\text{fem}}^{\text{II}} + \boldsymbol{K}_{\text{b}}^{\text{I}} \end{bmatrix} \begin{Bmatrix} \boldsymbol{u}_{\text{fem}}^{\text{F}} \\ \boldsymbol{u}_{\text{fem}}^{\text{I}} \end{Bmatrix} = \begin{Bmatrix} \boldsymbol{f}_{\text{fem}}^{\text{F}} \\ \boldsymbol{F}_{\text{b}}^{\text{I}} \end{Bmatrix} \tag{8.75}$$

由于矩阵 $\boldsymbol{K}_{\text{b}}^{\text{I}}$ 的非对称和稠密的特性，最终耦合系统方程的对称性和稀疏性遭到破坏。因此，引入一个对称的增强矩阵 $\boldsymbol{K}_{\text{enh}}$ 到耦合系统方程中，则可以得到

$$\begin{bmatrix} \boldsymbol{K}_{\text{fem}}^{\text{FF}} & \boldsymbol{K}_{\text{fem}}^{\text{FI}} \\ \boldsymbol{K}_{\text{fem}}^{\text{IF}} & \boldsymbol{K}_{\text{fem}}^{\text{II}} + \boldsymbol{K}_{\text{enh}} \end{bmatrix} \begin{Bmatrix} \boldsymbol{u}_{\text{fem}}^{\text{F}} \\ \boldsymbol{u}_{\text{fem}}^{\text{I}} \end{Bmatrix} = \begin{Bmatrix} \boldsymbol{f}_{\text{fem}}^{\text{F}} \\ \boldsymbol{F}_{\text{b}}^{\text{I}} \end{Bmatrix} + \begin{bmatrix} \boldsymbol{0} & \boldsymbol{0} \\ \boldsymbol{0} & -\boldsymbol{K}_{\text{b}}^{\text{I}} + \boldsymbol{K}_{\text{enh}} \end{bmatrix} \begin{Bmatrix} \boldsymbol{u}_{\text{fem}}^{\text{F}} \\ \boldsymbol{u}_{\text{fem}}^{\text{I}} \end{Bmatrix} \tag{8.76}$$

进而得到如下的迭代求解公式：

$$\bar{\boldsymbol{K}} \boldsymbol{U}_{n+1} = \bar{\boldsymbol{F}} + \bar{\boldsymbol{K}}_{\text{c}} \boldsymbol{U}_n \tag{8.77}$$

其中，

$$\bar{\boldsymbol{K}}=\begin{bmatrix}\boldsymbol{K}_{\text{fem}}^{\text{FF}} & \boldsymbol{K}_{\text{fem}}^{\text{FI}}\\ \boldsymbol{K}_{\text{fem}}^{\text{IF}} & \boldsymbol{K}_{\text{fem}}^{\text{II}}+\boldsymbol{K}_{\text{enh}}\end{bmatrix},\quad \bar{\boldsymbol{K}}_{\text{c}}=\begin{bmatrix}\boldsymbol{0} & \boldsymbol{0}\\ \boldsymbol{0} & -\boldsymbol{K}_{\text{b}}^{\text{I}}+\boldsymbol{K}_{\text{enh}}\end{bmatrix},\quad \bar{\boldsymbol{F}}=\begin{Bmatrix}\boldsymbol{f}_{\text{fem}}^{\text{F}}\\ \boldsymbol{F}_{\text{b}}^{\text{I}}\end{Bmatrix} \tag{8.78}$$

初始位移矢量 $\boldsymbol{U}_0$ 可以选择为零矢量。迭代过程直到误差 $\varepsilon=\dfrac{\left\|\boldsymbol{U}_{n+1}-\boldsymbol{U}_n\right\|}{\left\|\boldsymbol{U}_{n+1}\right\|}$ 小于给定的精度才终止。增强矩阵的选择严重影响迭代过程的性能。一般来说，增强矩阵有三种选择[26]：①单位矩阵 $\boldsymbol{K}_{\text{enh}}=\boldsymbol{I}$；② $\boldsymbol{K}_{\text{enh}}=\operatorname{diag}\left(\boldsymbol{K}_{\text{b}}^{\text{I}}\right)$（$\boldsymbol{K}_{\text{b}}^{\text{I}}$ 矩阵的对角线部分)；③ $\boldsymbol{K}_{\text{enh}}=\left[\boldsymbol{K}_{\text{b}}^{\text{I}}+\left(\boldsymbol{K}_{\text{b}}^{\text{I}}\right)^{\text{T}}\right]\Big/2$。基于收敛定理[27]，对称迭代耦合方法的收敛条件为

$$\rho\left(\bar{\boldsymbol{K}}^{-1}\bar{\boldsymbol{K}}_{\text{c}}\right)<1 \tag{8.79}$$

式中，ρ 是矩阵 $\bar{\boldsymbol{K}}^{-1}\bar{\boldsymbol{K}}_{\text{c}}$ 的谱半径。

8.5 数值算例

在本节中，我们通过几个数值例子来研究所提出的等几何有限元-边界元耦合分析方法的稳定性和准确性。除非另有说明，$(p+1)\times(q+1)\times(r+1)$ 个高斯点用于 IGAFEM 中的单元积分。在自适应积分方案中，目标误差水平设置为 $e_{\text{int}}=10^{-8}$，每个积分子单元中指定的最大高斯阶次为 6。在 M-PSEM 的实现中，Q 点是根据沿着母单元的每条线上的高斯积分点来选择的。改进幂级数展开法的验证算例详见文献[28]。

8.5.1 非齐次边界条件

与传统的拉格朗日基函数相比，NURBS 基函数一般情况下不满足 Kronecker-delta 性质，因此，控制点处的未知场变量不具有插值特性，且非齐次本质边界条件不能直接施加，本章采用最小二乘技术[29]来提取控制点处的控制变量。

该方法的基本思想是寻找合适的控制变量，使下列方程的值最小

$$\begin{aligned}\boldsymbol{J}_{\text{u}}&=\frac{1}{2}\int_{\Gamma_{\text{u}}}(\boldsymbol{u}-\bar{\boldsymbol{u}})^2\,\mathrm{d}\Gamma_{\text{u}}\\ \boldsymbol{J}_{\text{t}}&=\frac{1}{2}\int_{\Gamma_{\text{t}}}(\boldsymbol{t}-\bar{\boldsymbol{t}})^2\,\mathrm{d}\Gamma_{\text{t}}\end{aligned} \tag{8.80}$$

将式(8.9)和式(8.17)代入式(8.80)中可得

$$\begin{aligned} \boldsymbol{J}_{\mathrm{u}} &= \frac{1}{2}\int_{\Gamma_{\mathrm{u}}}\left(\boldsymbol{R}\boldsymbol{u}_{\mathrm{cp}} - \overline{\boldsymbol{u}}\right)^2 \mathrm{d}\Gamma_{\mathrm{u}} \\ \boldsymbol{J}_{\mathrm{t}} &= \frac{1}{2}\int_{\Gamma_{\mathrm{t}}}\left(\boldsymbol{R}\boldsymbol{t}_{\mathrm{cp}} - \overline{\boldsymbol{t}}\right)^2 \mathrm{d}\Gamma_{\mathrm{t}} \end{aligned} \tag{8.81}$$

其中，$\boldsymbol{u}_{\mathrm{cp}}$ 和 $\boldsymbol{t}_{\mathrm{cp}}$ 分别是位移边界 Γ_{u} 和面力边界 Γ_{t} 上控制点处的控制变量；$\overline{\boldsymbol{u}}$ 和 $\overline{\boldsymbol{t}}$ 则为边界上已知的位移和面力矢量。

分别取 $\boldsymbol{J}_{\mathrm{u}}$ 和 $\boldsymbol{J}_{\mathrm{t}}$ 对 $\boldsymbol{u}_{\mathrm{cp}}$ 和 $\boldsymbol{t}_{\mathrm{cp}}$ 的偏导数，得到

$$\begin{aligned} \frac{\partial \boldsymbol{J}_{\mathrm{u}}}{\partial \boldsymbol{u}_{\mathrm{cp}}} &= \int_{\Gamma_{\mathrm{u}}}\left(\boldsymbol{R}^{\mathrm{T}}\boldsymbol{R}\boldsymbol{u}_{\mathrm{cp}} - \boldsymbol{R}^{\mathrm{T}}\overline{\boldsymbol{u}}\right)\mathrm{d}\Gamma_{\mathrm{u}}=0 \\ \frac{\partial \boldsymbol{J}_{\mathrm{t}}}{\partial \boldsymbol{t}_{\mathrm{cp}}} &= \int_{\Gamma_{\mathrm{t}}}\left(\boldsymbol{R}^{\mathrm{T}}\boldsymbol{R}\boldsymbol{t}_{\mathrm{cp}} - \boldsymbol{R}^{\mathrm{T}}\overline{\boldsymbol{t}}\right)\mathrm{d}\Gamma_{\mathrm{t}}=0 \end{aligned} \tag{8.82}$$

因此，通过求解以下方程，可以得到未知的控制变量 $\boldsymbol{u}_{\mathrm{cp}}$ 和 $\boldsymbol{t}_{\mathrm{cp}}$

$$\begin{aligned} \boldsymbol{A}_{\mathrm{u}} \cdot \boldsymbol{u}_{\mathrm{cp}} &= \boldsymbol{b}_{\mathrm{u}} \\ \boldsymbol{A}_{\mathrm{t}} \cdot \boldsymbol{t}_{\mathrm{cp}} &= \boldsymbol{b}_{\mathrm{t}} \end{aligned} \tag{8.83}$$

式中，$\boldsymbol{A}_{\mathrm{u}} = \int_{\Gamma_{\mathrm{u}}}\boldsymbol{R}^{\mathrm{T}}\boldsymbol{R}\mathrm{d}\Gamma_{\mathrm{u}}$，$\boldsymbol{b}_{\mathrm{u}} = \int_{\Gamma_{\mathrm{u}}}\boldsymbol{R}^{\mathrm{T}}\overline{\boldsymbol{u}}\mathrm{d}\Gamma_{\mathrm{u}}$，$\boldsymbol{A}_{\mathrm{t}} = \int_{\Gamma_{\mathrm{t}}}\boldsymbol{R}^{\mathrm{T}}\boldsymbol{R}\mathrm{d}\Gamma_{\mathrm{t}}$，$\boldsymbol{b}_{\mathrm{t}} = \int_{\Gamma_{\mathrm{t}}}\boldsymbol{R}^{\mathrm{T}}\overline{\boldsymbol{t}}\mathrm{d}\Gamma_{\mathrm{t}}$。

8.5.2　受剪力作用的悬臂梁

在随后的两个数值算例中，将研究对称迭代耦合方法的精度和有效性。考虑如图 8.6 所示的具有长度 L=4m、宽度 D=1m 和高度 H=1m 的三维悬臂梁模型。悬臂梁左端固定，右端受到合力 P=100Pa 的抛物型分布的剪切面力作用。材料参数设置为：弹性模量 $E = 3\times10^7\,\mathrm{Pa}$，泊松比 $\nu = 0$。

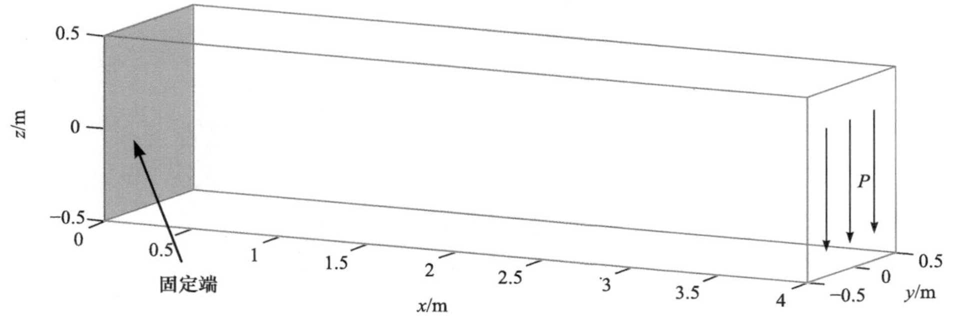

图 8.6　悬臂梁模型

该问题的位移解析解表示为[30]

$$u_x = \frac{Py}{6EI}\left[(6L-3x)x + 2(1+\nu)\left(y^2 - \frac{H^2}{4}\right)\right]$$
$$u_y = -\frac{P}{6EI}\left[(3L-x)x^2 + 2\nu y(L-x) + (4+5\nu)\frac{H^2 x}{4}\right] \tag{8.84}$$

式中，惯性矩 $I = \frac{H^3 D}{12}$。

应力分量的解析解为[30]

$$\sigma_{xx} = \frac{P(L-x)y}{I}$$
$$\sigma_{xy} = -\frac{P}{2I}\left(\frac{D^2}{4} - y^2\right) \tag{8.85}$$
$$\sigma_{yy} = 0$$

在本节中，使用一系列具有不同 NURBS 阶次(二次、三次和四次)的细化网格来研究对称迭代耦合方法的收敛行为。不同网格尺寸下的悬臂梁模型如图 8.7 所示，其中绿色和黄色区域分别代表耦合模型的有限元部分和边界元部分。注意

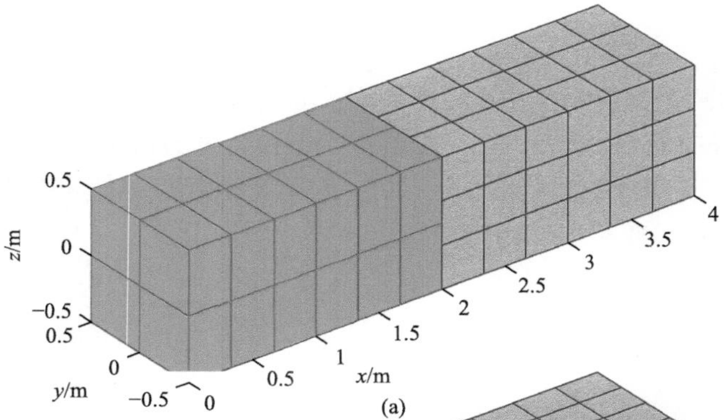

(a)

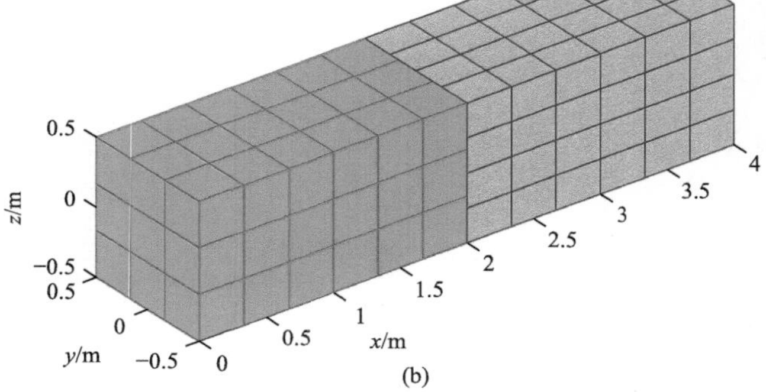

(b)

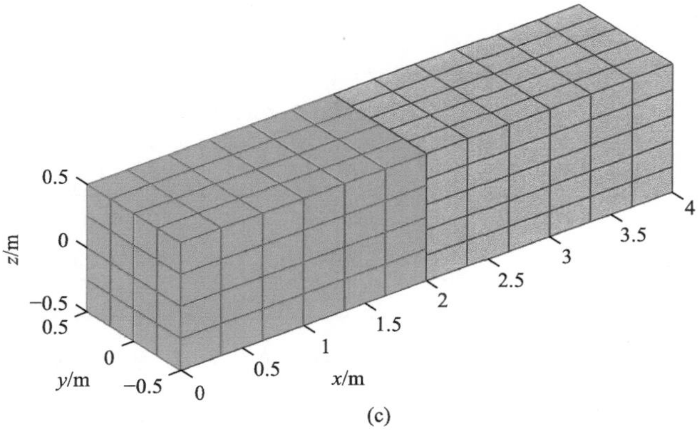

(c)

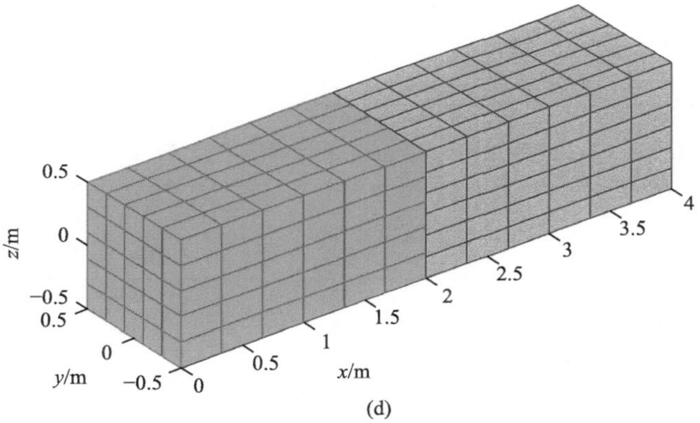

(d)

图 8.7　不同网格下的悬臂梁模型

(a) 网格 1；(b) 网格 2；(c) 网格 3；(d) 网格 4

有限元和边界元部分的细化过程是同时进行的，以保证一直处于非相适应的界面。可以看出，经过多次细化后，界面网格之间的差异会变得越来越小，并逐渐达到均匀。

在每个参数方向上 NURBS 阶次为 2～4 和网格 1～4 对应的对称迭代耦合方法中，不同增强矩阵选择下 $\bar{\boldsymbol{K}}^{-1}\bar{\boldsymbol{K}}_{\mathrm{c}}$ 的谱半径结果如表 8.1～表 8.3 所示。得出的主要收敛结论如下：①选择 $\boldsymbol{K}_{\mathrm{b}}^{\mathrm{I}}$ 矩阵的对称部分作为增强矩阵，保证了方法的收敛性和稳定性，其中 $\bar{\boldsymbol{K}}^{-1}\bar{\boldsymbol{K}}_{\mathrm{c}}$ 矩阵的谱半径远小于 1；②单元细化和阶次升高都会导致谱半径略有增加。这种现象出现的可能原因是，随着 NURBS 阶次的增加，每个 NURBS 形函数支持的单元数量也会随之增加。为了更清晰地比较，图 8.8 展示了当增强矩阵的选择为 $\boldsymbol{K}_{\mathrm{enh}}=\boldsymbol{I}$ 和 $\boldsymbol{K}_{\mathrm{enh}}=\mathrm{diag}\left(\boldsymbol{K}_{\mathrm{b}}^{\mathrm{I}}\right)$ 时，对称迭代耦合方法的误差随迭代次数的变化趋势，从图中可以清楚地观察到发散的趋势。

表 8.1 在增强矩阵 $\boldsymbol{K}_{\text{enh}}=\left[\boldsymbol{K}_{\text{b}}^{\text{I}}+\left(\boldsymbol{K}_{\text{b}}^{\text{I}}\right)^{\text{T}}\right]/2$ 情况下，$\bar{\boldsymbol{K}}^{-1}\bar{\boldsymbol{K}}_{\text{c}}$ 的谱半径

(p,q)	$\boldsymbol{K}_{\text{enh}}=\left[\boldsymbol{K}_{\text{b}}^{\text{I}}+\left(\boldsymbol{K}_{\text{b}}^{\text{I}}\right)^{\text{T}}\right]/2$			
	网格 1	网格 2	网格 3	网格 4
(2,2)	0.0732	0.0642	0.0666	0.0783
(3,3)	0.0834	0.0782	0.0817	0.0842
(4,4)	0.0737	0.0904	0.0972	0.0998

表 8.2 在增强矩阵 $\boldsymbol{K}_{\text{enh}}=\boldsymbol{I}$ 情况下，$\bar{\boldsymbol{K}}^{-1}\bar{\boldsymbol{K}}_{\text{c}}$ 的谱半径

(p,q)	$\boldsymbol{K}_{\text{enh}}=\boldsymbol{I}$			
	网格 1	网格 2	网格 3	网格 4
(2,2)	3.7808	3.8732	3.8779	3.8695
(3,3)	3.9726	3.9955	4.0129	4.0044
(4,4)	4.0586	4.1081	4.1039	4.1014

表 8.3 在增强矩阵 $\boldsymbol{K}_{\text{enh}}=\operatorname{diag}\left(\boldsymbol{K}_{\text{b}}^{\text{I}}\right)$ 情况下，$\bar{\boldsymbol{K}}^{-1}\bar{\boldsymbol{K}}_{\text{c}}$ 的谱半径

(p,q)	$\boldsymbol{K}_{\text{enh}}=\operatorname{diag}\left(\boldsymbol{K}_{\text{b}}^{\text{I}}\right)$			
	网格 1	网格 2	网格 3	网格 4
(2,2)	1.2019	1.0520	1.0198	0.9849
(3,3)	1.4733	1.3982	1.3464	1.2662
(4,4)	1.7077	1.6888	1.6253	1.5396

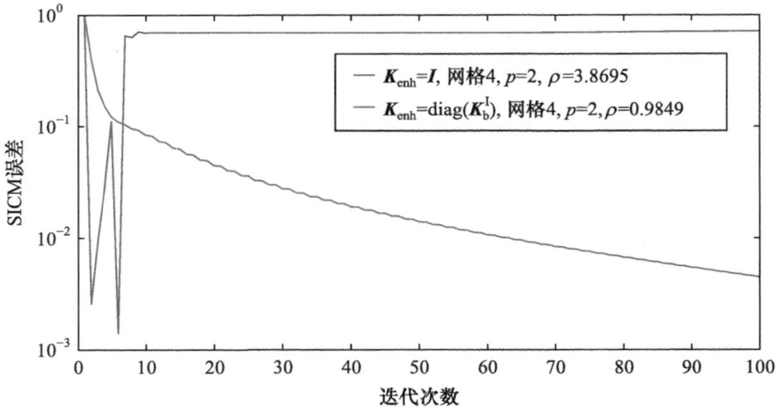

图 8.8 对称迭代耦合方法的误差随迭代次数的变化趋势

接下来，当选择 $\boldsymbol{K}_{\text{enh}}=\left[\boldsymbol{K}_{\text{b}}^{\text{I}}+\left(\boldsymbol{K}_{\text{b}}^{\text{I}}\right)^{\text{T}}\right]/2$ 作为增强矩阵时，研究达到指定精度所需的迭代次数。在不同的 NURBS 阶次和网格密度下，达到指定精度所需的迭

代次数曲线如图 8.9 所示。从图中可以看出，在所有情况下，只需要几次迭代就可以收敛到指定的精度。此外，提高 NURBS 阶次时对迭代次数的影响大于网格细化时的影响，即增加 NURBS 基函数阶次可以加快收敛速度。此外，在三次 NURBS 基函数和网格 4 的情况下，y 方向上的位移分布云图和 σ_{xx} 应力分量云图分别如图 8.10 和图 8.11 所示，计算结果与解析解非常吻合。

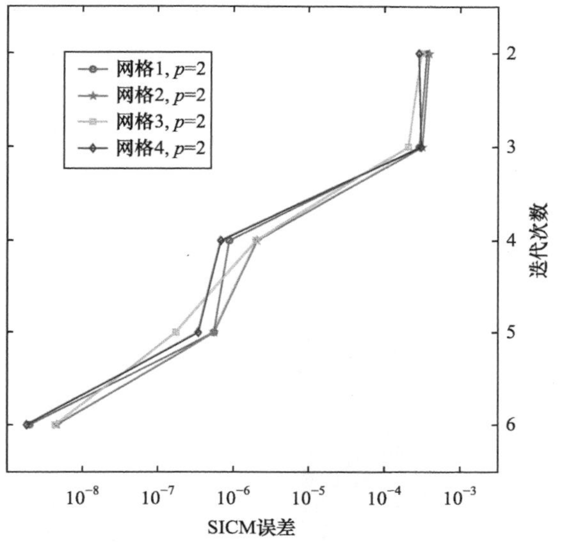

(a)

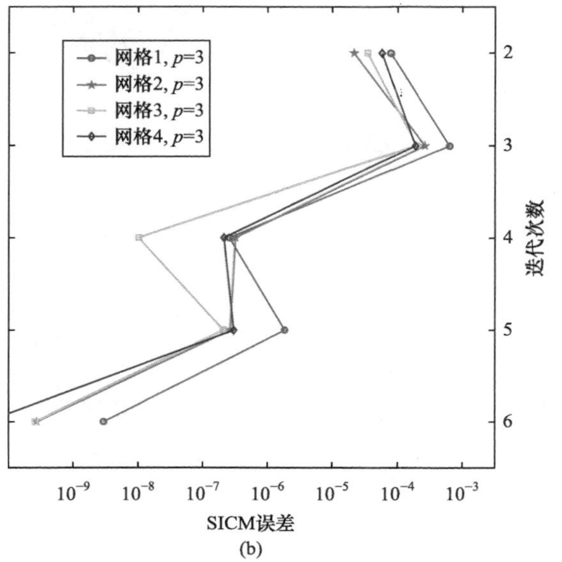

(b)

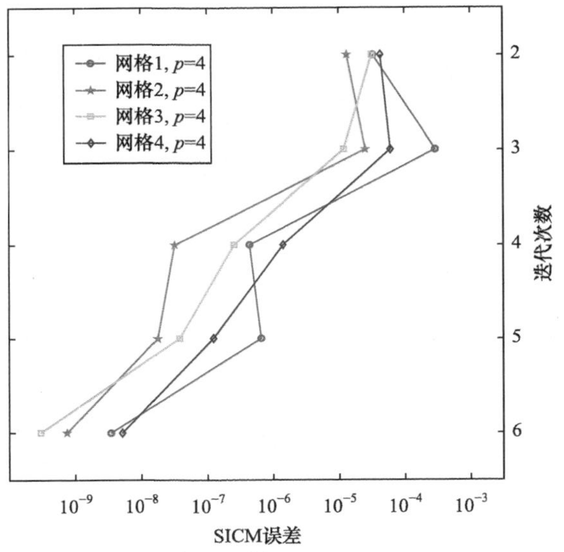

(c)

图 8.9　在不同的 NURBS 阶次和网格密度下，达到指定精度所需的迭代次数

(a) 二次 NURBS；(b) 三次 NURBS；(c) 四次 NURBS

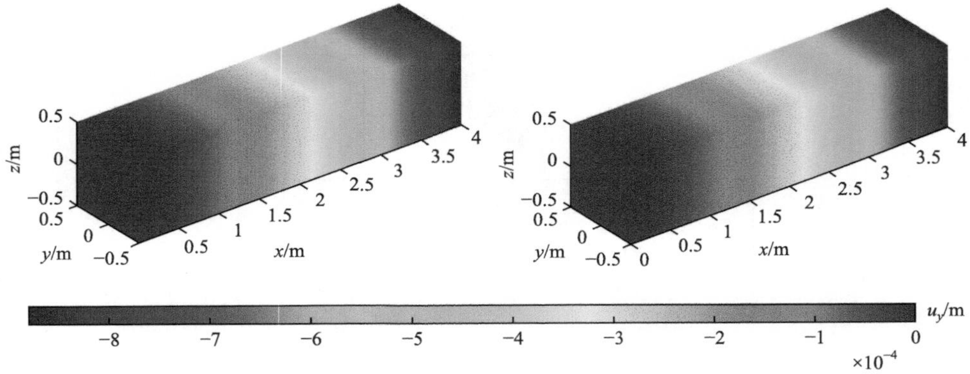

(a)　　(b)

图 8.10　y 方向位移分布云图

(a) 数值结果；(b) 解析解

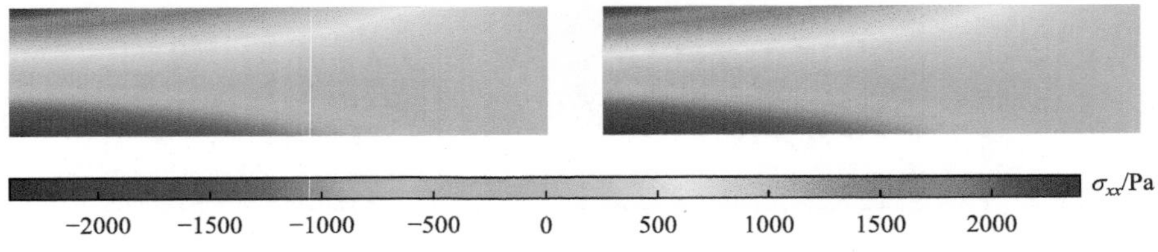

(a)　　(b)

图 8.11　σ_{xx} 分布云图

(a) 数值结果；(b) 解析解

8.5.3 受内压的圆柱体

考虑具有内径为 a 和外径为 b 的后壁圆筒受均匀内压 P 的作用。在极坐标形式下，位移和应力的解析解表示为[30]

$$\sigma_{rr}=\frac{a^2P}{b^2-a^2}\left(1-\frac{b^2}{r^2}\right),\quad \sigma_{\theta\theta}=\frac{a^2P}{b^2-a^2}\left(1+\frac{b^2}{r^2}\right)$$
$$u_r=\frac{1}{E}\frac{a^2P}{b^2-a^2}\left[(1-\nu)r+\frac{b^2(1+\nu)}{r}\right] \tag{8.86}$$

其中，r 和 θ 代表极坐标下的位移和角度。

根据对称性，仅考虑空心圆柱的四分之一模型，并施加相关的对称边界条件。在本算例中，模型的尺寸和材料参数设置为：圆柱内半径 a=1m、外半径 b=2m、圆柱高度 H=5m、弹性模量 E=1Pa、泊松比 $\nu=0.3$ 和压力载荷 P=1Pa。边界元部分由六个相同阶次的边兼容的 NURBS 片构成。通过固定圆柱底面 z 方向上的位移来消除刚体运动。采用 8.5.1 节给出的方法来施加非均匀内压边界条件。

首先，研究等几何有限元–边界元耦合分析方法的收敛性。不同细分次数下的网格如图 8.12 所示，其中绿色和黄色区域分别代表耦合模型有限元部分和边界元部分。位移相对误差的 L_2 范数定义为

$$e_{L_2}=\frac{\left\|\boldsymbol{u}_{\mathrm{ext}}(\boldsymbol{x})-\boldsymbol{u}_{\mathrm{h}}(\boldsymbol{x})\right\|}{\left\|\boldsymbol{u}_{\mathrm{ext}}(\boldsymbol{x})\right\|}=\frac{\left\{\int_{\Omega}\left[\boldsymbol{u}_{\mathrm{ext}}(\boldsymbol{x})-\boldsymbol{u}_{\mathrm{h}}(\boldsymbol{x})\right]^2\mathrm{d}\Omega\right\}^{\frac{1}{2}}}{\left\{\int_{\Omega}\left[\boldsymbol{u}_{\mathrm{ext}}(\boldsymbol{x})\right]^2\mathrm{d}\Omega\right\}^{\frac{1}{2}}} \tag{8.87}$$

式中，$\boldsymbol{u}_{\mathrm{ext}}(\boldsymbol{x})$ 和 $\boldsymbol{u}_{\mathrm{h}}(\boldsymbol{x})$ 分别代表位移的精确解和数值解。

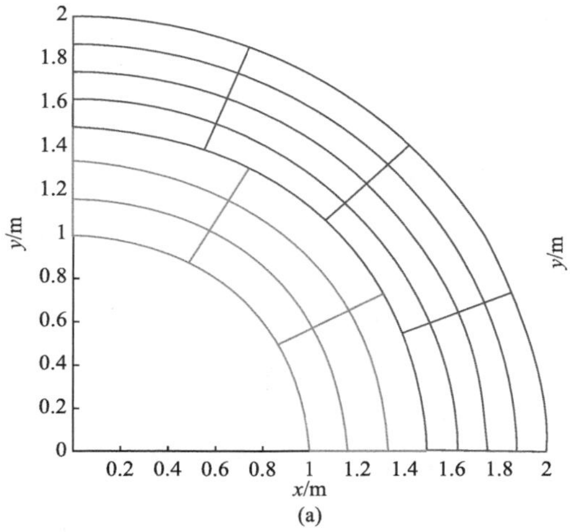

(a)

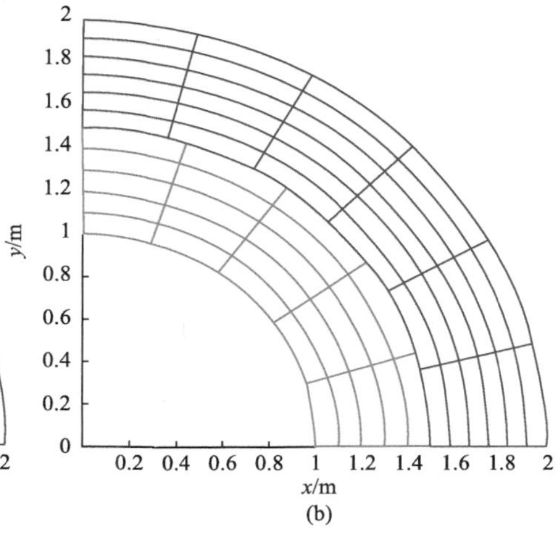

(b)

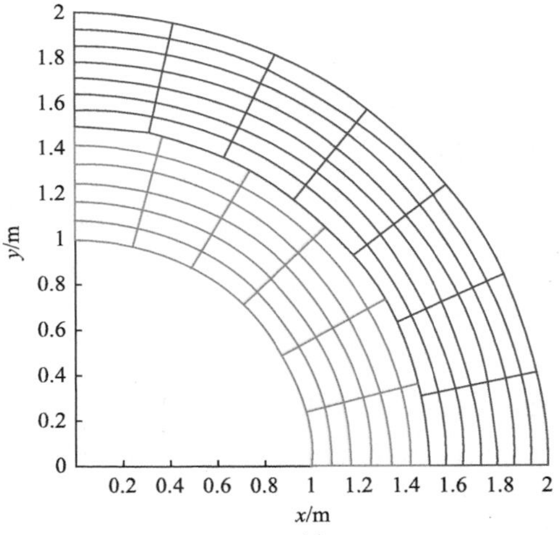

(c)

图 8.12 等几何圆柱体模型

(a) 网格 2；(b) 网格 4；(c) 网格 5

图 8.13 给出了在不同细化网格和 NURBS 阶次下位移相对误差 L_2 范数的收敛曲线。此外，图 8.13 中还显示了仅用 IGAFEM 求解的收敛曲线。结果表明，虽然 NURBS 阶次的升高对耦合模型的计算精度影响不大，但耦合模型的位移误差还是要小于 IGAFEM 的结果，可能原因是半解析形式的 BEM 产生了更高的精度。

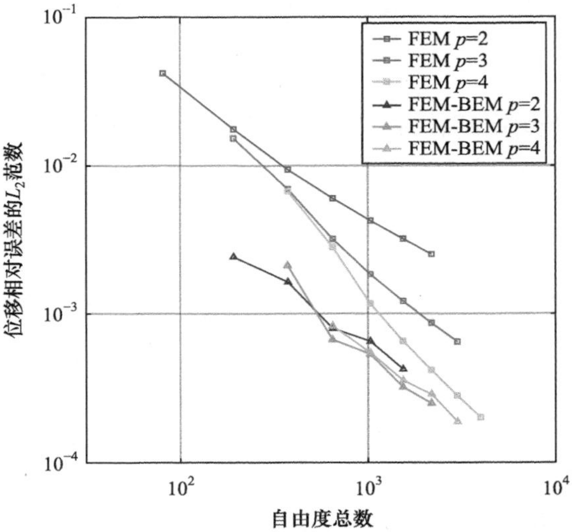

图 8.13 耦合模型和 IGAFEM 模型的收敛曲线

在接下来的分析中，采用三次 NURBS 和网格 4 的模型进行研究，其三维视图如图 8.14 所示。NURBS 模型的有限元子域中包含 125 个三次 NURBS 体单元和 1536 个位移自由度，而边界元子域中包含 216 个三次 NURBS 曲面单元和 1158 个位移自由度。与有限元和边界元子域相关的耦合界面位移自由度数分别为 192 和 243。在迭代耦合方法中 243 个边界元自由度将通过转换矩阵由 192 个有限元自由度进行表示，然后在全局耦合系统方程中进行静态压缩。

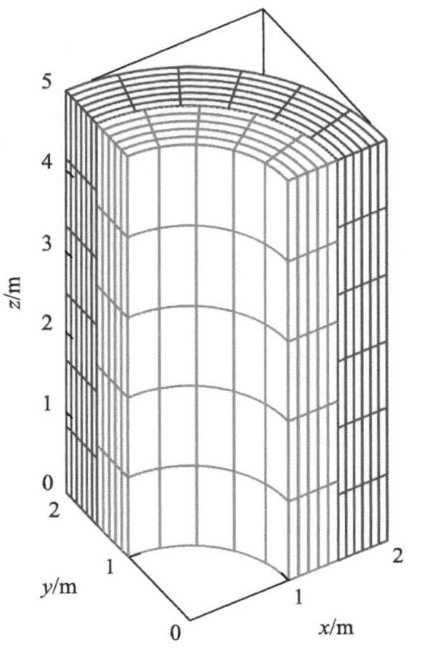

图 8.14　空心圆柱的 NURBS 耦合模型
其中红色和蓝色分别代表有限元和边界元网格

图 8.15 描绘了属于不同子域的控制点和耦合界面上的虚拟节点线的示意图。在虚拟相适应界面上控制点是一对一的约束关系。与耦合界面上连续和平衡条件约束相关的转换矩阵 $\mathcal{D}_{\text{fem}}$、$\mathcal{D}_{\text{bem}}$ 和 $\boldsymbol{M}$ 的结构如图 8.16 所示。我们可以看到，由于 NURBS 基函数的局部支撑特性，这些约束矩阵都是呈稀疏的和带状的。图 8.17 显示了对应某个源点的圆柱表面非奇异边界元单元内自适应积分网格的划分过程。

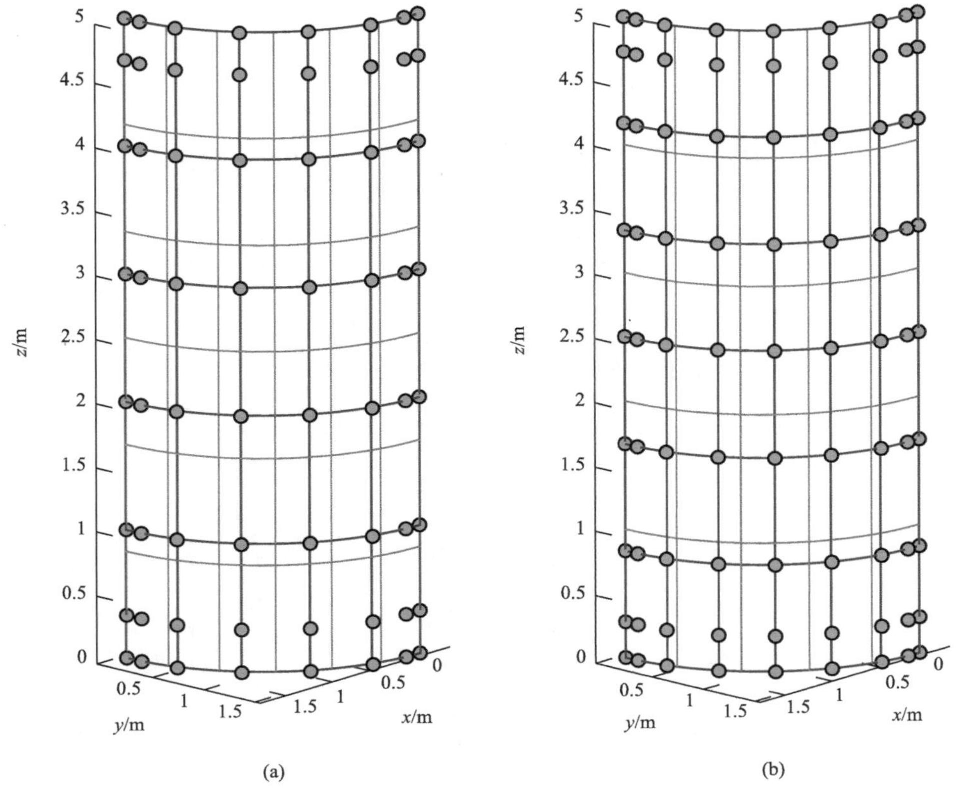

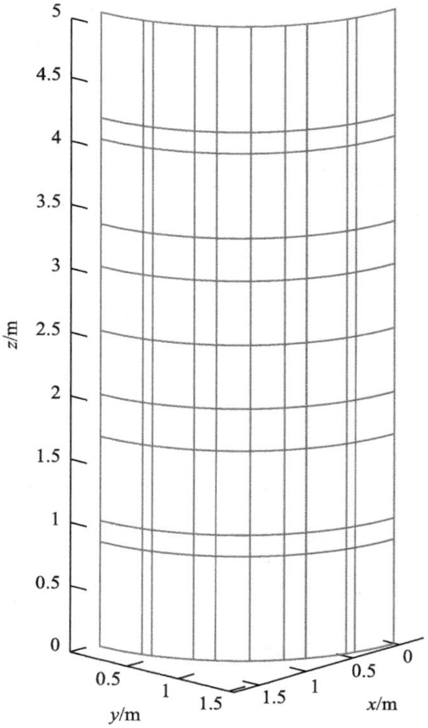

(c)

图 8.15 不同子域的控制点和耦合界面上的虚拟节点线

灰色线和绿色填充点分别代表虚拟节点线和控制点：(a) FE 部分；(b) BE 部分；(c) 虚拟相适应界面

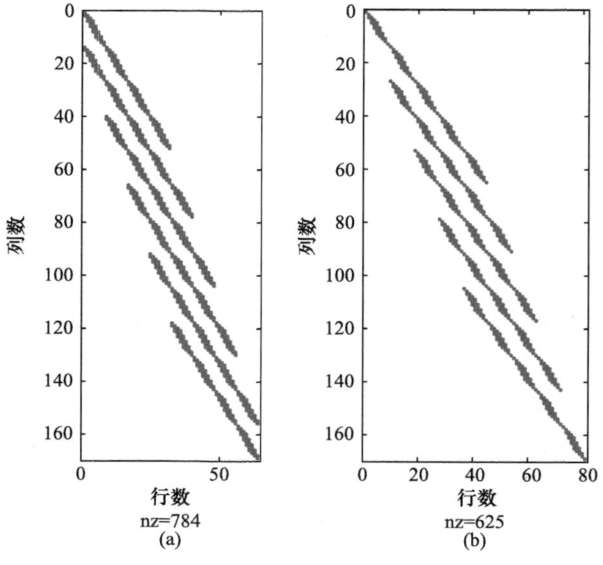

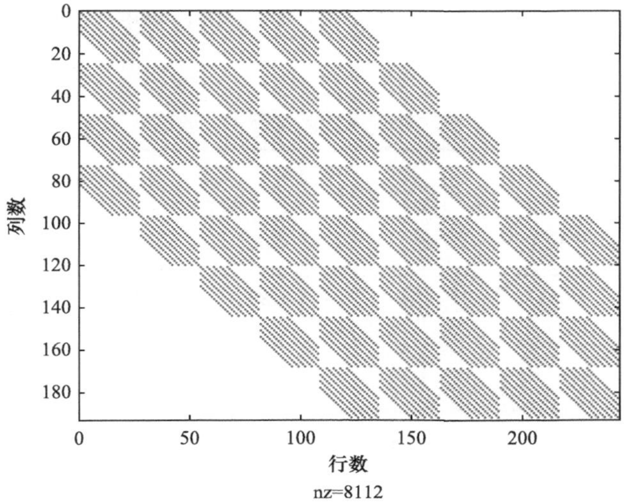

(c)

图 8.16　与界面约束相关的转换矩阵

nz 是矩阵中非零元素的个数：(a) $\mathcal{D}_{\text{fem}}$ ；(b) $\mathcal{D}_{\text{bem}}$ ；(c) $\boldsymbol{M}$

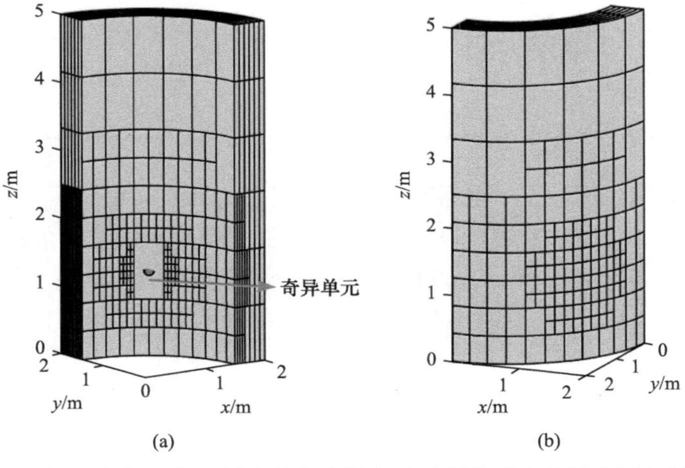

图 8.17　根据源点位置(红色填充点代表源点)非奇异单元内自适应积分划分过程

(a) 内边界；(b) 外边界

在不同阶次 NURBS 和增强矩阵选择下，收敛矩阵 $\bar{\boldsymbol{K}}^{-1}\bar{\boldsymbol{K}}_{\text{c}}$ 的谱半径变化如表 8.4 所示。结果表明：①随着 NURBS 基函数阶次的增加，谱半径略有增加；②增强矩阵为 $\boldsymbol{K}_{\text{enh}}=\left[\boldsymbol{K}_{\text{b}}^{\text{I}}+\left(\boldsymbol{K}_{\text{b}}^{\text{I}}\right)^{\text{T}}\right]\Big/2$ 的选择是最优的，能够稳定地确保迭代的收敛。此外，在对称增强矩阵 $\boldsymbol{K}_{\text{enh}}=\left[\boldsymbol{K}_{\text{b}}^{\text{I}}+\left(\boldsymbol{K}_{\text{b}}^{\text{I}}\right)^{\text{T}}\right]\Big/2$ 的选择下，达到指定精度所需

的迭代次数如表 8.5 所示。结果表明，增加 NURBS 的阶次可以加快迭代的收敛速度。

表 8.4 不同 NURBS 阶次和增强矩阵选择下，收敛矩阵 $\bar{K}^{-1}\bar{K}_c$ 的谱半径

(p,q)	增强矩阵类型		
	$K_{enh}=\left[K_b^I+\left(K_b^I\right)^T\right]/2$	$K_{enh}=I$	$K_{enh}=\mathrm{diag}\left(K_b^I\right)$
(2,2)	0.1455	0.9994	1.0910
(3,3)	0.1581	0.9997	1.3394
(4,4)	0.1759	0.9998	1.5587

表 8.5 在增强矩阵为 $K_{enh}=\left[K_b^I+\left(K_b^I\right)^T\right]/2$ 情况下，指定精度所需的迭代次数

(p,q)	ε						
	10^{-2}	10^{-3}	10^{-4}	10^{-5}	10^{-6}	10^{-7}	10^{-8}
(2,2)	2	2	4	4	5	6	7
(3,3)	2	2	4	4	6	6	8
(4,4)	2	2	3	4	5	5	7

径向位移和径向应力的数值解与解析解的比较如图 8.18 所示。数值解与解析解吻合得很好。图 8.19(a)描绘了径向位移的数值解与解析解之间的相对误差的 L_2 范数的分布云图。同时，有限元和边界元部分耦合界面上的误差水平也分别显示

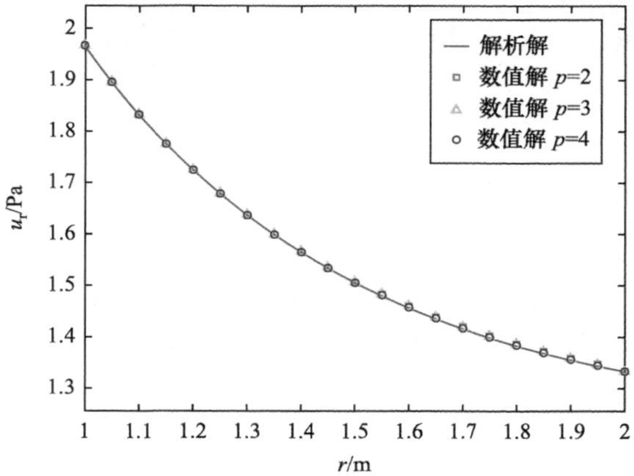

(a)

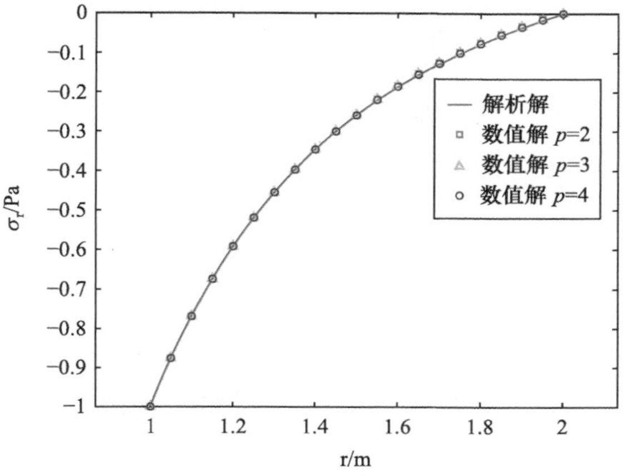

(b)

图 8.18　数值解与解析解的比较

(a) 径向位移；(b) 径向应力

在图 8.19(b)和(c)中。可以观察到，耦合界面两侧的相对误差水平在同一量级。

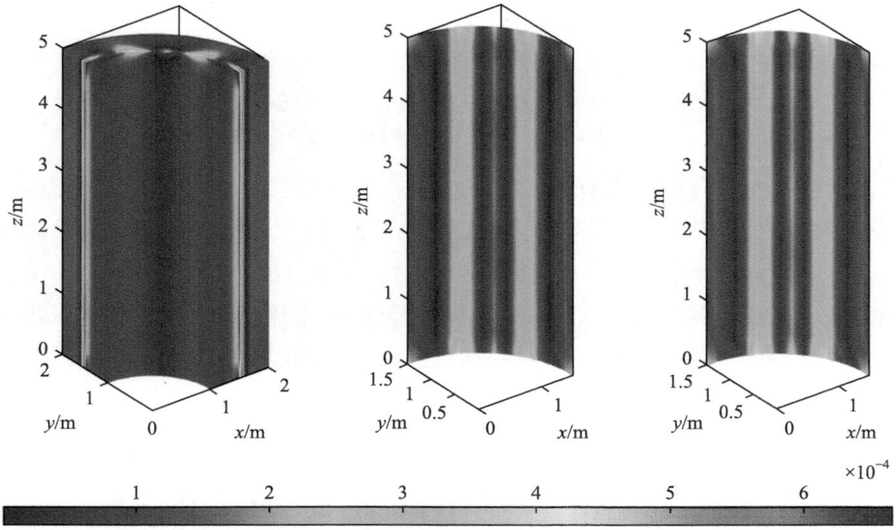

图 8.19　数值解与解析解之间的相对误差的 L_2 范数的分布云图

(a) 径向位移的数值解与解析解之间的相对误差；(b) 有限元部分界面上的数值解与解析解之间的相对误差；(c) 边界元部分界面上的径向位移的数值解与解析解之间的相对误差

8.5.4 马蹄状 U 形管

在后续的两个算例中研究了等几何有限元-边界元耦合分析方法对多个耦合

界面和不完全重合界面的适用性。在该算例中，考虑在其两个顶面受到相等和方向相反的位移作用的实心马蹄铁模型，该模型包含两个非相适应的耦合界面。使用IGAFEM进行的类似计算可在文献[14]中找到。有限元和边界元子域的NURBS网格如图8.20(a)所示，其中由红色网格线包含的区域和蓝色网格线包含的区域分别代表有限元和边界元部分。

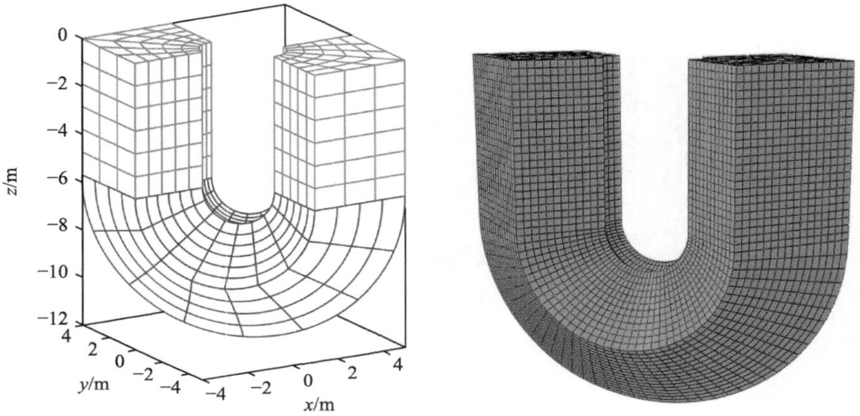

(a) (b)

图 8.20 实心马蹄铁的 NURBS 和 ABAQUS 网格

(a) NURBS 网格；(b) ABAQUS 网格

该NURBS耦合模型的有限元子域中包含288个二次NURBS体单元和2592个位移自由度，而边界元子域中包含288个二次NURBS曲面单元和1326个位移自由度。材料参数设置为：弹性模量 $E=3\times10^{7}\,\text{Pa}$ 和泊松比 $\nu=0.3$。初始施加的相等和方向相反的位移沿 y 方向应用于模型的两个顶面。为了防止刚体运动，以及形成反对称的加载条件，约束顶面点(−4, 4, 0)和点(5, −5, 0)处沿 x 方向和 z 方向的位移。

包含两个耦合界面的边界元系统方程可以写为

$$\begin{bmatrix} \boldsymbol{H}_{\text{bem}}^{\text{I}^1\text{I}^1} & \boldsymbol{H}_{\text{bem}}^{\text{I}^1\text{B}} & \boldsymbol{H}_{\text{bem}}^{\text{I}^1\text{I}^2} \\ \boldsymbol{H}_{\text{bem}}^{\text{BI}^1} & \boldsymbol{H}_{\text{bem}}^{\text{BB}} & \boldsymbol{H}_{\text{bem}}^{\text{BI}^2} \\ \boldsymbol{H}_{\text{bem}}^{\text{I}^2\text{I}^1} & \boldsymbol{H}_{\text{bem}}^{\text{I}^2\text{B}} & \boldsymbol{H}_{\text{bem}}^{\text{I}^1\text{I}^2} \end{bmatrix} \begin{Bmatrix} \boldsymbol{u}_{\text{bem}}^{\text{I}^1} \\ \boldsymbol{u}_{\text{bem}}^{\text{B}} \\ \boldsymbol{u}_{\text{bem}}^{\text{I}^2} \end{Bmatrix} = \begin{bmatrix} \boldsymbol{G}_{\text{bem}}^{\text{I}^1\text{I}^1} & \boldsymbol{G}_{\text{bem}}^{\text{I}^1\text{B}} & \boldsymbol{G}_{\text{bem}}^{\text{I}^1\text{I}^2} \\ \boldsymbol{G}_{\text{bem}}^{\text{BI}^1} & \boldsymbol{G}_{\text{bem}}^{\text{BB}} & \boldsymbol{G}_{\text{bem}}^{\text{BI}^2} \\ \boldsymbol{G}_{\text{bem}}^{\text{I}^2\text{I}^1} & \boldsymbol{G}_{\text{bem}}^{\text{I}^2\text{B}} & \boldsymbol{G}_{\text{bem}}^{\text{I}^1\text{I}^2} \end{bmatrix} \begin{Bmatrix} \boldsymbol{t}_{\text{bem}}^{\text{I}^1} \\ \boldsymbol{t}_{\text{bem}}^{\text{B}} \\ \boldsymbol{t}_{\text{bem}}^{\text{I}^2} \end{Bmatrix} \tag{8.88}$$

式中，上角标 I^1 和 I^2 分别代表耦合模型左边和右边的耦合界面。与有限元系统矩阵耦合后，压缩相应的自由度，并施加边界元子域的边界条件后，最终全局耦合

系数方程表示为

$$
\begin{bmatrix}
\boldsymbol{K}_{\text{fem}}^{\text{FF}^1} & \boldsymbol{K}_{\text{fem}}^{\text{FI}^1} & \mathbf{0} & \mathbf{0} & \mathbf{0} & \mathbf{0} & \mathbf{0} \\
\boldsymbol{K}_{\text{fem}}^{\text{IF}^1} & \boldsymbol{K}_{\text{fem}}^{\text{II}^1} & \boldsymbol{M}^1 & \mathbf{0} & \mathbf{0} & \mathbf{0} & \mathbf{0} \\
\mathbf{0} & \boldsymbol{H}_{\text{bem}}^{\text{I}^1\text{I}^1}\boldsymbol{T}_{\text{bf}}^1 & -\boldsymbol{G}_{\text{bem}}^{\text{I}^1\text{I}^1} & \boldsymbol{\mathcal{A}}_{\text{bem}}^{\text{I}^1\text{B}} & -\boldsymbol{G}_{\text{bem}}^{\text{I}^1\text{I}^2} & \boldsymbol{H}_{\text{bem}}^{\text{I}^1\text{I}^2}\boldsymbol{T}_{\text{bf}}^2 & \mathbf{0} \\
\mathbf{0} & \boldsymbol{H}_{\text{bem}}^{\text{BI}^1}\boldsymbol{T}_{\text{bf}}^1 & -\boldsymbol{G}_{\text{bem}}^{\text{BI}^1} & \boldsymbol{\mathcal{A}}_{\text{bem}}^{\text{BB}} & -\boldsymbol{G}_{\text{bem}}^{\text{BI}^2} & \boldsymbol{H}_{\text{bem}}^{BI^2}\boldsymbol{T}_{\text{bf}}^2 & \mathbf{0} \\
\mathbf{0} & \boldsymbol{H}_{\text{bem}}^{\text{I}^2\text{I}^1}\boldsymbol{T}_{\text{bf}}^1 & -\boldsymbol{G}_{\text{bem}}^{\text{I}^2\text{I}^1} & \boldsymbol{\mathcal{A}}_{\text{bem}}^{\text{I}^2\text{B}} & -\boldsymbol{G}_{\text{bem}}^{\text{I}^2\text{I}^2} & \boldsymbol{H}_{\text{bem}}^{\text{I}^2\text{I}^2}\boldsymbol{T}_{\text{bf}}^2 & \mathbf{0} \\
\mathbf{0} & \mathbf{0} & \mathbf{0} & \mathbf{0} & \mathbf{0} & \boldsymbol{K}_{\text{fem}}^{\text{FF}^2} & \boldsymbol{K}_{\text{fem}}^{\text{FI}^2} \\
\mathbf{0} & \mathbf{0} & \mathbf{0} & \mathbf{0} & \boldsymbol{M}^2 & \boldsymbol{K}_{\text{fem}}^{\text{IF}^2} & \boldsymbol{K}_{\text{fem}}^{\text{II}^2}
\end{bmatrix}
\begin{Bmatrix}
\boldsymbol{u}_{\text{fem}}^{\text{F}^1} \\
\boldsymbol{u}_{\text{fem}}^{\text{I}^1} \\
\boldsymbol{t}_{\text{bem}}^{\text{I}^1} \\
\boldsymbol{\mathcal{X}}_{\text{bem}}^{\text{B}} \\
\boldsymbol{t}_{\text{bem}}^{\text{I}^2} \\
\boldsymbol{u}_{\text{fem}}^{\text{F}^2} \\
\boldsymbol{u}_{\text{fem}}^{\text{I}^2}
\end{Bmatrix}
=
\begin{Bmatrix}
\boldsymbol{f}_{\text{fem}}^{\text{F}^1} \\
\mathbf{0} \\
\boldsymbol{\mathcal{B}}_{\text{bem}}^{\text{I}^1\text{B}}\boldsymbol{\mathcal{Y}}_{\text{bem}}^{\text{B}} \\
\boldsymbol{\mathcal{B}}_{\text{bem}}^{\text{BB}}\boldsymbol{\mathcal{Y}}_{\text{bem}}^{\text{B}} \\
\boldsymbol{\mathcal{B}}_{\text{bem}}^{\text{I}^2\text{B}}\boldsymbol{\mathcal{Y}}_{\text{bem}}^{\text{B}} \\
\boldsymbol{f}_{\text{fem}}^{\text{F}^2} \\
\mathbf{0}
\end{Bmatrix}
\tag{8.89}
$$

本节将数值计算结果与 ABAQUS 软件结果进行了比较。相关有限元模型由 8 结点线性六面体拉格朗日网格组成，并包含 53352 个结点和 48578 个单元，其单元网格如图 8.20(b)所示。有限元子域的初始控制点和节点向量分别如图 8.21(a)和表 8.6 所示。边界元子域的初始控制点和节点向量分别如图 8.21(b)和表 8.7 所示。最终全局耦合系统矩阵的结构如图 8.22 所示。矩阵的左上和右下部分是 IGAFEM 形成的对称刚度矩阵，而中间部分是 IGABEM 系数矩阵乘以相关变换矩阵生成的完全填充矩阵，其中压缩了与界面相关的边界元子域中的位移自由度。图 8.23 展示了对应某个源点的模型表面非奇异边界元单元中自适应积分网格的划分过程。在单元自适应划分过程中，只有靠近源点的部分单元才需要更多的积分点，这既节省了计算成本，又提高了计算效率。

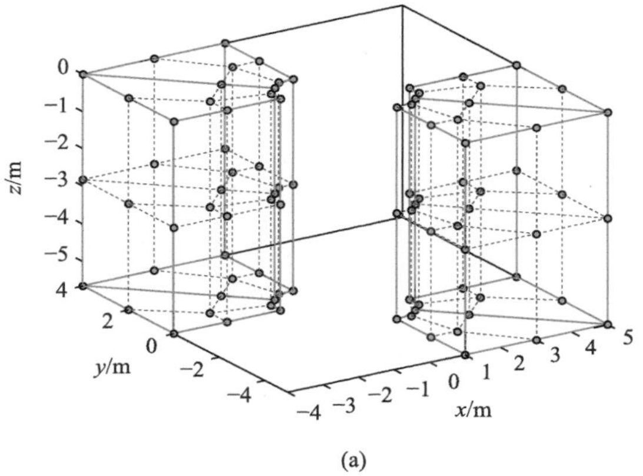

(a)

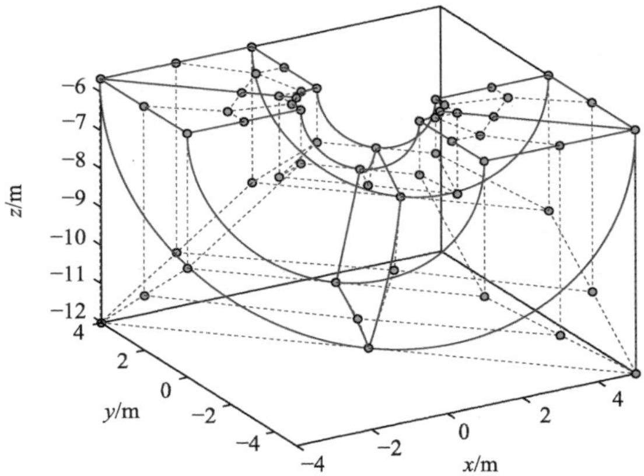

(b)

图 8.21 有限元子域和边界元子域的初始控制点
(a) 有限元子域；(b) 边界元子域

表 8.6 有限元子域的 NURBS 阶次和节点向量

参数方向	阶次	节点矢量
ξ	2	$\boldsymbol{\Xi}=\{0,0,0,0.5,0.5,1,1,1\}$
η	2	$\boldsymbol{\Psi}=\{0,0,0,1,1,1\}$
ζ	2	$\boldsymbol{\mathcal{M}}=\{0,0,0,1,1,1\}$

表 8.7 边界元子域的 NURBS 阶次和节点向量

片序号	参数方向	阶次	节点矢量
1	ξ	2	$\boldsymbol{\Xi}=\{0,0,0,1,1,1\}$
	η	2	$\boldsymbol{\Psi}=\{0,0,0,0.5,0.5,1,1,1\}$
2	ξ	1	$\boldsymbol{\Xi}=\{0,0,1,1\}$
	η	2	$\boldsymbol{\Psi}=\{0,0,0,0.5,0.5,1,1,1\}$
3	ξ	1	$\boldsymbol{\Xi}=\{0,0,1,1\}$
	η	2	$\boldsymbol{\Psi}=\{0,0,0,0.5,0.5,1,1,1\}$
4	ξ	2	$\boldsymbol{\Xi}=\{0,0,0,0.5,0.5,1,1,1\}$
	η	2	$\boldsymbol{\Psi}=\{0,0,0,0.5,0.5,1,1,1\}$

续表

片序号	参数方向	阶次	节点矢量
5	ξ	2	$\boldsymbol{\Xi}=\{0,0,0,0.5,0.5,1,1,1\}$
	η	2	$\boldsymbol{\Psi}=\{0,0,0,1,1,1\}$
6	ξ	2	$\boldsymbol{\Xi}=\{0,0,0,0.5,0.5,1,1,1\}$
	η	2	$\boldsymbol{\Psi}=\{0,0,0,1,1,1\}$

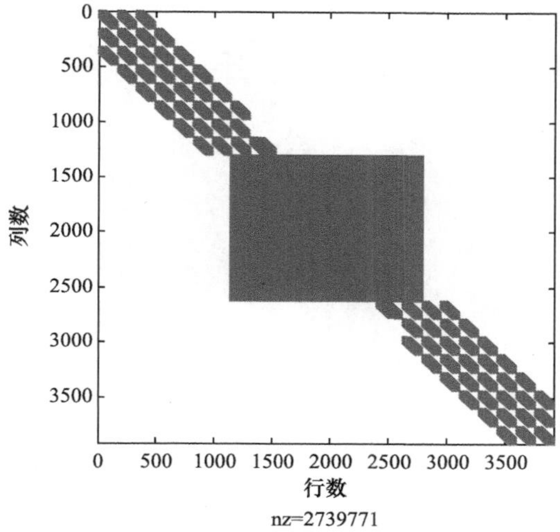

图 8.22 全局耦合系统矩阵的结构
nz 是矩阵中非零元素的个数

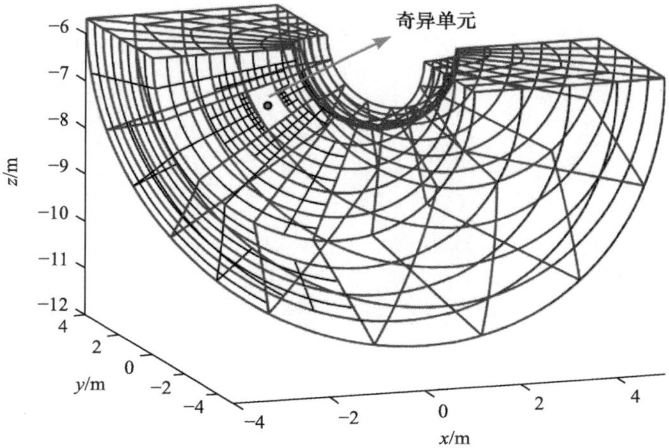

图 8.23 与源点(红色填充点)相关的非奇异边界元单元内自适应积分网格的划分过程

整个耦合模型表面上的σ_{zz}应力计算结果与 ABAQUS 软件计算的参考解的云图分别如图 8.24(a)和(b)所示。此外，非对称载荷条件下的局部应力σ_{yy}与 ABAQUS 软件计算的参考解的云图分别显示在图 8.25(a)和(b)中。从 MATLAB 和 ABAQUS 图例中的数据可以看出，计算结果与参考结果相吻合。

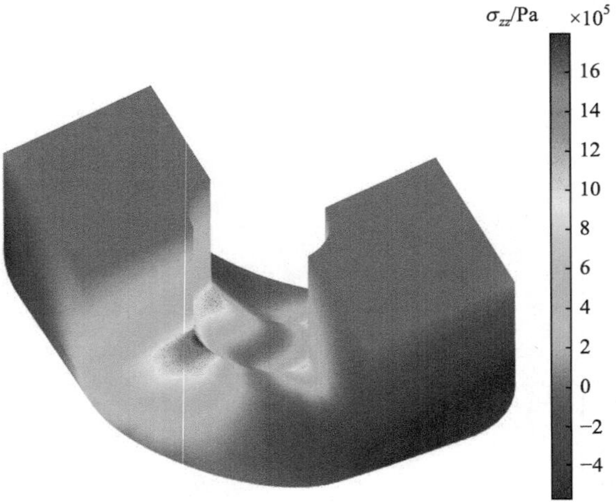

(a)

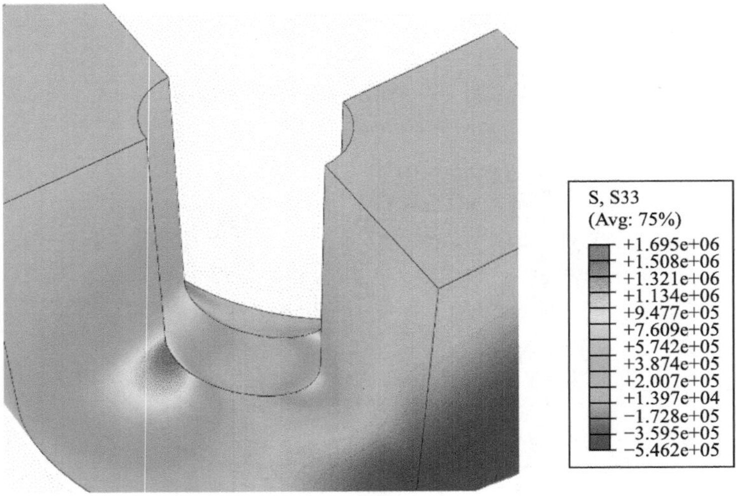

(b)

图 8.24 σ_{zz}应力云图

(a) 数值结果；(b) ABAQUS 结果

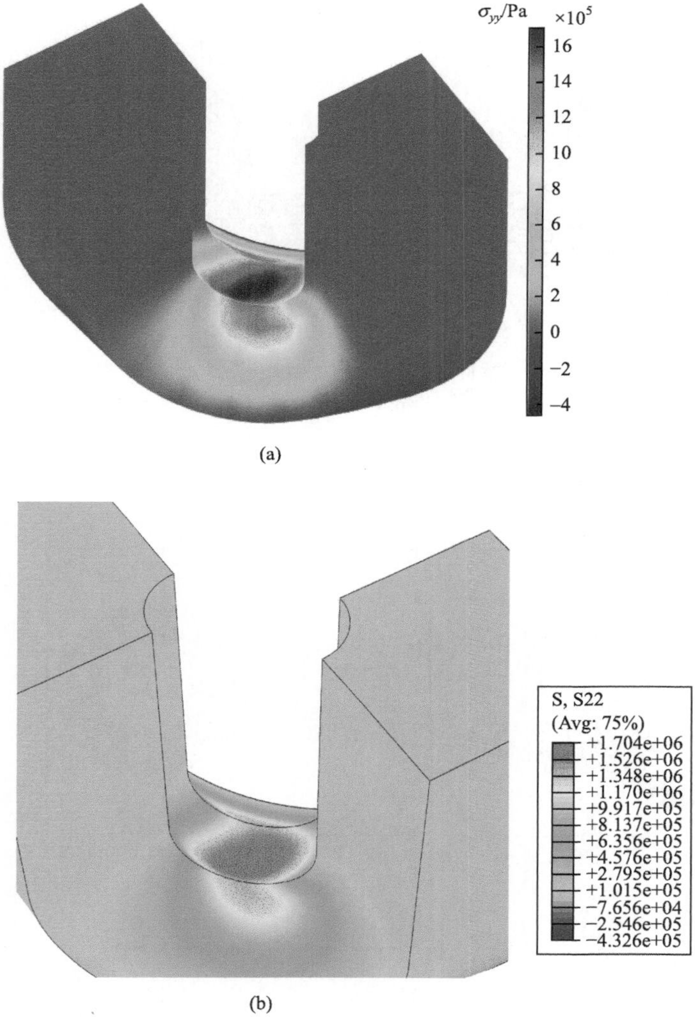

图 8.25　σ_{yy} 应力云图

(a) 数值结果；(b) ABAQUS 结果

8.5.5　三维连杆

考虑如图 8.26(a)所示的三维连杆模型，其中深灰色和浅灰色区域分别代表有限元和边界元子域。模型几何来源于文献[31]。该模型的主视图和俯视图分别如图 8.26(b)和(c)所示。材料参数设置为：弹性模量 $E = 2\times10^5$ Pa 和泊松比 $\nu = 0.3$。耦合模型的左端完全固定，右端环形内表面承受沿 z 方向均匀横向载荷 $P = -100$N。

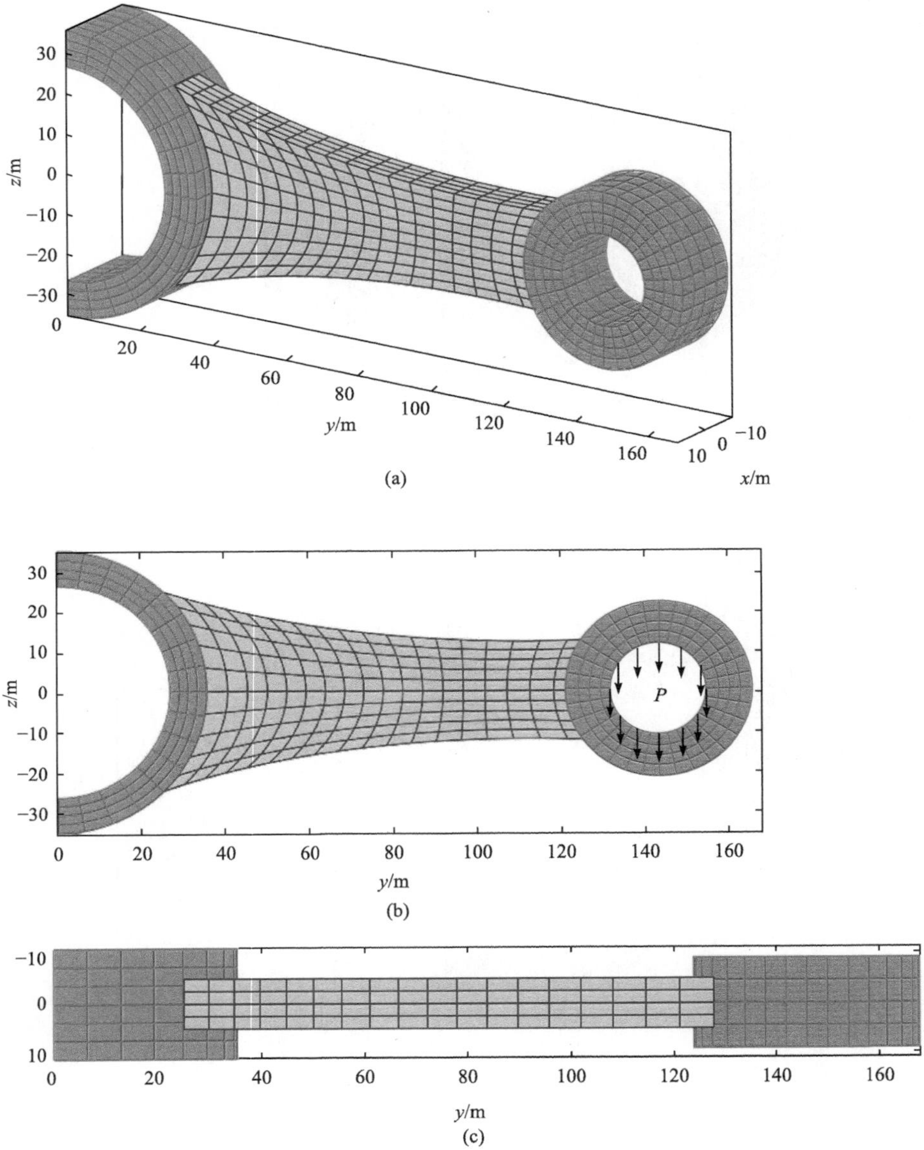

图 8.26　三维连杆模型

(a) NURBS 网格；(b) 主视图；(c) 俯视图

该 NURBS 耦合模型的左端有限元子域中包含 384 个二次 NURBS 体单元和 2736 个位移自由度，右端有限元子域中包含 512 个二次 NURBS 体单元和 3888 个位移自由度，以及边界元子域中包含 584 个二次 NURBS 曲面单元和 2160 个位移自由度。有限元子域的初始控制点和节点向量分别如图 8.27、表 8.8 和表 8.9

所示。边界元子域的初始控制点和节点向量分别如图 8.28 和表 8.10 所示。

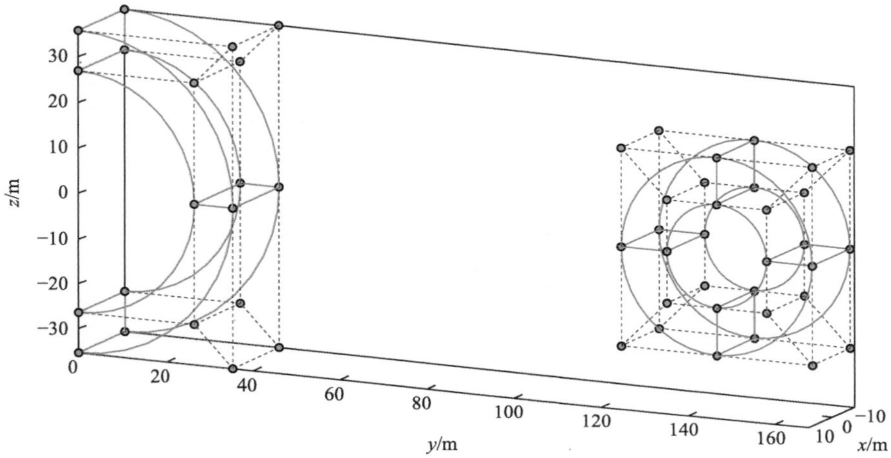

图 8.27　有限元子域的初始控制点

表 8.8　左端有限元子域的 NURBS 阶次和节点向量

参数方向	阶次	节点矢量
ξ	2	$\boldsymbol{\Xi}=\{0,0,0,0.5,0.5,1,1,1\}$
η	1	$\boldsymbol{\Psi}=\{0,0,1,1\}$
ξ	1	$\mathcal{M}=\{0,0,1,1\}$

表 8.9　右端有限元子域的 NURBS 阶次和节点向量

参数方向	阶次	节点矢量
ξ	2	$\boldsymbol{\Xi}=\{0,0,0,0.25,0.25,0.5,0.5,0.75,0.75,1,1,1\}$
η	1	$\boldsymbol{\Psi}=\{0,0,1,1\}$
ζ	1	$\mathcal{M}=\{0,0,1,1\}$

表 8.10　边界元子域的 NURBS 阶次和节点向量

片编号	参数方向	阶次	节点矢量
1～6	ξ	2	$\boldsymbol{\Xi}=\{0,0,0,1,1,1\}$
	η	2	$\boldsymbol{\Psi}=\{0,0,0,1,1,1\}$

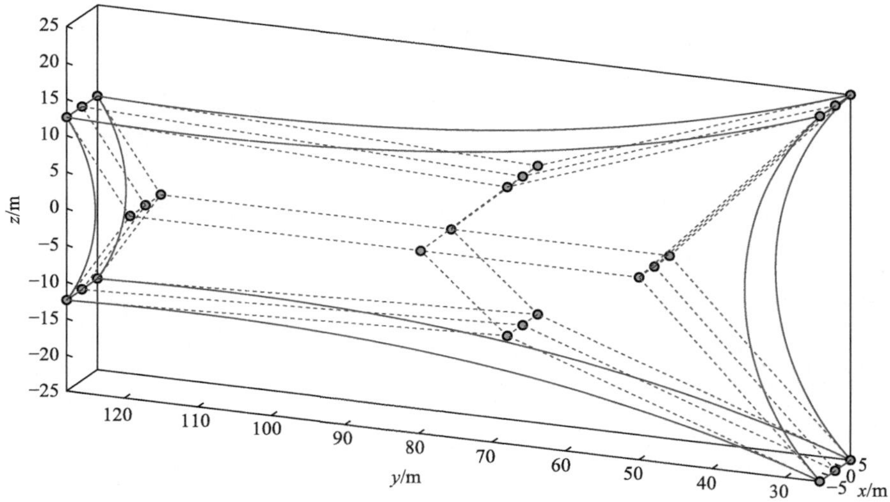

图 8.28 边界元子域的初始控制点

该模型包括两个非相适应且不完全重合的耦合界面。在本章中，通过构造相应的 C^0 连续的虚拟节点线来实施界面约束。具体实现过程如下：①采用等几何点的反求算法将边界元耦合界面上的角点投影到相应的有限元耦合界面上，得到相应的有限元子域的参数坐标；②插入上述参数坐标的节点，直到其重复性等于有限元界面上 NURBS 基函数的阶次，以获得 C^0 连续的虚拟节点线。图 8.29 和图 8.30 展示了详细的实施过程。此外，耦合转换矩阵 $\mathcal{D}_{\mathrm{fem}}$ 的结构也展示在两幅图中，其非零元素所在的列编号即表示支撑该耦合区域的控制点编号。最终全局耦合系统矩阵的结构如图 8.31 所示。

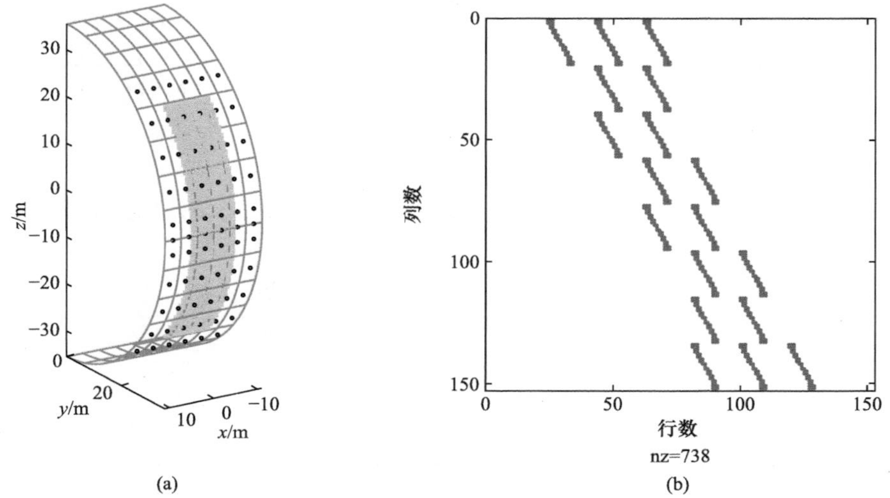

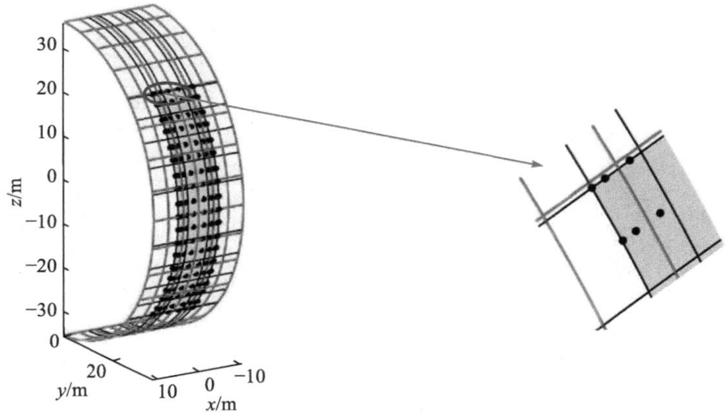

(c)

图 8.29　左侧有限元部分不完全重合界面的实施过程

(a) 左侧耦合界面和受影响的控制点；(b) 与左侧有限元部分相关矩阵 $\mathcal{D}_{\text{fem}}$ 的结构；(c) 虚拟插入的节点线(黑线)和相应的控制点 $\tilde{\boldsymbol{P}}$ (实心黑点)

将数值计算结果与 ABAQUS 软件结果进行了比较。有限元模型由 10 结点四面体拉格朗日网格组成，并包含 212374 个结点和 146787 个单元，其单元网格图如图 8.32 所示。整个耦合模型表面 z 方向上的位移计算结果与 ABAQUS 软件计算的参考解的云图分别如图 8.33(a)和(b)所示。从 MATLAB 和 ABAQUS 图例中的数据可以看出，计算结果与参考结果吻合得很好。

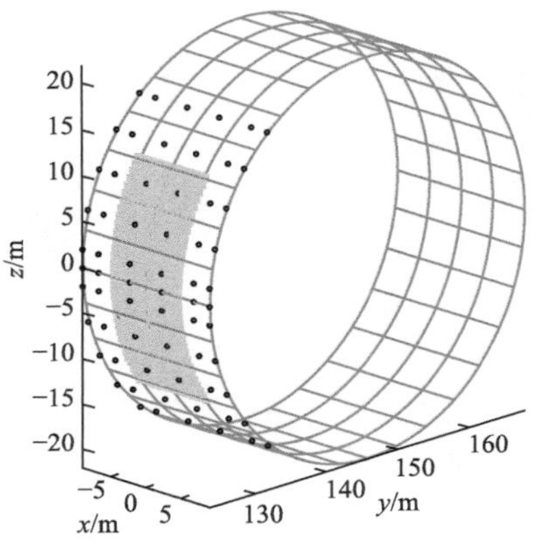

(a)

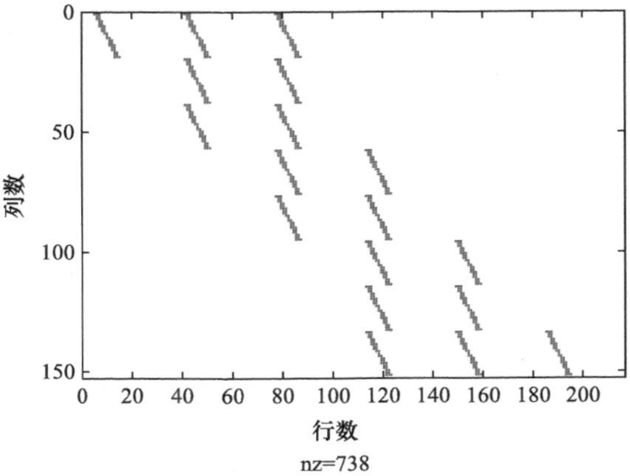

(b)

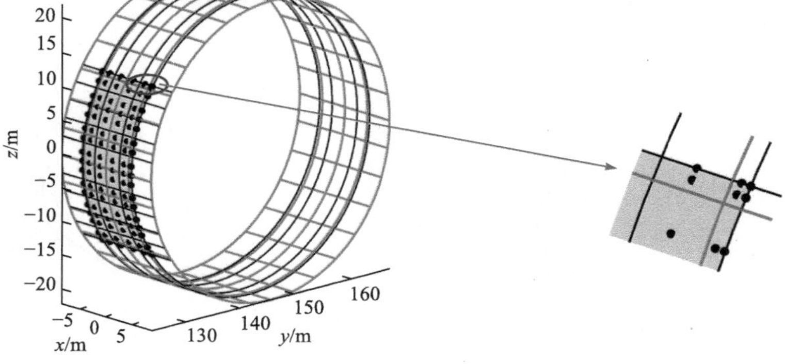

(c)

图 8.30 右侧有限元部分不完全重合界面的实施过程

(a) 右侧耦合界面和受影响的控制点；(b) 与右侧有限元部分相关矩阵 $\mathcal{D}_{\text{fem}}$ 的结构；(c) 虚拟插入的节点线(黑线)和相应的控制点 $\tilde{\boldsymbol{P}}$ (实心黑点)

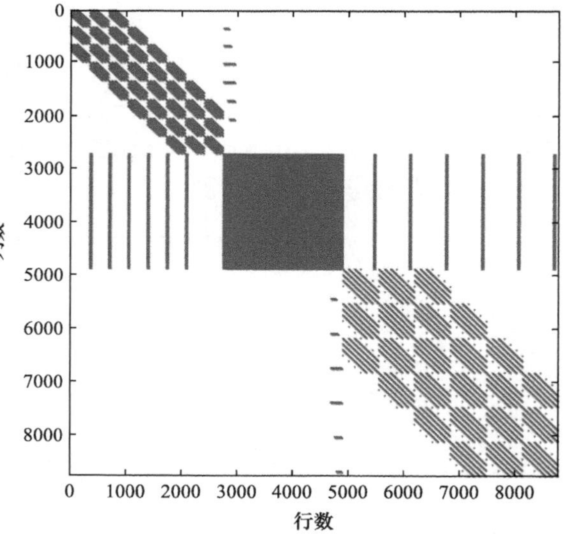

图 8.31 全局耦合系统矩阵结构图

图 8.32 三维连杆的 ABAQUS 模型

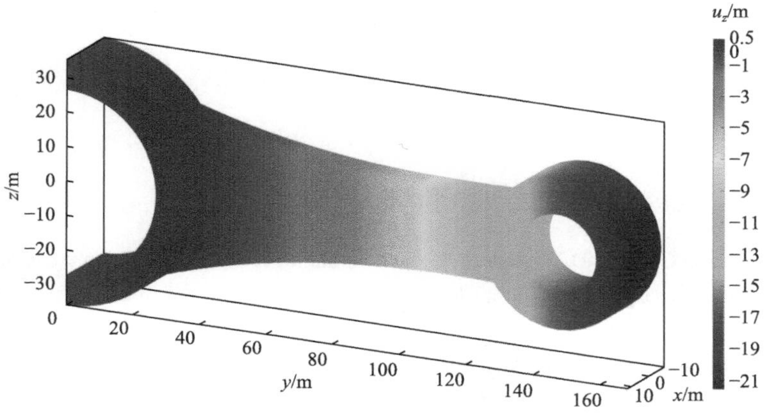

(a)

(b)

图 8.33　*z* 方向位移云图
(a) 数值结果；(b) ABAQUS 结果

8.6 小　　结

本章研究了三维弹性力学问题等几何有限元–边界元耦合算法中非相适应界面和对称迭代求解问题。针对非相适应耦合界面，提出了一套与求解问题类型无关的虚拟节点插入技术，有效地建立了非相适应(或不完全重合)界面上的位移连续条件，其实施仅依赖于界面两侧的 NURBS 信息且易于编程；提出了改进的幂级数展开法，克服了传统方法在处理高阶 NURBS 基函数($\geqslant 3$)奇异积分时出现的 Vandermonde 矩阵接近奇异的问题，推广了幂级数展开法在等几何领域中的适用范围；采用基于高斯积分误差公式和四叉树逐层分解技术的自适应积分法，处理等几何边界元子域中的非奇异积分，达到了求解精度和效率的平衡；探究了在不同的与等几何边界元子域相关的增强矩阵选择下，对称迭代耦合方法的收敛性和准确性。结果表明，当增强矩阵选择为$\boldsymbol{K}_{\text{enh}}=\left[\boldsymbol{K}_{\text{b}}^{\text{I}}+\left(\boldsymbol{K}_{\text{b}}^{\text{I}}\right)^{\text{T}}\right]\Big/2$时，对称迭代耦合方法的收敛性和准确性均是最优的。

参 考 文 献

[1] Zienkiewicz O C, Kelly D W, Bettess P. The coupling of the finite element method and boundary solution procedures[J]. International Journal for Numerical Methods in Engineering, 1977, 11(2): 355-375.

[2] Coda H B, Venturini W S, Aliabadi M H. A general 3D BEM/FEM coupling applied to elastodynamic continua/frame structures interaction analysis[J]. International Journal for Numerical Methods in Engineering, 1999, 46(5): 695-712.

[3] Elleithy W, Grzhibovskis R. An adaptive domain decomposition coupled finite element-boundary element method for solving problems in elasto-plasticity[J]. International Journal for Numerical Methods in Engineering, 2009, 79(8): 1019-1040.

[4] Aour B, Rahmani O, Nait-Abdelaziz M. A coupled FEM/BEM approach and its accuracy for solving crack problems in fracture mechanics[J]. International Journal of Solids and Structures, 2007, 44(7): 2523-2539.

[5] Mobasher M E, Waisman H. Adaptive modeling of damage growth using a coupled FEM/BEM approach[J]. International Journal for Numerical Methods in Engineering, 2016, 105(8): 599-619.

[6] Soares D, Jr. Acoustic modelling by BEM-FEM coupling procedures taking into account explicit and implicit multi-domain decomposition techniques[J]. International Journal for Numerical Methods in Engineering, 2009, 78(9): 1076-1093.

[7] Zhao W, Chen L, Chen H, et al. Topology optimization of exterior acoustic-structure interaction systems using the coupled FEM-BEM method[J]. International Journal for Numerical Methods in Engineering, 2019, 119(5): 404-431.

[8] Liu Z, Majeed M, Cirak F, et al. Isogeometric FEM-BEM coupled structural-acoustic analysis of shells using subdivision surfaces[J]. International Journal for Numerical Methods in Engineering, 2018, 113(9): 1507-1530.

[9] Heltai L, Kiendl J, Desimone A, et al. A natural framework for isogeometric fluid-structure interaction based on BEM-shell coupling[J]. Computer Methods in Applied Mechanics and Engineering, 2017, 316: 522-546.

[10] Yildizdag M E, Ardic I T, Kefal A, et al. An isogeometric FE-BE method and experimental investigation for the hydroelastic analysis of a horizontal circular cylindrical shell partially filled with fluid[J]. Thin-Walled Structures, 2020, 151: 106755.

[11] May S, Kästner M, Müller S, et al. A hybrid IGAFEM/IGABEM formulation for two-dimensional stationary magnetic and magneto-mechanical field problems[J]. Computer Methods in Applied Mechanics and Engineering, 2014, 273: 161-180.

[12] Maestre J, Pallares J, Cuesta I, et al. A 3D isogeometric BE-FE analysis with dynamic

remeshing for the simulation of a deformable particle in shear flows[J]. Computer Methods in Applied Mechanics and Engineering, 2017, 326: 70-101.

[13] Cottrell J A, Hughes T J R, Bazilevs Y. Isogeometric Analysis: Toward Integration of CAD and FEA[M]. New York: John Wiley & Sons, Inc., 2009.

[14] Hughes T J R, Cottrell J A, Bazilevs Y. Isogeometric analysis: CAD, finite elements, NURBS, exact geometry and mesh refinement[J]. Computer Methods in Applied Mechanics and Engineering, 2005, 194(39): 4135-4195.

[15] Zienkiewicz O C, Taylor R L, Zhu J Z. The Finite Element Method: Its Basis and Fundamentals[M]. 7th ed. United Kingdom: Butterworth-Heinemann, 2013.

[16] Beer G, Marussig B, Duenser C. The Isogeometric Boundary Element Method[M]. Switzerland: Springer Nature Switzerland AG, 2020.

[17] Simpson R N, Bordas S P A, Trevelyan J, et al. A two-dimensional isogeometric boundary element method for elastostatic analysis[J]. Computer Methods in Applied Mechanics and Engineering, 2012, 209-212: 87-100.

[18] Gao X W. An effective method for numerical evaluation of general 2D and 3D high order singular boundary integrals[J]. Computer Methods in Applied Mechanics and Engineering, 2010, 199(45): 2856-2864.

[19] Gong Y P, Dong C Y, Qin X C. An isogeometric boundary element method for three dimensional potential problems[J]. Journal of Computational and Applied Mathematics, 2017, 313: 454-468.

[20] Gao X W. The radial integration method for evaluation of domain integrals with boundary-only discretization[J]. Engineering Analysis with Boundary Elements, 2002, 26(10): 905-916.

[21] Bu S, Davies T G. Effective evaluation of non-singular integrals in 3D BEM[J]. Advances in Engineering Software, 1995, 23(2): 121-128.

[22] Gao X W, Davies T G. Adaptive integration in elasto-plastic boundary element analysis[J]. Journal of the Chinese Institute of Engineers, 2000, 23(3): 349-356.

[23] Coox L, Greco F, Atak O, et al. A robust patch coupling method for NURBS-based isogeometric analysis of non-conforming multipatch surfaces[J]. Computer Methods in Applied Mechanics and Engineering, 2017, 316: 235-260.

[24] Borden M J, Scott M A, Evans J A, et al. Isogeometric finite element data structures based on Bézier extraction of NURBS[J]. International Journal for Numerical Methods in Engineering, 2011, 87(1-5): 15-47.

[25] Ma J, Le M. A new method for coupling of boundary element method and finite element method[J]. Applied Mathematical Modelling, 1992, 16(1): 43-46.

[26] Dong C Y. An iterative FE-BE coupling method for elastostatics[J]. Computers & Structures, 2001, 79(3): 293-299.

[27] Stoer J, Bulirsch R. Introduction to Numerical Analysis[M]. Switzerland: Springer Nature Switzerland AG, 2002.

[28] 董春迎, 公颜鹏, 孙芳玲. 等几何边界元法[M]. 北京: 科学出版社, 2023.

[29] Nguyen V P, Anitescu C, Bordas S P A, et al. Isogeometric analysis: an overview and computer

implementation aspects[J]. Mathematics and Computers in Simulation, 2015, 117: 89-116.

[30] Timoshenko S P, Goodier J N. Theory of Elasticity[M]. New York: McGraw-Hill, 1969.

[31] Nguyen V P, Kerfriden P, Brino M, et al. Nitsche's method for two and three dimensional NURBS patch coupling[J]. Computational Mechanics, 2014, 53(6): 1163-1182.

第 9 章

混合维度实体-壳结构的等几何有限元-边界元耦合分析

9.1 引　言

壳体作为一种典型的降维结构，由于其轻量化和良好的抗弯刚度等优点而具有最佳的承载性能，广泛应用于航空航天、潜艇、汽车，以及其他国防和民用工业领域，如飞机机翼、风力涡轮机叶片、压力容器和机身蒙皮等。开展对壳体的有效和准确的数值分析具有重要意义。壳体的几何结构可用嵌入三维空间中的双参数曲面来描述，并在厚度方向给出合理的变形假设。然而，在某些情况下，由于几何、载荷和材料行为的复杂性，简化的壳理论模拟其力学行为并不能满足所有的要求，因此需要借助三维实体单元来准确获取某些区域的完整响应。因此，开展实体-壳耦合问题的等几何耦合算法研究具有重要的实际意义。

实现混合维度实体-壳耦合分析的主要思想是建立低维单元类型(壳)和高维单元类型(实体)之间耦合界面上控制变量之间的联系，即沿耦合边的位移协调条件，以及面力和壳体应力合力(法向力矩、扭矩和剪力)的平衡条件。大量的研究主要集中在有限元实体单元与有限元壳单元的耦合分析上。依据壳单元内转角和横向位移表示的边界合反力与三维实体边界上的面力在界面虚位移上所做的虚功相等，建立混合维度耦合模型的多点约束方程。采用等几何有限元-边界元耦合方法分析混合维度实体-壳耦合问题，因实体区域由等几何边界元法进行模拟，因此整个耦合模型的数值分析只需要提供边界的 CAD 网格信息，这在实现耦合分析与 CAD 系统的紧密联系方面具有很大的潜力。

另一方面，在耦合界面两侧的壳体单元和实体单元的自由度数量和类型是不同的，耦合的关键是有效地建立不同类型单元的控制变量之间的混合维度耦合约束。基于此，本章将采用等几何有限元-边界元耦合方法来实现混合维度实体-壳结构的耦合分析，其中采用等几何 Reissner-Mindlin 壳单元对壳体进行建模，采用基于配点的等几何边界元法对实体部分进行模拟。该耦合方法只需提供整个耦合模型边界的CAD网格信息，易于实现等几何耦合算法与CAD系统的紧密融合。

9.2 等几何 Reissner-Mindlin 壳公式

9.2.1 壳体曲面的微分几何

壳体参考面(也称作壳体中性面)可由两个面内自然曲线坐标 ξ^1 和 ξ^2 进行描述。在此处我们约定拉丁字母的取值范围为 $\{1,2,3\}$，希腊字母的取值范围为 $\{1,2\}$。给定壳中性面的位置矢量 $\boldsymbol{r}\left(\xi^1,\xi^2\right)$，则协变基矢量为

$$\boldsymbol{g}_\alpha = \frac{\partial \boldsymbol{r}}{\partial \xi^\alpha} \tag{9.1}$$

中性面在该点的法线为

$$\boldsymbol{n} = \frac{\boldsymbol{g}_1 \times \boldsymbol{g}_2}{\left\| \boldsymbol{g}_1 \times \boldsymbol{g}_2 \right\|} \tag{9.2}$$

其中，$\|\cdot\|$ 代表欧拉范数。

因此，壳体曲面第一基本形式的分量可以表示为[1]

$$g_{\alpha\beta} = \boldsymbol{g}_\alpha \cdot \boldsymbol{g}_\beta \tag{9.3}$$

第二基本形式的分量可以表示为[1]

$$b_{\alpha\beta} = \boldsymbol{g}_{\alpha,\beta} \cdot \boldsymbol{n} \tag{9.4}$$

式中，$\boldsymbol{g}_{\alpha,\beta} = \dfrac{\partial^2 \boldsymbol{r}}{\partial \xi^\alpha \partial \xi^\beta}$。

根据式(9.3)，我们可以得到逆变基矢量为

$$\boldsymbol{g}^\alpha = g^{\alpha\beta} \cdot \boldsymbol{g}_\beta \tag{9.5}$$

其中，$g^{\alpha\beta} = \left[g_{\alpha\beta} \right]^{-1}$。

壳体在 ξ^1 和 ξ^2 两个方向上的物理曲率表示为

$$\frac{1}{R_1} = \frac{b_{11}}{g_{11}}, \qquad \frac{1}{R_2} = \frac{b_{22}}{g_{22}} \tag{9.6}$$

式中，R_1 和 R_2 代表曲率半径。当所考虑的法向量指向曲率中心时，这些值为正。

9.2.2 壳体的位移描述

图 9.1 所示为物理空间中壳体模型的示例。本章中涉及的壳体均具有恒定的厚度 h。壳体上任意一点处的位置矢量 $\boldsymbol{R}_{\mathrm{pv}}$ 描述为

$$\boldsymbol{R}_{\mathrm{pv}}\left(\xi^1,\xi^2,\zeta\right)=\boldsymbol{r}\left(\xi^1,\xi^2\right)+\frac{h}{2}\zeta\boldsymbol{n}\left(\xi^1,\xi^2\right) \tag{9.7}$$

其中，$\boldsymbol{r}\left(\xi^1,\xi^2\right)$是该点在壳体中性面上正交投影点处的位置向量；$\zeta$ 是沿法线方向定义的厚度坐标，其取值范围为$[-1,1]$；$\boldsymbol{n}\left(\xi^1,\xi^2\right)$为单位法向量，可由式(9.2)求得。

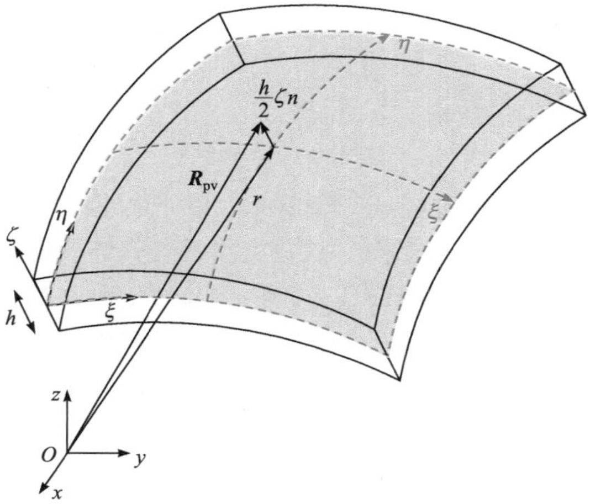

图 9.1　物理空间中的壳体模型

红色虚线包围的浅灰色区域为中性面

由于壳体厚度方向上一阶横向剪切变形的运动学假设，壳体变形后法线不一定再与中性面垂直，而是绕法线旋转了一定的角度。因此，变形后壳体上任意一点处的位置矢量表示为

$$\overline{\boldsymbol{R}}_{\mathrm{pv}}\left(\xi^1,\xi^2,\zeta\right)=\overline{\boldsymbol{r}}\left(\xi^1,\xi^2\right)+\frac{h}{2}\zeta\boldsymbol{\mathcal{R}}\boldsymbol{n}\left(\xi^1,\xi^2\right) \tag{9.8}$$

式中，头标“–”代表变形之后的量。通过旋转轴矢量 $\boldsymbol{\theta}$ 定义的 Rodrigues 旋转张量 $\boldsymbol{\mathcal{R}}$ 定义为[2]

$$\begin{gathered}\boldsymbol{\mathcal{R}}=\boldsymbol{I}+c_1\boldsymbol{\Omega}+c_2\boldsymbol{\Omega}^2\\ c_1=\frac{\sin\theta}{\theta},\quad c_2=\frac{1-\cos\theta}{\theta^2},\quad \boldsymbol{\Omega}=\begin{bmatrix}0 & -\theta_3 & \theta_2\\ \theta_3 & 0 & -\theta_1\\ -\theta_2 & \theta_1 & 0\end{bmatrix},\quad \theta=\|\boldsymbol{\theta}\|\end{gathered} \tag{9.9}$$

其中，$\boldsymbol{I}$ 为三阶单位矩阵。

在小应变和小旋转的假设条件下，Rodrigues 旋转张量 $\boldsymbol{\mathcal{R}}$ 可以简化为 $\boldsymbol{\mathcal{R}}=\boldsymbol{I}+\boldsymbol{\Omega}$。因此，壳体上任意一点处的位置矢量表示为

$$\begin{aligned}\boldsymbol{u}\left(\xi^1,\xi^2,\zeta\right)&=\overline{\boldsymbol{R}}_{\mathrm{pv}}\left(\xi^1,\xi^2,\zeta\right)-\boldsymbol{R}_{\mathrm{pv}}\left(\xi^1,\xi^2,\zeta\right)\\&=\overline{\boldsymbol{r}}\left(\xi^1,\xi^2\right)-\boldsymbol{r}\left(\xi^1,\xi^2\right)+\frac{h}{2}\zeta(\boldsymbol{\mathcal{R}}-\boldsymbol{I})\boldsymbol{n}\left(\xi^1,\xi^2\right)\\&=\boldsymbol{U}\left(\xi^1,\xi^2\right)+\frac{h}{2}\zeta\boldsymbol{\theta}\left(\xi^1,\xi^2\right)\times\boldsymbol{n}\left(\xi^1,\xi^2\right)\end{aligned}\tag{9.10}$$

式中，$\boldsymbol{U}\left(\xi^1,\xi^2\right)$和$\boldsymbol{\theta}\left(\xi^1,\xi^2\right)$分别是该点在壳体中性面上正交投影点处的位移矢量和旋转矢量。在式(9.10)中，关系式$(\boldsymbol{\mathcal{R}}-\mathbf{I})\boldsymbol{n}\simeq\boldsymbol{\Omega}\boldsymbol{n}=\boldsymbol{\theta}\times\boldsymbol{n}$对于所有小旋转的情况都成立。

9.2.3 壳体的等几何离散公式

本章选择等几何 Reissner-Mindlin 壳单元[3,4]进行模拟，它的每个控制点处有六个自由度，包括三个平移和三个旋转，这些变量都与壳体曲线坐标系无关，其有助于接下来要讨论的混合维度耦合约束的实施。等几何分析中的参数坐标ξ和η与壳体曲线坐标系中参数ξ^1和ξ^2是一致的，而不再需要任何额外的坐标转换。

变形前壳体中性面的精确几何描述可由对应控制点和 NURBS 基函数插值得到，并表示为

$$\boldsymbol{r}(\xi,\eta)=\sum_{i=1}^{n\times m}R_i(\xi,\eta)\boldsymbol{P}_i\tag{9.11}$$

其中，$R_i(\xi,\eta)$是双变量的 NURBS 基函数；n 和 m 为两个参数方向上 NURBS 基函数的阶次；$\boldsymbol{P}_i$ 是相关的控制点。

通过使用相同的 NURBS 基函数来逼近未知场，中性面上离散的位移场和旋转场描述为

$$\begin{aligned}\boldsymbol{U}\left(\xi^1,\xi^2\right)&=\sum_{i=1}^{n\times m}R_i(\xi,\eta)\boldsymbol{U}_i\\\boldsymbol{\theta}\left(\xi^1,\xi^2\right)&=\sum_{i=1}^{n\times m}R_i(\xi,\eta)\boldsymbol{\theta}_i\end{aligned}\tag{9.12}$$

式中，$\boldsymbol{U}_i=\left\{U_{ix},U_{iz},U_{iy}\right\}^{\mathrm{T}}$和$\boldsymbol{\theta}_i=\left\{\theta_{ix},\theta_{iz},\theta_{iy}\right\}^{\mathrm{T}}$分别为控制点$\boldsymbol{P}_i$处的定义在整体笛卡儿坐标系下的位移和转角自由度。

因此，根据式(9.11)和式(9.12)，壳体上任意一点处位置矢量和位移矢量的离散化形式表示为

$$\begin{aligned}\boldsymbol{R}(\xi,\eta,\zeta)&=\sum_{i=1}^{n\times m}R_i(\xi,\eta)\boldsymbol{P}_i+\frac{h}{2}\zeta\boldsymbol{n}(\xi,\eta)\\\boldsymbol{u}(\xi,\eta,\zeta)&=\sum_{i=1}^{n\times m}R_i(\xi,\eta)\boldsymbol{U}_i+\frac{h}{2}\zeta\sum_{i=1}^{n\times m}R_i(\xi,\eta)\boldsymbol{\theta}_i\times\boldsymbol{n}(\xi,\eta)\end{aligned}\tag{9.13}$$

可以注意到，由于等几何分析中壳体的精确几何描述，每个积分点处的法线都可以精确地计算，而不再是传统方法中通过结点处的法线值近似插值而得。

从物理空间到参数空间的雅可比变换矩阵表示为[4]

$$\begin{aligned}\boldsymbol{J}=\boldsymbol{R}_{,\Xi}&=\sum_{i=1}^{n\times m}R_{i,\Xi}\boldsymbol{P}_i+\frac{h}{2}(\zeta\boldsymbol{n})_{,\Xi}\quad(\Xi=\xi,\eta,\zeta)\\&=\left(\sum_{i=1}^{n\times m}R_{i,\xi}\boldsymbol{P}_i+\frac{h}{2}\zeta\boldsymbol{n}_{,\xi};\ \sum_{i=1}^{n\times m}R_{i,\eta}\boldsymbol{P}_i+\frac{h}{2}\zeta\boldsymbol{n}_{,\eta};\ \frac{h}{2}\zeta\boldsymbol{n}\right)\end{aligned}\tag{9.14}$$

根据式(9.1)和式(9.2)，我们可以在每个计算点处构造一个局部正交坐标系 $\boldsymbol{L}$，其表示为

$$\boldsymbol{L}=\left(\boldsymbol{t}_1=\boldsymbol{g}_1/\|\boldsymbol{g}_1\|;\ \boldsymbol{t}_2=\boldsymbol{n}\times\boldsymbol{t}_1;\ \boldsymbol{n}\right)\tag{9.15}$$

考虑均质、各向同性和线弹性材料，其本构行为仅需用弹性模量 E 和泊松比 ν 就能描述。在局部坐标系中使用平面应力假设，即 $\sigma_{33}^{\text{loc}}=0$，其中上标“loc”表示局部坐标系。因此，局部坐标系下的材料本构矩阵 $\boldsymbol{D}^{\text{loc}}$ 定义为

$$\boldsymbol{D}^{\text{loc}}=\frac{E}{1-\nu^2}\begin{bmatrix}1&\nu&0&0&0\\\nu&1&0&0&0\\0&0&(1-\nu)/2&0&0\\0&0&0&k_s(1-\nu)/2&0\\0&0&0&0&k_s(1-\nu)/2\end{bmatrix}\tag{9.16}$$

其中，k_s 为剪切修正因子，其值为5/6。

利用整体坐标系和局部坐标系之间的变换关系，我们可以得到整体坐标系下的材料本构矩阵 $\boldsymbol{D}^{\text{glo}}$ 为

$$\boldsymbol{D}^{\text{glo}}=\boldsymbol{Q}^{\text{T}}\boldsymbol{D}^{\text{loc}}\boldsymbol{Q}\tag{9.17}$$

式中，上标“glo”表示全局坐标系；$\boldsymbol{Q}$ 是转换矩阵，其表达式为

$$\boldsymbol{Q}=\begin{bmatrix}L_{11}^2&L_{12}^2&L_{13}^2&L_{11}L_{12}&L_{12}L_{13}&L_{13}L_{11}\\L_{21}^2&L_{22}^2&L_{23}^2&L_{21}L_{22}&L_{22}L_{23}&L_{23}L_{21}\\2L_{11}L_{21}&2L_{12}L_{22}&2L_{13}L_{23}&L_{11}L_{22}+L_{12}L_{21}&L_{12}L_{23}+L_{13}L_{22}&L_{13}L_{21}+L_{11}L_{23}\\2L_{21}L_{31}&2L_{22}L_{32}&2L_{23}L_{33}&L_{21}L_{32}+L_{22}L_{31}&L_{22}L_{33}+L_{23}L_{32}&L_{23}L_{31}+L_{21}L_{33}\\2L_{31}L_{11}&2L_{32}L_{12}&2L_{33}L_{13}&L_{31}L_{12}+L_{32}L_{11}&L_{32}L_{13}+L_{33}L_{12}&L_{33}L_{11}+L_{31}L_{13}\end{bmatrix}\tag{9.18}$$

将式(9.13)代入文献[4]中的几何方程，可得到整体坐标系下的线性应变矩阵

$\boldsymbol{B}^{\mathrm{glo}}$ 为

$$\boldsymbol{B}^{\mathrm{glo}}=\left[\boldsymbol{B}_1^{\boldsymbol{u}}\boldsymbol{B}_1^{\boldsymbol{\theta}} \quad \cdots \quad \boldsymbol{B}_i^{\boldsymbol{u}}\boldsymbol{B}_i^{\boldsymbol{\theta}} \quad \cdots \quad \boldsymbol{B}_{n\times m}^{\boldsymbol{u}}\boldsymbol{B}_{n\times m}^{\boldsymbol{\theta}}\right] \tag{9.19}$$

其中，

$$\boldsymbol{B}_i^{\boldsymbol{u}}=\begin{bmatrix} R_{i,x} & 0 & 0 \\ 0 & R_{i,y} & 0 \\ 0 & 0 & R_{i,z} \\ R_{i,y} & R_{i,x} & 0 \\ 0 & R_{i,z} & R_{i,y} \\ R_{i,z} & 0 & R_{i,x} \end{bmatrix}, \quad \boldsymbol{B}_i^{\boldsymbol{\theta}}=\begin{bmatrix} 0 & (\tilde{R}_i n_z)_{,x} & -(\tilde{R}_i n_y)_{,x} \\ -(\tilde{R}_i n_z)_{,y} & 0 & (\tilde{R}_i n_x)_{,y} \\ (\tilde{R}_i n_y)_{,z} & -(\tilde{R}_i n_x)_{,z} & 0 \\ -(\tilde{R}_i n_z)_{,x} & (\tilde{R}_i n_z)_{,y} & (\tilde{R}_i n_x)_{,x}-(\tilde{R}_i n_y)_{,y} \\ (\tilde{R}_i n_y)_{,y}-(\tilde{R}_i n_z)_{,z} & -(\tilde{R}_i n_x)_{,y} & (\tilde{R}_i n_x)_{,z} \\ (\tilde{R}_i n_y)_{,x} & (\tilde{R}_i n_z)_{,z}-(\tilde{R}_i n_x)_{,x} & -(\tilde{R}_i n_y)_{,z} \end{bmatrix} \tag{9.20}$$

式中，$\tilde{R}_i=\dfrac{h}{2}\zeta R_i$。形函数和单位法向量对物理坐标的一阶导数为

$$\begin{aligned} &\left\{R_{i,x},\ R_{i,y},\ R_{i,z}\right\}=\boldsymbol{J}^{-1}\left\{R_{i,\xi},\ R_{i,\eta},\ 0\right\} \\ &\left\{\tilde{R}_{i,x},\ \tilde{R}_{i,y},\ \tilde{R}_{i,z}\right\}=\boldsymbol{J}^{-1}\left\{\frac{h}{2}\zeta R_{i,\xi},\ \frac{h}{2}\zeta R_{i,\eta},\ \frac{h}{2}R_i\right\} \\ &\left[\boldsymbol{n}_{,x}\quad \boldsymbol{n}_{,y}\quad \boldsymbol{n}_{,z}\right]=\boldsymbol{J}^{-1}\left[\boldsymbol{n}_{,\xi}\quad \boldsymbol{n}_{,\eta}\quad \boldsymbol{0}\right] \end{aligned} \tag{9.21}$$

其中，单位法向量对自然曲线坐标的导数为

$$\frac{\partial \boldsymbol{n}}{\partial \xi^{\alpha}}=\frac{\boldsymbol{g}_{3,\alpha}\cdot\overline{\boldsymbol{g}}_3-\boldsymbol{g}_3\cdot\overline{\boldsymbol{g}}_{3,\alpha}}{\overline{\boldsymbol{g}}_3^2}, \quad \frac{\partial \boldsymbol{n}}{\partial \zeta}=\boldsymbol{0} \tag{9.22}$$

式中，

$$\begin{aligned} &\boldsymbol{g}_3=\boldsymbol{g}_1\times\boldsymbol{g}_2,\quad \boldsymbol{g}_{3,\alpha}=\boldsymbol{g}_{1,\alpha}\times\boldsymbol{g}_2+\boldsymbol{g}_1\times\boldsymbol{g}_{2,\alpha} \\ &\overline{\boldsymbol{g}}_3=\left\|\boldsymbol{g}_1\times\boldsymbol{g}_2\right\|,\quad \overline{\boldsymbol{g}}_{3,\alpha}=\frac{\boldsymbol{g}_3\cdot\boldsymbol{g}_{3,\alpha}}{\overline{\boldsymbol{g}}_3} \end{aligned} \tag{9.23}$$

在本质边界 Γ_{u} 上指定位移边界条件为 $\boldsymbol{u}=\overline{\boldsymbol{u}}$，在自然边界 Γ_{t} 上指定面力边界条件为 $\boldsymbol{t}=\overline{\boldsymbol{t}}$，则壳体求解方程的弱形式为[4]

$$\int_{\Omega}\varepsilon(\delta\boldsymbol{u})^{\mathrm{T}}\boldsymbol{D}^{\mathrm{glo}}\varepsilon(\boldsymbol{u})\mathrm{d}\Omega=\int_{\Omega}(\delta\boldsymbol{u})^{\mathrm{T}}\overline{\boldsymbol{p}}\mathrm{d}\Omega+\int_{\Gamma_{\mathrm{t}}}(\delta\boldsymbol{u})^{\mathrm{T}}\overline{\boldsymbol{t}}\mathrm{d}\Gamma_{\mathrm{t}} \tag{9.24}$$

其中，$\boldsymbol{\varepsilon}(\boldsymbol{u})$ 是应变矢量；$\overline{\boldsymbol{p}}$ 是作用在壳体中性面上的面力矢量；位移矢量 $\boldsymbol{u}$ 由整体坐标系中的三个平移自由度和三个转角自由度组成；而面力矢量 $\overline{\boldsymbol{t}}$ 包含沿全局坐标方向的三个面力分量，以及绕由式(9.15)定义的局部坐标轴 $\boldsymbol{t}_1$ 和 $\boldsymbol{t}_2$ 的两个弯矩分量。

利用 NURBS 基函数逼近未知位移场后，得到如下离散后的代数方程：

$$\boldsymbol{K}\boldsymbol{u}=\boldsymbol{f} \tag{9.25}$$

其中，$\boldsymbol{K}$ 和 $\boldsymbol{f}$ 分别为全局刚度矩阵和外力矢量，并表示为

$$\begin{aligned}\boldsymbol{K}&=\int_{\Omega}\left(\boldsymbol{B}^{\mathrm{glo}}\right)^{\mathrm{T}}\boldsymbol{D}^{\mathrm{glo}}\boldsymbol{B}^{\mathrm{glo}}\mathrm{d}\Omega\\ \boldsymbol{f}&=\int_{\Omega}\boldsymbol{R}^{\mathrm{T}}\boldsymbol{b}\mathrm{d}\Omega+\int_{\Gamma_{\mathrm{t}}}\boldsymbol{R}^{\mathrm{T}}\overline{\boldsymbol{t}}\,\mathrm{d}\Gamma_{\mathrm{t}}\end{aligned} \tag{9.26}$$

旋转向量 $\boldsymbol{\theta}_i$ 定义旋转轴的方向，其大小为绕该轴的旋转角度。此外，在整体坐标系中定义的三个转角分量是线性相关的，其引起的刚度矩阵奇异性可以通过人为地向式(9.25)引入一个与法向量相关的旋转刚度子矩阵来解决。刚度矩阵的更新部分可以写为[3]

$$\boldsymbol{K}_{ii}^{\theta\theta}=\boldsymbol{K}_{ii}^{\theta\theta}+sk\,\boldsymbol{n}\otimes\boldsymbol{n} \tag{9.27}$$

其中，s 的取值范围为 $10^{-6}\sim10^{-4}$；k 是矩阵 $\boldsymbol{K}_{ii}^{\theta\theta}$ 对角线部分的最大值。此外，使用高阶 NURBS 基函数可以有效地避免 Reissner-Mindlin 假设引起的剪切自锁问题，其详细的讨论可参考文献[5]。

9.3　混合维度耦合实施

图 9.2 显示了混合维度耦合问题的一个示例，其中耦合域 $\boldsymbol{\Omega}$ 包含边界元实体子域 Ω^{B} (含耦合表面 Γ_{c})和有限元壳体子域 Ω^{F} (含耦合边 S_{c})。在接下来的推导中，约定上标 F 、B 和 I 分别表示有限元子域、边界元子域和耦合界面，下标 c、b 和 f 分别代表界面、边界元和有限元子域内排除界面的其他部分。采用大写粗体字母 $\boldsymbol{U}$ 和 $\boldsymbol{\theta}$ 表示有限元子域的位移和旋转矢量，而小写粗体字母 $\boldsymbol{u}$ 和 $\boldsymbol{t}$ 表示边界元子域的位移和面力矢量。实体-壳耦合问题的目标是建立耦合界面上不同类型单元的控制变量之间的联系，即位移的协调条件，以及沿耦合边缘的面力和壳体应力合力(法向力矩、扭矩和剪力)的平衡条件。

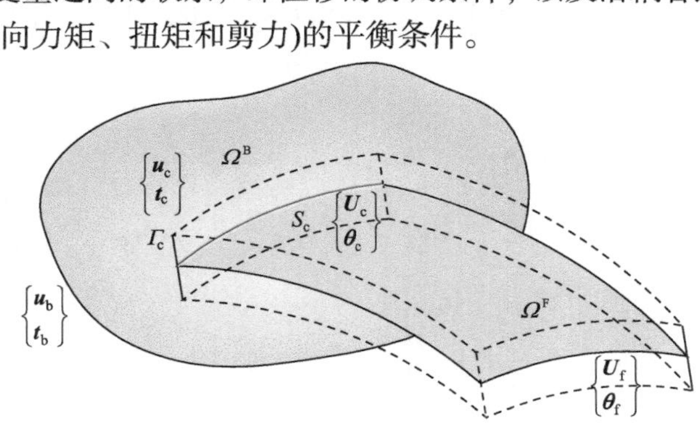

图 9.2　固体(边界元部分)-壳(有限元部分)耦合模型

9.3.1 刚度矩阵形式的边界元子域方程

忽略体积力情况下，三维弹性体的位移边界积分方程表示为[6]

$$c_{ij}(\boldsymbol{P}_{\mathrm{s}})u_j(\boldsymbol{P}_{\mathrm{s}})+\int_{\Gamma}T_{ij}(\boldsymbol{P}_{\mathrm{s}},\boldsymbol{Q})u_j(\boldsymbol{Q})\mathrm{d}\Gamma=\int_{\Gamma}U_{ij}(\boldsymbol{P}_{\mathrm{s}},\boldsymbol{Q})t_j(\boldsymbol{Q})\mathrm{d}\Gamma \tag{9.28}$$

其中，$\boldsymbol{P}_{\mathrm{s}}$ 和 $\boldsymbol{Q}$ 分别称作源点和场点；U_{ij} 和 T_{ij} 分别是线弹性材料的位移和面力基本解，它们的表达式可在文献[6]中找到；$c_{ij}(\boldsymbol{P}_{\mathrm{s}})$ 是与源点处边界几何形状相关的自由项。

通过等几何分析对式(9.28)进行离散，需使用参数空间中的 Greville 横坐标[7]获得配点的位置，并使用 NURBS 基函数来近似位移和面力变量。在循环所有的配点之后，得到如下线性系统方程：

$$\boldsymbol{H}\boldsymbol{u}=\boldsymbol{G}\boldsymbol{t} \tag{9.29}$$

其中，$\boldsymbol{H}$ 和 $\boldsymbol{G}$ 是影响系数矩阵；$\boldsymbol{u}$ 和 $\boldsymbol{t}$ 分别是包含整个边界的全局位移和面力矢量。

将自由度划分为与耦合界面相关的自由度和其余自由度两个部分，式(9.29)可以扩展为

$$\begin{bmatrix}\boldsymbol{H}_{\mathrm{cc}} & \boldsymbol{H}_{\mathrm{cb}}\\ \boldsymbol{H}_{\mathrm{bc}} & \boldsymbol{H}_{\mathrm{bb}}\end{bmatrix}\begin{Bmatrix}\boldsymbol{u}_{\mathrm{c}}\\ \boldsymbol{u}_{\mathrm{b}}\end{Bmatrix}=\begin{bmatrix}\boldsymbol{G}_{\mathrm{cc}} & \boldsymbol{G}_{\mathrm{cb}}\\ \boldsymbol{G}_{\mathrm{bc}} & \boldsymbol{G}_{\mathrm{bb}}\end{bmatrix}\begin{Bmatrix}\boldsymbol{t}_{\mathrm{c}}\\ \boldsymbol{t}_{\mathrm{b}}\end{Bmatrix} \tag{9.30}$$

在施加边界元部分的边界条件后，式(9.30)变为

$$\begin{bmatrix}\boldsymbol{H}_{\mathrm{cc}} & \mathcal{A}_{\mathrm{cb}}\\ \boldsymbol{H}_{\mathrm{bc}} & \mathcal{A}_{\mathrm{bb}}\end{bmatrix}\begin{Bmatrix}\boldsymbol{u}_{\mathrm{c}}\\ \mathcal{X}_{\mathrm{b}}\end{Bmatrix}=\begin{bmatrix}\boldsymbol{G}_{\mathrm{cc}} & \mathcal{B}_{\mathrm{cb}}\\ \boldsymbol{G}_{\mathrm{bc}} & \mathcal{B}_{\mathrm{bb}}\end{bmatrix}\begin{Bmatrix}\boldsymbol{t}_{\mathrm{c}}\\ \mathcal{Y}_{\mathrm{b}}\end{Bmatrix} \tag{9.31}$$

其中，矢量 $\mathcal{X}_{\mathrm{b}}$ 包含所有的未知位移和面力；矢量 $\mathcal{Y}_{\mathrm{b}}$ 包含所有已知的位移和面力。根据舒尔补理论[8]，将式(9.31)改写为

$$\begin{bmatrix}\boldsymbol{H}_{\mathrm{cc}}-\mathcal{A}_{\mathrm{cb}}(\mathcal{A}_{\mathrm{bb}})^{-1}\boldsymbol{H}_{\mathrm{bc}} & \boldsymbol{0}\\ \boldsymbol{H}_{\mathrm{bc}} & \mathcal{A}_{\mathrm{bb}}\end{bmatrix}\begin{Bmatrix}\boldsymbol{u}_{\mathrm{c}}\\ \mathcal{X}_{\mathrm{b}}\end{Bmatrix}=\begin{bmatrix}\boldsymbol{G}_{\mathrm{cc}}-\mathcal{A}_{\mathrm{cb}}(\mathcal{A}_{\mathrm{bb}})^{-1}\boldsymbol{G}_{\mathrm{bc}} & \mathcal{B}_{\mathrm{cb}}-\mathcal{A}_{\mathrm{cb}}(\mathcal{A}_{\mathrm{bb}})^{-1}\mathcal{B}_{\mathrm{bb}}\\ \boldsymbol{G}_{\mathrm{bc}} & \mathcal{B}_{\mathrm{bb}}\end{bmatrix}\begin{Bmatrix}\boldsymbol{t}_{\mathrm{c}}\\ \mathcal{Y}_{\mathrm{b}}\end{Bmatrix} \tag{9.32}$$

由式(9.32)的第一行，得到

$$\begin{aligned}&\left[\boldsymbol{H}_{\mathrm{cc}}-\mathcal{A}_{\mathrm{cb}}(\mathcal{A}_{\mathrm{bb}})^{-1}\boldsymbol{H}_{\mathrm{bc}}\right]\boldsymbol{u}_{\mathrm{c}}-\left[\mathcal{B}_{\mathrm{cb}}-\mathcal{A}_{\mathrm{cb}}(\mathcal{A}_{\mathrm{bb}})^{-1}\mathcal{B}_{\mathrm{bb}}\right]\mathcal{Y}_{\mathrm{b}}\\ &=\left[\boldsymbol{G}_{\mathrm{cc}}-\mathcal{A}_{\mathrm{cb}}(\mathcal{A}_{\mathrm{bb}})^{-1}\boldsymbol{G}_{\mathrm{bc}}\right]\boldsymbol{t}_{\mathrm{c}}\end{aligned} \tag{9.33}$$

因此，耦合界面上的面力和位移之间的关系可以记为

$$\boldsymbol{t}_{\mathrm{c}}=\mathcal{M}\boldsymbol{u}_{\mathrm{c}}-\mathcal{N} \tag{9.34}$$

其中，

$$\begin{aligned}\mathcal{M}&=\left[\boldsymbol{G}_{\mathrm{cc}}-\mathcal{A}_{\mathrm{cb}}\left(\mathcal{A}_{\mathrm{bb}}\right)^{-1}\boldsymbol{G}_{\mathrm{bc}}\right]^{-1}\left[\boldsymbol{H}_{\mathrm{cc}}-\mathcal{A}_{\mathrm{cb}}\left(\mathcal{A}_{\mathrm{bb}}\right)^{-1}\boldsymbol{H}_{\mathrm{bc}}\right]\\ \mathcal{N}&=\left[\boldsymbol{G}_{\mathrm{cc}}-\mathcal{A}_{\mathrm{cb}}\left(\mathcal{A}_{\mathrm{bb}}\right)^{-1}\boldsymbol{G}_{\mathrm{bc}}\right]^{-1}\left[\mathcal{B}_{\mathrm{cb}}-\mathcal{A}_{\mathrm{cb}}\left(\mathcal{A}_{\mathrm{bb}}\right)^{-1}\mathcal{B}_{\mathrm{bb}}\right]\mathcal{Y}_{\mathrm{b}}\end{aligned}\tag{9.35}$$

值得注意的是，在这里我们改进了第 8 章中的实现方法。第 8 章中是通过高斯消去法部分地消去式(9.31)中的子矩阵 $\mathcal{A}_{\mathrm{cb}}$，其过程是相对低效且耗时的。

将未知矢量 $\mathcal{X}_{\mathrm{b}}$ 进一步展开，将式(9.31)重新写为

$$\begin{bmatrix}\boldsymbol{H}_{\mathrm{cc}} & \mathcal{A}_{\mathrm{c},\boldsymbol{u}_{\mathrm{b}k}} & \mathcal{A}_{\mathrm{c},\boldsymbol{t}_{\mathrm{b}k}}\\ \boldsymbol{H}_{\boldsymbol{u}_{\mathrm{b}k},\mathrm{c}} & \mathcal{A}_{\boldsymbol{u}_{\mathrm{b}k},\boldsymbol{u}_{\mathrm{b}k}} & \mathcal{A}_{\boldsymbol{u}_{\mathrm{b}k},\boldsymbol{t}_{\mathrm{b}k}}\\ \boldsymbol{H}_{\boldsymbol{t}_{\mathrm{b}k},\mathrm{c}} & \mathcal{A}_{\boldsymbol{t}_{\mathrm{b}k},\boldsymbol{u}_{\mathrm{b}k}} & \mathcal{A}_{\boldsymbol{t}_{\mathrm{b}k},\boldsymbol{t}_{\mathrm{b}k}}\end{bmatrix}\begin{Bmatrix}\boldsymbol{u}_{\mathrm{c}}\\ \boldsymbol{u}_{\mathrm{b}k}\\ \boldsymbol{t}_{\mathrm{b}k}\end{Bmatrix}=\begin{bmatrix}\boldsymbol{G}_{\mathrm{cc}} & \mathcal{B}_{\mathrm{c},\boldsymbol{t}_k} & \mathcal{B}_{\mathrm{c},\boldsymbol{u}_k}\\ \boldsymbol{G}_{\boldsymbol{t}_k,\mathrm{c}} & \mathcal{B}_{\boldsymbol{t}_k,\boldsymbol{t}_k} & \mathcal{B}_{\boldsymbol{t}_k,\boldsymbol{u}_k}\\ \boldsymbol{G}_{\boldsymbol{u}_k,\mathrm{c}} & \mathcal{B}_{\boldsymbol{u}_k,\boldsymbol{t}_k} & \mathcal{B}_{\boldsymbol{u}_k,\boldsymbol{u}_k}\end{bmatrix}\begin{Bmatrix}\boldsymbol{t}_{\mathrm{c}}\\ \boldsymbol{t}_k\\ \boldsymbol{u}_k\end{Bmatrix}\tag{9.36}$$

其中，$\boldsymbol{u}_{\mathrm{b}k}$ 和 $\boldsymbol{t}_{\mathrm{b}k}$ 分别代表除了耦合界面之外的边界上未知的位移和面力；$\boldsymbol{u}_k$ 和 $\boldsymbol{t}_k$ 则分别代表对应的已知量。

将式(9.34)代入式(9.36)中，消除掉未知的面力 $\boldsymbol{t}_{\mathrm{c}}$，可得

$$\begin{bmatrix}\boldsymbol{H}_{\mathrm{cc}}-\boldsymbol{G}_{\mathrm{cc}}\mathcal{M} & \mathcal{A}_{\mathrm{c},\boldsymbol{u}_{\mathrm{b}k}} & \mathcal{A}_{\mathrm{c},\boldsymbol{t}_{\mathrm{b}k}}\\ \boldsymbol{H}_{\boldsymbol{u}_{\mathrm{b}k},\mathrm{c}}-\boldsymbol{G}_{\boldsymbol{t}_k,\mathrm{c}}\mathcal{M} & \mathcal{A}_{\boldsymbol{u}_{\mathrm{b}k},\boldsymbol{u}_{\mathrm{b}k}} & \mathcal{A}_{\boldsymbol{u}_{\mathrm{b}k},\boldsymbol{t}_{\mathrm{b}k}}\\ \boldsymbol{H}_{\boldsymbol{t}_{\mathrm{b}k},\mathrm{c}}-\boldsymbol{G}_{\boldsymbol{u}_k,\mathrm{c}}\mathcal{M} & \mathcal{A}_{\boldsymbol{t}_{\mathrm{b}k},\boldsymbol{u}_{\mathrm{b}k}} & \mathcal{A}_{\boldsymbol{t}_{\mathrm{b}k},\boldsymbol{t}_{\mathrm{b}k}}\end{bmatrix}\begin{Bmatrix}\boldsymbol{u}_{\mathrm{c}}\\ \boldsymbol{u}_{\mathrm{b}k}\\ \boldsymbol{t}_{\mathrm{b}k}\end{Bmatrix}=\begin{Bmatrix}\mathcal{R}_{\mathrm{c}}\\ \mathcal{R}_{\boldsymbol{u}_{\mathrm{b}k}}\\ \mathcal{R}_{\boldsymbol{t}_{\mathrm{b}k}}\end{Bmatrix}\tag{9.37}$$

其中，

$$\begin{cases}\mathcal{R}_{\mathrm{c}}=-\boldsymbol{G}_{\mathrm{cc}}\mathcal{N}+\mathcal{B}_{\mathrm{c},\boldsymbol{t}_k}\boldsymbol{t}_k+\mathcal{B}_{\mathrm{c},\boldsymbol{u}_k}\boldsymbol{u}_k\\ \mathcal{R}_{\boldsymbol{u}_{\mathrm{b}k}}=-\boldsymbol{G}_{\boldsymbol{t}_k,\mathrm{c}}\mathcal{N}+\mathcal{B}_{\boldsymbol{t}_k,\boldsymbol{t}_k}\boldsymbol{t}_k+\mathcal{B}_{\boldsymbol{t}_k,\boldsymbol{u}_k}\boldsymbol{u}_k\\ \mathcal{R}_{\boldsymbol{t}_{\mathrm{b}k}}=-\boldsymbol{G}_{\boldsymbol{u}_k,\mathrm{c}}\mathcal{N}+\mathcal{B}_{\boldsymbol{u}_k,\boldsymbol{t}_k}\boldsymbol{t}_k+\mathcal{B}_{\boldsymbol{u}_k,\boldsymbol{u}_k}\boldsymbol{u}_k\end{cases}\tag{9.38}$$

由式(9.37)中的第三行，得到未知面力 $\boldsymbol{t}_{\mathrm{b}k}$ 与未知位移的关系

$$\boldsymbol{t}_{\mathrm{b}k}=\left(\mathcal{A}_{\boldsymbol{t}_{\mathrm{b}k},\boldsymbol{t}_{\mathrm{b}k}}\right)^{-1}\mathcal{R}_{\boldsymbol{t}_{\mathrm{b}k}}-\left(\mathcal{A}_{\boldsymbol{t}_{\mathrm{b}k},\boldsymbol{t}_{\mathrm{b}k}}\right)^{-1}\left(\boldsymbol{H}_{\boldsymbol{t}_{\mathrm{b}k},\mathrm{c}}-\boldsymbol{G}_{\boldsymbol{u}_k,\mathrm{c}}\mathcal{M}\right)\boldsymbol{u}_{\mathrm{c}}-\left(\mathcal{A}_{\boldsymbol{t}_{\mathrm{b}k},\boldsymbol{t}_{\mathrm{b}k}}\right)^{-1}\mathcal{A}_{\boldsymbol{t}_{\mathrm{b}k},\boldsymbol{u}_{\mathrm{b}k}}\boldsymbol{u}_{\mathrm{b}k}\tag{9.39}$$

将式(9.39)代入式(9.37)中，并静力压缩 $\boldsymbol{t}_{\mathrm{b}k}$，即得到刚度矩阵形式的边界元子域方程为

$$\begin{bmatrix}\boldsymbol{K}_{\mathrm{c,c}} & \boldsymbol{K}_{\mathrm{c,b}k}\\ \boldsymbol{K}_{\mathrm{b}k,\mathrm{c}} & \boldsymbol{K}_{\mathrm{b}k,\mathrm{b}k}\end{bmatrix}\begin{Bmatrix}\boldsymbol{u}_{\mathrm{c}}\\ \boldsymbol{u}_{\mathrm{b}k}\end{Bmatrix}=\begin{Bmatrix}\boldsymbol{f}_{\mathrm{c}}\\ \boldsymbol{f}_{\mathrm{b}k}\end{Bmatrix}\tag{9.40}$$

其中，

$$
\begin{aligned}
&\boldsymbol{K}_{\mathrm{c,c}} = \boldsymbol{H}_{\mathrm{cc}} - \boldsymbol{G}_{\mathrm{cc}}\mathcal{M} - \mathcal{A}_{\mathrm{c},\boldsymbol{t}_{\mathrm{b}k}}\left(\mathcal{A}_{\boldsymbol{t}_{\mathrm{b}k},\boldsymbol{t}_{\mathrm{b}k}}\right)^{-1}\left(\boldsymbol{H}_{\boldsymbol{t}_{\mathrm{b}k},\mathrm{c}} - \boldsymbol{G}_{\boldsymbol{u}_k,\mathrm{c}}\mathcal{M}\right)\\
&\boldsymbol{K}_{\mathrm{c,b}k} = \mathcal{A}_{\mathrm{c},\boldsymbol{u}_{\mathrm{b}k}} - \mathcal{A}_{\mathrm{c},\boldsymbol{t}_{\mathrm{b}k}}\left(\mathcal{A}_{\boldsymbol{t}_{\mathrm{b}k},\boldsymbol{t}_{\mathrm{b}k}}\right)^{-1}\mathcal{A}_{\boldsymbol{t}_{\mathrm{b}k},\boldsymbol{u}_{\mathrm{b}k}}\\
&\boldsymbol{K}_{\mathrm{b}k,\mathrm{c}} = \boldsymbol{H}_{\boldsymbol{u}_{\mathrm{b}k},\mathrm{c}} - \boldsymbol{G}_{\boldsymbol{t}_k,\mathrm{c}}\mathcal{M} - \mathcal{A}_{\boldsymbol{u}_{\mathrm{b}k},\boldsymbol{t}_{\mathrm{b}k}}\left(\mathcal{A}_{\boldsymbol{t}_{\mathrm{b}k},\boldsymbol{t}_{\mathrm{b}k}}\right)^{-1}\left(\boldsymbol{H}_{\boldsymbol{t}_{\mathrm{b}k},\mathrm{c}} - \boldsymbol{G}_{\boldsymbol{u}_k,\mathrm{c}}\mathcal{M}\right)\\
&\boldsymbol{K}_{\mathrm{b}k,\mathrm{b}k} = \mathcal{A}_{\boldsymbol{u}_{\mathrm{b}k},\boldsymbol{u}_{\mathrm{b}k}} - \mathcal{A}_{\boldsymbol{u}_{\mathrm{b}k},\boldsymbol{t}_{\mathrm{b}k}}\left(\mathcal{A}_{\boldsymbol{t}_{\mathrm{b}k},\boldsymbol{t}_{\mathrm{b}k}}\right)^{-1}\mathcal{A}_{\boldsymbol{t}_{\mathrm{b}k},\boldsymbol{u}_{\mathrm{b}k}}\\
&\boldsymbol{f}_{\mathrm{c}} = \mathcal{R}_{\mathrm{c}} - \mathcal{A}_{\mathrm{c},\boldsymbol{t}_{\mathrm{b}k}}\left(\mathcal{A}_{\boldsymbol{t}_{\mathrm{b}k},\boldsymbol{t}_{\mathrm{b}k}}\right)^{-1}\mathcal{R}_{\boldsymbol{t}_{\mathrm{b}k}}\\
&\boldsymbol{f}_{\mathrm{b}k} = \mathcal{R}_{\boldsymbol{u}_{\mathrm{b}k}} - \mathcal{A}_{\boldsymbol{u}_{\mathrm{b}k},\boldsymbol{t}_{\mathrm{b}k}}\left(\mathcal{A}_{\boldsymbol{t}_{\mathrm{b}k},\boldsymbol{t}_{\mathrm{b}k}}\right)^{-1}\mathcal{R}_{\boldsymbol{t}_{\mathrm{b}k}}
\end{aligned}
\tag{9.41}
$$

式(9.40)中隐含了面力平衡条件，一旦求解出未知位移，就可以在后处理步骤中使用式(9.34)和式(9.39)计算出未知面力。因此，固体和壳体之间的耦合约束只需要建立耦合界面上的位移连续关系。

9.3.2 直接运动耦合约束方法

根据式(9.9)和式(9.13)，壳体上任意一点处的位移场可由其自由度表示为

$$
\begin{cases}
u_x = U_x - \dfrac{h}{2}\zeta n_y\theta_z + \dfrac{h}{2}\zeta n_z\theta_y\\
u_y = U_y + \dfrac{h}{2}\zeta n_x\theta_z - \dfrac{h}{2}\zeta n_z\theta_x\\
u_z = U_z - \dfrac{h}{2}\zeta n_x\theta_y + \dfrac{h}{2}\zeta n_y\theta_x
\end{cases}
\tag{9.42}
$$

边界元子域耦合界面 $\varGamma_{\mathrm{c}}$ 上的位移场可以从相关控制点处的控制变量由 NURBS 基函数插值得到，并表示为

$$
u_j^{\mathrm{B}} = \sum_{k=1}^{n_{\mathrm{surf}}} R_k^{\mathrm{B}}\left(\xi^{\mathrm{B}},\eta^{\mathrm{B}}\right)u_{jk}^{\mathrm{B}} \quad \left(j = x,y,z\right)
\tag{9.43}
$$

其中，n_{surf} 是边界元子域耦合界面 $\varGamma_{\mathrm{c}}$ 上的控制点总数。采用如下的矢量记号：

$$
\boldsymbol{R}^{\mathrm{B}} = \left\{R_1^{\mathrm{B}},R_2^{\mathrm{B}},\cdots,R_{n_{\mathrm{surf}}}^{\mathrm{B}}\right\}, \quad \bar{\boldsymbol{u}}_j^{\mathrm{B}} = \left\{u_{j1}^{\mathrm{B}},u_{j2}^{\mathrm{B}},\cdots,u_{jn_{\mathrm{surf}}}^{\mathrm{B}}\right\}^{\mathrm{T}}
\tag{9.44}
$$

式(9.43)可以写为如下矩阵形式：

$$
u_j^{\mathrm{B}} = \boldsymbol{R}^{\mathrm{B}}\cdot\bar{\boldsymbol{u}}_j^{\mathrm{B}}
\tag{9.45}
$$

同样地，有限元子域耦合界面 S_{c} 的整体平移和旋转场也可以从相关控制点处的控制变量由 NURBS 基函数插值得到，并表示为

$$
\begin{aligned}
U_j^{\mathrm{F}} &= \sum_{k=1}^{n_{\mathrm{edge}}} R_k^{\mathrm{F}}\left(\xi^*, \eta^{\mathrm{F}}\right) U_{jk}^{\mathrm{F}} \\
\theta_j^{\mathrm{F}} &= \sum_{k=1}^{n_{\mathrm{edge}}} R_k^{\mathrm{F}}\left(\xi^*, \eta^{\mathrm{F}}\right) \theta_{jk}^{\mathrm{F}} \quad (j = x, y, z)
\end{aligned} \tag{9.46}
$$

其中，n_{edge} 是有限元子域耦合界面 S_{c} 上的控制点总数；ξ^* 是沿壳体耦合边的已知参数坐标。采用如下的矢量记号：

$$
\begin{aligned}
&\boldsymbol{R}^{\mathrm{F}} = \left\{R_1^{\mathrm{F}}, R_2^{\mathrm{F}}, \cdots, R_{n_{\mathrm{edge}}}^{\mathrm{F}}\right\}, \quad \bar{\boldsymbol{U}}_j^{\mathrm{F}} = \left\{U_{j1}^{\mathrm{F}}, U_{j2}^{\mathrm{F}}, \cdots, U_{jn_{\mathrm{edge}}}^{\mathrm{F}}\right\}^{\mathrm{T}} \\
&\bar{\boldsymbol{\theta}}_j^{\mathrm{F}} = \left\{\theta_{j1}^{\mathrm{F}}, \theta_{j2}^{\mathrm{F}}, \cdots, \theta_{jn_{\mathrm{edge}}}^{F}\right\}^{\mathrm{T}} \quad (j = x, y, z)
\end{aligned} \tag{9.47}
$$

式(9.46)可以写为如下矩阵形式：

$$
\begin{aligned}
U_j^{\mathrm{F}} &= \boldsymbol{R}^{\mathrm{F}} \cdot \bar{\boldsymbol{U}}_j^{\mathrm{F}} \\
\theta_j^{\mathrm{F}} &= \boldsymbol{R}^{\mathrm{F}} \cdot \bar{\boldsymbol{\theta}}_j^{\mathrm{F}} \quad (j = x, y, z)
\end{aligned} \tag{9.48}
$$

由于 NURBS 基函数的非插值特性，控制点不一定位于曲面上。因此，采用边界元子域耦合界面上的配点(参数空间中 Greville 横坐标的映射)来构造直接运动耦合约束条件。配点和控制点之间的关系可以表示为

$$
\boldsymbol{P}^{\mathrm{col}} = \mathcal{M}_{\mathrm{col}} \cdot \boldsymbol{P}^{\mathrm{ctr}} \tag{9.49}
$$

其中，$\mathcal{M}_{\mathrm{col}}$ 是由 NURBS 形函数组成的变换矩阵。

同样地，式(9.49)也适用于配点处的位移场

$$
\begin{Bmatrix} \boldsymbol{u}_x^{\mathrm{B}} \\ \boldsymbol{u}_y^{\mathrm{B}} \\ \boldsymbol{u}_z^{\mathrm{B}} \end{Bmatrix}^{\mathrm{col}} = \underbrace{\begin{bmatrix} \mathcal{M}_{\mathrm{col}} & & \\ & \mathcal{M}_{\mathrm{col}} & \\ & & \mathcal{M}_{\mathrm{col}} \end{bmatrix}}_{\mathcal{M}_{\mathrm{col}}^{\mathrm{g}}} \begin{Bmatrix} \bar{\boldsymbol{u}}_x^{\mathrm{B}} \\ \bar{\boldsymbol{u}}_y^{\mathrm{B}} \\ \bar{\boldsymbol{u}}_z^{\mathrm{B}} \end{Bmatrix}^{\mathrm{ctr}} \tag{9.50}
$$

由式(9.42)和式(9.50)，并结合式(9.45)和式(9.48)，直接运动耦合约束条件表示为

$$
\begin{Bmatrix} \bar{\boldsymbol{u}}_x^{\mathrm{B}} \\ \bar{\boldsymbol{u}}_y^{\mathrm{B}} \\ \bar{\boldsymbol{u}}_z^{\mathrm{B}} \end{Bmatrix}^{\mathrm{ctr}} = \underbrace{\left(\mathcal{M}_{\mathrm{col}}^{\mathrm{g}}\right)^{-1} \cdot \begin{bmatrix} \boldsymbol{R}^{\mathrm{F}} & \boldsymbol{0} & \boldsymbol{0} & \boldsymbol{0} & \dfrac{h}{2}\zeta n_z \boldsymbol{R}^{\mathrm{F}} & -\dfrac{h}{2}\zeta n_y \boldsymbol{R}^{\mathrm{F}} \\ \boldsymbol{0} & \boldsymbol{R}^{\mathrm{F}} & \boldsymbol{0} & -\dfrac{h}{2}\zeta n_z \boldsymbol{R}^{\mathrm{F}} & \boldsymbol{0} & \dfrac{h}{2}\zeta n_x \boldsymbol{R}^{\mathrm{F}} \\ \boldsymbol{0} & \boldsymbol{0} & \boldsymbol{R}^{\mathrm{F}} & \dfrac{h}{2}\zeta n_y \boldsymbol{R}^{\mathrm{F}} & -\dfrac{h}{2}\zeta n_x \boldsymbol{R}^{\mathrm{F}} & \boldsymbol{0} \end{bmatrix}}_{\mathbf{Tr}_{\mathrm{bf}}^{\mathrm{g}}} \begin{Bmatrix} \bar{\boldsymbol{U}}_x^{\mathrm{F}} \\ \bar{\boldsymbol{U}}_y^{\mathrm{F}} \\ \bar{\boldsymbol{U}}_z^{\mathrm{F}} \\ \bar{\boldsymbol{\theta}}_x^{\mathrm{F}} \\ \bar{\boldsymbol{\theta}}_y^{\mathrm{F}} \\ \bar{\boldsymbol{\theta}}_z^{\mathrm{F}} \end{Bmatrix} \tag{9.51}
$$

遍历边界元子域耦合界面 Γ_c 上的所有配点，最终得到全局界面耦合矩阵为

$$\mathbf{Tr}_{\mathrm{bf}} = \sum \mathbf{Tr}_{\mathrm{bf}}^{\mathrm{g}} \tag{9.52}$$

9.3.3 基于界面虚功相等的弱耦合方法

建立弱耦合方程的基本原理是界面两侧的壳体和三维实体在相同的界面位移下所做的功相等。壳体和实体耦合界面上相关的变量及其假设的正方向分别如图 9.3 和图 9.4 所示。

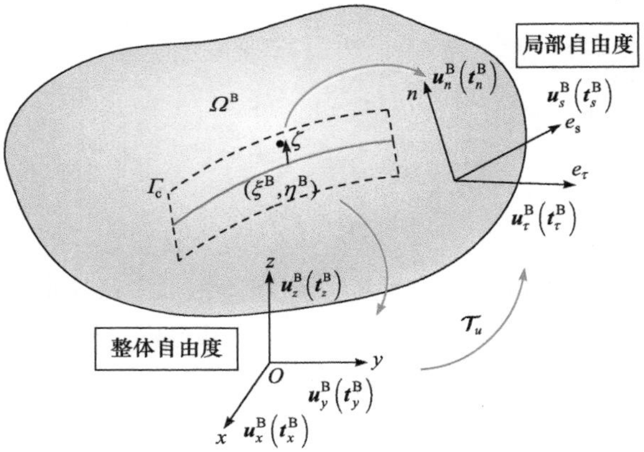

图 9.3 与边界元子域耦合界面相关的所有变量和相应的坐标变换

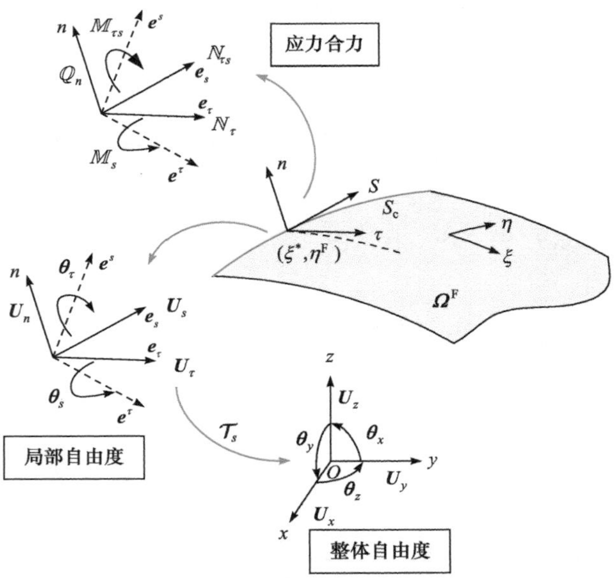

图 9.4 与有限元子域耦合界面相关的所有变量和相应的坐标变换

边界元子域在耦合界面上所做的功在整体坐标系中表示为

$$\mathcal{W}_{\mathrm{c}}^{\mathrm{bem}}=\int_{\Gamma_{\mathrm{c}}}\left(t_x^{\mathrm{B}}u_x^{\mathrm{B}}+t_y^{\mathrm{B}}u_y^{\mathrm{B}}+t_z^{\mathrm{B}}u_z^{\mathrm{B}}\right)\mathrm{d}\Gamma_{\mathrm{c}} \tag{9.53}$$

壳体耦合边上每个计算点处的局部坐标系如图 9.4 所示，相关的方向向量表示为

$$\boldsymbol{e}_\tau=\frac{\boldsymbol{g}_1}{\left\|\boldsymbol{g}_1\right\|},\quad \boldsymbol{e}_s=\frac{\boldsymbol{g}_2}{\left\|\boldsymbol{g}_2\right\|} \tag{9.54}$$

相关的法线矢量通过式(9.2)计算。

用于描述壳体变形的局部自由度由沿局部坐标轴的平移自由度$\left(U_\tau,U_s,U_n\right)$和描述截面旋转的转角自由度$\left(\theta_\tau,\theta_s\right)$组成。应注意的是，$\theta_\tau$和$\theta_s$分别是围绕$\boldsymbol{e}^s$和$\boldsymbol{e}^\tau$的旋转角度。$\boldsymbol{e}^s$和$\boldsymbol{e}^\tau$的计算公式如下：

$$\boldsymbol{e}^s=\frac{\boldsymbol{g}^2}{\left\|\boldsymbol{g}^2\right\|},\quad \boldsymbol{e}^\tau=\frac{\boldsymbol{g}^1}{\left\|\boldsymbol{g}^1\right\|} \tag{9.55}$$

规定绕旋转轴逆时针旋转的转角为正。因此，壳体的局部自由度和整体自由度之间的转换关系为

$$\begin{Bmatrix}U_x\\U_y\\U_z\\\theta_x\\\theta_y\\\theta_z\end{Bmatrix}=\underbrace{\begin{bmatrix}\boldsymbol{e}_{\tau x}&\boldsymbol{e}_{sx}&n_x&0&0\\\boldsymbol{e}_{\tau y}&\boldsymbol{e}_{sy}&n_y&0&0\\\boldsymbol{e}_{\tau z}&\boldsymbol{e}_{sz}&n_z&0&0\\0&0&0&\boldsymbol{e}_x^s&\boldsymbol{e}_x^\tau\\0&0&0&\boldsymbol{e}_y^s&\boldsymbol{e}_y^\tau\\0&0&0&\boldsymbol{e}_z^s&\boldsymbol{e}_z^\tau\end{bmatrix}}_{\mathcal{T}_s}\begin{Bmatrix}U_\tau\\U_s\\U_n\\\theta_\tau\\\theta_s\end{Bmatrix}=\mathcal{T}_s\begin{Bmatrix}U_\tau\\U_s\\U_n\\\theta_\tau\\\theta_s\end{Bmatrix} \tag{9.56}$$

耦合公式中涉及的壳体应力合力包括：面内薄膜力$\mathcal{N}_\tau$和$\mathcal{N}_{\tau s}$、剪力$\mathcal{Q}_n$以及弯矩$\mathcal{M}_s$和$\mathcal{M}_{\tau s}$。这些应力合力的正负号约定如图 9.4 所示。因此，耦合界面上壳体所做的功在局部坐标系中表示为

$$\mathcal{W}_{\mathrm{c}}^{\mathrm{fem}}=\int_{S_{\mathrm{c}}}\left(\mathcal{N}_\tau U_\tau^{\mathrm{F}}+\mathcal{N}_{\tau s}U_s^{\mathrm{F}}+\mathcal{Q}_nU_n^{\mathrm{F}}+\mathcal{M}_{\tau s}\theta_\tau^{\mathrm{F}}+\mathcal{M}_s\theta_s^{\mathrm{F}}\right)\mathrm{d}S_{\mathrm{c}} \tag{9.57}$$

将式(9.53)中整体坐标系下边界元子域在耦合界面上所做的功转换到壳体的局部坐标系中，并表示为

$$\mathcal{W}_{\mathrm{c}}^{\mathrm{bem}}=\int_{\Gamma_{\mathrm{c}}}\left(t_\tau^{\mathrm{B}}u_\tau^{\mathrm{B}}+t_s^{\mathrm{B}}u_s^{\mathrm{B}}+t_n^{\mathrm{B}}u_n^{\mathrm{B}}\right)\mathrm{d}\Gamma_{\mathrm{c}} \tag{9.58}$$

式(9.58)中自由度的转换关系表示为

$$\begin{Bmatrix} u_\tau^{\mathrm{B}} \\ u_s^{\mathrm{B}} \\ u_n^{\mathrm{B}} \end{Bmatrix} = \underbrace{\begin{bmatrix} \boldsymbol{e}_{\tau x} & \boldsymbol{e}_{\tau y} & \boldsymbol{e}_{\tau z} \\ \boldsymbol{e}_{sx} & \boldsymbol{e}_{sy} & \boldsymbol{e}_{sz} \\ n_x & n_y & n_z \end{bmatrix}}_{\mathcal{T}_u} \begin{Bmatrix} u_x^{\mathrm{B}} \\ u_y^{\mathrm{B}} \\ u_z^{\mathrm{B}} \end{Bmatrix} = \mathcal{T}_u \begin{Bmatrix} u_x^{\mathrm{B}} \\ u_y^{\mathrm{B}} \\ u_z^{\mathrm{B}} \end{Bmatrix} \tag{9.59}$$

引入 Reissner-Mindlin 壳体截面上与应力合力有关的先验应力分布假设，即面内薄膜力在厚度上均匀分布，弯曲和扭转力矩产生的法向应力呈线性变化，横向剪应力随厚度呈抛物线变化。根据耦合界面上的内力平衡条件，得到以下方程：

$$\begin{aligned} t_\tau^{\mathrm{B}} &= \sigma_{\tau\tau}^{\mathrm{F}} = \frac{\mathcal{N}_\tau}{h} + \frac{6\mathcal{M}_s}{h^2}\left(\zeta - \zeta^0\right) \\ t_s^{\mathrm{B}} &= \sigma_{\tau s}^{\mathrm{F}} = \frac{\mathcal{N}_{\tau s}}{h} - \frac{6\mathcal{M}_{\tau s}}{h^2}\left(\zeta - \zeta^0\right) \\ t_n^{\mathrm{B}} &= \sigma_{\tau n}^{\mathrm{F}} = \frac{3\mathcal{Q}_n}{2h}\left(1 - \zeta^2\right) \end{aligned} \tag{9.60}$$

式中，h 是壳体的厚度；ζ^0 是确保在纯弯曲或扭转情况下壳体中性面处产生零应力状态的偏心率，其值为[9]

$$\zeta^0 = \frac{h}{6R_{\mathrm{para}}} \tag{9.61}$$

其中，R_{para} 是与壳体几何形状相关的耦合界面参数方向上的物理曲率，其值由式(9.6)求得。

弱耦合公式以功相等的形式表示为

$$\mathcal{W}_{\mathrm{c}}^{\mathrm{fem}} = \mathcal{W}_{\mathrm{c}}^{\mathrm{bem}} \tag{9.62}$$

将式(9.60)代入式(9.58)中，得到

$$\begin{aligned} \mathcal{W}_{\mathrm{c}}^{\mathrm{bem}} &= \int_{\Gamma_{\mathrm{c}}} \left(t_\tau^{\mathrm{B}} u_\tau^{\mathrm{B}} + t_s^{\mathrm{B}} u_s^{\mathrm{B}} + t_n^{\mathrm{B}} u_n^{\mathrm{B}}\right) \mathrm{d}\Gamma_{\mathrm{c}} \\ &= \int_{\Gamma_{\mathrm{c}}} \left\{ \left[\frac{\mathcal{N}_\tau}{h} + \frac{6\mathcal{M}_s}{h^2}\left(\zeta - \zeta^0\right)\right] u_\tau^{\mathrm{B}} + \left[\frac{\mathcal{N}_{\tau s}}{h} - \frac{6\mathcal{M}_{\tau s}}{h^2}\left(\zeta - \zeta^0\right)\right] u_s^{\mathrm{B}} + \frac{3\mathcal{Q}_n}{2h}\left(1 - \zeta^2\right) u_n^{\mathrm{B}} \right\} \mathrm{d}\Gamma_{\mathrm{c}} \\ &= \int_{\Gamma_{\mathrm{c}}} \left[\frac{\mathcal{N}_\tau}{h} u_\tau^{\mathrm{B}} + \frac{\mathcal{N}_{\tau s}}{h} u_s^{\mathrm{B}} + \frac{3\mathcal{Q}_n}{2h}\left(1 - \zeta^2\right) u_n^{\mathrm{B}} + \frac{6\mathcal{M}_s}{h^2}\left(\zeta - \zeta^0\right) u_\tau^{\mathrm{B}} - \frac{6\mathcal{M}_{\tau s}}{h^2}\left(\zeta - \zeta^0\right) u_s^{\mathrm{B}} \right] \mathrm{d}\Gamma_{\mathrm{c}} \end{aligned} \tag{9.63}$$

比较式(9.57)和式(9.63)中应力项前的系数，得到最终的耦合积分方程为

$$
\begin{aligned}
&\int_{S_c} \mathbb{N}_\tau U_\tau^{\mathrm{F}} \mathrm{d}S_c = \int_{\Gamma_c} \frac{\mathbb{N}_\tau}{h} u_\tau^{\mathrm{B}} \mathrm{d}\Gamma_c \\
&\int_{S_c} \mathbb{N}_{\tau s} U_s^{\mathrm{F}} \mathrm{d}S_c = \int_{\Gamma_c} \frac{\mathbb{N}_{\tau s}}{h} u_s^{\mathrm{B}} \mathrm{d}\Gamma_c \\
&\int_{S_c} \mathbb{Q}_n U_n^{\mathrm{F}} \mathrm{d}S_c = \int_{\Gamma_c} \frac{3\mathbb{Q}_n}{2h} \left(1-\zeta^2\right) u_n^{\mathrm{B}} \mathrm{d}\Gamma_c \\
&\int_{S_c} \mathbb{M}_{\tau s} \theta_\tau^{\mathrm{F}} \mathrm{d}S_c = \int_{\Gamma_c} -\frac{6\mathbb{M}_{\tau s}}{h^2} \left(\zeta-\zeta^0\right) u_s^{\mathrm{B}} \mathrm{d}\Gamma_c \\
&\int_{S_c} \mathbb{M}_s \theta_s^{\mathrm{F}} \mathrm{d}S_c = \int_{\Gamma_c} \frac{6\mathbb{M}_s}{h^2} \left(\zeta-\zeta^0\right) u_\tau^{\mathrm{B}} \mathrm{d}\Gamma_c
\end{aligned}
\tag{9.64}
$$

随后，推导式(9.64)通过 NURBS 形函数插值得到的离散形式。边界元子域耦合界面 Γ_c 上的位移场可以从相关控制点处的控制变量由 NURBS 基函数插值得到，并表示为

$$
\begin{aligned}
u_j^{\mathrm{B}} &= \sum_{k=1}^{n_{\mathrm{surf}}} R_k^{\mathrm{B}}\left(\xi^{\mathrm{B}},\eta^{\mathrm{B}}\right) u_{jk}^{\mathrm{B}} \quad (j=\tau,s,n) \\
t_j^{\mathrm{B}} &= \sum_{k=1}^{n_{\mathrm{surf}}} R_k^{\mathrm{B}}\left(\xi^{\mathrm{B}},\eta^{\mathrm{B}}\right) t_{jk}^{\mathrm{B}} \quad (j=\tau,s,n)
\end{aligned}
\tag{9.65}
$$

其中，n_{surf} 是边界元子域耦合界面 Γ_c 上的控制点总数。采用如下的矢量记号：

$$
\begin{aligned}
&\boldsymbol{R}^{\mathrm{B}} = \left\{R_1^{\mathrm{B}}, R_2^{\mathrm{B}}, \cdots, R_{n_{\mathrm{surf}}}^{\mathrm{B}}\right\}, \quad \hat{\boldsymbol{u}}_j^{\mathrm{B}} = \left\{u_{j1}^{\mathrm{B}}, u_{j2}^{\mathrm{B}}, \cdots, u_{jn_{\mathrm{surf}}}^{\mathrm{B}}\right\}^{\mathrm{T}} \\
&\hat{\boldsymbol{t}}_j^{\mathrm{B}} = \left\{t_{j1}^{\mathrm{B}}, t_{j2}^{\mathrm{B}}, \cdots, t_{jn_{\mathrm{surf}}}^{\mathrm{B}}\right\}^{\mathrm{T}} \quad (j=\tau,s,n)
\end{aligned}
\tag{9.66}
$$

式(9.65)可以写为如下的矩阵形式：

$$
\begin{aligned}
u_j^{\mathrm{B}} &= \boldsymbol{R}^{\mathrm{B}} \cdot \hat{\boldsymbol{u}}_j^{\mathrm{B}} \\
t_j^{\mathrm{B}} &= \boldsymbol{R}^{\mathrm{B}} \cdot \hat{\boldsymbol{t}}_j^{\mathrm{B}}
\end{aligned}
\tag{9.67}
$$

同样地，在局部坐标系下，有限元子域耦合界面 S_c 上的平移和旋转场也可以从相关控制点处的控制变量由 NURBS 基函数插值得到，并表示为

$$
\begin{aligned}
U_j^{\mathrm{F}} &= \sum_{k=1}^{n_{\mathrm{edge}}} R_k^{\mathrm{F}}\left(\xi^*,\eta^{\mathrm{F}}\right) U_{jk}^{\mathrm{F}} \quad (j=\tau,s,n) \\
\theta_j^{\mathrm{F}} &= \sum_{k=1}^{n_{\mathrm{edge}}} R_k^{\mathrm{F}}\left(\xi^*,\eta^{\mathrm{F}}\right) \theta_{jk}^{\mathrm{F}} \quad (j=\tau,s)
\end{aligned}
\tag{9.68}
$$

其中，n_{edge} 是有限元子域耦合界面 S_c 上的控制点总数；ξ^* 是沿壳体耦合边的已知参数坐标。采用如下的矢量记号：

$$
\begin{aligned}
&\boldsymbol{R}^{\mathrm{F}}=\left\{R_1^{\mathrm{F}},R_2^{\mathrm{F}},\cdots,R_{n_{\mathrm{edge}}}^{\mathrm{F}}\right\},\quad \hat{\boldsymbol{U}}_j^{\mathrm{F}}=\left\{U_{j1}^{\mathrm{F}},U_{j2}^{\mathrm{F}},\cdots,U_{jn_{\mathrm{edge}}}^{\mathrm{F}}\right\}^{\mathrm{T}}\quad (j=\tau,s,n)\\
&\hat{\boldsymbol{\theta}}_j^{\mathrm{F}}=\left\{\theta_{j1}^{\mathrm{F}},\theta_{j2}^{\mathrm{F}},\cdots,\theta_{jn_{\mathrm{edge}}}^{F}\right\}^{\mathrm{T}}\quad (j=\tau,s)
\end{aligned}
\tag{9.69}
$$

式(9.68)可以写为如下的矩阵形式:

$$
\begin{aligned}
U_j^{\mathrm{F}}&=\boldsymbol{R}^{\mathrm{F}}\cdot\hat{\boldsymbol{U}}_j^{\mathrm{F}}\quad (j=\tau,s,n)\\
\theta_j^{\mathrm{F}}&=\boldsymbol{R}^{\mathrm{F}}\cdot\hat{\boldsymbol{\theta}}_j^{\mathrm{F}}\quad (j=\tau,s)
\end{aligned}
\tag{9.70}
$$

在式(9.64)的数值实施中，壳体耦合边上积分点处的应力合力(薄膜力、剪切力和弯矩)也通过单元控制点处的应力合力值由 NURBS 形函数插值求得，并表示为

$$
\begin{aligned}
&\mathbb{N}_j=\sum_{k=1}^{n_{\mathrm{edge}}}R_k^{\mathrm{F}}\left(\xi^*,\eta^{\mathrm{F}}\right)\mathbb{N}_{jk}\quad (j=\tau,\tau s)\\
&\mathbb{Q}_n=\sum_{k=1}^{n_{\mathrm{edge}}}R_k^{\mathrm{F}}\left(\xi^*,\eta^{\mathrm{F}}\right)\mathbb{Q}_{nk},\quad \mathbb{M}_j=\sum_{k=1}^{n_{\mathrm{edge}}}R_k^{\mathrm{F}}\left(\xi^*,\eta^{\mathrm{F}}\right)\mathbb{M}_{jk}\quad (j=\tau s,s)
\end{aligned}
\tag{9.71}
$$

采用如下的矢量记号:

$$
\begin{aligned}
&\hat{\mathbb{N}}_j=\left\{\mathbb{N}_{j1},\mathbb{N}_{j2},\cdots,\mathbb{N}_{jn_{\mathrm{edge}}}\right\}^{\mathrm{T}}\quad (j=\tau,\tau s)\\
&\hat{\mathbb{Q}}_n=\left\{\mathbb{Q}_{n1},\mathbb{Q}_{n2},\cdots,\mathbb{Q}_{nn_{\mathrm{edge}}}\right\}^{\mathrm{T}}\\
&\hat{\mathbb{M}}_j=\left\{\mathbb{M}_{j1},\mathbb{M}_{j2},\cdots,\mathbb{M}_{jn_{\mathrm{edge}}}\right\}^{\mathrm{T}}\quad (j=\tau s,s)
\end{aligned}
\tag{9.72}
$$

式(9.71)可以写为如下的矩阵形式:

$$
\begin{aligned}
\mathbb{N}_j&=\boldsymbol{R}^{\mathrm{F}}\cdot\hat{\mathbb{N}}_j\\
\mathbb{Q}_n&=\boldsymbol{R}^{\mathrm{F}}\cdot\hat{\mathbb{Q}}_n\\
\mathbb{M}_j&=\boldsymbol{R}^{\mathrm{F}}\cdot\hat{\mathbb{M}}_j
\end{aligned}
\tag{9.73}
$$

将式(9.67)、式(9.70)和式(9.73)代入式(9.64)中，可以得到

$$
\begin{aligned}
&\hat{\mathbb{N}}_\tau{}^{\mathrm{T}}\cdot\underbrace{\left[\int_{S_{\mathrm{c}}}\left(\boldsymbol{R}^{\mathrm{F}}\right)^{\mathrm{T}}\cdot\boldsymbol{R}^{\mathrm{F}}\mathrm{d}S_{\mathrm{c}}\right]}_{\mathcal{M}_{\mathrm{F}}}\cdot\hat{\boldsymbol{U}}_\tau^{\mathrm{F}}=\hat{\mathbb{N}}_\tau{}^{\mathrm{T}}\cdot\underbrace{\left[\int_{\Gamma_{\mathrm{c}}}\frac{1}{h}\left(\boldsymbol{R}^{\mathrm{F}}\right)^{\mathrm{T}}\cdot\boldsymbol{R}^{\mathrm{B}}\mathrm{d}\Gamma_{\mathrm{c}}\right]}_{\mathcal{M}_{\mathrm{B}}^{\mathbb{N}}}\cdot\hat{\boldsymbol{u}}_\tau^{\mathrm{B}}\\
&\hat{\mathbb{N}}_{\tau s}{}^{\mathrm{T}}\cdot\underbrace{\left[\int_{S_{\mathrm{c}}}\left(\boldsymbol{R}^{\mathrm{F}}\right)^{\mathrm{T}}\cdot\boldsymbol{R}^{\mathrm{F}}\mathrm{d}S_{\mathrm{c}}\right]}_{\mathcal{M}_{\mathrm{F}}}\cdot\hat{\boldsymbol{U}}_s^{\mathrm{F}}=\hat{\mathbb{N}}_{\tau s}{}^{\mathrm{T}}\cdot\underbrace{\left[\int_{\Gamma_{\mathrm{c}}}\frac{1}{h}\left(\boldsymbol{R}^{\mathrm{F}}\right)^{\mathrm{T}}\cdot\boldsymbol{R}^{\mathrm{B}}\mathrm{d}\Gamma_{\mathrm{c}}\right]}_{\mathcal{M}_{\mathrm{B}}^{\mathbb{N}}}\cdot\hat{\boldsymbol{u}}_s^{\mathrm{B}}
\end{aligned}
$$

$$\hat{\mathcal{Q}}_n^{\mathrm{T}} \cdot \underbrace{\left[\int_{s_c}\left(\boldsymbol{R}^{\mathrm{F}}\right)^{\mathrm{T}} \cdot \boldsymbol{R}^{\mathrm{F}} \mathrm{d}S_{\mathrm{c}}\right]}_{\mathcal{M}_{\mathrm{F}}} \cdot \hat{\boldsymbol{U}}_n^{\mathrm{F}} = \hat{\mathcal{Q}}_n^{\mathrm{T}} \cdot \underbrace{\left[\int_{\Gamma_{\mathrm{c}}} \frac{3\left(1-\zeta^2\right)}{2h}\left(\boldsymbol{R}^{\mathrm{F}}\right)^{\mathrm{T}} \cdot \boldsymbol{R}^{\mathrm{B}} \mathrm{d}\Gamma_{\mathrm{c}}\right]}_{\mathcal{M}_{\mathrm{B}}^{\mathcal{Q}}} \cdot \hat{\boldsymbol{u}}_n^{\mathrm{B}} \tag{9.74}$$

$$\hat{\mathcal{M}}_{\tau s}^{\mathrm{T}} \cdot \underbrace{\left[\int_{s_c}\left(\boldsymbol{R}^{\mathrm{F}}\right)^{\mathrm{T}} \cdot \boldsymbol{R}^{\mathrm{F}} \mathrm{d}S_{\mathrm{c}}\right]}_{\mathcal{M}_{\mathrm{F}}} \cdot \hat{\boldsymbol{\theta}}_\tau^{\mathrm{F}} = -\hat{\mathcal{M}}_{\tau s}^{\mathrm{T}} \cdot \underbrace{\left[\int_{\Gamma_{\mathrm{c}}} \frac{6\left(\zeta-\zeta^0\right)}{h^2}\left(\boldsymbol{R}^{\mathrm{F}}\right)^{\mathrm{T}} \cdot \boldsymbol{R}^{\mathrm{B}} \mathrm{d}\Gamma_{\mathrm{c}}\right]}_{\mathcal{M}_{\mathrm{B}}^{\mathcal{M}}} \cdot \hat{\boldsymbol{u}}_s^{\mathrm{B}}$$

$$\hat{\mathcal{M}}_{s}^{\mathrm{T}} \cdot \underbrace{\left[\int_{s_c}\left(\boldsymbol{R}^{\mathrm{F}}\right)^{\mathrm{T}} \cdot \boldsymbol{R}^{\mathrm{F}} \mathrm{d}S_{\mathrm{c}}\right]}_{\mathcal{M}_{\mathrm{F}}} \cdot \hat{\boldsymbol{\theta}}_s^{\mathrm{F}} = \hat{\mathcal{M}}_{s}^{\mathrm{T}} \cdot \underbrace{\left[\int_{\Gamma_{\mathrm{c}}} \frac{6\left(\zeta-\zeta^0\right)}{h^2}\left(\boldsymbol{R}^{\mathrm{F}}\right)^{\mathrm{T}} \cdot \boldsymbol{R}^{\mathrm{B}} \mathrm{d}\Gamma_{\mathrm{c}}\right]}_{\mathcal{M}_{\mathrm{B}}^{\mathcal{M}}} \cdot \hat{\boldsymbol{u}}_\tau^{\mathrm{B}}$$

应注意的是，$\mathcal{M}_{\mathrm{F}}$ 始终是方阵且其逆矩阵一定存在，这是因为支持壳体耦合边的 NURBS 基函数之间是线性无关的。将式(9.74)中相应的应力合力变量删除后，在局部坐标系下，耦合界面上有限元和边界元子域自由度之间的关系可以表示为

$$\begin{Bmatrix} \hat{\boldsymbol{U}}_\tau^{\mathrm{F}} \\ \hat{\boldsymbol{U}}_s^{\mathrm{F}} \\ \hat{\boldsymbol{U}}_n^{\mathrm{F}} \\ \hat{\boldsymbol{\theta}}_\tau^{\mathrm{F}} \\ \hat{\boldsymbol{\theta}}_s^{\mathrm{F}} \end{Bmatrix} = \underbrace{\begin{bmatrix} \mathcal{M}_{\mathrm{F}}^{-1} \cdot \mathcal{M}_{\mathrm{B}}^{\mathcal{N}} & \mathbf{0} & \mathbf{0} \\ \mathbf{0} & \mathcal{M}_{\mathrm{F}}^{-1} \cdot \mathcal{M}_{\mathrm{B}}^{\mathcal{N}} & \mathbf{0} \\ \mathbf{0} & \mathbf{0} & \mathcal{M}_{\mathrm{F}}^{-1} \cdot \mathcal{M}_{\mathrm{B}}^{\mathcal{Q}} \\ \mathbf{0} & -\mathcal{M}_{\mathrm{F}}^{-1} \cdot \mathcal{M}_{\mathrm{B}}^{\mathcal{M}} & \mathbf{0} \\ \mathcal{M}_{\mathrm{F}}^{-1} \cdot \mathcal{M}_{\mathrm{B}}^{\mathcal{M}} & \mathbf{0} & \mathbf{0} \end{bmatrix}}_{\mathbf{Tr}_{\mathrm{fb}}^{1}} \begin{Bmatrix} \hat{\boldsymbol{u}}_\tau^{\mathrm{B}} \\ \hat{\boldsymbol{u}}_s^{\mathrm{B}} \\ \hat{\boldsymbol{u}}_n^{\mathrm{B}} \end{Bmatrix} = \mathbf{Tr}_{\mathrm{fb}}^{l} \begin{Bmatrix} \hat{\boldsymbol{u}}_\tau^{\mathrm{B}} \\ \hat{\boldsymbol{u}}_s^{\mathrm{B}} \\ \hat{\boldsymbol{u}}_n^{\mathrm{B}} \end{Bmatrix} \tag{9.75}$$

将式(9.75)中的耦合约束关系转换到整体坐标系下，并表示为

$$\mathbf{Tr}_{\mathrm{fb}}^{\mathrm{g}} = \mathcal{T}_s^{\mathrm{g}} \cdot \mathbf{Tr}_{\mathrm{fb}}^{1} \cdot \mathcal{T}_u^{\mathrm{g}} \tag{9.76}$$

其中，$\mathcal{T}_s^{\mathrm{g}}$ 和 $\mathcal{T}_u^{\mathrm{g}}$ 是根据支持有限元和边界元子域界面的控制点数量，对矩阵 $\mathcal{T}_s$ 和 $\mathcal{T}_u$ 进行扩展后的形式表示为

$$\mathcal{T}_s^{\mathrm{g}} = \begin{bmatrix} \mathcal{T}_s & & & \\ & \mathcal{T}_s & & \\ & & \ddots & \\ & & & \mathcal{T}_s \end{bmatrix}_{6n_{\mathrm{edge}} \times 5n_{\mathrm{edge}}}, \quad \mathcal{T}_u^{\mathrm{g}} = \begin{bmatrix} \mathcal{T}_u & & & \\ & \mathcal{T}_u & & \\ & & \ddots & \\ & & & \mathcal{T}_u \end{bmatrix}_{3n_{\mathrm{surf}} \times 3n_{\mathrm{surf}}} \tag{9.77}$$

其中，$\mathcal{T}_s$ 和 $\mathcal{T}_u$ 分别是由式(9.56)和式(9.59)定义的。

因此，整体坐标系下，在有限元和边界元子域的耦合界面上，与每个计算点

相关的控制点处自由度之间的关系为

$$\begin{aligned}&\left\{\bar{U}_x \quad \bar{U}_y \quad \bar{U}_z \quad \bar{\theta}_x \quad \bar{\theta}_y \quad \bar{\theta}_z\right\}^{\mathrm{T}}\\&=\mathbf{Tr}_{\mathrm{fb}}^{\mathrm{g}}\left\{\bar{\boldsymbol{u}}_x^{\mathrm{B}} \quad \bar{\boldsymbol{u}}_y^{\mathrm{B}} \quad \bar{\boldsymbol{u}}_z^{\mathrm{B}}\right\}^{\mathrm{T}}\end{aligned} \tag{9.78}$$

由于式(9.74)中积分矩阵 $\mathcal{M}_{\mathrm{B}}^{N}$、$\mathcal{M}_{\mathrm{B}}^{Q}$ 和 $\mathcal{M}_{\mathrm{B}}^{M}$ 的求解均涉及与壳体参数相关的形函数 $\boldsymbol{R}^{\mathrm{F}}$ 的计算，需要将边界元子域耦合面上参数坐标为 $\left(\xi^{\mathrm{B}},\eta^{\mathrm{B}}\right)$ 的点投影到有限元子域的耦合边上，并获取对应壳体上的参数坐标 $\left(\xi^{*},\eta^{\mathrm{F}}\right)$。实施过程为：首先计算参数坐标 $\left(\xi^{\mathrm{B}},\eta^{\mathrm{B}}\right)$ 在边界元子域耦合面上的物理坐标；然后由最近点算法计算其在壳体中性面上投影点的物理坐标；最后利用等几何分析中点的反求算法确定参数坐标 $\left(\xi^{*},\eta^{\mathrm{F}}\right)$。

循环耦合界面上的所有单元和每个单元中的积分点后，根据单元连接矩阵组装形成的最终全局耦合约束矩阵为

$$\mathbf{Tr}_{\mathrm{fb}}=\sum_{\text{单元}}\sum_{\text{积分点}}\mathbf{Tr}_{\mathrm{fb}}^{\mathrm{g}} \tag{9.79}$$

9.3.4 耦合系统的控制方程

在获得界面位移耦合条件后，将有限元部分的自由度划分为与耦合界面相关的自由度和其余自由度两个部分，则壳体的控制方程式(9.25)可以扩展为

$$\begin{bmatrix}\boldsymbol{K}_{\mathrm{fem}}^{\mathrm{FF}} & \boldsymbol{K}_{\mathrm{fem}}^{\mathrm{FI}}\\ \boldsymbol{K}_{\mathrm{fem}}^{\mathrm{IF}} & \boldsymbol{K}_{\mathrm{fem}}^{\mathrm{II}}\end{bmatrix}\begin{Bmatrix}\boldsymbol{u}_{\mathrm{fem}}^{\mathrm{F}}\\ \boldsymbol{u}_{\mathrm{fem}}^{\mathrm{I}}\end{Bmatrix}=\begin{Bmatrix}\boldsymbol{f}_{\mathrm{fem}}^{\mathrm{F}}\\ \boldsymbol{f}_{\mathrm{fem}}^{\mathrm{I}}\end{Bmatrix}+\begin{Bmatrix}\boldsymbol{0}\\ \boldsymbol{f}_{\mathrm{c}}^{\mathrm{F}}\end{Bmatrix} \tag{9.80}$$

边界元子域耦合界面上的未知等效控制力矢量可以用未知界面上面力表示为

$$\boldsymbol{f}_{\mathrm{c}}^{\mathrm{B}}=\mathbb{L}_{\mathrm{c}}^{\mathrm{B}}\boldsymbol{t}_{\mathrm{c}} \tag{9.81}$$

其中，$\mathbb{L}_{\mathrm{c}}^{\mathrm{B}}$ 为转换矩阵，其表达式为

$$\mathbb{L}_{\mathrm{c}}^{\mathrm{B}}=\int_{\Gamma_{\mathrm{c}}}\left(\boldsymbol{R}_{\mathrm{c}}^{\mathrm{B}}\right)^{\mathrm{T}}\boldsymbol{R}_{\mathrm{c}}^{\mathrm{B}}\mathrm{d}\Gamma_{\mathrm{c}} \tag{9.82}$$

其中，$\boldsymbol{R}_{\mathrm{c}}^{\mathrm{B}}$ 是关联的 NURBS 形函数矩阵，表示为

$$\boldsymbol{R}_{\mathrm{c}}^{\mathrm{B}}=\begin{bmatrix}{}^{1}R_{\mathrm{c}}^{\mathrm{B}} & 0 & 0 & \cdots & {}^{n_{\mathrm{b}}}R_{\mathrm{c}}^{\mathrm{B}} & 0 & 0\\ 0 & {}^{1}R_{\mathrm{c}}^{\mathrm{B}} & 0 & \cdots & 0 & {}^{n_{\mathrm{b}}}R_{\mathrm{c}}^{\mathrm{B}} & 0\\ 0 & 0 & {}^{1}R_{\mathrm{c}}^{\mathrm{B}} & \cdots & 0 & 0 & {}^{n_{\mathrm{b}}}R_{\mathrm{c}}^{\mathrm{B}}\end{bmatrix} \tag{9.83}$$

其中，n_{b} 是支持边界元子域耦合界面 $\varGamma_{\mathrm{c}}$ 上的控制点总数。

将式(9.34)和式(9.81)代入式(9.40)可得更新后的刚度矩阵形式的边界元子域方程为

$$\begin{bmatrix} \bar{\boldsymbol{K}}_{\mathrm{c,c}} & \boldsymbol{K}_{\mathrm{c,b}k} \\ \boldsymbol{K}_{\mathrm{b}k,\mathrm{c}} & \boldsymbol{K}_{\mathrm{b}k,\mathrm{b}k} \end{bmatrix} \begin{Bmatrix} \boldsymbol{u}_{\mathrm{c}} \\ \boldsymbol{u}_{\mathrm{b}k} \end{Bmatrix} = \begin{Bmatrix} \bar{\boldsymbol{f}}_{\mathrm{c}} \\ \boldsymbol{f}_{\mathrm{b}k} \end{Bmatrix} + \begin{Bmatrix} \boldsymbol{f}_{\mathrm{c}}^{\mathrm{B}} \\ \boldsymbol{0} \end{Bmatrix} \tag{9.84}$$

其中，

$$\begin{cases} \bar{\boldsymbol{K}}_{\mathrm{c,c}} = \boldsymbol{K}_{\mathrm{c,c}} + \mathbb{L}_{\mathrm{c}}^{\mathrm{B}}\boldsymbol{\mathcal{M}} \\ \bar{\boldsymbol{f}}_{\mathrm{c}} = \boldsymbol{f}_{\mathrm{c}} + \mathbb{L}_{\mathrm{c}}^{\mathrm{B}}\boldsymbol{\mathcal{N}} \end{cases} \tag{9.85}$$

结合式(9.80)和式(9.84)，并考虑式(9.52)中的耦合条件 $\boldsymbol{u}_{\mathrm{c}} = \mathbf{Tr}_{\mathrm{bf}}\boldsymbol{u}_{\mathrm{fem}}^{\mathrm{I}}$ 时，最终简化的耦合系统方程为

$$\begin{bmatrix} \boldsymbol{K}_{\mathrm{fem}}^{\mathrm{FF}} & \boldsymbol{K}_{\mathrm{fem}}^{\mathrm{FI}} & \boldsymbol{0} \\ \boldsymbol{K}_{\mathrm{fem}}^{\mathrm{IF}} & \boldsymbol{K}_{\mathrm{fem}}^{\mathrm{II}} + \left(\mathbf{Tr}_{\mathrm{bf}}\right)^{\mathrm{T}} \bar{\boldsymbol{K}}_{\mathrm{c,c}}\mathbf{Tr}_{\mathrm{bf}} & \left(\mathbf{Tr}_{\mathrm{bf}}\right)^{\mathrm{T}} \boldsymbol{K}_{\mathrm{c,b}k} \\ \boldsymbol{0} & \boldsymbol{K}_{\mathrm{b}k,\mathrm{c}}\mathbf{Tr}_{\mathrm{bf}} & \boldsymbol{K}_{\mathrm{b}k,\mathrm{b}k} \end{bmatrix} \begin{Bmatrix} \boldsymbol{u}_{\mathrm{fem}}^{\mathrm{F}} \\ \boldsymbol{u}_{\mathrm{fem}}^{\mathrm{I}} \\ \boldsymbol{u}_{\mathrm{b}k} \end{Bmatrix} = \begin{Bmatrix} \boldsymbol{f}_{\mathrm{fem}}^{\mathrm{F}} \\ \boldsymbol{f}_{\mathrm{fem}}^{\mathrm{I}} + \left(\mathbf{Tr}_{\mathrm{bf}}\right)^{\mathrm{T}} \bar{\boldsymbol{f}}_{\mathrm{c}} \\ \boldsymbol{f}_{\mathrm{b}k} \end{Bmatrix} \tag{9.86}$$

另一种情况下，考虑式(9.79)中的耦合条件 $\boldsymbol{u}_{\mathrm{fem}}^{\mathrm{I}} = \mathbf{Tr}_{\mathrm{fb}}\boldsymbol{u}_{\mathrm{c}}$ 时，最终简化的耦合系统方程为

$$\begin{bmatrix} \boldsymbol{K}_{\mathrm{fem}}^{\mathrm{FF}} & \boldsymbol{K}_{\mathrm{fem}}^{\mathrm{FI}}\mathbf{Tr}_{\mathrm{fb}} & \boldsymbol{0} \\ \left(\mathbf{Tr}_{\mathrm{fb}}\right)^{\mathrm{T}} \boldsymbol{K}_{\mathrm{fem}}^{\mathrm{IF}} & \left(\mathbf{Tr}_{\mathrm{fb}}\right)^{\mathrm{T}} \boldsymbol{K}_{\mathrm{fem}}^{\mathrm{II}}\mathbf{Tr}_{\mathrm{fb}} + \bar{\boldsymbol{K}}_{\mathrm{c,c}} & \boldsymbol{K}_{\mathrm{c,b}k} \\ \boldsymbol{0} & \boldsymbol{K}_{\mathrm{b}k,\mathrm{c}} & \boldsymbol{K}_{\mathrm{b}k,\mathrm{b}k} \end{bmatrix} \begin{Bmatrix} \boldsymbol{u}_{\mathrm{fem}}^{\mathrm{F}} \\ \boldsymbol{u}_{\mathrm{c}} \\ \boldsymbol{u}_{\mathrm{b}k} \end{Bmatrix} = \begin{Bmatrix} \boldsymbol{f}_{\mathrm{fem}}^{\mathrm{F}} \\ \left(\mathbf{Tr}_{\mathrm{fb}}\right)^{\mathrm{T}} \boldsymbol{f}_{\mathrm{fem}}^{\mathrm{I}} + \bar{\boldsymbol{f}}_{\mathrm{c}} \\ \boldsymbol{f}_{\mathrm{b}k} \end{Bmatrix} \tag{9.87}$$

压缩自由度 $\boldsymbol{u}_{\mathrm{c}}$ 或 $\boldsymbol{u}_{\mathrm{fem}}^{\mathrm{I}}$ 的值可在求解耦合系统方程式(9.86)和式(9.87)后，通过在相应的界面耦合条件下，替换相应的非独立自由度而求出。另外，式(9.86)和式(9.87)成立的隐含的耦合界面上的平衡条件表示为

$$\begin{gathered} \boldsymbol{f}_{\mathrm{c}}^{\mathrm{F}} + \left(\mathbf{Tr}_{\mathrm{bf}}\right)^{\mathrm{T}} \boldsymbol{f}_{\mathrm{c}}^{\mathrm{B}} = \boldsymbol{0} \\ \left(\mathbf{Tr}_{\mathrm{fb}}\right)^{\mathrm{T}} \boldsymbol{f}_{\mathrm{c}}^{\mathrm{F}} + \boldsymbol{f}_{\mathrm{c}}^{\mathrm{B}} = \boldsymbol{0} \end{gathered} \tag{9.88}$$

9.4 数 值 算 例

本节将通过相关的四个数值算例，说明所提出方法的收敛性、准确性和可靠性。除非另有说明，有限元壳单元中的数值积分使用 $(p+1)\times(q+1)$ 个高斯点，其中 p 和 q 分别是参数 ξ 和 η 方向上 NURBS 基函数的阶次。边界元部分中非奇异

积分的计算使用 20×20 个高斯点。采用第 8 章提出的改进幂级数展开法(M-PSEM)计算 NURBS 基函数下的各类奇异积分。

9.4.1 三维圆环体耦合模型

首先验证所提出的刚度矩阵形式边界元子域方程的准确性和可靠性。考虑如图 9.5 所示的三维圆环体耦合模型内径 a=4m、外径 b=5m 和厚度 h=1m，左端完全固定，在其右端施加沿 x 轴负方向的表面剪力作用。相关的材料参数设置为：弹性模量 $E=10^5\,\text{Pa}$，泊松比 $\nu=0$，其用以阻止横向压缩效应。

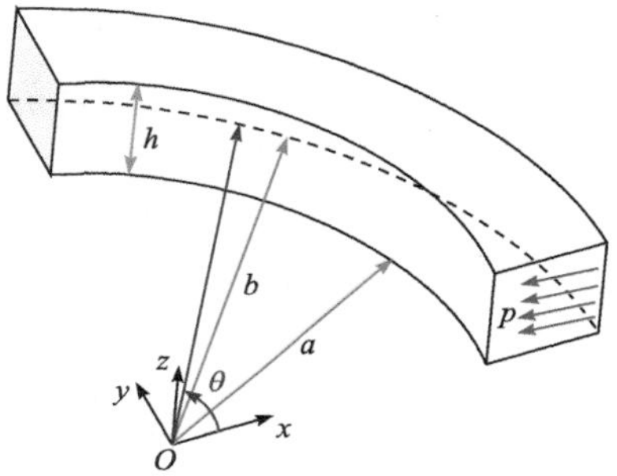

图 9.5 左端完全固定和在其右端施加表面剪力作用的三维圆环体耦合模型

在极坐标形式下，应力分量的解析解为[10]

$$
\begin{aligned}
\sigma_{rr} &= \frac{p}{N_0}\sin\theta\left(r+\frac{a^2b^2}{r^3}-\frac{a^2+b^2}{r}\right)\\
\sigma_{\theta\theta} &= \frac{p}{N_0}\sin\theta\left(3r-\frac{a^2b^2}{r^3}-\frac{a^2+b^2}{r}\right)\\
\sigma_{r\theta} &= -\frac{p}{N_0}\cos\theta\left(r+\frac{a^2b^2}{r^3}-\frac{a^2+b^2}{r}\right)
\end{aligned}
\tag{9.89}
$$

其中，$\theta\in\left[0,\dfrac{\pi}{2}\right]$；几何常数 N_0 定义为

$$
N_0=a^2-b^2+\left(a^2+b^2\right)\ln\frac{b}{a} \tag{9.90}
$$

整体坐标系中的应力分量可通过以下坐标变换规则获得：

$$
\begin{aligned}
\sigma_{xx} &= \sigma_{rr}\cos^2\theta+\sigma_{\theta\theta}\sin^2\theta-2\sigma_{r\theta}\sin\theta\cos\theta\\
\sigma_{yy} &= \sigma_{rr}\sin^2\theta+\sigma_{\theta\theta}\cos^2\theta+2\sigma_{r\theta}\sin\theta\cos\theta\\
\sigma_{xy} &= \left(\sigma_{rr}-\sigma_{\theta\theta}\right)\sin\theta\cos\theta+\sigma_{r\theta}\cos 2\theta
\end{aligned}
\tag{9.91}
$$

在极坐标形式下，位移分量的解析解为[10]

$$u_r = -\frac{2D}{E}\theta\cos\theta + \frac{\sin\theta}{E}\left[D(1-\nu)\ln r + A(1-3\nu)r^2 + \frac{B(1+\nu)}{r^2}\right] + \frac{D\pi}{E}\cos\theta$$

$$u_\theta = \frac{2D}{E}\theta\sin\theta - \frac{\cos\theta}{E}\left[A(5+\nu)r^2 + \frac{B(1+\nu)}{r^2} - D(1-\nu)\ln r\right] + \frac{D(1+\nu)}{E}\cos\theta - \frac{D\pi}{E}\sin\theta \tag{9.92}$$

其中相关的参数定义为

$$A = \frac{p}{2N_0},\quad B = -\frac{pa^2b^2}{2N_0},\quad D = -\frac{p}{N_0}\left(a^2+b^2\right) \tag{9.93}$$

固定端处的本质边界条件按照式(9.92)给出的位移解施加，并根据式(9.89)中的应力解施加自由端的剪力载荷，其值为

$$p(r) = \frac{p}{N_0}\left(r + \frac{a^2b^2}{r^3} - \frac{a^2+b^2}{r}\right) \tag{9.94}$$

由于 NURBS 基函数的非插值特性，非均匀边界条件可通过最小二乘技术[9]施加。通过第 8 章中提出的虚拟节点插入技术，建立非相适应界面上的位移耦合约束条件。如图 9.6 所示，圆环体被划分为两个子区域，并包含一个非相适应的界面。深灰和浅灰区域分别代表耦合模型的有限元和边界元部分。有限元子域由 $9\times7\times7$ 个双三次 NURBS 体网格离散，而边界元子域由 8×5 或 5×5 个双三次 NURBS 曲面网格离散。

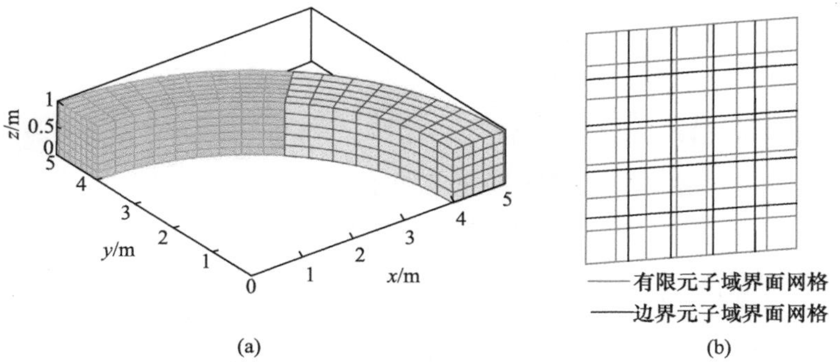

图 9.6　三维圆环体耦合模型
(a) 网格；(b) 非相适应界面

沿指定路径($0\leqslant\theta\leqslant\frac{\pi}{2}, r=\frac{a+b}{2}$)的位移和应力解与解析解的比较绘制在图 9.7 中。此外，沿 x 方向上的位移分布和径向应力与解析解的比较分别如图 9.8 和图 9.9 所示。可从图中观察到数值解与解析解之间良好的一致性，且在耦合界面

上有限元和边界元子域获得的位移和应力值也非常一致。

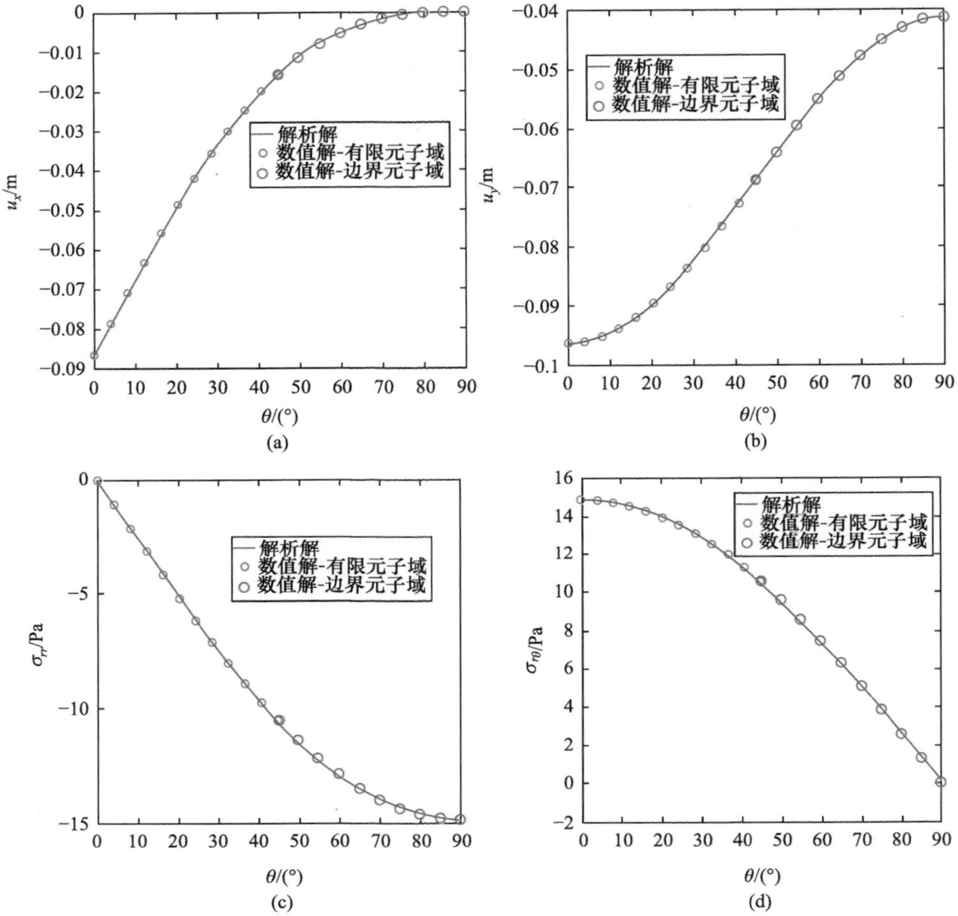

图 9.7 沿指定路径位移和应力数值解以及精确解

(a) u_x；(b) u_y；(c) σ_{rr}；(d) $\sigma_{r\theta}$

最终系统耦合矩阵的结构如图 9.10 所示。图中右下部分是 IGABEM 生成的稠密的系数矩阵，而左上部分是由 IGAFEM 形成的对称刚度矩阵乘以相关变换矩阵而得，此处压缩了与界面相关的有限元子域中的位移自由度。

界面上位移误差的绝对值定义为

$$\text{err} = \left| u_{\text{h}}^{\text{B}}(\boldsymbol{x}) - u_{\text{h}}^{\text{F}}(\boldsymbol{x}) \right| \tag{9.95}$$

其中，$u_{\text{h}}^{\text{F}}(\boldsymbol{x})$ 和 $u_{\text{h}}^{\text{B}}(\boldsymbol{x})$ 分别是有限元和边界元子域求出的界面位移，其误差分布云图如图 9.11 所示。可以观察到，误差水平 err 在10^{-9}数量级，表明了所提出方

法的准确性。

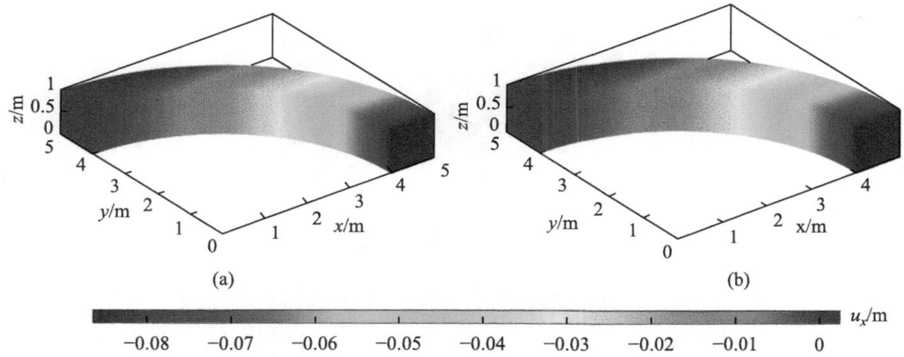

图 9.8　x 方向上的位移分布

(a) 数值解；(b) 解析解

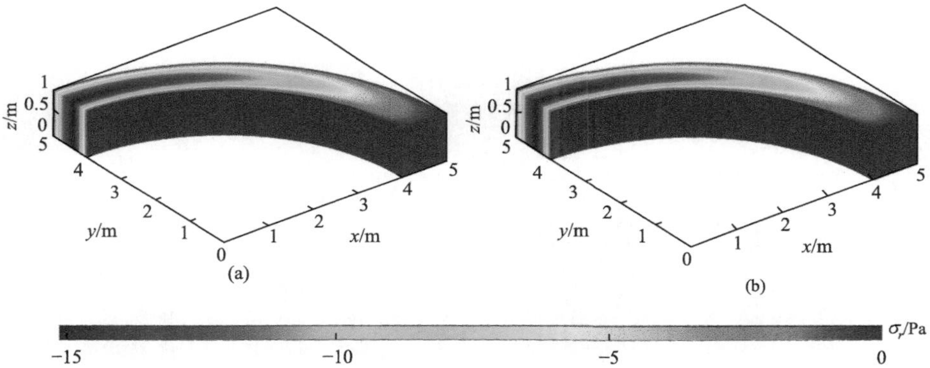

图 9.9　径向应力分布

(a) 数值解；(b) 解析解

随后，研究了非相适应离散耦合模型的收敛性。位移相对误差的 L_2 范数定义为

$$e_{L_2}=\frac{\left\|\boldsymbol{u}_{\text{ext}}(\boldsymbol{x})-\boldsymbol{u}_{\text{h}}(\boldsymbol{x})\right\|}{\left\|\boldsymbol{u}_{\text{ext}}(\boldsymbol{x})\right\|}=\frac{\left\{\int_{\Omega}\left[\boldsymbol{u}_{\text{ext}}(\boldsymbol{x})-\boldsymbol{u}_{\text{h}}(\boldsymbol{x})\right]^2\mathrm{d}\Omega\right\}^{\frac{1}{2}}}{\left\{\int_{\Omega}\left[\boldsymbol{u}_{\text{ext}}(\boldsymbol{x})\right]^2\mathrm{d}\Omega\right\}^{\frac{1}{2}}} \tag{9.96}$$

其中，Ω 是整个耦合模型的表面；$\boldsymbol{u}_{\text{ext}}(\boldsymbol{x})$ 和 $\boldsymbol{u}_{\text{h}}(\boldsymbol{x})$ 分别为位移矢量的精确解和数值解。

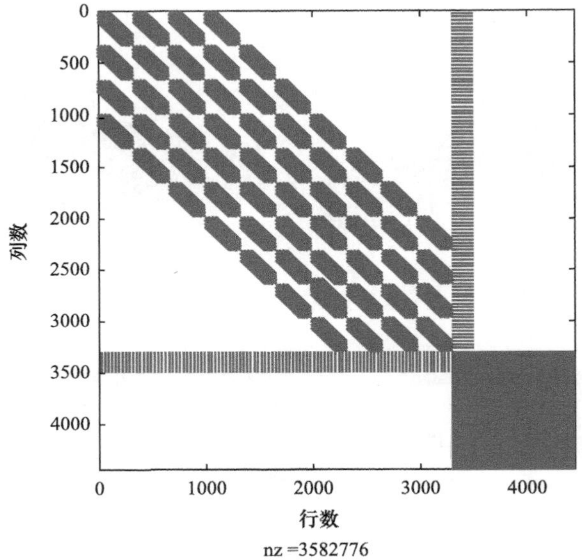

图 9.10 最终系统耦合矩阵的结构

nz 是矩阵中非零元素的个数

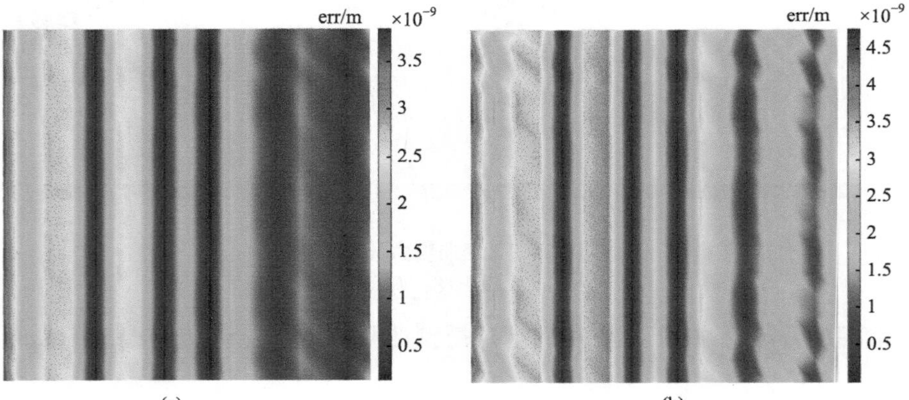

图 9.11 界面上位移误差绝对值的分布

(a) u_x 误差的分布；(b) u_y 误差的分布

初始最粗糙网格如图 9.12(a)所示。有限元和边界元子域的单元分别使用 $(j+3)\times(j+2)\times(j+2)$ 和 $(j+1)\times j\times j$ 的细分规则进行连续细分，其中 $j=1,2,3,\cdots,10$。$j=3$ 和 $j=5$ 时的网格分别如图 9.12(b)和(c)所示。我们假设有限元和边界元子域中所有参数方向上都使用相同阶次的 NURBS 基函数。图 9.13 给出了不同网格和 NURBS 阶次下的位移相对误差 L_2 范数的收敛结果。不同 NURBS 阶次下的结果都表现出了良好的收敛性，且随着 NURBS 基函数阶次的增加，收敛速度会提高。

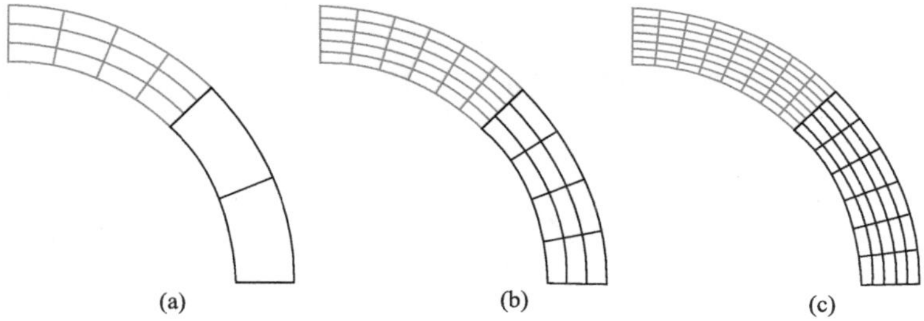

图 9.12　收敛分析的 h 细化

(a) 最粗糙网格，有限元，4×3×3 NURBS 体单元，边界元，2×1×1 NURBS 曲面单元；(b) 有限元，6×5×5 NURBS 体单元，边界元，4×3×3 NURBS 曲面单元；(c) 有限元，8×7×7 NURBS 体单元，边界元，6×5×5 NURBS 曲面单元

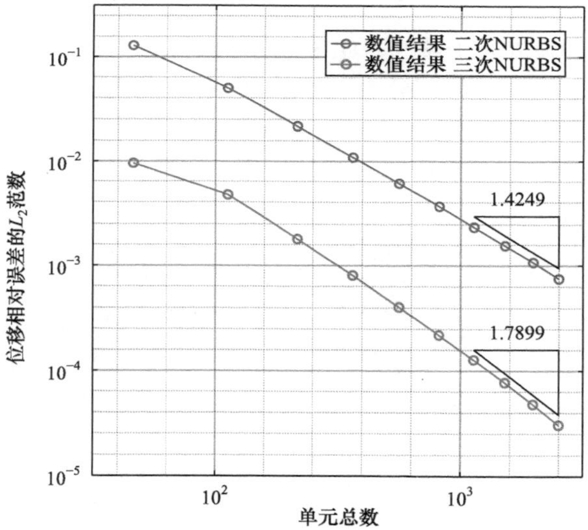

图 9.13　二次与三次 NURBS 基函数下耦合模型的收敛曲线

9.4.2　悬臂实体-平壳耦合模型

在后续数值算例中，考虑实体和壳体的混合维度耦合问题。实体部分采用 IGABEM 建模，只需对物体的表面进行离散，避免了实体内部体网格的划分。壳体采用等几何 Reissner-Mindlin 壳单元建模和分析。考虑如图 9.14 所示悬臂实体-平壳耦合模型，其右端受均匀剪力 $t=-10\text{N}/\text{m}$ 作用，且模型被均匀分为两部分。模型的相关几何参数为：长度 L=320m、宽度 b=80m 和厚度 h=20m。该模型的材料参数为：弹性模量 E=1000Pa、泊松比 $\nu=0.3$。根据悬臂梁理论，挠度 w 表示为

$$w=\frac{tb}{6EI}(x-L)^3-\frac{tbL^2}{2EI}x+\frac{tbL^3}{6EI} \tag{9.97}$$

式中，$I=\dfrac{bh^3}{12}$ 为惯性矩。需要注意的是，在悬臂梁理论求解中并没有考虑泊松比的影响。

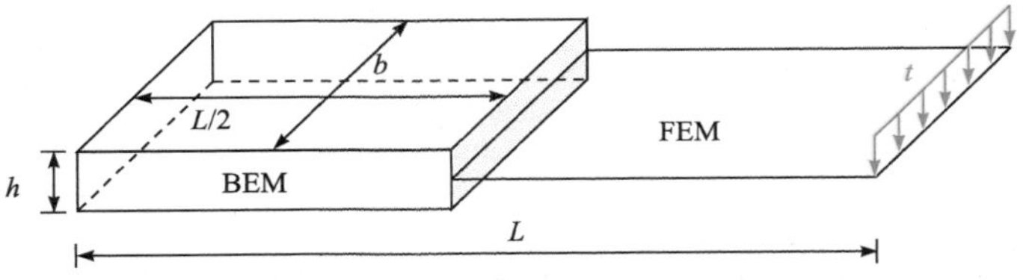

图 9.14　右端受均匀剪力作用的悬臂实体-平壳耦合模型

模型有限元和边界元子域的 NURBS 离散网格如图 9.15 所示。有限元部分由 32 个壳单元组成，而边界元部分由 136 个 NURBS 曲面单元构成。该操作的优点是避免了在使用三维 IGAFEM 分析时需要生成内部体网格的复杂过程，尤其是当模型的几何由多个 NURBS 片组成时。

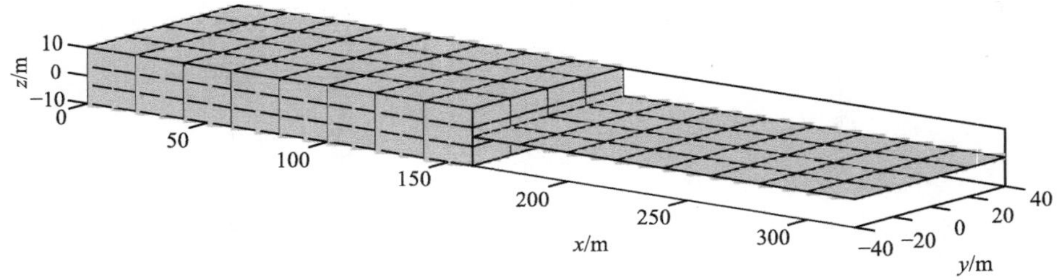

图 9.15　悬臂实体-平壳耦合模型的有限元和边界元子域 NURBS 离散网格

选择在五阶 NURBS 基函数下，由 48×32 个等几何 Reissner-Mindlin 壳单元构成整体模型的计算结果作为参考解，用以数值比较。研究并分析了 9.3.2 节和 9.3.3 节中提出的两种耦合方法：第一种是基于壳体位移假设和基于配点的直接运动耦合方法；第二种是基于壳体和三维实体在界面上所做功相等的弱耦合方法。二次和三次 NURBS 基函数下边界元子域耦合界面上配点的分布，以及其在壳体耦合边上投影点的分布如图 9.16 所示。图 9.17 显示了不同 NURBS 阶次和不同耦合方法下沿路径($0\leqslant x\leqslant320, y=0, z=0$)上模型的挠度曲线，以及利用梁理论计算的参考解。

数值结果表明：①在界面两侧的相同位置处，基于配点的直接耦合方法得到的位移值完全相等，而弱耦合方法得到的结果具有一定的误差；②与参考解相比，弱耦合方法得到的结果具有较高的精度，而基于配点的直接耦合方法存在较大的

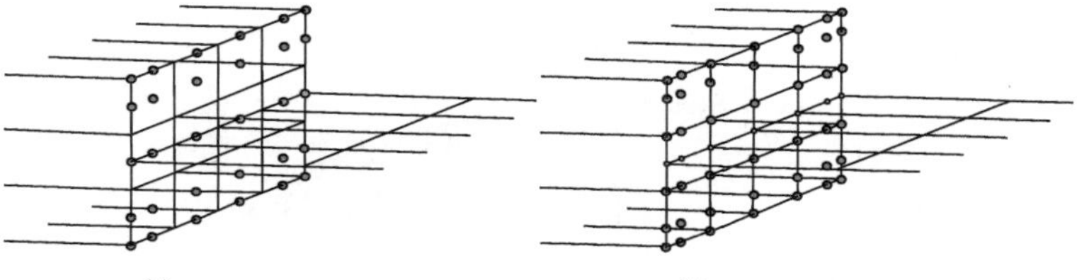

图 9.16 不同阶次 NURBS 基函数下边界元子域耦合界面上配点的分布，以及其在壳体耦合边上投影点的分布

(a) 二次 NURBS；(b) 三次 NURBS

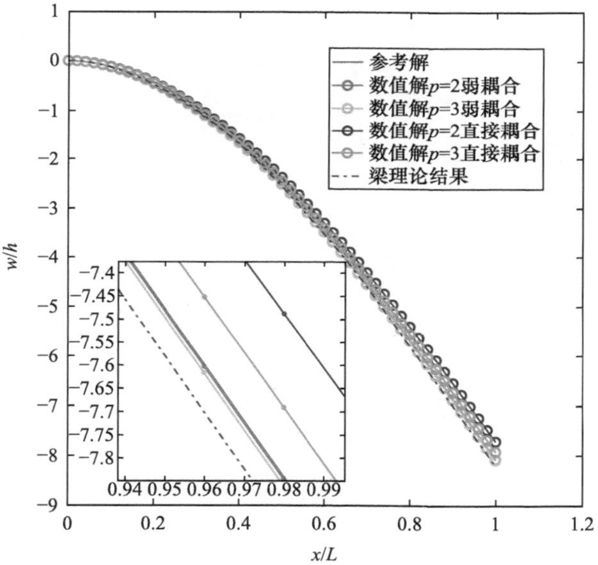

图 9.17 不同 NURBS 阶次和不同耦合方法下沿路径($0\leqslant x\leqslant 320, y=0, z=0$)上的挠度曲线

偏差；③在其他条件相同的情况下，增加 NURBS 基函数的阶次可以提高数值结果的精度。当使用二次 NURBS 基函数和弱耦合方法时，路径上模型的挠度和 von Mises 应力分布以及与参考解的比较如图 9.18 所示。可以清楚地看到，数值结果与参考解吻合得很好。

最后，考虑边界元子域耦合界面厚度方向上的单元数量对耦合精度的影响。不同边界元子域耦合界面厚度方向上的单元分布如图 9.19 所示。不同耦合方法、不同 NURBS 基函数阶次和不同厚度方向上单元数量的数值结果如图 9.20 所示。可以得出以下结论：① 对于基于配点的直接耦合方法，增加单元数量会使数值结果逐渐偏离参考解，可能的原因是，随着单元数量的增加，不断增加的配点过度加强了界面上的耦合约束；②对于弱耦合方法，增加单元数可以提高数值结果的精度，使其逐渐收敛到参考解。

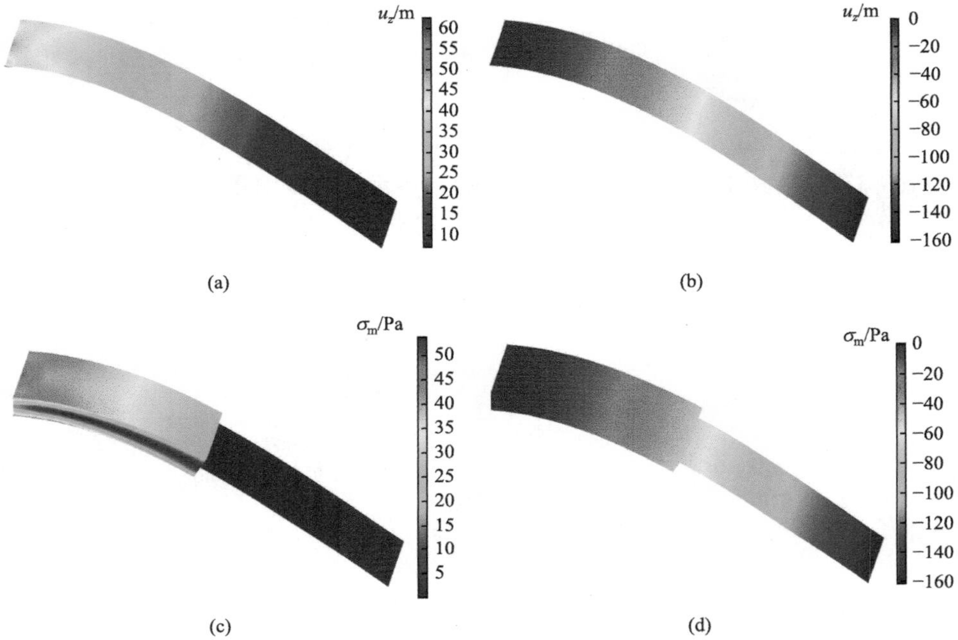

图 9.18　挠度和 von Mises 应力分布与参考解的比较

(a) 挠度的参考解；(b) 挠度的数值结果；(c) von Mises 应力的参考解；(d) von Mises 应力的数值结果

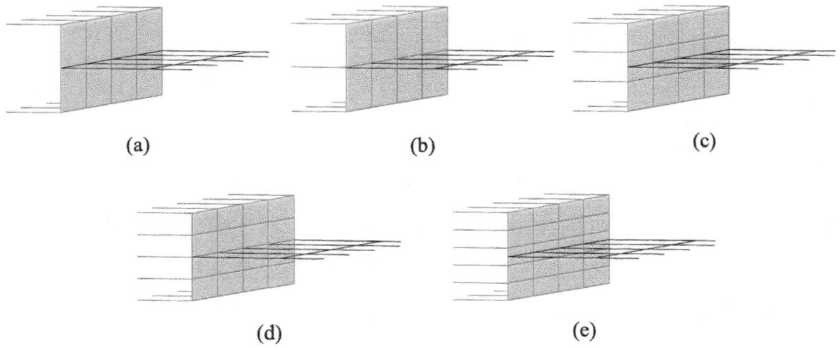

图 9.19　不同边界元子域耦合界面厚度方向上的单元分布

(a) 一个单元；(b) 两个单元；(c) 三单元；(d) 四个单元；(e) 五个单元

综上所述，基于功相等的弱耦合方法，在精度和准确性上都优于基于配点的直接耦合方法，其具有更好的鲁棒性。因此，在接下来的算例中，我们使用弱耦合方法来建立界面上的位移连续条件。

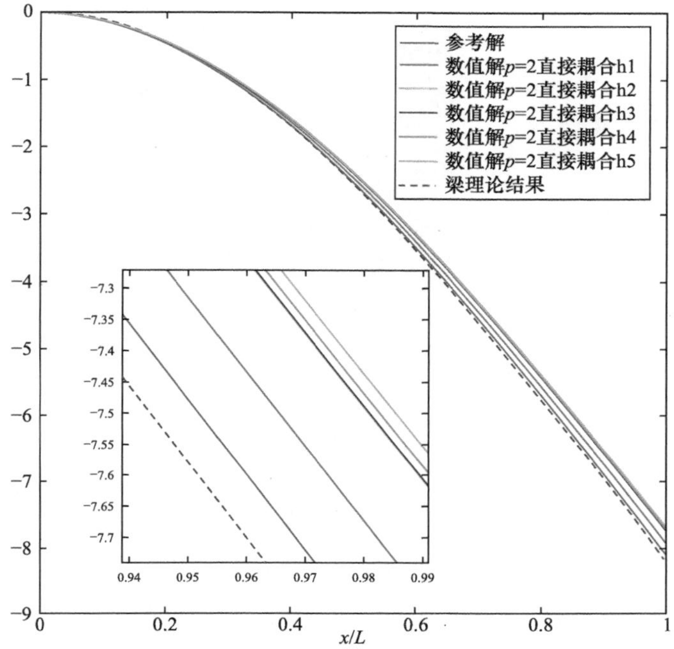
参考解
数值解p=2直接耦合h1
数值解p=2直接耦合h2
数值解p=2直接耦合h3
数值解p=2直接耦合h4
数值解p=2直接耦合h5
梁理论结果
w/h
x/L
(a)

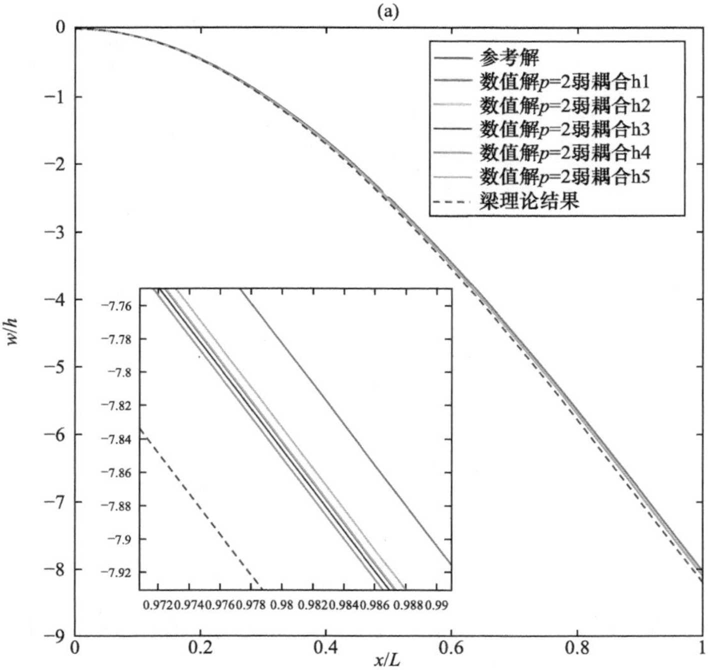
参考解
数值解p=2弱耦合h1
数值解p=2弱耦合h2
数值解p=2弱耦合h3
数值解p=2弱耦合h4
数值解p=2弱耦合h5
梁理论结果
w/h
x/L
(b)

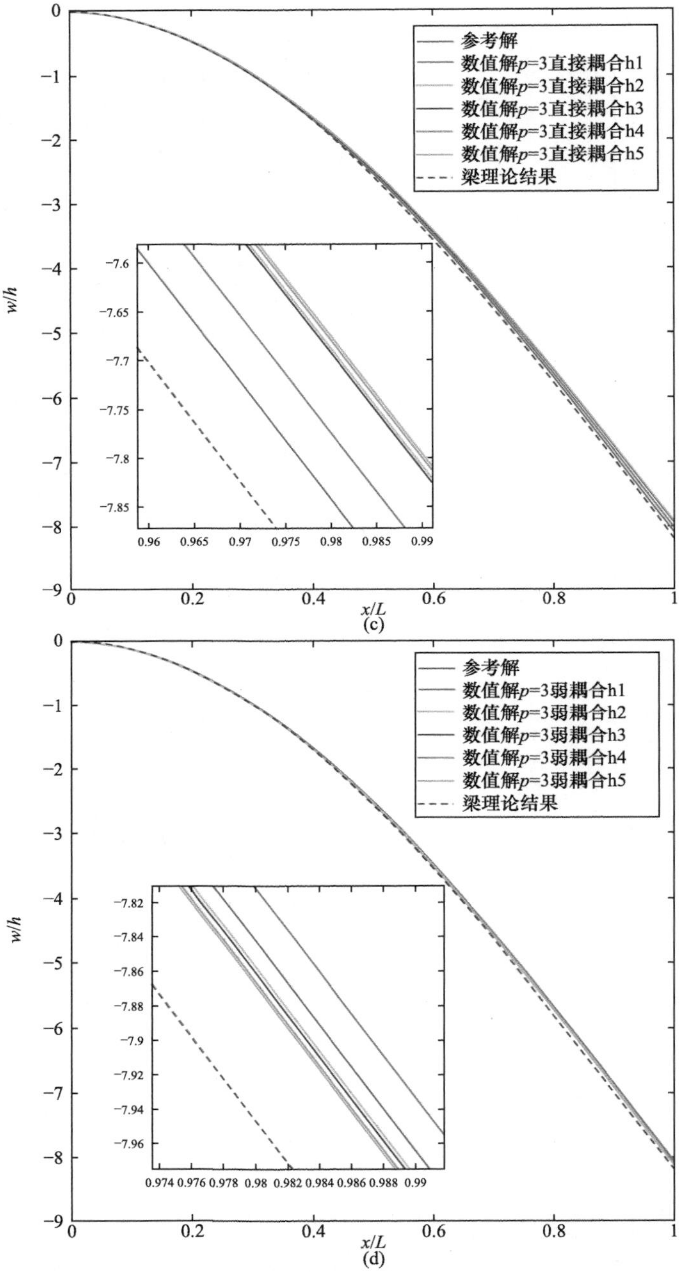

图 9.20 在不同耦合方法、不同 NURBS 基函数阶次和不同厚度方向上不同单元数的情况下，沿路径($0 \leqslant x \leqslant 320, y = 0, z = 0$)上点的挠度曲线

(a) 二次 NURBS 和直接耦合方法；(b) 二次 NURBS 和弱耦合方法；(c) 三次 NURBS 和直接耦合方法；(d) 三次 NURBS 和弱耦合方法

9.4.3　3/4 圆柱实体-曲壳耦合模型

本算例研究了式(9.61)中偏移量对耦合精度的影响。考虑如图 9.21 所示的左端固支右端受到均匀拉力作用的等几何空心圆柱耦合模型。均匀拉力的幅值为 t=1N/m。材料性能由弹性模量 E=1000Pa 和泊松比 $\nu = 0.3$ 定义。模型的几何参数包括：内径 $R_{\min} = 4\text{m}$ 、外径 $R_{\max} = 5\text{m}$ 和长度 $L = 30\text{m}$ 。

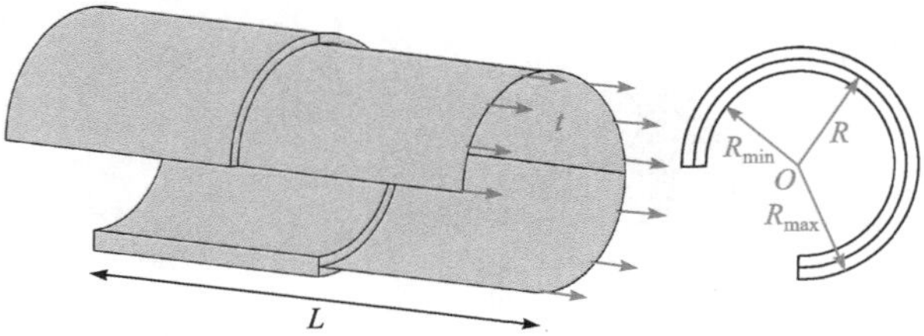

图 9.21　等几何空心圆柱耦合模型

该模型平均分为两部分：左侧部分由 IGABEM 计算，而右侧部分由等几何 Reissner-Mindlin 壳单元模拟。左侧边界元子域包含 312 个二次 NURBS 曲面单元和 1392 个位移自由度，而右侧有限元子域包含 96 个二次 NURBS 壳单元和 960 个广义自由度。耦合模型的 NURBS 网格和边界元子域的配点分布如图 9.22 所示。

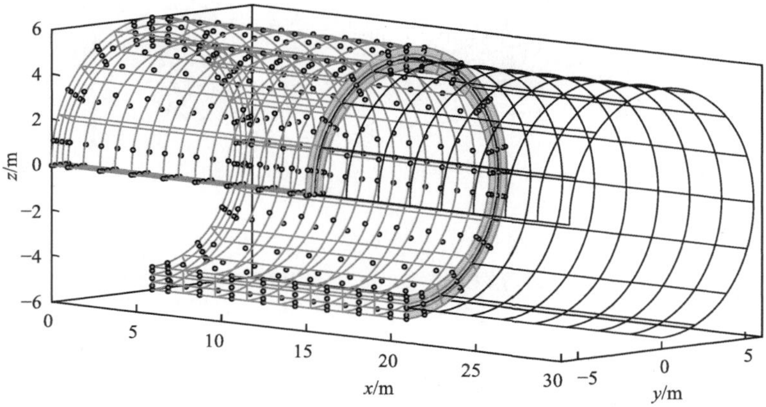

图 9.22　耦合模型的 NURBS 网格和边界元子域的配点分布

图 9.23 和图 9.24 分别显示了考虑和不考虑偏移量影响下的位移分量 u_x 和应力分量 σ_{xx} 的分布云图。如果不考虑偏心率的影响，很容易地观察到错误的位移和应力。此外，偏移量对应力产生的影响大于对位移产生的影响，甚至在应力的计算中导致超过 50%的误差。根据式(9.61)中的偏移量的计算表达式，可知偏心率的影响主要取决于壳体曲率半径与其厚度的比值。当不考虑偏移的影响时，比值越大，误差越大(本算例中比率为 0.0303)。值得注意的是，在 9.4.2 节的数值算例中，

平壳的曲率半径为无限大，故其偏移量为零。因此，在曲面几何的耦合分析中，偏移量的影响不容忽视。

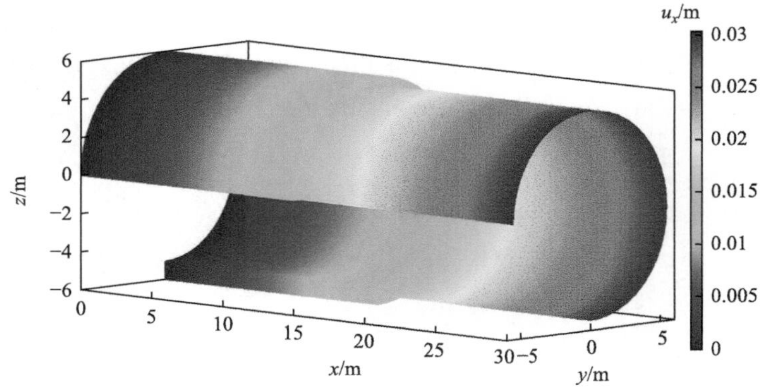

(a)

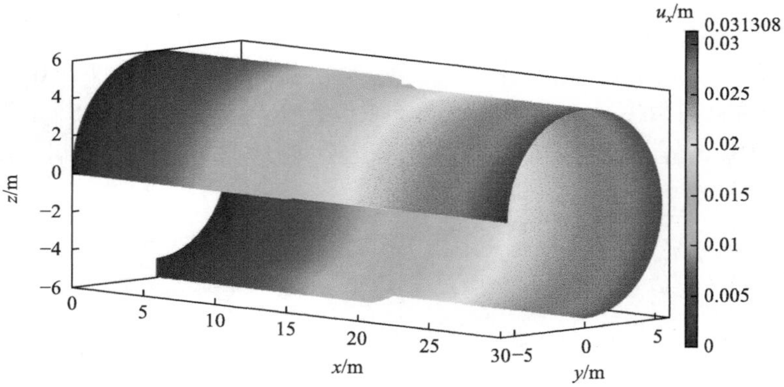

(b)

图 9.23 位移分量 u_x 的云图

(a) 考虑偏移量的影响；(b) 不考虑偏移量的影响

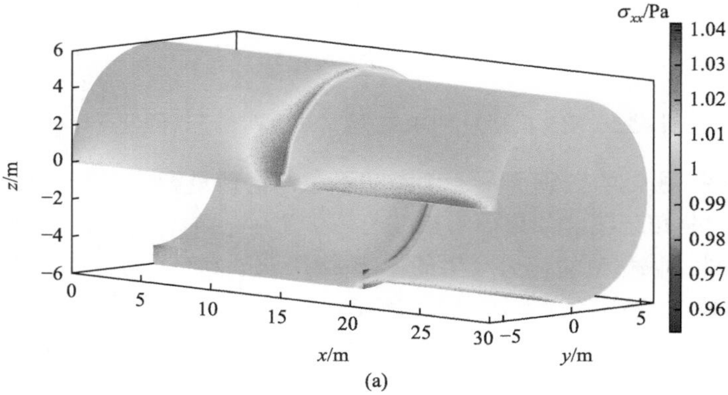

(a)

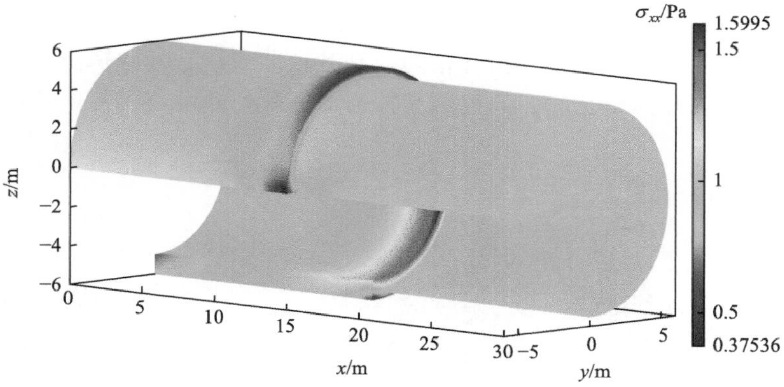

(b)

图 9.24 应力分量 σ_{xx} 的云图

(a) 考虑偏移量的影响；(b) 不考虑偏移量的影响

9.4.4 叶轮叶片耦合模型

考虑如图 9.25 所示的复杂叶轮叶片耦合模型。该模型由一个轮毂和八个叶片组成。轮毂由 IGABEM 建模，而叶片由等几何 Reissner-Mindlin 壳单元模拟。为了避免 Mindlin 壳分析中的剪切自锁效应，采用双三次 NURBS 基函数对叶片进行几何构造和耦合分析。文献[5]中指出，对于半径与厚度之比小于 1000(本例中的比率为 50)的情况，三次 NURBS 形函数可以避免剪切自锁效应。

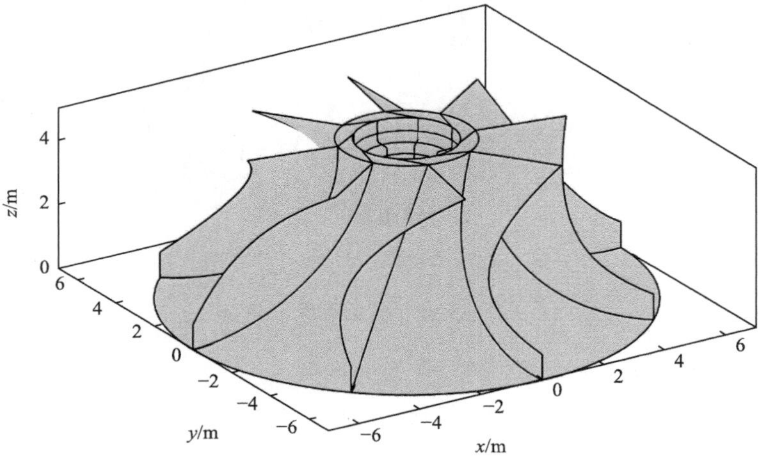

图 9.25 复杂叶轮叶片耦合模型

通过调整叶片 NURBS 几何的控制点及其权重，可以灵活地控制叶型曲线的参数，实现对叶轮的参数化设计。有限元和边界元子域的 NURBS 网格如图 9.26 所示，其中红色和黑色网格线分别表示有限元和边界元部分。整个模型由 13 个

NURBS 片组成，其中轮毂由四个边兼容的 NURBS 片组成。如图 9.27 所示，轮毂耦合面和叶片包覆面可分别由叶根曲线和叶顶曲线绕 z 轴旋转而成。在构造好任意一个叶片的几何形状后，通过绕 z 轴的圆环阵列即可获得其他叶片的实例。

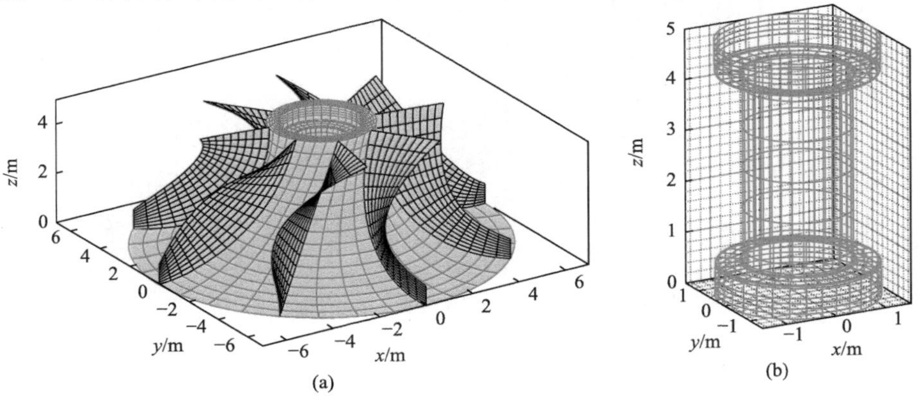

图 9.26　叶轮叶片耦合模型的 NURBS 网格

(a) 整体模型网格；(b) 轮毂的内表面网格

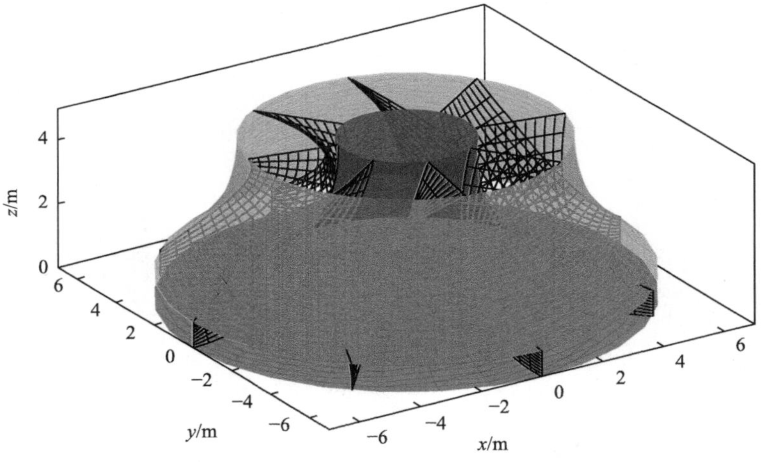

图 9.27　轮毂耦合面和叶片包覆面

该模型的材料为不锈钢，其参数为：弹性模量 $E = 1.9\times10^{11}\text{Pa}$ 、泊松比 $\nu = 0.305$ 。叶片的厚度设置为 $h = 0.1\text{m}$ 。通过将轮毂内表面上的所有位移分量设置为零来施加本质边界条件。通过沿叶片表面法线方向施加 $\boldsymbol{t} = 100000\boldsymbol{n}\text{Pa}$ 的面力，模拟叶片的负载情况。该耦合模型的有限元子域由 1152 个三次 NURBS 壳单元和 10944 个广义自由度组成，而边界元子域由 1920 个三次 NURBS 曲面单元和 11520 个位移自由度构成，最终耦合自由度总数为 15360。轮毂整个表面配点在物理空间中的分布如图 9.28 所示。

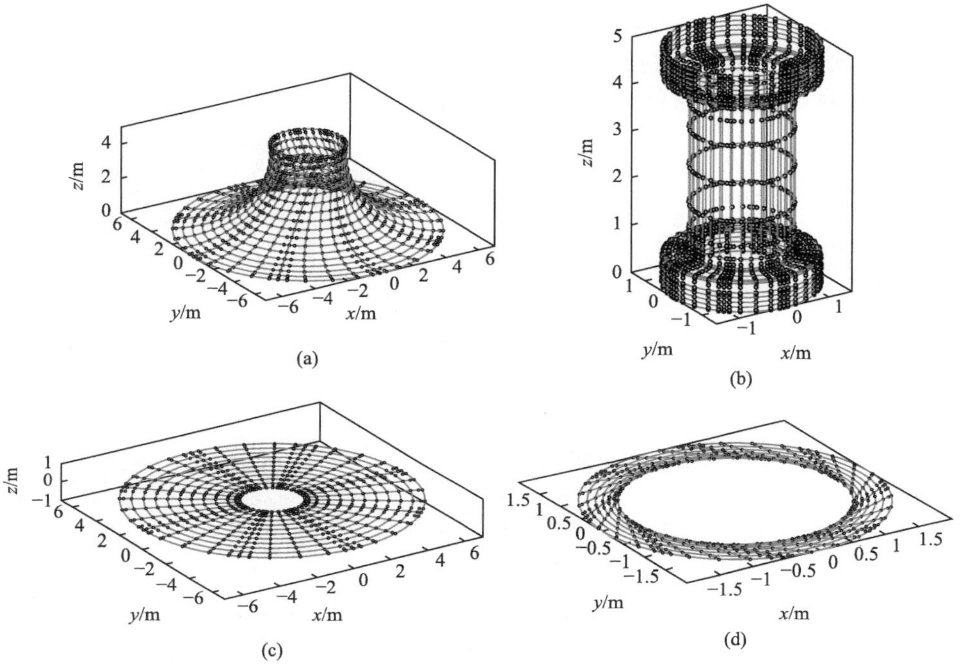

图 9.28　轮毂物理空间中的配点位置

(a) 轮毂外表面的配点；(b) 轮毂内表面的配点；(c) 轮毂底面的配点；(d) 轮毂顶面的配点

由于轮毂上的耦合面是旋转曲面，并为了确保壳体在物理空间中的厚度相等，其耦合边在参数空间$\left((\xi,\eta),0\leqslant\xi\leqslant1,0\leqslant\eta\leqslant1\right)$中的投影不再是平行于节点线的直线(图 9.29)，而且由这些投影线包围的区域不再是矩形区域。我们采用标准

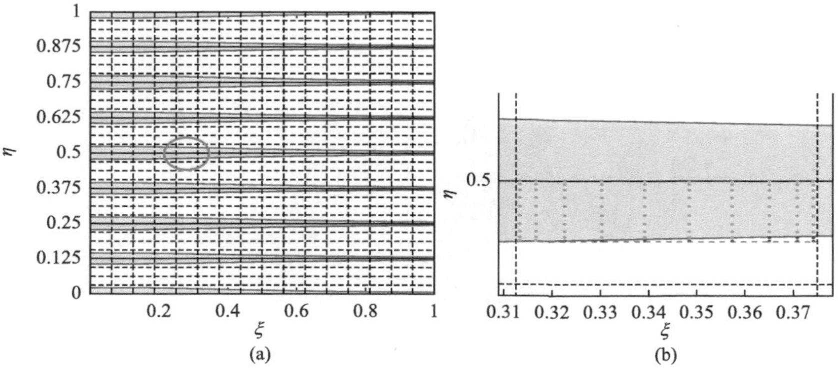

图 9.29　叶片几何形状在参数空间中的投影

黑色虚线表示非零节点区间(单元)；(a) 参数空间中所有叶片的投影，其中橙色、黑色和蓝色线分别代表壳体上、中和下边缘的映射；(b) 积分点在指定单元中的分布(绿色填充点用于计算，而蓝色填充点被舍弃)

高斯求积方法计算相关参数单元中在局部子矩形区域内的耦合积分，并舍弃耦合区域外的积分点。在上述过程实施后，物理空间中耦合界面上叶片沿其厚度方向的上下边缘轮廓如图 9.30 所示。

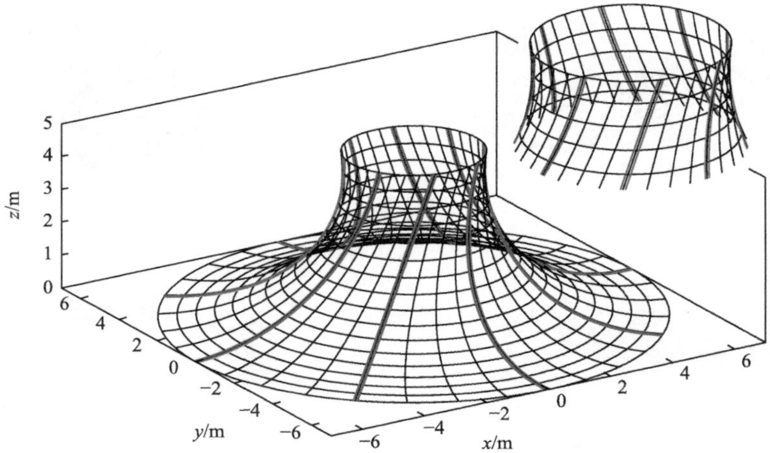

图 9.30 物理空间中耦合界面上叶片上下边缘的轮廓

图 9.31 显示了未变形的结构和在放大因子为 50 时的 y 方向上的位移分布。von Mises 应力分布云图如图 9.32 所示。应注意的是，对数可视化的比例被用于清楚地显示应力的平滑变化。因为叶片和轮毂底面之间的连接区域并不是平滑过渡的，所以最大应力出现在这些区域附近，其可以通过倒圆角的形式来缓解。此外，轮毂和叶片的 von Mises 应力分布云图也在图 9.33 和图 9.34 中单独绘制，应注意到八个叶片的 von Mises 应力分布并不完全一致。

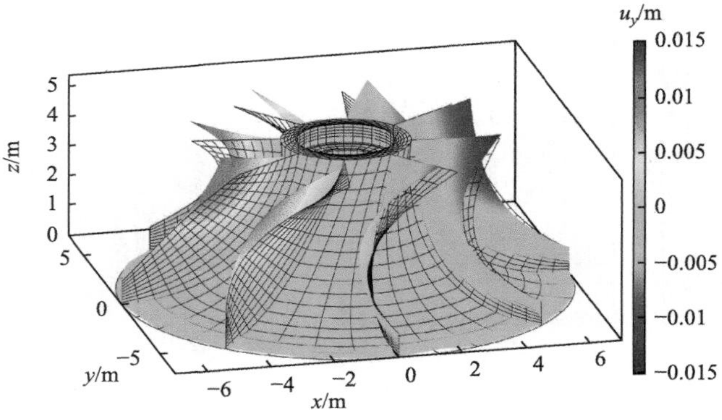

(a)

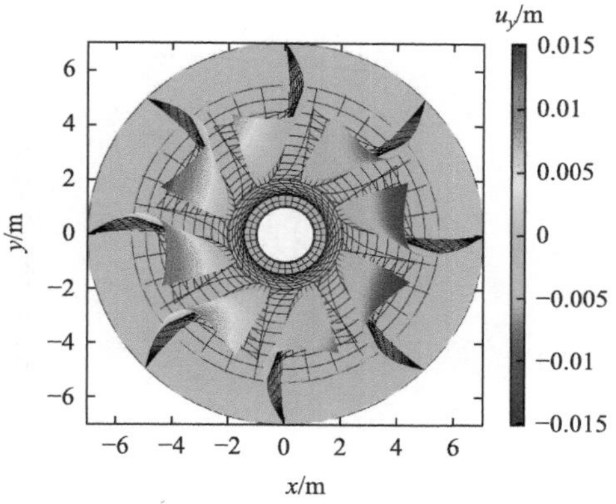

(b)

图 9.31　y 方向上的位移分布云图，放大因子为 50

(a) 整个模型；(b) 俯视图

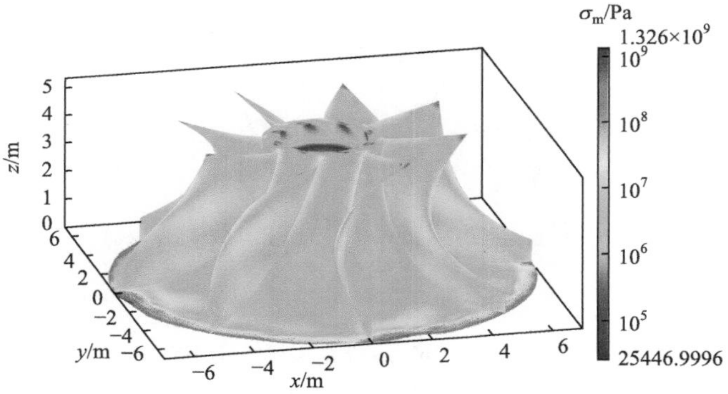

(a)

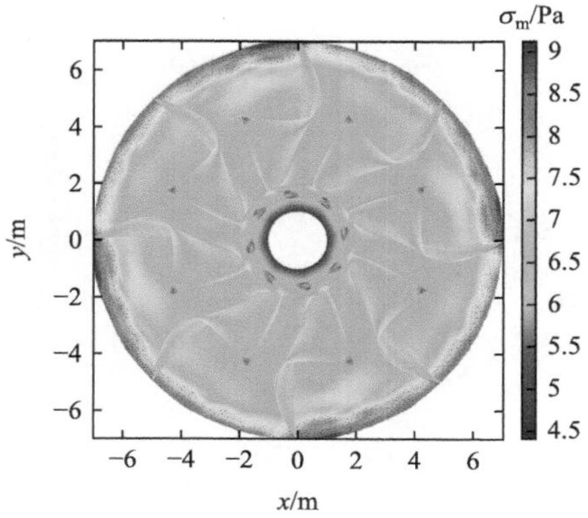

(b)

图 9.32　von Mises 应力分布云图

(a) 整个模型；(b) 俯视图

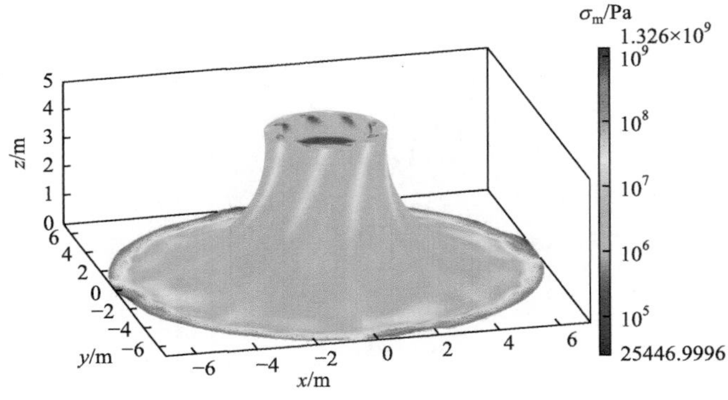

(a)

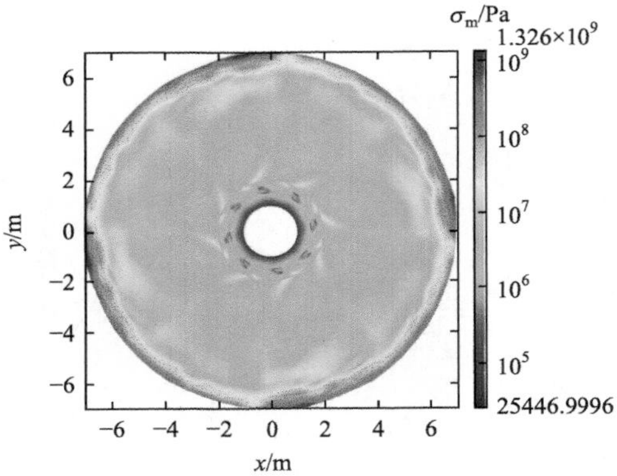

(b)

图 9.33　轮毂的 von Mises 应力分布云图

(a) 整个模型；(b) 俯视图

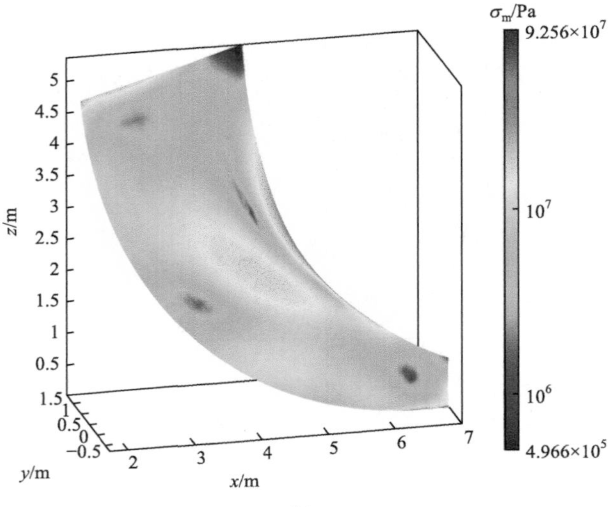

(a)

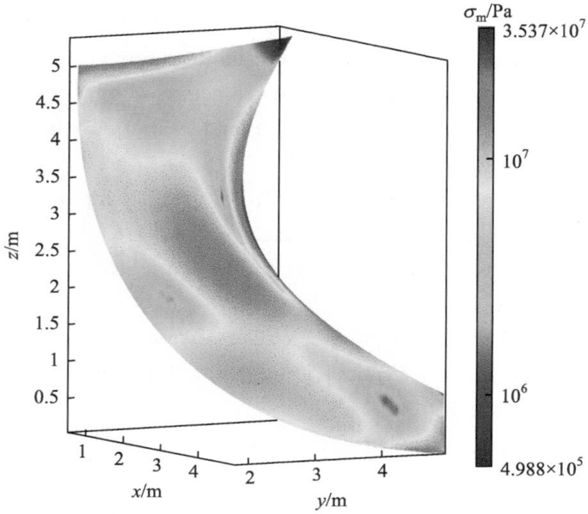

(b)

图 9.34 叶片的 von Mises 应力分布云图

(a) 耦合边缘位于 xOz 和 yOz 平面内的叶片；(b) 其他位置处的叶片

9.5 小 结

本章建立了混合维度实体-壳耦合问题的等几何有限元-边界元耦合算法，其实体部分采用等几何边界元模拟，能避免实体内部体网格的划分，壳体区域采用等几何 Reissner-Mindlin 壳单元进行分析。因此耦合分析仅需提供整个模型边界的 CAD 网格信息，易于实现等几何耦合方法与 CAD 系统的紧密融合；推导了刚度矩阵形式的等几何边界元子域的系数方程，简化了实体-壳耦合界面上约束条件的实施，即只需显式地建立界面上的位移协调关系；提出和探究了基于壳体变形假设的直接运动耦合方法和基于界面上虚功相等的弱耦合方法的收敛性和稳定性。研究结果表明，弱耦合方法具有更好的精度、灵活性和稳定性，当考虑曲壳与实体耦合，以及较大的曲率半径与厚度比值时，微小偏心率对耦合力矩合力的影响不能被忽略，否则会出现不可接受的误差。

参考文献

[1] Chapelle D, Bathe K J. Fundamental considerations for the finite element analysis of shell structures[J]. Computers & Structures, 1998, 66(1): 19-36.

[2] Dornisch W, Müller R, Klinkel S. An efficient and robust rotational formulation for isogeometric Reissner-Mindlin shell elements[J]. Computer Methods in Applied Mechanics and Engineering, 2016, 303: 1-34.

[3] Benson D J, Bazilevs Y, Hsu M C, et al. Isogeometric shell analysis: the Reissner-Mindlin shell[J]. Computer Methods in Applied Mechanics and Engineering, 2010, 199(5): 276-289.

[4] Adam C, Bouabdallah S, Zarroug M, et al. Improved numerical integration for locking treatment in isogeometric structural elements. Part Ⅱ: Plates and shells[J]. Computer Methods in Applied Mechanics and Engineering, 2015, 284: 106-137.

[5] Wu Y H, Dong C Y, Yang H S. Isogeometric FE-BE coupling approach for structural-acoustic interaction[J]. Journal of Sound and Vibration, 2020, 481: 115436.

[6] Beer G, Marussig B, Duenser C. The Isogeometric Boundary Element Method[M]. Switzerland: Springer Nature Switzerland AG, 2020.

[7] Simpson R N, Bordas S P A, Trevelyan J, et al. A two-dimensional isogeometric boundary element method for elastostatic analysis[J]. Computer Methods in Applied Mechanics and Engineering, 2012, 209-212: 87-100.

[8] Björck A. Numerical Methods in Matrix Computations[M]. Switzerland: Springer International Publishing, 2015.

[9] Shim K W, Monaghan D J, Armstrong C G. Mixed dimensional coupling in finite element stress analysis[J]. Engineering with Computers, 2002, 18(3): 241-252.

[10] Timoshenko S P, Goodier J N. Theory of Elasticity[M]. New York: McGraw-Hill, 1969.